The Internet and American Business

History of Computing

I. Bernard Cohen and William Aspray, editors

John Agar, *The Government Machine: A Revolutionary History of the Computer*

William Aspray and Paul E. Ceruzzi, *The Internet and American Business*

William Aspray, *John von Neumann and the Origins of Modern Computing*

Charles J. Bashe, Lyle R. Johnson, John H. Palmer, and Emerson W. Pugh, *IBM's Early Computers*

Martin Campbell-Kelly, *From Airline Reservations to Sonic the Hedgehog: A History of the Software Industry*

Paul E. Ceruzzi, *A History of Modern Computing*

I. Bernard Cohen, *Howard Aiken: Portrait of a Computer Pioneer*

I. Bernard Cohen and Gregory W. Welch, editors, *Makin' Numbers: Howard Aiken and the Computer*

John Hendry, *Innovating for Failure: Government Policy and the Early British Computer Industry*

Michael Lindgren, *Glory and Failure: The Difference Engines of Johann Müller, Charles Babbage, and Georg and Edvard Scheutz*

David E. Lundstrom, *A Few Good Men from Univac*

René Moreau, *The Computer Comes of Age: The People, the Hardware, and the Software*

Arthur L. Norberg, *Computers and Commerce: A Study of Technology and Management at Eckert-Mauchly Computer Company, Engineering Research Associates, and Remington Rand, 1946–1957*

Emerson W. Pugh, *Building IBM: Shaping an Industry and Its Technology*

Emerson W. Pugh, *Memories That Shaped an Industry*

Emerson W. Pugh, Lyle R. Johnson, and John H. Palmer, *IBM's 360 and Early 370 Systems*

Kent C. Redmond and Thomas M. Smith, *From Whirlwind to MITRE: The R&D Story of the SAGE Air Defense Computer*

Alex Roland with Philip Shiman, *Strategic Computing: DARPA and the Quest for Machine Intelligence, 1983–1993*

Raúl Rojas and Ulf Hashagen, editors, *The First Computers—History and Architectures*

Dorothy Stein, *Ada: A Life and a Legacy*

John Vardalas, *The Computer Revolution in Canada: Building National Technological Competence, 1945–1980*

Maurice V. Wilkes, *Memoirs of a Computer Pioneer*

The Internet and American Business

edited by William Aspray and Paul E. Ceruzzi

The MIT Press
Cambridge, Massachusetts
London, England

For information about special quantity discounts, please e-mail special_sales@mitpress.mit.edu

This book was set in Stone Serif and Stone Sans on 3B2 by Asco Typesetters, Hong Kong.
Printed and bound in the United States of America.

Library of Congress Cataloging-in-Publication Data

The Internet and American business / edited by William Aspray, Paul E. Ceruzzi.
    p.   cm. — (History of computing)
Includes bibliographical references and index.
ISBN 978-0-262-01240-9 (hardcover : alk. paper)
1. Internet—United States—Economic aspects. 2. Electronic commerce—United States.
3. Internet industry—United States. 4. Internet—United States—Social aspects. 5. Information technology—United States—Economic aspects. I. Aspray, William. II. Ceruzzi, Paul E.
HE7583.U6I57   2008
384.3'30973—dc22   2007005460

10 9 8 7 6 5 4 3 2 1

Contents

Preface   vii
Acknowledgments   ix

I Introduction   1

1 Introduction   3
William Aspray and Paul E. Ceruzzi

2 The Internet before Commercialization   9
Paul E. Ceruzzi

II Internet Technologies Seeking a Business Model   45

3 Innovation and the Evolution of Market Structure for Internet Access in the United States   47
Shane Greenstein

4 Protocols for Profit: Web and E-mail Technologies as Product and Infrastructure   105
Thomas Haigh

5 The Web's Missing Links: Search Engines and Portals   159
Thomas Haigh

6 The Rise, Fall, and Resurrection of Software as a Service: Historical Perspectives on the Computer Utility and Software for Lease on a Network   201
Martin Campbell-Kelly and Daniel D. Garcia-Swartz

III Commerce in the Internet World   231

7 Discovering a Role Online: Brick-and-Mortar Retailers and the Internet   233
Ward Hanson

8   Small Ideas, Big Ideas, Bad Ideas, Good Ideas: "Get Big Fast" and Dot-Com
Venture Creation   259
David A. Kirsch and Brent Goldfarb

IV   Industry Transformation and Selective Adoption   277

9   Internet Challenges for Media Businesses   279
Christine Ogan and Randal A. Beam

10   Internet Challenges for Nonmedia Industries, Firms, and Workers: Travel
Agencies, Realtors, Mortgage Brokers, Personal Computer Manufacturers, and
Information Technology Services Professionals   315
Jeffrey R. Yost

11   Resistance Is Futile? Reluctant and Selective Users of the Internet   351
Nathan Ensmenger

V   New Technology—Old and New Business Uses   389

12   New Wine in Old and New Bottles: Patterns and Effects of the Internet on
Companies   391
James W. Cortada

13   Communities and Specialized Information Businesses   423
Atsushi Akera

VI   Newly Created or Amplified Problems   449

14   File Sharing and the Music Industry   451
William Aspray

15   Eros Unbound: Pornography and the Internet   491
Blaise Cronin

VII   Lessons Learned, Future Opportunities   539

16   Market and Agora: Community Building by Internet   541
Wolfgang Coy

17   Conclusions   557
William Aspray and Paul E. Ceruzzi

List of Contributors   565
Index   569

Preface

This book is a historical study of the effect of the Internet, a recent advance in communications and computing technology, on traditional business practices in the United States. We begin with a brief look at the technological evolution of the Internet. We then quickly move to uncharted territory, in which we treat the existence of a robust and powerful computer network as a given, and examine what has happened as a result of the Internet's creation.

More than ten years have passed since the Silicon Valley company Netscape had its initial public offering, an event that many argue triggered the gold rush of Internet commercialization. It has now been about five years since the "Internet bubble" burst, leaving behind a well-told story of bankruptcies, lawsuits, and plunging stock valuations. A look at U.S. business history reveals that this is not the first time such bubbles have occurred. The overbuilding of railroads in the nineteenth century and electric power networks in the early twentieth century followed a similar pattern. Huge fortunes were amassed, scoundrels went to jail, newspapers ran daily stories of malfeasance, and investors lost personal savings. But after all that happened, the nation was left with an infrastructure on which it could trade and prosper. Parts of that infrastructure survive and are in heavy use today. The same is true with the Internet. The bursting of the bubble in 2000–2001 left behind a mesh of high-capacity fiber-optic telecommunications lines, a network of switching nodes, routers, switching, and database software, and a host of other components that are the twenty-first-century counterpart to the nineteenth-century rail network or the twentieth-century electric power network. A present generation of businesses can use this infrastructure. The technology is still evolving and advancing rapidly, but enough has stabilized for commercial interests to concentrate on their business model, not on chasing or building the technological base. The major components are in place, and one can turn to a stable group of suppliers of hardware and software, such as Cisco, Sun, Oracle, and IBM. That story is the prologue. The emergence of those suppliers set the stage for the main thrust of this book: the practical commercial use of this remarkable technical accomplishment.

In developing this book, we consciously made an attempt to balance the focus of the daily press on a few well-known facets of this story, such as Amazon's success in online selling or Google's success in selling advertisements on its pages. These stories leave out a large swath of territory. Some companies use the Internet to replace proprietary networks they had previously built for internal use; the consumer would never see these, but they are no less important. Others combine the Internet with traditional business practices to come up with entirely new business models—to an extent Amazon is such a company, but there are many others that have done this in ways that are less well-known. The cliché one often hears is that the Internet reduces "friction": the inefficiencies in transactions that are an inevitable result of parties to a transaction having incomplete, inaccurate, or out-of-date information. In practice that effect manifests itself in different ways, at different stages of a business transaction, and we propose to examine this phenomenon at these various points.

Related to that search for balance is our desire to give sufficient coverage to the threats that the Internet poses to traditional businesses. To continue the analogy above, the reduction in friction harms those who have made a business out of exploiting inefficiency. The following chapters are a first attempt to get an understanding of a dynamic and exciting period of American business history. We hope to have captured some of that excitement in the chapters that follow.

Acknowledgments

The preparation of this book was a cooperative effort that spanned continents, and involved a number of persons and institutions, including but not limited to those who have written the following chapters and their host institutions. We wish first of all to acknowledge the Smithsonian Institution and Indiana University for their support of the editors' efforts, and for allowing us the time to complete this study. We specifically acknowledge the support of the Smithsonian's National Air and Space Museum, Division of Space History, and Indiana University School of Informatics for their support of travel expenses for the editors to meet, and the latter specifically for its support of a conference held in summer 2006 at the Deutsches Museum in Munich, where many of the authors and others were able to gather and discuss the themes of this study. We wish to thank Oskar Blumtritt, Hartmut Petzold, and Helmut Trischler of the Deutsches Museum for organizing and hosting that meeting, and for their insightful comments on early drafts of the chapters. We also thank Robert Kahn of the Corporation for National Research Initiatives, Jonathan Coopersmith of Texas A&M University, Tom Misa of the Charles Babbage Institute, and David Walden for their critical reading and comments on early drafts.

I Introduction

1 Introduction

William Aspray and Paul E. Ceruzzi

[People] say about the Web: One third retail, one third porn, and one third lies, all of our baser nature in one quick stop.

Stephen L. Carter, *The Emperor of Ocean Park*

When we think of the Internet, we generally think of Amazon, Google, Hotmail, Napster, MySpace, and a host of other sites for buying products, searching for information, downloading entertainment, chatting with friends, or posting photographs. If we examine the historical literature about the Internet, however, it is hardly an exaggeration to say that none of these topics is covered. This book aims to fix this problem.

It is not as though historians and academically minded journalists have not paid attention to the Internet. Yet their focus has mainly been on the origins of the Internet in the 1960s and 1970s through the efforts of the U.S. Department of Defense's Advanced Research Projects Agency (ARPA) to build a network to connect defense researchers and later military sites to one another. When not focused on ARPA, this historical literature has addressed the use of these networks in the 1980s mainly by the scientific community to communicate with one another, share data, and gain remote access to powerful computing facilities.

Although the literature is much broader, let us here consider only the three best-known books on Internet history.[1] Arthur Norberg and Judy O'Neill's *Transforming Computer Technology* (Baltimore, MD: Johns Hopkins University Press, 1996) provides a careful historical analysis of the early years of ARPA's Information Processing Technology Office, and in a detailed case study, describes and analyzes ARPA's role in the development of networking technology and the ARPANET. Janet Abbate's *Inventing the Internet* (Cambridge, MA: MIT Press, 1999) traces the history of the technical developments, and the social shaping of people and institutions on these technical developments, from ARPA's development of basic technology to the creation of the World Wide Web. Katie Hafner's *Where Wizards Stay Up Late* (New York: Simon and Schuster, 1998) takes a readable but more journalistic approach to the same topics as covered in the Norberg/O'Neill and Abbate books—focusing more on personalities and anecdotes to present a strong narrative drive.

There is also a mass-market business and journalistic literature, which covers the companies and individuals making news in the Internet world.[2] These are valuable additions to our understanding of the development of the Internet as a commercial venture. Altogether too often, however, these tend to be either heroic accounts of how an individual or company has changed the world, or an antiheroic one purporting to show the colossal blunders made by these individuals or companies. Most of them have a narrow focus, and many do not build on the background and ways of understanding offered by the academic literature on the history of technology, science and technology studies, business history, economic history, and gender and cultural history.

This book looks at the historical development of the Internet as a commercial entity and the impact it has had on American business. This history begins in 1992, only fifteen years ago, when the U.S. Congress first permitted the Internet to be used by people besides academic, government, or military users. This book is informed by the academic literature mentioned above, while remaining as accessible to the general reader as the numerous popular histories that are in print. While we use this academic literature to inform our account, this book will not concentrate on the current academic debates about the theories and methods of postmodern history. Our focus is on understanding how in the past fifteen years the Internet added a commercial dimension to what was already a social and scientific phenomenon, and the way that added dimension affected commerce in the United States. We believe that this book is suitable for a general audience or classroom use, and hope that it will inspire historians to write more academic monographs that explore more deeply many of the topics we merely survey.

We begin with a chapter that reviews the early history of the Internet, with a different emphasis from the three books mentioned above. The focus is not so much on the development of the military network ARPANET or the scientific network NSFNET. Instead, chapter 2 traces seven lines of development that were critical if the Internet was to become commercialized.

The second section of the book (chapters 3 to 6) examines the basic technologies of the Internet: Internet service provision, e-mail, Web browsers, search engines, portals, and computer utilities and software services. The focus is not so much on the development of the technologies themselves, or how they work, as on the business models that make it possible for companies to supply these technologies to millions of users.

The third section of the book (chapters 7 and 8) explores the development of commerce on the Internet. It includes a description of the thousands of new companies—both the successes and failures—formed to sell directly to consumers over the Internet.

This coverage includes a more general discussion of the dot-com boom and crash. This section also discusses how traditional brick-and-mortar companies have responded to the Internet to sell to consumers. For these companies, the Internet presented a great opportunity, but it has also created the challenges of major new investments, changes in business strategy, and new competition.

The fourth section (chapters 9 to 11) considers one set of problems introduced by the commercialization of the Internet—namely, the tumultuous change and harm it caused to certain traditional industries. One chapter analyzes changes to the media industries: news, magazine, television, and radio. Another chapter presents examples of changes to several nonmedia industries: travel agencies, realtors, mortgage brokers, and personal computer manufacturers. The final chapter in this section uses the medical and academic professions to study the elective use of the Internet. This section reveals how many users do not fall into the simple categories of total adopters or Luddites in their use of the Internet, as some of the popular literature suggests. Instead, different professions tend to adopt the use of certain technologies associated with the Internet and not others, depending on how these technologies fit into their established work habits and bases of authority.

The fifth section (chapters 12 and 13) considers two very different examples of Internet applications—one traditional, and the other entirely novel. The traditional one examines the use of the Internet to replace, supplement, and alter traditional means for information transactions within a company as well as between a company and its suppliers and sales network. The second application involves using the Internet to create communities. This application is not entirely novel, as the chapter shows the deep roots of the social uses of the Internet in American society, but it does reveal how this new technology changed the character of the communities thus created.

The sixth section (chapters 14 and 15) considers another kind of problem introduced by the commercial Internet: providing a highly effective way of promoting and disseminating behaviors that are illegal, or at least widely regarded as socially undesirable. One chapter looks at the copyright violations associated with sharing music on the Internet, and the other one discusses Internet pornography.

The final section reminds the reader of the noncommercial forces that remain powerful as the Internet evolved, continuing the notion of the Internet as a community-building medium. The final chapter revisits the common themes discussed across multiple chapters and briefly speculates on possible future paths of evolution.

The writing of this book presented great challenges. The amount of material to be covered was enormous. We fought continuously throughout the writing to keep the chapter length under control. Each chapter could easily have been the subject of its own book, and we will not be surprised if such books appear in the coming years. Most of the contributors to this book are historians, and they are used to having chronological distance from their topic. That distance gives historians a perspective that enables them to identify which developments have lasting importance and determine which contextual factors really matter. Almost all of our chapters come up to the present, though, and it is a temptation we have had to resist to keep changing our texts as new developments unfold. We have not had the opportunity for the dust to settle to get a better impression of the landscape. Another challenge has been that the authors of this book have not had the "bricks of history" available—the focused journal papers, monographs, and specialized studies conducted by other scholars—as they attempt to write a general overview. Out of necessity, the authors have relied heavily on news reports and magazine feature articles, which most academic scholarship tends to avoid.

It is always difficult to provide the coherence in an edited volume that one finds in a single-author work. We have taken a number of steps to address that challenge. Many of the contributors have known one another for many years, and there has been a high level of dialogue between the authors as they have worked on their chapters: sorting out boundaries between chapters, discussing which themes are in common, agreeing on who will provide certain background information, and deciding what approach to take. Each chapter has typically been critiqued by three or four of the other contributors. The two editors have nudged the contributors toward a common set of themes (which are discussed in the final chapter). More than half of the contributors met together for three days of intensive conversation about their initial drafts. This meeting was held in Munich, where German, other European, and other American historians of technology, technologists, and sociologists were invited to participate. Although it may seem odd that a book that focuses primarily on U.S. developments would be discussed at a conference in Munich, in fact the participants found that the German perspective was a great help as the authors and editors examined the assumptions implicit in their drafts.

There was important value added, however, by preparing this book as an edited volume. The topics covered are wide-ranging, and it is unlikely that any single author would be as well prepared to undertake this project as our collective group. The authors of the chapter on the media industries, for example, are professors of journalism and mass communication. The chapter on Internet commerce is written by a business school professor who has made an extensive analysis of failed dot-com companies. The author of the chapter on brick-and-mortar companies has taught this subject in the Stanford business school since the time of the dot-com bubble. And so on for the expertise of our other various chapter authors, who included historians of technology, science and technology scholars, and business consultants.

As the editors of this book, we feel fortunate to have assembled such a capable group. Every one of the authors is well-known in their academic discipline. Six of the contributors have been recipients of the prestigious Tomash Fellowship for study in the history of computing. Five of the contributors hold distinguished professorships in their universities. The younger scholars are among the most promising in their profession. Biographies of the individual authors can be found at the back of the book.

We hope our readers will enjoy and be informed by these accounts, and will let us know where we have gone astray.

Notes

1. In recent years a large body of historical literature has appeared, at all levels of technical, political, and social detail. Among those consulted by the authors, one of the better sources is "A Brief History of the Internet," published electronically by the Internet Society, at ⟨http://www.isoc.org/internet/history/brief.shtml⟩.

2. A search on the term "History of the Internet" on Amazon.com reveals a list of several thousand titles. To these would be added the numerous online histories, including the one cited above.

2 The Internet before Commercialization

Paul E. Ceruzzi

One popular history of the Internet begins by quoting a pioneer's hopes that historians would dispel the myth, already prevalent, that the Internet was invented "to protect national security in the face of a nuclear attack."[1] That such a myth has arisen is no surprise; the speed at which the Internet became a part of American society complicates an attempt to write an objective history of it. The steamboat, the mass-produced automobile, and interchangeable parts had myths about their invention, too. With careful and patient scholarship, a more complete picture of those technologies emerged. After such research, historians often realize that the myths in fact have a lot of truth to them—that, for example, Henry Ford truly was the inventor of mass production, even if others before him had produced inexpensive automobiles in large numbers. The Internet's connection to surviving a nuclear war is not entirely wrong, either. The Internet that we use today grew out of a project funded by the U.S. Department of Defense's Advanced Research Projects Agency (ARPA), a division set up in 1958 in the wake of the Soviet Union's orbiting of Sputnik the year before.[2] Nearly every history of the Internet, including the one cited above, argues that the network that ARPA built, the ARPANET, was the ancestor of today's Internet. So while the story of the Internet's origins in surviving a nuclear war may be fanciful, no one questions its connection to the cold war. The critics of that creation myth do not deny the cold war origins; they want to provide a context, to show that the forces that drove the creation of computer networks were more complex than implied by that story. More than fifteen years have passed since the end of the cold war, perhaps enough time to allow us to place the Internet and its cold war origins in context.

This chapter chronicles the steps that took a theoretical, experimental, military-funded network, begun in the 1960s, to the commercialized and private Internet of today. Most readers are intimately familiar with this network. It is an indication of its maturity that one can make good use of it without understanding how it works in detail. Individuals can purchase and drive a high-performance sports car without knowing much about the engine, suspension, or drive train that give the car such performance. Likewise, a user of the Internet—even a sophisticated "power user"—may not know what lies under its hood, either. This chapter does not explain the technology of what lies under the Internet's hood but it does attempt to tell how those components got there.

The ARPANET was conceived in the mid-1960s, achieved rudimentary operation by 1970, and was demonstrated to an audience of specialists at a conference in October 1972. For the next ten years it was further developed, as more computers were connected to it, and as users found ways to make productive use of it. There are a number of well-written histories, including several that are available on the World Wide Web, that cover that pioneering era.[3] This chapter mainly deals with the events that followed the adoption of the current Internet protocols in 1983, to about 1995, when the Internet shed its government and academic roots, and became fully a commercial entity. But it begins with a brief outline of the major events before 1983 in order to set the stage for what followed.

One can find the roots of computer networking far back into the nineteenth century, when the Morse telegraph inaugurated communications using electricity. When the electronic digital computer was invented in the late 1930s and 1940s, some of its creators recognized that such a machine could assist in communicating information, but their immediate goal was to create a machine that would aid mathematical calculating—hence the name itself, *computer*, borrowed from the name given to people who performed calculations by hand or with desk calculators. In the United States at that time, local and long-distance communications were handled by AT&T, a regulated monopoly, as well as Western Union and other related companies. The U.S. telephone network was perhaps the finest in the world. In 1940, George Stibitz of AT&T's Bell Laboratories demonstrated what may have been the first remote access to a calculating machine, for a mathematics conference held in New Hampshire (the calculator remained at Bell Labs in New York). But remote access to computing machines was not a priority in the years that followed, and connecting one computer to another was even less desired.

One should resist any attempt to find a single point of origin of computer networks. Historical research thus far has identified a host of actors, agencies, and locations where the roots of today's Internet may be found. Among historians there is a consensus that much credit belongs to J. C. R. Licklider, a scientist working at the Cambridge, Massachusetts research firm Bolt Beranek and Newman (BBN), who in the early 1960s began writing influential papers and leading discussions on the value of using computers interactively with human beings, and as a communications as well as a calculating device.[4]

In 1962, Licklider became the director of ARPA's Information Processing Techniques Office (IPTO), and with access to generous amounts of funding he directed an effort to realize that vision. To summarize a rich and detailed story, Larry Roberts, who had joined ARPA in 1966, learned the following year of a method of communication called *packet switching*, which although untested in any serious system, offered numerous advantages over the way that AT&T passed information on its network. Packet switching was conceived, and the term coined, by Donald Davies, a British physicist. The concept had also been independently developed by Paul Baran, a RAND Corporation mathematician. (Baran was specifically looking for ways for the United States to communicate effectively during a nuclear crisis with the Soviet Union, and thus it was his work that gave rise to the mythical, but not entirely inaccurate, story of the Internet's origins.) In essence, packet switching divided a message into blocks, called *packets*, each of which was separately put in a kind of electronic envelope that gave its destination address plus other information. In contrast to traditional phone circuits, these packets could be sent separately to the destination, by different routes and at different times, and reassembled once they got there. Although at first glance the overhead required to route packets would seem to be absurdly high, in practice, and after years of careful mathematical analysis and design of the system (much of that done by Professor Leonard Kleinrock of UCLA), packet switching was found to be a far superior method of transmitting computer data.

To continue the analogy with a letter sent through the postal service, note that one must obey conventions when addressing a letter as to the placement of the destination address, return address, postal code, postage, and so on. These conventions hold true regardless of the contents of the envelope—a critical distinction that also holds true for the Internet. The information that enveloped an electronic packet also had to obey certain conventions, regardless of the contents of the packet. These conventions were called *protocols*, from the Greek word meaning the leaf glued to a scroll that identified its contents (who first applied this word to packets is not known).

By 1967, Robert Taylor was head of IPTO, and the project to network ARPA-funded computers, now under the name ARPANET, continued. Again to telescope a rich history, the first connection was made at UCLA in 1969; by 1971, fifteen nodes were connected across the United States. In 1972 Taylor and Robert Kahn, a mathematician who had moved from BBN to ARPA, sensed a lot of skepticism among computer scientists and communications experts about the growing network. To counter that, they arranged a demonstration at the International Conference on Computer Communication in Washington, DC, from October 24 to 26.[5] By most accounts the demonstration was a success. Much work remained to be done, but one could say that the ARPANET, and with it the notion of practical computer-to-computer networking, had arrived.[6]

In 1983, the network switched from an initial scheme for connecting computers to a set of so-called protocols that remain in use to this day. Those protocols, the Transmission-Control Protocol/Internet Protocol (TCP/IP), are fundamental to the working of the Internet, and because of their persistence over the following decades it is appropriate to begin this narrative in 1983, when ARPA decreed that its network would use them.[7] This chapter ends in 1995, when the U.S. government relinquished ownership of the Internet backbone, and when access to the Internet through a program called the World Wide Web became commonplace. That is a span of only a dozen years—a much shorter period than it took to build other comparable systems, such as the Interstate Highway System—but in the compressed world known as "Internet time," it was a long interval indeed. In that interval the basic components of a private, commercial Internet were put in place.

Those basic components include the following:

• A set of transcontinental data channels called *backbones* (plural, paradoxically), operated by telecommunications companies such as Sprint or Verizon. Initially using copper wire, microwave, or satellites, now these are nearly all fiber-optic cables.
• Local, regional, and metropolitan-area networks attached to the backbones, serving most of the country. Among them are Internet service providers (ISPs): companies that make the final connection from the Internet to business clients and end users.
• Routers, switches, and other specialized hardware that route traffic among the backbones and ISPs.
• Software—specialized switching programs, algorithms, and standards and rules—that route the traffic. These include the fundamental protocols of TCP/IP, but many others as well.
• A political system of governance and management, including the Internet Corporation for Assigned Names and Numbers (ICANN), the Internet Society, and others. These entities do not "operate" the Internet in a command-and-control capacity, but their existence, however controversial, is necessary for the network to function.
• Software and hardware for the end user. The hardware may be a personal computer or laptop, cell phone, personal digital assistant, or specialized device. The software includes Web browsers, instant-messaging and e-mail programs, file-sharing programs, and others.
• Finally, a social and business model, which establishes norms about who is using the Internet and what they do there, and allows those who own and operate its segments to make money from their investment. This is the most difficult aspect to define, but understanding it is crucial to understanding the following chapters.

The development and promulgation of TCP/IP, and the associated theories of routing, error control, and packet configuration, were the contributions of the ARPANET to the above list, and one may forgive those who, because of these, give the ARPANET such a privileged place in the history. The TCP/IP protocols are still in use, and to use the jargon of the field, they have "scaled up" remarkably well: they work for a network many orders of magnitude larger than the one they were written for. Other switching algorithms and software developed under ARPA's direction, less familiar to the layperson, have also evolved and continue to serve. The Internet initially was conceived as a "network of networks"—of which the ARPANET was to be only one. What emerged is a global network that relies on an almost universal use of TCP/IP.

Other components of the ARPANET have evolved more radically. The initial copper wire and microwave backbone, operating at fifty thousand bits per second, has given way to fiber-optic lines with orders of magnitude greater speed. The end-user equipment has also changed. ARPANET linked mainframes and minicomputers such as the IBM System/360, Control Data 6600, or the Digital Equipment Corporation PDP-10 and PDP-11. Few of these are still running anywhere, much less connected to the Internet, although large mainframes and supercomputers are part of the network today. Routers and other switching hardware are supplied by companies including Sun, Cisco, IBM, and other less well-known suppliers. Most users connect to the Internet through a desktop or laptop computer. The preferred terminal of the ARPANET was the Teletype Model ASR-33, an electromechanical device whose greatest legacy was the @ sign on its keyboard—adopted for e-mail and now the icon of the networked world.[8] At the beginning of the transition to the Internet, when this chapter begins, the preferred terminal was the Digital Equipment Corporation VT-100: a monochrome, text-only monitor and keyboard. As of this writing it seems that the laptop, accessed wirelessly in a café, is the preferred method for connection, but that may change.

The hierarchy of backbones, regional networks, metropolitan networks, and ISPs was not part of the initial ARPANET, which connected its computers more in a gridlike topology. By the late 1980s that topology evolved to this hierarchical one, which has survived to the present day (with some changes, discussed later). Likewise, the router, other switching hardware, and the associated protocols have a lineage that can be traced back to the mid-1980s.

The governance of the Internet has had a most interesting evolution. The Department of Defense, as expected, managed and had authority over the ARPANET. In 1983, the military established a network for its internal use (MILNET), which it managed closely. It turned over the management of a network for other uses to the National Science Foundation (NSF). Even then, however, the Department of Defense retained the critical role of assigning names and addresses to Internet users. Since about 1995, the U.S. Department of Commerce has held authority for Internet governance including the assignment of names, although Commerce waited until 2005 to state this formally.[9] This chapter, which concerns events between 1983 and 1995, will therefore focus on the NSF's role and its relationship to the Department of Defense. Like the TCP/IP and other protocols, Internet governance retained a continuity as it evolved and scaled up in size.

That leaves the social and cultural component. The difference here between the ARPANET and the commercialized Internet is great. Subsequent chapters will describe a host of applications that take advantage of the Internet's unique abilities to transmit, process, and store information of all types. Almost none of that, not even e-mail, was present in the early ARPANET. Its successor, the NSFNET, likewise had few of these applications. That suggests an obvious question: if the social and economic uses of the Internet did not come from its Department of Defense and NSF ancestors, where did they come from? I discuss this question first, and then return to the other components listed above.

Personal Computer Services

One clue may be found in a remarkable book that appeared in successive editions during the early years of the personal computer, from about 1977 to 1990.[10] The thesis of *The Complete Handbook of Personal Computer Communications,* by Alfred Glossbrenner, was that if you were among the "tens of millions" of people who had purchased a personal computer but had not yet connected it to one of the many online services then available, you were missing the most important dimension of the personal computer revolution.[11] In the book, the author listed dozens—out of what he estimated as over thirteen hundred—different online databases, which a personal computer user could access via a modem, communications software, and a telephone. Most of these databases had arisen in the earlier, mainframe era, and grew out of "remote job entry" and time-sharing services intended to bring mainframe processing to off-site locations. These began as specialized services, available only to select corporate or professional customers, and they were usually expensive. Their evolution in the service of business customers is discussed in a subsequent chapter. What is of interest here is that the services also became available to individuals, beginning in the late 1970s, thanks to the advent of the inexpensive personal computer. The personal computer "changed everything. The electronic information industry took off like a rocket and has been blasting at full warp speed ever since."[12] Glossbrenner observed two interrelated phenomena. One was the obvious increase in customers for these services, brought on by the low costs of owning a personal computer. The second was the broadening of the services offered, from the specialized data or job entry, to more general services providing news, consumer information, financial advice and services, movie and restaurant reviews, and so forth. In a positive feedback cycle, as more and more people accessed these services, the providers saw a reason to offer more types of information, which in turn spurred more customers. Glossbrenner's observation was echoed by both Alan Kay, of the Xerox Palo Alto Research Center, who stated that "a computer is a communications device first, second, and third," and Stewart Brand, of the *Whole Earth Catalog*, who remarked that "'telecommunicating' is our founding domain."[13] At the time these three were making those statements, the personal computer was *not* first and foremost a communications device but rather was used primarily for games, word processing, and calculations utilizing spreadsheet programs. Glossbrenner, Kay, and Brand were evangelists who felt a mission to will this world into being.

If the ARPANET was not open to the public, and if its successor the NSFNET did not allow commercial use, how then did individuals access these commercial services? The primary way was for an individual to buy a modem for their personal computer, plug it into a telephone line, dial a local number, and connect at speeds typically in the three hundred to twelve hundred bits per second range. If the service was not located in the local calling zone—and this was typically the case—the user had two choices. They could dial a long-distance number and pay the charges on top of whatever charges there were for the service—not an enticing proposition given the pricing of long-distance calls in those days. Or they could dial a local number that the provider established in most area codes, with the provider making a connection to its mainframes via a commercial data network. In the United States, by tradition, one usually paid a flat rate for local calls regardless of how long the call lasted, and the telephone company made no distinction between a voice or data call. Since most people had a telephone anyway, this method of connecting to a service appeared to be "free." It was not free, of course, but the provider usually bundled the cost of the telephone connection into the overall cost of the service. There was a variety of pricing schemes. CompuServe, for example, charged $19.95 for an initial setup and one hour of connection to its service, plus an additional $15.00 an hour at twelve hundred baud during business hours, dropping to $12.50 an hour during evenings and weekends (the term baud is roughly equivalent to, though not the same as, bits per second; I shall use the latter term for the rest of this chapter). The Source charged a $10.00 monthly fee, plus a fee per hour of connect time depending on the speed and so on.[14] At the top of the line was Nexis, which charged $50.00 a month, plus $9.00 to $18.00 per search during business hours.[15]

Most of these services maintained a central database on mainframe computers. The distributed model that is the hallmark of the Internet was only found in a few, isolated exceptions. For instance, CompuServe's computers were in Columbus, Ohio; The Source's computers were in Vienna, Virginia.[16] The topology of these services was that of a hub and spokes, harking back to the time-sharing services from which they descended.

For the provider, the cost of setting up local telephone numbers was not trivial. The developers of the ARPANET came up with the notion of packet switching, a more economical way to move computer data than the techniques used for voice calls. While ARPANET was not open to commercial use, providers could use one of several commercial services that shipped data by packet switching. In the 1980s the two most popular were Tymnet of San Jose, California, and Telenet of Reston, Virginia. (By 1990, these were subsidiaries of McDonnell Douglas and U.S. Sprint, respectively.) It was no coincidence that the first president of Telenet was Larry Roberts, who had been instrumental in getting the ARPANET going, while Tymnet counted among its employees Douglas Engelbart, the inventor of the mouse and a pioneer in online information access.[17]

Writing in the mid-1980s, Glossbrenner and Brand listed about a dozen major services available to individuals, with comparisons of their pricing schemes, details on how to connect, the types of services offered, and so on. Both singled out The Source as best exemplifying the spirit of this new world. From looking back on descriptions of it many years later, in a world where the commercialized Internet prevails and The Source is long gone, one has an impression that it was "the source" of the social and business model of computer networking. It was not the only service to offer features like discussion forums, for example, but it seemed to have the best illusion that its subscribers were entering into a social world that was uniquely defined by interactive computing. Later on, these concepts would take on other names, especially *cyberspace*. But the concepts were present here, in rudimentary form. When in 1989 CompuServe purchased the assets of The Source Telecomputing Corporation, Glossbrenner lamented its passing. He observed, "The Source was more than just another online system. It was the system that started it all. It made the mistakes and experienced the successes that paved the way for those who followed. Above all, it contributed a *vision* of what an online utility could and should be."[18]

These personal computer services provided the model for the commercialized Internet a decade later, but I do not wish to press the analogy too far. For example, none of them had an effective and secure way to handle credit card transactions—the foundation of Internet commerce. One service required a user to call a voice telephone number and give the credit card information to a person at the other end, after going online and identifying what to buy. That did not succeed. Recall that the idea of ordering goods from a catalog by calling a toll-free number and giving one's credit card information over the phone was itself a novel idea at the time, and it took a while before many consumers felt comfortable doing that. Security and consumer confidence were serious issues that had to be addressed by the early World Wide Web browsers before that final piece could be put into place. That was a crucial exception, but otherwise it is uncanny how much of the modern social use of the Internet could be found in these services.

The Source was the brainchild of the Virginia entrepreneur William von Meister, biographies of whom describe a person whose ideas always seemed to run far ahead of an ability to implement them. Just as The Source was being established, in 1979, von Meister was pushed out.[19] Undeterred, he founded another service, hoping to deliver songs to consumers over telecommunication links—another idea that was twenty years ahead of its time and the topic of a later chapter of this book. Then he founded yet another, this one called Control Video Corporation, whose intent was to allow owners of the Atari 2600 computer to play games interactively with others in cyberspace.[20] The service was launched in the early 1980s, just as the market for computer games was crashing. Years later, von Meister's ideas would become the bedrock of commercial traffic on the Internet, but he did not live to see that, succumbing to illness at age fifty-three in 1995.

The history of computing has many stories of failed enterprises and visionaries who could not put their ideas into practice. This one stands out for what happened after von Meister left Control Video. Among the people attracted to his vision was a young marketer from Pizza Hut named Steve Case. Case was not a computer expert but discovered the potential of the online world after logging on to The Source on a Kaypro computer from his apartment in Wichita, Kansas. He joined Control Video and kept it going through a slough of poor sales, bad timing, angry creditors, and other perils that no company could have expected to survive. With the financial backing from a Washington, DC, entrepreneur named Jim Kimsey, and with technical help from Marc Seriff, who had worked on the ARPANET, in 1985 a new company was formed, Quantum Computer Services, which initially also hoped to deliver online games to home computer owners.[21] A few years later it was renamed America Online (AOL), with Case as its CEO; by the late 1990s it connected more individuals—as many as thirty million— to the Internet than any other service.

Besides CompuServe and AOL, one other service from that era deserves a brief mention. In 1984 a service named Prodigy was founded with joint support from CBS, IBM, and Sears. A few years later it was relaunched, with IBM and Sears the primary backers. It established local telephone numbers for most U.S. metropolitan areas, and through Sears offered a package of personal computers, modems, and communications software to help get consumers started. Where it differed from the other services mentioned above was its early use of computer graphics at a time when personal computers ran the text-only DOS operating system and modems had very low speeds. Prodigy software preloaded a graphics interface on to the user's computer to get around those barriers. The graphics were there not just to make the service more attractive; they were there to provide advertisers a rich environment to place ads before Prodigy subscribers. The ads were intended to offset most of the costs, the assumption being that consumers would not mind if this meant their monthly subscription costs were low. Thus Prodigy anticipated one of the business models of the World Wide Web that was developed later on, when personal computers could more easily handle graphics. Later Web-based graphics did not adopt the Prodigy standard, but Prodigy's work was among the influences of today's graphics-rich, ad-filled—some would say ad-cluttered—World Wide Web.

What doomed Prodigy was its users' heavy use of e-mail and discussion forums, which used a lot of connect time but did not generate enough ad views. Reflecting the corporate images of Sears and IBM, the service also attempted to censor the discussions, which led to a hostile reaction from subscribers. By contrast, AOL had a more freewheeling policy in its chat rooms, where it appointed volunteers to monitor the content. These volunteers—who not coincidentally included people who were physically disabled or otherwise shut-in—worked diligently to guide chat room discussions while having only a light touch of censorship. They became an electronic version of

Dorothy Parker at the Algonquin Hotel's Round Table, only the table could accommodate a few million people. Parker did not live to see the emergence of these chat rooms, but if she had, she would have recognized the racy content of most of them—not much different from what she and her friends gossiped about at the Algonquin in the 1920s. Prodigy transformed itself into an ISP in the mid-1990s, and was eventually acquired by a regional telephone company. Its history, along with that of AOL and CompuServe, shows the perils and opportunities of being a pioneer.

During the years these services were being established, they were not connected to the ARPANET or its successors, although they used commercial packet-switched services to connect subscribers to their computers. Their connection to the ARPANET was the transfer of people like Seriff. The classic 1981 book on computer architecture, *Computer Structures*, by Daniel Siewiorek, Gordon Bell, and Allen Newell, devotes several chapters to a detailed discussion of the ARPANET, but Brand's pathbreaking *Whole Earth Software Catalog*, published in 1984, does not mention that network at all.[22] Neither does Glossbrenner in the early editions of his book. A chapter in the 1990 edition of *The Complete Handbook* looks at electronic mail, which Glossbrenner predicts will soon be a service that everyone will want. But he laments how the various e-mail systems are incompatible, and that mail originating in one network cannot be delivered to another. He mentions a service, now forgotten, called DASnet, which promised to provide interoperation. Among the twenty-seven separate e-mail systems it claimed to connect, one—listed among all the others—was called Internet.[23] That is the only place in the book where the term is used.

The Internet Emerges

The Internet that Glossbrenner was referring to was the ARPANET and other networks that were connected to it, using TCP/IP and a device called a *gateway* (more on that later). After 1985 the term Internet became more common, although one major survey called it the "ARPA-Internet." In other instances, the word was written in lower case to denote any network of networks, including private ones, which were not necessarily connected to the ARPANET or used its protocols.[24]

That listing of the twenty-seven different e-mail services tells us why the Internet prevailed. It was designed from the start to connect heterogeneous networks. Few, if any, of the other networks had an interest in that—often the opposite was the case, where the network's administrators desired to preserve exclusivity. Just as the personal networks enjoyed a positive feedback cycle of more customers leading to more services, the Internet likewise enjoyed a positive feedback cycle. Its openness and interoperability made it possible for different networks to connect to it, and as they did, more and more networks *wanted* to connect to it.[25]

Although the Internet grew out of a closed network, the details of its design were debated and established in the open, via an innocent-sounding set of documents called Request for Comments (RFC). The protocols were described among these requests and published in technical journals. Computer vendors could adopt the protocols without paying a royalty or obtaining a license, in contrast to IBM's Systems Network Architecture or Digital Equipment Corporation's DECnet, which were guarded by their suppliers. A decision by ARPA's Information Processing Techniques Office, then under the direction of Robert Kahn, to fund the incorporation of TCP/IP into the Unix operating system accelerated this feedback loop. Unix was developed at AT&T's Bell Labs, which owned a trademark on the name, but AT&T did not restrain others from using it. Because it lent itself to modification and experimentation, Unix was popular among academic computer science departments. Beginning in 1980, ARPA supported a project at the University of California at Berkeley to develop a version of Unix with the protocols written into it. Other universities could get the program and a license at low cost. The programming effort was led by a young computer scientist named Bill Joy, and the version of Unix with TCP/IP in it was called 4.2 Berkeley Software Distribution (4.2BSD), released in August 1983.[26] Kahn thought of Joy as a "vector" for TCP/IP, like a mosquito carrying a virus from one host (a computer running Unix) to another. I have not asked Joy what he thinks of the metaphor, but he did succeed in spreading the protocols in his travels throughout the academic world and then the commercial one when he became an employee of Sun Microsystems. Thus by 1983, when ARPA-NET switched to TCP/IP, preparatory work had already been done to make that transition effective.

Academic Networks

As mentioned, the for-profit private networks were not connected to the ARPANET. Likewise, even on university campuses where there was a connection, access to the ARPANET was limited. That led to discontent at many schools, where even members of computer science departments found themselves shut out of what they recognized was an exciting new development in their field. Hence, one of the first external networks to connect to the ARPANET was built for the computer scientists, especially the have-nots in universities that had no Department of Defense contracts. Beginning in 1981, with funding from the NSF along with technical support from Kahn and Vinton Cerf at ARPA, CSNET was inaugurated and grew steadily thereafter. It contracted with Telenet to handle traffic, but it also developed a low-cost, dial-up telephone connection for those colleges that could not afford Telenet's fees.[27] The network used TCP/IP to make the connection. CSNET lasted only a decade, but it has an important place in history for at least three reasons. It was one of the first networks to address the issue of cost—a concern faced by smaller colleges and universities that had modest budgets and no access to Department of Defense monies. It pioneered in interconnecting different types of networks: the commercial Telenet, the ARPANET, and a network based on simple dial-up telephone access. And it was an early test of TCP/IP's capabilities at a time when other competing protocols were being touted and much of the ARPANET itself was using the older protocol.

Researchers at other campuses developed similar networks, some of which reached a high level of sophistication. Eventually, these too would be connected to the Internet as they succumbed to the advantages of interoperability. But that was not their goal from the start, and for some the Internet connection would not be made until much later. Often their designs were quite different from ARPANET's, and that also made interconnection difficult. An influential survey published in 1990 by John Quarterman lists about a dozen, of which two deserve a brief mention: BITNET and Usenet.[28]

BITNET was established beginning in 1981 as a way to connect IBM System/370 mainframes, which in those days were the mainstay of academic computing facilities.[29] It used leased telecommunications lines and communicated by an ingenious use of a feature—some might call it a *hack*—that let the mainframe read a message as if it were a deck of punched cards being read from a remote location. The first two installations to be connected were the City University of New York and Yale University. The service was accessible to students and faculty from the humanities as well as the sciences, and it soon became popular. Just as CSNET made networking available to computer scientists who had no access to military funds, so did BITNET provide access to the many others on a campus or research institution who otherwise were denied access.[30] One of its most popular features was the Listserv: a discussion forum similar to those provided by the Source and CompuServe; these became especially popular among students and faculty in the humanities. BITNET had some drawbacks: for example, it had a treelike topology, with the root at the City University of New York, and thus lacked the robustness of the Internet. But IBM mainframes were reliable workhorses, and users could depend on the network. BITNET merged with CSNET in 1987, which itself ceased operation in 1991. Listserv software is now widely available on the Internet for any operating system, and most Web services like Yahoo! offer a variant of it.

Usenet began the same way in 1980, as an interconnection between computers at Duke University and the University of North Carolina. Whereas BITNET ran on IBM mainframes running the VM operating system, Usenet linked computers that ran Unix. The software that it used was called the Unix-to-Unix CoPy program (UUCP, sometimes written in lowercase uucp as was the Unix custom). This program originated at Bell Labs, where Unix was developed, using dial-up telephone connections. UUCP was not easy to use, and it often seemed that Unix programmers wanted it that way. Nor was Usenet connected to the Internet. It is of historical importance for its impact on people who occupied the other end of the spectrum from BITNET—that is, programmers. Because of the significance of the BSD, Unix was the favored operating system for the switching hardware that the Internet needed as it grew. Through the 1980s, most Internet nodes used Digital Equipment Corporation VAX computers and Sun workstations, all running Berkeley Unix. The people writing the software for the Internet had to be fluent in Unix and UUCP, and they spent a lot of time discussing programming and related technical topics on Usenet groups. Usenet was well-known for these groups, which tended to be more freewheeling than those on BITNET. Some groups are still active, although they have been the subject of harsh attacks of spam. The reader may get a flavor of these groups by accessing them through Google, which absorbed Usenet around 2000. Finally, UUCP was the inspiration for the name and more of a company called UUNet, which played a crucial role in the transition of the Internet to a commercial entity—this will be discussed later.

In January 1983, all ARPANET nodes were mandated to use the TCP/IP protocols. The transition from the older Network Control Protocol (NCP) was painful, but by June all hosts had made the change. The network was split into two parts: MILNET was to be used for secure military operations, while the original ARPANET now referred to a research network. Both were managed by the Department of Defense. Traffic of a restricted nature was allowed to flow between them, using a machine called a gateway (no relation to the commercial computers by that name). A gateway was in essence a minicomputer configured to pass messages from one network to another, making the necessary adjustments to the packets as needed. Software running on the gateway also contained routing information that determined where a packet was to go and how it was to get there. The gateway evolved into a machine called a router, as the function of determining where to send packets became more and more important. Like the notion of packet switching itself and TCP/IP, the router was, and remains, a key element in constituting the Internet.

In 1977, well before this interconnection, experimental connections were established among ARPANET, a packet radio system in the San Francisco Bay Area, and a European network via satellite. This has led to confusion over when the Internet really began. These 1977 experiments clearly were first, and they joined networks far less similar than MILNET and ARPANET. Nonetheless, the experiments were still part of the "closed world" of military contracting, even if the network's design was open.[31] That began to change only after 1983, as more networks external to the national security establishment were connected. Many would like to identify a moment of invention that corresponds to Samuel Morse's "What hath God wrought?" but the Internet was not invented that way. My choice of 1983 as a starting point can be legitimately challenged, yet it is not an arbitrary choice.

The National Science Foundation

Thus far I have examined three of the seven components that make up the commercial Internet. The Source, CompuServe, Usenet, and BITNET inspired the social model of an online community. From the ARPA came packet switching (developed elsewhere but heavily ARPA supported), interoperability, the Internet protocols, and the informal, open development methods including the RFC. ARPA-supported research also led to the "plumbing" of the Internet: the specialized software and hardware that evolved into the router and the switching programs running on them. I shall next look at three more components: the topology of fiber-optic backbones, the regional networks and ISPs connected to them, and the political system of Internet governance. ARPA had a role in the emergence of these three as well, but their direct ancestor was the NSF.

The NSF, like ARPA, followed an indirect route toward building these pieces of infrastructure. Its immediate goal was to provide access to supercomputers: expensive devices that were to be installed in a few places, with NSF support. Supercomputers are sometimes compared to Formula One race cars: high performance, expensive, hand built, and for limited use. A better comparison is with an astronomical observatory. University astronomy departments have their own telescopes, and individuals with modest resources and skills can do observational astronomy in their backyards, but the cutting edges of astronomy require large installations that can only be funded by governments or wealthy foundations. Likewise, for certain problems at the edges of science, only supercomputers will do.

The United States was the acknowledged world leader in the design and production of these machines, but in the 1980s scientists faced a conundrum. Most of these computers were bought by the U.S. defense and related agencies: the CIA, the National Security Agency, the weapons laboratories at Los Alamos and Livermore, and so on. Thus, the NSF found itself reviewing proposals from American scientists, at American universities, asking for money to travel to the United Kingdom, where some of the few supercomputers not doing classified work were located. The NSF was already supporting a low-cost network for computer scientists—the CSNET, described above—and now it would become involved in a network with supercomputers in mind.

The NSF began work in 1984, and by 1986, under the direction of Dennis Jennings, had linked five supercomputer centers affiliated with universities (Cornell, the University of Illinois, the University of California at San Diego, Princeton, and Carnegie-Mellon) plus the National Center for Atmospheric Research in Boulder, Colorado. The initial link was at fifty-six kilobits per second. That was slow, but computer output in those days was mostly tables of numbers, not rich graphics. Jennings and his cohorts knew that higher speeds were necessary, and had plans for upgrading it from the start. The group made three crucial decisions that would affect later history. The first was to adopt the protocols, TCP/IP, that were promulgated by ARPA—this seems obvious in hindsight, but recall the competing proprietary protocols offered by IBM and others, and it was not evident that this network would interconnect.[32] The second decision was to make it a general-purpose network, available to researchers, not just to a specific discipline. In other words, it *would* interconnect not just supercomputers but with the ARPANET, CSNET, and other networks as appropriate. The third decision was to build the links in the form of a backbone, running across the country, to which not only these five centers but also local and regional networks would connect.[33] These three decisions, as much as any, effected the transition of networking from the ARPANET to the Internet.

That an agency devoted to funding pure science would be involved with engineering a network was not a complete surprise. In June 1984, President Ronald Reagan nominated Eric Bloch to be the NSF director. Bloch was trained as a physicist in Europe, and had a long and distinguished career at IBM, where he was instrumental in directing many IBM innovations including the circuitry for its System/360 computers. Few challenged his qualifications as a scientist, although some grumbled about the choice of a nonacademic scientist to head the NSF. A couple of years later, Bloch appointed C. Gordon Bell to be the assistant director in charge of computer science and computer engineering. Bell also had a distinguished career, at Carnegie-Mellon University and the Digital Equipment Corporation, where he helped design the computers that were the main nodes of the ARPANET—the PDP-11 and PDP-10—and the computer that ran most installations of Berkeley Unix—the VAX. Bloch was a scientist who did engineering for IBM; Bell was an engineer, who once told me that he considered himself a "hardware hacker."

In a 1995 interview, Bell hinted that the NSF knew that the network it was supporting was not going to be restricted to supercomputer access. "[The NSFNET] was proposed to be used for supercomputers. Well, all the networkers knew it wasn't supercomputers. There was no demand. We knew that supercomputers needed bandwidth, they needed to communicate, but when you really force people to use them they would prefer their own machines."[34]

For Bell, what was important was to get a network running, even at a slow speed, and leverage the support the network got from its users to push for higher speeds. In 1987, the NSF awarded a contract to replace the original backbone with a new one, at a speed known as T1: 1.5 million bits per second (Mbps). In 1992, the NSFNET was upgraded to 45 Mbps—T3. As fiber-optic technology matured, it displaced older copper and microwave channels, and set off on a course of exponential speed increases—an equivalent to Moore's Law in silicon chips (Moore's Law is an empirical observation that silicon chip density has doubled every eighteen months since the mid-1960s). Around 1983 the first fiber cable was laid between New York and Washington, and in December 1988 the first transatlantic fiber-optic cable was completed, offering speeds up to 560 Mbps at a much lower cost than equivalent microwave radio circuits. By the late

1990s backbones were measured at multiples of T3 speeds; for example, Optical Carrier–3 (OC-3) was equivalent to three T3 lines, and so on. At the turn of the millennium OC-192 speeds were common. In 2005, Verizon was offering some of its residential customers a fiber line into the home at speeds close to T3. By comparison, first-generation commercial satellites were launched in the 1970s with onboard transponders equivalent to a few T3s each. Land-based microwave relays have an equivalent bandwidth—the fastest microwave channel has the capacity of a half-dozen T3 lines or so. Clearly the move to fiber-optic communications is another driver of the commercial Internet.[35]

The 1987 the NSF contract was awarded to two corporations and a consortium, which offered to build the network for $14 million, plus another $5 million supplied by the state of Michigan. According to the official history, "IBM would provide the hardware and software for the packet-switching network and network management, while MCI would provide the transmission circuits for the NSFNET backbone." The Michigan Education Research Information Triad (MERIT), a consortium of Michigan universities, was to perform the "overall engineering, management and operation of the project." MERIT would also "be responsible for developing user support and information services."[36] Many histories of that era focus on MERIT for its management skills as well as its role in establishing a partnership among the government, academia, and industry, which was a defining characteristic of the Internet's creation. IBM's role was also of interest. One would expect IBM to support only a network that used its own, proprietary Systems Network Architecture, but IBM supported the decision to use TCP/IP. And it constructed switching nodes using clusters of the IBM RT computers that ran Berkeley Unix, not an IBM operating system. This was a critical application for the RT computer. The RT did not become a big seller for IBM, but a more powerful version of it did later on. Initially the VAX was the preferred computer for Internet nodes, but the VAX became obsolete in the 1990s. Many of them were replaced by large IBM RS/6000 computers, using the same architecture as the RT.[37]

Most interesting was the role of MCI, the company that supplied the backbone for the NSF. To this day, MCI (since 2006 a subsidiary of Verizon) is the principal carrier of Internet backbone traffic. And that began with this contract.

The "M" in MCI stands for "microwave." MCI descended from a company founded in 1963 to provide a private microwave link between Chicago and Saint Louis. AT&T, in those days a powerful monopoly, fought MCI at every step, and it was not until 1971, after an expenditure of millions in lobbyists' and lawyers' fees, that MCI was able to offer this service. In 1968 it moved to Washington, DC—according to legend, so that its lawyers could walk to the headquarters of the Federal Communications Commission, where they spent most of their time. MCI eventually prevailed, becoming the first company not part of AT&T to offer long-distance service to consumers. It was the first of many such competitors after the court-ordered breakup of AT&T in

1984. Having fought that battle, and become famous in the consumer and business world for its tenacity and spunk, MCI quietly changed course. At the very moment when nearly all of its revenue came from long-distance voice traffic, MCI looked into a crystal ball and saw that voice traffic was becoming a low-profit commodity, while packet-switched data traffic was growing exponentially. Again quoting from the NSF's history of its network, "MCI had always been known as a voice company, and if MCI was going to learn about data, there was no better opportunity than this project."[38] The company also learned of the new paradigm from Vint Cerf, who it employed from 1982 through 1986. No doubt Cerf continued his role as an evangelist for the Internet and its protocols, as he and Kahn had done while at ARPA.

The contractors delivered a T1 backbone to the NSF by mid-1988. By 1990 the Net connected about two hundred universities as well as other government networks, including those operated by NASA and the Department of Energy. It was growing rapidly. BITNET and Usenet established connections, as did many other domestic and several international networks. With these connections in place, the original ARPANET became obsolete and was decommissioned in 1990. CSNET shut down the following year. Any remaining nodes on those networks were switched over to the NSFNET where necessary.

In this way, the NSF-supported backbone formed the basis of the Internet: a network of heterogeneous networks. Between 1987 and 1990 the name gained currency, although outside of academia the phenomenon was little noticed. The popular press at the time was obsessed with Microsoft and its chair Bill Gates, and ignored news of networking activities. An unforeseen event in November 1988, however, brought the Internet into the public consciousness. On November 2, Robert T. Morris introduced a file on to the Internet that rapidly spread through the network and effectively disabled it. Later called a *worm*, it exploited weaknesses in Berkeley Unix and some quirks in the instruction set of the VAX. Morris was a graduate student at Cornell; his father, also named Robert, was a high-ranking mathematician at the National Security Agency. Morris was caught and convicted of violating the 1986 Computer Fraud and Abuse Act. He was sentenced to probation, community service, and a light fine.

The popular press did pick up this story and in general did a poor job reporting it. At that time, few people understood how this network fit into society; fewer still could estimate the financial cost of damaging it. One figure that was reported gave a sense of how big the network had become: the Morris worm crippled an estimated six thousand computers—only a fraction of those connected. Computer programmers studied and debated the code that Morris wrote. Some hailed him as a hero who revealed the inherent weaknesses in the network; others noted the clever tricks he used in his attack. The press rarely mentioned that his father had published articles, in open scholarly literature, on how easy it was to guess passwords and otherwise break into systems. One Unix programmer looked at the code and criticized it for being sloppy, although he was clearly impressed by the program. After he was caught, Morris claimed that he did not intend to bring down the Internet but did so because of a bug in his program. The presence of lines of code that intentionally tried to cover his tracks suggests that he was being disingenuous. Some called for much harsher punishment. The president of the Association for Computing Machinery was not among them; he noted that the younger Morris had accomplished what his father had tried but failed to do: call attention to the vulnerability of Unix and networks. MIT must have felt the same: Morris is currently (as of 2006) employed there as a professor with tenure.[39]

For this narrative, the Morris worm is significant for two reasons. First, it brought the Internet into the public's consciousness, even if in a negative light. The popular press conveyed a sense that being connected to this network, although fraught with peril, was a sign that you were someone important. Second, the worm had implications for the commercial traffic that the Internet would later handle. Consumers would not buy and sell things, nor would businesses place sensitive corporate data on the Internet, if it were vulnerable to an attack by a bright graduate student. Defending against worms, viruses, "phishing," spoofing, and so forth, has characterized the Internet ever since. As Morris understood and exploited so well, its vulnerability is directly related to its philosophy of open standards and interoperability.

Topology

The topology of the NSF network was different from the ARPANET's. Its backbone crossed the country and linked only fourteen nodes at T1 speeds, whereas ARPANET had over sixty nodes in its 1979 configuration. Below these nodes were sets of regional networks, below them campuswide networks, and below them local networks that might span an office or academic department. It was a treelike structure, with physical connections to the backbone limited to a few branches. That structure allowed the network to grow without becoming unwieldy. It also represented a departure from the initial notion of a distributed network designed to survive a nuclear attack. In practice, however, the topology is quite robust. The term backbone is a misnomer: there are currently several backbones in operation, each connected to the others at selected points, and each capable of picking up traffic if one were attacked or disabled.

This topology dovetailed with an innovation not foreseen by the ARPANET pioneers—namely, the invention and spread of local area networks (LANs). These networks, especially Ethernet, linked workstations and personal computers together at high speeds, and typically spanned an office or campus department. Outside of the home, where Internet connections did not come until later, most users thus accessed the Internet through a desktop computer or workstation connected to Ethernet, which in turn was connected to a regional network, and then to the NSF backbone or its successors.

Other government networks were connected at two Federal Internet Exchange sites: FIX-East and FIX-West in College Park, Maryland, and at Moffett Field, California. The universities were connected through regional networks—for example, MERIT in Michigan, NorthWestNet based in Seattle, SURANET serving the southeastern United States, NYSERNET based in New York, and so on.[40]

Commercial networks were also allowed to connect to the backbone, but that traffic was constrained by what the NSF called an Acceptable Use Policy. The Congress could not allow the NSF to support a network that others were using to make profits from. The policy read, in part:

General Principle
NSFNET services are provided to support open research and education in and among U.S. research and instructional institutions, plus research arms of for-profit firms when engaged in open scholarly communication and research. Use for other purposes is not acceptable.

Specifically Acceptable Uses
...
Communication and exchange for professional development, to maintain currency, or to debate issues in a field or subfield of knowledge....
Announcements of new products or services for use in research or instruction, but not advertising of any kind.
Communication incidental to otherwise acceptable use, except for illegal or specifically unacceptable use.

Unacceptable Use
Use for for-profit activities....
Extensive use for private or personal business....

This statement applies to use of the NSF backbone only. NSF expects that connecting networks will formulate their own use and policies.[41]

Even a reader with no background in the law can sense that those words were chosen carefully. From a current perspective, where e-mail is the most common form of communication, it is interesting to see how the NSF wrestled with the issue of the content of mail. Later chapters will discuss how the academic world has struggled with the modern variant of this conundrum—for instance, having students download music or movies from campus networks. Those who work in corporations face similar policies, which typically allow employees to send occasional personal e-mail, look at news sites, or do some shopping, but the policies prohibit extensive personal use, or visiting pornographic or gambling sites. In 1988, the NSF allowed MCI to connect its MCI Mail service to the network, initially for "research purposes"—to explore the feasibility of connecting a commercial mail service.[42] Cerf had worked on developing MCI Mail while he was employed there. The MCI connection gave its customers access to the growing Internet, and not long after, CompuServe and Sprint got a similar connection.

Glossbrenner, in the 1990 edition of his book, devotes a whole chapter to MCI Mail, calling it "'the nation's electronic mail system.' People may have accounts on many different systems, but *everybody* is on MCI."[43] Whether or not these connections served a research purpose, they were popular.

The political need for the Acceptable Use Policy was obvious. There was also a technical need for it. The topology of the NSF network, the result of a technical decision, encouraged regional networks to connect to the NSF's backbone. That allowed entrepreneurs to establish private networks and begin selling network services, hoping to use a connection to the backbone to gain a national and eventually global reach. The NSF hoped that its Acceptable Use Policy would encourage commercial entities to fund the creation of other backbones with Internet connectivity. That would allow the NSF to focus on its mission of funding scientific research, while bringing in capital from private investors to fund further upgrades to the network. That transition did happen; by 1995, all Internet backbone services were operated by commercial entities. But the transition was awkward. Given the place of the Internet in modern society, it is hard to criticize the government's role in creating it. What criticism there is has centered on two actions the NSF took between 1989 and 1995. The first is how the backbone evolved from a government to a commercial resource. (The second, the establishment of the domain name registry, will be discussed later.)

In 1989 two creators of the nonprofit, New York–area regional network (NYSERNET) founded a for-profit company called Performance Systems International for the purpose of selling TCP/IP network services to commercial customers. Led by William L. Schrader, the company was better able to raise capital since it was not constrained by the regional network's contractual relationship with the government. Schrader moved the company to northern Virginia, where it became known as PSINet (pronounced "Pee-Ess-*Eye*-Net"). Some back in New York complained that Schrader was making a profit from technology he developed under government support, but in any event, PSINet was successful, its stock riding the Internet bubble into the 1990s. In 1999 it bought the naming rights for the Baltimore Ravens stadium, which left a few football fans confused about its pronunciation. Like many Washington-area tech firms, PSINet did not survive the bursting of the bubble and declared bankruptcy in May 2001.[44]

PSINet was joined by another for-profit company that had similar roots. In the mid-1980s, Rich Adams was working for a Department of Defense contractor, where he had access to the ARPANET. He devised a way to make a connection to the Usenet community and its UUCP, and found that many of his colleagues in the Unix world clamored for a similar connection. In 1987 he quit his job, founded a nonprofit company, and began working out of his home in suburban Virginia. He called the company UUNet (pronounced "*You*-You-Net"), a variant of UUCP, although Adams said the name meant nothing.[45] By 1988, the service was funneling vast amounts of traffic between Unix networks and the Internet, all through a single computer located in Arlington,

Virginia. In 1990 UUNet switched to for-profit status, as PSINet did, after being criticized for its tax-exempt status. With venture capital funding it began to grow, and went public in 1995. John Sidgmore, a young but experienced manager, was brought in as its CEO, and its revenues increased dramatically. Its stock soared, largely based on a deal it had with MCI to supply bandwidth for Microsoft's entrée into networking. Subsequent events will be covered in later chapters, but like PSINet, UUNet's stock crashed after the bursting of the Internet bubble. It ended up as a part of MCI (for a while WorldCom; after 2006 Verizon). The primary beneficiary of all this activity was MCI. The combination of MCI's role as a supplier of the NSF backbone, the early connection of MCI Mail to the Internet, plus MCI's alliance with UUNet, made MCI "the world's largest Internet access provider."[46] It may still be, even with all the turmoil those companies have gone through since 1995. UUNet never bought the rights to name a sports stadium, but today the MCI division of Verizon in Ashburn, Virginia, is located on UUNet Drive.

In 1991, PSINet, UUNet, and other commercial providers agreed to set up a connection among one another, in effect creating a network independent of the NSF and not subject to government restrictions. The Commercial Internet eXchange (CIX) allowed a free exchange of data among its subscribers, who paid a modest initiation fee to join. At first CIX was located at a PSINet facility in the Washington, DC, area; it later moved to Silicon Valley. In 1992, a connection was made with the NSFNET through Advanced Network and Services, Inc. (ANS; discussed next), thus linking this traffic to the Internet, although this was subject to the Acceptable Use Policy.

In September 1990, the consortium that built the NSF backbone announced the formation of a nonprofit corporation, ANS, which would give the operation of the backbone a measure of independence from MCI, IBM, and MERIT. It also spun off a for-profit arm in May 1991. This led to criticism, this time among for-profit companies that charged that this company would have privileged access to the backbone. There followed a series of complex events, including congressional hearings and a series of NSF-sponsored conferences at Harvard's Kennedy School of Government, all concerning the need to turn over the backbone to the private sector while preserving a national resource for academic research.

One outcome of that activity was the relaxation of the Acceptable Use Policy. The milestone was contained in the language of an amendment to the Scientific and Technology Act of 1992, which pertained to the authorization of the NSF. Its sponsor was Congressperson Rick Boucher of Virginia's Ninth Congressional District. The legislation passed and was signed into law by President George H. W. Bush on November 23, 1992.[47] Paragraph (g) reads in full:

In carrying out subsection (a) (4) of this section, the Foundation is authorized to foster and support access by the research and education communities to computer networks which may be used substantially for purposes in addition to research and education in the sciences and engineering, if the additional uses will tend to increase the overall capabilities of the networks to support such research and education activities.[48]

The key phrase was "in addition to." With those three words, the commercial Internet was born. And with apologies to Winston Churchill, never before have so few words meant so much to so many.

Whenever the role of Congress comes up in a discussion of the Internet, another politician, not Boucher, is mentioned. As a candidate for president, in 1999 Vice President Al Gore told Wolf Blitzer of CNN that "during my service in the United States Congress, I took the initiative in creating the Internet." The press, led ironically by the online version of *Wired* magazine, ridiculed him. A press release typically distorted his words to a claim to have "invented" the Internet, which he never said. Gore was a senator from Tennessee when the Internet emerged out of ARPANET and the NSF. His father, Al Gore Sr., had also been a senator and in the 1950s had championed what became the Interstate Highway System. His son did not coin the term *information superhighway* but he popularized that term while in the Senate and may have had his father in mind. In the early 1980s, Gore proposed federal funding for a national computer network, primarily for the scientific community. The bill (informally called the Gore Bill during its gestation) went through a number of changes, was passed as the High Performance Computing Act of 1991 (Public Law 102–194), and was signed into law by President George H. W. Bush. It was seen at the time as the first step in implementing a "High Performance Computing and Communications Initiative" that many federal agencies would contribute to and benefit from.[49] Gore envisioned a National Research and Education Network, which would have a backbone orders of magnitude faster than the existing NSF backbone and would be the "successor" to the Internet.[50]

By the time the bill was signed into law, events had overtaken the idea. The Internet became its own successor. The emerging Internet was more and more a link of LANs of personal computers and workstations, and e-mail was a primary source of traffic. Supercomputers were important, but less so. Opening up the backbone to commercial traffic liberated venture capital, which drove the Internet after 1993. A segment of the academic community lobbied for a closed, research-oriented network (under the name of Internet2), but the Internet had moved on. William Wulf, Bell's successor as assistant director for computing research and education at the NSF in the late 1980s, saw this transition. In 1993, as the commercialized Internet was bursting out of the NSF shell, Wulf summarized the feeling among academics, saying, "I don't think any of us know where this thing is going anymore, but there's something exciting happening, and it's big."[51] Subsequent chapters will show the accuracy of that statement.

Gore's role in championing a network echoes that of his father. His father proposed a network of superhighways that would have been funded by general tax revenues, but an expansion of the federal bureaucracy was unacceptable to the Eisenhower administration. The highways were built only after President Dwight David Eisenhower brokered an agreement to fund them from a dedicated tax on fuels, monies that go into a highway trust fund. Thus "Ike," not Senator Gore, got credit for the Interstate Highway System. Al Gore Jr. deserves credit for helping create the Internet, but the network he championed was not the Internet we know today. One must keep this in mind when speaking of the government's role in fostering this or any similar technology. There is no such thing as "the government." There are a number of competing agencies that have different agendas, sources of funding, and access to legislative or executive power. Histories of the Internet rightly credit the ARPA as the source of the protocols; what few histories mention is that other agencies of the government were, at the same time, promoting an alternative suite of protocols, call the Open Systems Interconnection (OSI).[52] I have not found evidence that Gore favored one protocol suite over another. The National Bureau of Standards (later the National Institute of Standards and Technology) supported the OSI, as did other agencies in the United States and abroad. Well into the 1980s, one reads accounts of how the TCI/IP will somehow either be absorbed into the OSI or be replaced by it. Neither happened. The National Institute of Standards and Technology kept supporting the OSI until the mid-1990s, when it realized that its assumption that TCP/IP was only an interim standard was naive.

With the relaxation of the Acceptable Use Policy, the NSF's role in maintaining the backbone receded. In 1993, the agency brokered an arrangement to replace its backbone with the very high speed Backbone Network Service, operated by MCI, for research use. This would be one of several commercially operated backbones, any one of which would be available for regional networks to connect to. These backbones would be connected at only a limited number of points, called Network Access Points (NAPs). Initially there were only four: two on the East Coast, one in Chicago, and one in the Silicon Valley area. These points became heavy concentrations of ISPs, and were called "a sensible alternative to the system that preceded it—a series of expensive point-to-point connections between the backbones of the various Internet providers."[53] (In fact, they quickly became congested choke points and were soon supplemented by other connections.)

These sites became places where ISPs connected to one another as well, in an arrangement known as *peering*, whereby each ISP agreed not to charge the other for traffic traveling through it as long as the traffic was even over the long term. The ISPs—with names like Digex, Earthlink, BBN Planet, and others—gradually replaced the nonprofit regional networks that had previously formed the second tier of the NSF's Internet topology. Along with fiber companies, they dug trenches in metropolitan areas and laid high-speed rings of fiber, the most popular called Synchronous Optical Network (SONET), which served customers in the region. For a time before the crash of 2000, city streets were full of trenching machines laying fiber. In the frenzy to lay cable, companies sometimes cut the fiber of another provider as they laid their own cables. SONET has a ring structure, so that if it is cut in one place the signal can get around the other way, but errant backhoes continue to be the bane of the Internet.

The NAPs in San Jose and the Washington, DC, area were operated by a company called Metropolitan Fiber Systems (MFS), and they took on the name Metropolitan Access Exchange (MAE)–West and MAE-East. (The similarity of the name of the MAE-West to the actress was a coincidence.) MAE-East became an especially busy hub. It absorbed the federal connections that had been at FIX-East in College Park, and at one time it switched over 50 percent of all Internet traffic—ironic in view of the popular notion of the Internet being a distributed network. Likewise, as the Internet expanded overseas, European providers found it cheaper and politically easier to connect their countries by routing traffic via undersea cables, across the Atlantic to an East Coast NAP, then back across the Atlantic again to another European country (hence the brief disruption of intra-European service after the September 11, 2001, attacks on the World Trade Center on Manhattan).

MAE-East occupied a walled-off section of a parking garage in an ordinary office building in Vienna, Virginia—a few blocks from AOL's headquarters at that time. It ran largely unattended, but those who went there to install or maintain a connection spoke of its lack of climate control, little security, and no backup power generators. MAE-West in San Jose, like MAE-East "one of the busiest thoroughfares in the Internet," was in an ordinary office building downtown. But these were, for a time, where the action was: "If you want to predict hot new trends in Internet infrastructure, you should poke around the MAE's."[54] In the late 1990s you could actually do that; today, a visitor "poking around" is likely to be interrogated by a security officer. MAE-East became legendary among network fanatics as the mysterious secret place where the whole Internet converged, an Internet equivalent to Area 51. MFS did not publicize its location, and most Web sites devoted to this folklore give the wrong address. By the time this tale arose, the MAEs lost their significance. The heavy concentration of switching was inefficient, and many ISPs established alternate connections elsewhere. And since the turn of the millennium, MCI (which absorbed MFS along with UUNet) dispersed MAE-East along a thirty-mile corridor between Vienna and Dulles Airport in Virginia, and MAE-West throughout Silicon Valley.

The Domain Name System

One final component needs to be discussed to finish my narrative of the transition to a commercial Internet, and this is the second place where the NSF's role has been criticized. We saw how the connection of private networks via CIX and later the NAPs created in effect an alternate Internet, not subject to the NSF or any direct governmental control. We also saw the rapid growth of personal services like AOL, after entrepreneurs such as Case began to aggressively market them to the public. That has led to claims that the modern world of cyberspace would have happened anyway, regardless of what the NSF did.[55] The fundamental reason why this was not so was the overwhelming advantage of the Internet's philosophy of inclusion and openness. Related to that was a need for a minimal level of governance, which allowed inputs from different parties and was not under the control of a for-profit corporation. When one looks at the recent history of AOL or especially AT&T, it is clear that having a company like that running the Internet would have been unwise. Nevertheless, the governance that did emerge, though not corporate, has been criticized for its lack of transparency.

Besides an agreement on the TCP/IP and other protocols, the Internet requires a common addressing scheme. The analogy is not with highways but with the telephone system, especially after AT&T was broken up in 1984. Anyone is free to set up a telephone system, and if one follows the proper electric standards, such a system can be connected to the U.S. telephone network. But if a person using it wants to call or be called by someone on the rest of the network, their area code and phone number had better not be the same as someone else's. Likewise a commercial Internet service, like Amazon.com, needs to preserve the exclusivity of its address. Amazon might have trouble selling books if, at random intervals, customers who tried to access it were connected instead to a women's roller derby team. Some entity has to control the allocation and registration of these addresses. As the Internet evolved, the control of this became almost identical with the governance of the Internet in general.[56]

Through the 1970s and into the early 1980s, the ARPANET managed this by having a simple file called HOSTS.TXT that contained the numerical address for each computer connected to the network. As new nodes were added, subtracted, or moved, the file was updated accordingly. A copy of it was stored on each computer, and those responsible for the management of the nodes were responsible for downloading the latest version on to their machines. HOSTS.TXT was a simple, small file at first, and the scheme worked well as long as the network was small. It had obvious drawbacks, including the need for managers to download and keep the latest file current, but the main problem was that it did not scale well. As the ARPANET grew, the size of HOSTS.TXT grew too, and the overhead involved in repeatedly downloading the file grew to unmanageable proportions.[57]

Beginning in August 1982, Zaw-Sing Su, Jonathan Postel, Paul Mokapetris, and others authored a series of RFCs that proposed replacing this system with a hierarchical naming convention, which came to be known as the Domain Name System (DNS). The hierarchy is familiar to most users today: for example, a server in the Computer Science Department of Big State University, an educational institution, might have the address ⟨server.cs.bigstate.edu.⟩ The advantage of this system is that a search need not go all the way to the *root* (.edu) unless a lower-level directory does not have the requested information. A computer in that department of that university could connect with all the other computers in that department without querying any computers outside the building, and it was reasonable to assume that a large percentage of queries to the address file would be of this type.

Initially the root of this tree was a node called .arpa; old ARPANET addresses were mapped on to it. That faded away as the Internet evolved, leaving a set of seven so-called top-level domains: .com, .edu, .mil, .net, .org, .gov, and .int. Most readers are familiar with these, except for the last one. That domain, for international treaty organizations, was later replaced by two-letter country codes as specified by the International Organization for Standardization's conventions—for instance, .fr for France. The addition included a domain for the United States: .us. The letters used to name hosts were restricted to a subset of ASCII and the twenty-six letters of the English alphabet. That decision seemed obvious at the time, when only a few nodes existed outside the United States, and when computers had little memory capacity. But that decision, made for technical reasons, would reverberate later on, when the Internet had a global reach and was vital to the commerce of countries that used not only the accented letters of the Roman alphabet but different alphabets altogether.

The addresses that the routers, nodes, and switches used were not names but thirty-two-bit binary numbers, which mapped on to the names. For example, the address of a server at the Smithsonian's National Air and Space Museum was, written in decimal, ⟨160.111.70.27⟩. Having that named ⟨nasm.si.edu⟩ made it much easier to remember, obviously. This decision, too, would later lead to criticism—say, from the hypothetical roller-derby team above, which might wonder why it could not use the term Amazon even if that was the registered name of its team.

The Internet began converting to this system in 1984 and was fully converted by 1987. It was a brilliant technical design. It allowed additions, deletions, and changes to addresses without having always to access the top-level root server. There was no single file that everyone had to download and keep up to date. Above all, it scaled well. In the world of the Internet, that is the highest compliment one can give to a system.[58]

From a social standpoint it worked, too, throughout the early, noncommercial phase of Internet growth. In late 1992, academic addresses (.edu) outnumbered commercial (.com) nodes by approximately 370,000 to 304,000. These two together vastly outnumbered all the others as well.[59] After the Acceptable Use Policy was amended, the balance shifted to dot-com, and Internet addresses began climbing into the millions. University information technology administrators often named their nodes after what caught their fancy—after departments, buildings, science-fictional characters, or comic book figures, such as "Calvin" and "Hobbes." Such laid-back naming was not suitable for the dot-com world. There, the names took on a monetary value, something not foreseen when the DNS was set up. During the inflation of the Internet bubble in the late 1990s, a secondary market arose whereby people traded domain names like

⟨sex.com⟩ or ⟨business.com⟩ for absurd values. By the time people became aware of the inequity of the scheme, the system was in place and working so well that it was difficult to change.

One place where the conversion was deficient in a technical sense was in the routing of e-mail from different systems connected to the Internet. Recall that this was an issue with the personal computer services that Glossbrenner surveyed; it was also an issue for the first few years after the DNS went into effect. For example, mail from an Internet user to someone running UUCP took on the form "user%host.UUCP@uunet.uu.net." Someone with an account on MCI Mail composed an address as "TO:full user name ENS: INTERNET MBX:user@host."[60] UUCP could sometimes be especially arcane, with a need to specify all the intermediate nodes, each separated by a "bang" (exclamation point) on the way to a destination. By 1991 these differences were resolved, and the familiar ⟨user@host.edu⟩ (or .com and so forth) became the norm for nearly all the mail services.

Those who devised the DNS did not foresee a time when dial-up connections for e-mail became common for the home, but that turned out not to be a problem. These home computers would not have a permanent domain name address but would be assigned a temporary one on the fly, each time a person logged on. Other protocols, operating in tandem with TCI/IP, handled the connection of these computers smoothly. Literature from the early 1990s spoke of two tiers of access: those with a full Internet connection, and those who dialed in and could only use it for e-mail and a few other simple applications. Some of this will be discussed in later chapters; in any event, that distinction faded as the World Wide Web became popular and broadband connections became more available.

There remained the issue of what entity would be the keeper of the registry of names and numbers. That was not an issue with the military-sponsored ARPANET, and when the NSFNET began it seemed logical that the NSF would take that responsibility. It did not quite work out that way. The Department of Defense kept control over the addresses of MILNET, and by implication, the dot-mil domain. But it also held control for a while over the rest of the NSFNET as well. The Department of Defense funded an agency called by the awkward name of Internet Network Information Center (InterNIC) to manage the registration. Those wishing to receive a domain name address sent a request to InterNIC, and if they qualified and no one else had that name, it was granted. There was no charge. Nor was there a test to determine whether that person was the best qualified to receive the name. Foreign countries often imposed restrictions on who could register under their two-letter country code, and thus many foreign businesses simply registered a dot-com address, if one was available. Conversely, some U.S. companies registered domains in foreign countries—the tiny Pacific nation of Tuvalu was popular because of its code: .tv. Registering names became a major source of revenue for Tuvalu.

InterNIC was at the Stanford Research Institute in northern California, but Jonathan Postel, who worked in Southern California, had a lot of personal control over this registry. Some informal accounts say that Postel kept the registry on a computer in his office in Marina Del Rey, and in effect ran the entire DNS with the help of an assistant. I have been unable to confirm this, but it is not as outlandish as it sounds. Shortly before his untimely death in 1998, Postel unilaterally asked the administrators of the servers keeping the registry (by then there were thirteen) to direct all their information to him; about half the administrators obeyed his request. In other words, he hijacked the Internet, if only briefly, to make a point about its governance.[61]

Only in 1993 did the NSF formally take over this job from the Department of Defense, as authorized by a clause in the High Performance Computing Act. What followed next was the most contentious decision about Internet policy, one that has been strongly criticized. The previous year saw control of the DNS transferred from the Stanford Research Institute to a Washington, DC, company called Government Systems, Inc. (GSI). Briefly the registry was located at a GSI office in Chantilly, Virginia, south of Dulles Airport. In January 1993, the NSF entered into a contract with another company, Network Solutions, Inc. (NSI), to manage the registry of five of the top-level domains: .com, .org, .net, .edu, and .gov. Registration was free, with the NSF assuming the costs.[62] According to an account of this transfer published by the legal community, "The origins of NSI are unclear. NSI's public documents state that NSI was founded in 1979. However, according to Jon Postel, NSI was a spin-off of GSI in 1992."[63] Two years later, the San Diego–based defense contractor Science Applications International Corporation (SAIC) bought NSI; soon after that, the company renegotiated its contract with the NSF to allow it to charge an annual fee for domain name registration. As dot-com addresses became valuable commodities, this change in policy suddenly made NSI a valuable company indeed. (In 2000, just before the bursting of the Internet bubble, SAIC sold the registry business to Verisign for an estimated $3 billion profit.)[64]

Postel was among those who criticized this arrangement. He and others proposed adding more top-level domains and letting other companies compete for the business. His suggestion was not adopted, and this arrangement with NSI lasted into the new millennium. The controversy simmered for the next few years, leading in 1998 to the creation of a new entity, ICANN, which since that time is regarded as the entity that governs the Internet for all practical purposes.

Formally, the Department of Commerce assumed the authority to manage the DNS, and by extension, the governance of the Internet. As of 2006 the controversies were far from settled. The issue of U.S. control has been especially nettlesome now that many countries rely on the Internet for their daily commercial and government operations. This dispute became public at the United Nations–sponsored World Summit on the Information Society, held in Tunis in November 2005. There, about seventy countries pressured the United States to turn over its oversight of the Internet to the United Nations. The United States refused. (It was during the preparations for this meeting that the Department of Commerce first formally stated its claim to oversight of the Internet.) Related to that issue is the perception that ICANN does not make its decisions in an open and fair manner, as the late Postel did in editing and promulgating the RFCs. In 2005, ICANN proposed an .xxx domain for adult (pornographic) Web sites, but it was forced to back down because of pressure from Christian groups, which lobbied the U.S. Congress. That was further evidence that the Internet was too much under the control of politicians. Also in 2005, a movement arose in parts of the world that used non-Roman alphabets, especially China and the Middle East, to set up a separate Internet not connected to the DNS. A similar movement also arose in Germany and the Netherlands, with the unexpected support of Paul Vixie, an Internet pioneer who manages one of the domain name servers in Palo Alto, California.[65] Postel had passed away by this time, but one can sense his spirit presiding over these activities. For the topics addressed in this book, these disputes are of more than academic interest, as they go to the heart of the ability of the Internet to perform commercial services.

When NSI received this contract, it established the A: Root server at its offices in Herndon, Virginia, near Dulles Airport, and twelve other servers, which mirrored the one in Herndon and provided a backup. Of these thirteen, five were in northern Virginia. Another two were in Maryland, three in California (including Vixie's F: Root server in Palo Alto), and one each in the United Kingdom, Japan, and Sweden. So there was not much dispersion of this critical resource. For technical reasons it was not practical to have more than thirteen servers maintaining the registry, but recently a technique called *anycasting* has allowed data from one to be replicated seamlessly on other computers. After the 2001 terrorist attacks on New York and the Pentagon, these files were further dispersed and protected. The location of the A: Root server is no longer disclosed to the public. ICANN has an office in California, but the corridor between Vienna, Virginia, and Dulles Airport remains the primary place where the domain name registry is kept and managed. As mentioned, both AOL and MCI have their headquarters in Ashburn, Virginia, and both Cerf and Kahn remain active in Internet activities and work in the region. MAE-East still exists in the corridor as well.

A visitor to northern Virginia's "Internet Alley" who expects to find a "master control" computer such as those seen in old science-fictional movies like *2001, a Space Odyssey* or *Colossus: The Forbin Project* will be disappointed. In 2003, I was privileged to be allowed a visit to one of the root servers at a nondescript office park near Dulles Airport. There I saw an IBM RS/6000 computer that stored the domain name registry; it had just replaced a similar machine from Sun. Next to it were racks of servers—they looked like pizza boxes—from Sun, Dell, IBM, and a few less familiar names. These helped distribute the registry to the ISPs that requested it. I saw no Cray or other supercomputer. I was told that the Root Zone File was very small, but it was receiving millions of queries a day. Twice a day, the A: Root server would send an update of the Root Zone File to the other twelve servers, which together comprise the "root." I saw what apparently was a master control console: a person operating an ordinary Dell personal computer, connected to a digital projector. The operator was projecting a map of the world, which showed the status of the root file at all the backup sites around the world. The map was written in PowerPoint. Such is the nature of the Internet.[66]

Denouement

By 1995, nearly all of components to comprise a commercialized Internet were in place.[67] To summarize, they were:

- Several high-speed backbones run by commercial entities (in the late 1990s these included UUNet, ANS, Sprint, BBN, and MCI; some of these merged or later spun off subsidiaries).
- The gradual replacement of the nonprofit regional networks by ISPs, including several that were dominant: AOL, CompuServe, Microsoft, Earthlink, Netzero, and AT&T.
- Below that level, the widespread adoption of Ethernet as a high-speed local network to link workstations and personal computers.
- The establishment of smaller-scale ISPs that provided dial-up access for individuals at home. The first of these may have been the World, based in the Boston area, followed by many others.
- The availability of personal computers that had good graphics capability (using Intel's Pentium processor), a built-in modem, communications software, and the Windows 95 Operating System with TCP/IP included.
- The end of the Acceptable Use Policy, and the end of the active participation of the NSF in making Internet policy, except for the DNS. The NSFNET was retired in 1995; by that time ARPANET, CSNET, and BITNET were all either retired or configured to run TCP/IP, and in effect become part of the Internet.
- The interconnection of most of the private networks to the Internet. The most significant of these was AOL, which initially provided an e-mail connection, and eventually a full connection.
- The establishment of a system of governance based on NSI's (later Verisign's) contract to maintain the domain name registry. ICANN would not be founded until 1998, but its predecessors were in place by 1995.
- Finally, the development of the World Wide Web, a program that allowed an easy on-ramp to the information superhighway.

The Internet of course did not stop evolving after 1995, but these pieces remained stable enough to provide a platform for the commercial activities that followed. The next two chapters will elaborate somewhat on this platform and its subsequent evolution; they are followed by examples of what was built on top of that.

Notes

1. Katie Hafner and Matthew Lyon, *Where Wizards Stay Up Late: The Origins of the Internet* (New York: Simon and Schuster, 1996), 10.

2. ARPA was founded in 1958 and later renamed the Defense Advanced Research Projects Agency (DARPA). Its name reverted to ARPA in 1993. The original acronym is used throughout this chapter.

3. In addition to the book by Hafner and Lyon, cited above, I recommend Janet Abbate, *Inventing the Internet* (Cambridge, MA: MIT Press, 1999), and Arthur Norberg and Judy E. O'Neill, *Transforming Computer Technology: Information Processing for the Pentagon, 1962–1986* (Baltimore, MD: Johns Hopkins University Press, 1996). It should not be surprising that one can also find excellent histories on the Internet itself, although the quality varies. Some are cited in this chapter, but rather than give a comprehensive survey of online histories, I suggest that the reader perform their own search under "Internet history."

4. J. C. R. Licklider, "Man-Computer Symbiosis," *IRE Transactions on Human Factors in Electronics* 1 (1960): 4–11; J. C. R. Licklider and Welden C. Clark, "On-Line Man-Computer Communication," *AFIPS Proceedings, Spring Joint Computer Conference* (1962): 113–128; and J. C. R. Licklider, with Robert C. Taylor, "The Computer as a Communication Device," *Science and Technology* 76 (1968): 21–31.

5. ARPA Network Information Center, "Scenarios for Using the ARPANET at the International Conference on Computer Communication, Washington, DC, October 24–26, 1972." Copy in the author's possession.

6. This information has been taken largely from Abbate, *Inventing the Internet*, and Norberg and O'Neill, *Transforming Computer Technology*.

7. April Marine, Susan Kirkpatrick, Vivian Neou, and Carol Ward, *Internet: Getting Started* (Englewood Cliffs, NJ: Prentice Hall, 1993), 156. They define a network protocol as "a rule defining how computers should interact."

8. Ray Tomlinson, address to the American Computer Museum, Bozeman, MT, April 28, 2000.

9. U.S. Department of Commerce, National Telecommunications and Information Administration, "Domain Names: U.S. Principles on the Internet's Domain Name and Addressing System," available at ⟨http://www.ntia.doc.gov/ntiahome/domainname/USDNSprinciples_06302005.htm⟩ (accessed May 9, 2006).

10. The IBM PC, running Microsoft's DOS operating system, was introduced in 1981 and became the standard in the years thereafter. Of all the competing personal computers that were being sold between 1977 and 1983, only a few, such as the Apple II, remained as an alternative to the IBM platform.

11. Alfred Glossbrenner, *The Complete Handbook of Personal Computer Communications* (New York: St. Martin's Press); the first edition was published in 1983, and the third edition in 1990. The quotation is from the third edition, xiv.

12. Ibid., 1983 edition, 8.

13. Stewart Brand, ed., *Whole Earth Software Catalog* (New York: Quantum Press/Doubleday, 1984), 139.

14. Ibid., 140 (table).

15. Ibid., 144.

16. CompuServe databases were handled by Digital Equipment Corporation PDP-10s and its successor computers, which used octal arithmetic to address data. CompuServe users thus had account numbers that contained the digits zero through seven but never eight or nine.

17. Abbate, *Inventing the Internet*, 80.

18. Glossbrenner, *Complete Handbook*, 1983 edition, 68.

19. Kara Swisher, *aol.com: How Steve Case Beat Bill Gates, Nailed the Netheads, and Made Millions in the War for the Web* (New York: Random House, 1998), chapter 2.

20. Ibid.

21. Ibid.; see also Glossbrenner, *Complete Handbook*, 1983 edition, chapter 10.

22. Daniel P. Siewiorek, C. Gordon Bell, and Allen Newell, *Computer Structures: Readings and Examples* (New York: McGraw-Hill, 1982), 387–438; Brand, *Whole Earth Software Catalog*, 138–157.

23. Glossbrenner, *Complete Handbook*, 255.

24. John S. Quarterman and Josiah C. Hoskins, "Computer Networks," *CACM* 29 (October 1986): 932–971.

25. This is sometimes expressed informally as Metcalfe's Law: the value of a network increases as the square of the number of nodes connected to it.

26. William Joy and John Gage, "Workstations in Science," *Science* (April 25, 1985): 467–470. Later in the 1980s it was upgraded to 4.3, to allow more interconnectivity to heterogeneous networks.

27. Douglas Comer, "The Computer Science Research Network CSNET: A History and Status Report," *CACM* 26 (October 1983): 747–753.

28. Quarterman and Hoskins, "Computer Networks"; see also John S. Quarterman, *The Matrix: Computer Networks and Conferencing Systems Worldwide* (Bedford, MA: Digital Press, 1990).

29. Quarterman states (in *The Matrix*) that the term is an acronym for "Because It's Time NETwork"; other accounts say it stood for "Because It's There NETwork," referring to the existence of these large IBM installations already in place.

30. The Smithsonian Institution, for example, had BITNET accounts for its employees long before most of them had an Internet connection.

31. The phrase is from Paul Edwards, *The Closed World: Computers and the Politics of Discourse in Cold War America* (Cambridge, MA: MIT Press, 1996).

32. The initial NSF backbone used networking software called fuzzball, which was not carried over to the T1 network.

33. "NSFNET—National Science Foundation Network," Living Internet, online resource, available at ⟨http://www.livinginternet.com/i/ii_nsfnet.htm⟩ (accessed November 10, 2005); see also Jay P. Kesan and Rajiv C. Shah, "Fool Us Once, Shame on You—Fool Us Twice, Shame on Us: What We Can Learn from the Privatization of the Internet Backbone Network and the Domain Name System," *Washington University Law Quarterly* 79 (2001): 106 ⟨http://papers.ssrn.com/sol3/papers .cfm?abstract_id=260834#paper%20Download⟩ (accessed electronically May 4, 2006).

34. C. Gordon Bell, interview with David Allison, April 1995, Smithsonian Institution, National Museum of American History, Oral History Collections, ⟨http://americanhistory.si.edu/collections/ comphist/bell.htm⟩ (accessed electronically May 4, 2006).

35. Bell, interview; see also David J. Whalen, "Communications Satellites: Making the Global Village Possible," unpublished manuscript, 2004; Shawn Young, "Why the Glut in Fiber Lines Remains Huge," *Wall Street Journal*, May 12, 2005, B1, B10; "Verizon FIOS: Quick Facts to Get You Up to Speed," brochure, ca. 2005, distributed to Verizon customers in the Washington, DC, area.

36. National Science Foundation, "NSFNET: A Partnership for High-Speed Networking," final report, 1987–1995 ⟨http://www.livinginternet.com/i/ii_nsfnet.htm⟩ (accessed electronically November 15, 2005).

37. The "R" in RT stood for "reduced," indicating that the computer's instruction set was smaller and simpler than that used by the VAX or System/360.

38. National Science Foundation, "NSFNET," 9.

39. Bryan Kocher, President's Letter, *CACM* 32/1 (1989): 3, 6; Robert Morris Web site, available at ⟨http://pdos.csail.mit.edu/~rtm/⟩ (accessed June 2006); Charles Schmidt and Tom Darby, "The What, Why, and How of the 1988 Internet Worm," available at ⟨http://snowplow.org/tom/ worm/worm.html⟩ (accessed June 2006).

40. National Science Foundation, "NSFNET," 24–27.

41. Ed Krol, *The Whole Internet User's Guide and Catalog* (Sebastopol, CA: O'Reilly, 1992), appendix C.

42. Robert Kahn, personal communication with the author September 4, 1999.

43. Glossbrenner, *Complete Handbook*, 1990 edition, 265.

44. Ellen McCarthy, "Prominent Tech Figure Returns," *Washington Post*, June 21, 2005, D4.

45. Kara Swisher, "Anticipating the Internet," *Washington Post*, business sec., May 6, 1996, 1, 15–16.

46. Patricia Sullivan, "Technology Executive John Sidgmore Dies at 52," *Washington Post*, December 12, 2003, B7.

47. Web site of Congressperson Rick Boucher, available at ⟨http://www.boucher.house.gov⟩ (accessed June 5, 2006).

48. 42 United States Code (USC) 1862, paragraph g.

49. U.S. National Research Council, Computer Science and Telecommunications Board, "Evolving the High Performance Computing and Communications Initiative to Support the Nation's Information Infrastructure" (Washington, DC: National Academy of Sciences, 1995), 1.

50. Al Gore, "Infrastructure for the Global Village," special issue, *Scientific American*, September 1991, 150–153.

51. Christopher Anderson, "The Rocky Road to a Data Highway," *Science* 260 (May 21, 1993): 1064–1065. Wulf at this time was a professor at the University of Virginia. When they both taught at Carnegie-Mellon University, Wulf and Bell were the two principal architects of the PDP-11 computer, one of the most common machines connected by the ARPANET.

52. Abbate, *Inventing the Internet*, chapter 5.

53. Stimson Garfinkel, "Where Streams Converge," *Hot Wired*, September 11, 1996 ⟨http://www.hotwired.com/packet/garfinkel/97/22/index29.html⟩ (accessed electronically December 12, 2006).

54. Ibid.

55. See, for example, Peter H. Salus, *Casting the Net: From ARPANET to INTERNET and Beyond* (Boston: Addison-Wesley Professional, 1995), 237; see also various Usenet discussions on "When the Internet Went Private," in alt.folklore.computers, ca. September 7, 2000 ⟨http:/www.garlic.com/~lynn/internet.htm#0⟩ (accessed via the Internet Archive, May 15, 2006).

56. Milton L. Mueller, *Ruling the Root: Internet Governance and the Taming of Cyberspace* (Cambridge, MA: MIT Press, 2002).

57. U.S. National Research Council, *Signposts in Cyberspace: The Domain Name System and Internet Navigation* (Washington, DC: National Academies Press, 2005), chapter 2.

58. Ibid., 47.

59. Marine et al., *Internet: Getting Started*, 195.

60. Ibid., 221; see also Glossbrenner, *Complete Handbook*, 1990 edition, chapters 20–21.

61. Katie Hafner, "Jonathan Postel Is Dead at 55; Helped Start and Run Internet," *New York Times*, October 18, 1988 ⟨http://www.myri.com/jon/nyt.html⟩ (accessed electronically December 16, 2005).

62. National Research Council, *Signposts in Cyberspace*, 75.

63. Kesan and Shah, "Fool Us Once," 171.

64. Ibid. See also National Research Council, *Signposts in Cyberspace*, 76; Steven Pearlstein, "A Beltway Bubble About to Burst?" *Washington Post*, June 20, 2003, E1.

65. Christopher Rhoads, "In Threat to Internet's Clout, Some Are Starting Alternatives," *Wall Street Journal*, January 19, 2006, A1, A7.

66. U.S. National Research Council, *The Internet under Crisis Conditions: Learning from September 11* (Washington, DC: National Academies Press, 2003). The Internet survived the attacks well.

67. Kesan and Shah, "Fool Us Once," passim.

II Internet Technologies Seeking a Business Model

3   Innovation and the Evolution of Market Structure for Internet Access in the United States

Shane Greenstein

How and why did the U.S. commercial Internet access market structure evolve during its first decade? Neither question has a ready answer. Within the United States, the first country to commercialize the Internet, business development did not follow a pre-scribed road map. Events that came after the National Science Foundation (NSF) priva-tized the Internet—such as the invention of World Wide Web and the commercial browser—created uncertainty about market value. In the midst of diffusion a major piece of legislation, the 1996 Telecommunications Act, altered many of the regulatory limits that shaped business decisions. Technological capabilities and business opera-tions coevolved as firms sought solutions to their business problems. Virtually all firms were open about their exploratory motives, taking a variety of approaches to commer-cialization, and refining business practices as the environment changed and they learned from experience.

These efforts built a network that made the information of the online world available to participants nationwide for a low cost. After a decade, more than half of U.S. house-holds were using the Internet (see figure 3.1), as were a greater fraction of business establishments. The access market alone generated revenues of over $23.4 billion in 2004—an impressive accomplishment since this does not include activities built on top of it, such as electronic retailing or auctions.[1]

In this chapter I describe the most salient features of the market structure and ana-lyze its evolution, focusing on developments in pricing, operational practices, and geographic coverage. One of my goals is to explain events in an accessible narrative—accessible, that is, to those who did not live through these complex events. A second goal is to further our understanding of how market structure shaped innovative behav-ior and visa versa.

I choose this latter emphasis for three reasons. First, it is appropriate for the era under scrutiny. During the first decade of the commercial Internet, many of the most basic facets of the commercial Internet were invented, tried in the marketplace, and re-fined. Second, it is rare to find such a rich and well-documented window into innova-tive behavior during the early years of a new market. These events will have value to

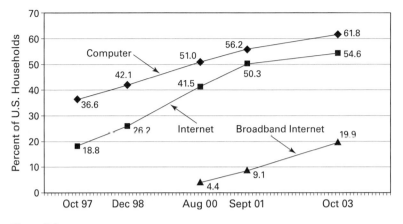

Figure 3.1

Percent of U.S. households with computers and Internet connections, selected years, 1997–2003. *Note*: 2001 and 2003 reflect Census-based weights, and earlier years use 1990 Census-based weights.

scholars of young industries. Third, contemporary policy debates often mischaracterize the lessons for innovative behavior from recent events. These lessons still are relevant. To paraphrase Bruce Owen, the players have reached only the fifth inning of a nine-inning game, and there is no rain delay in sight.[2]

This chapter seeks to make the relationship between market structure and innovative behavior more central to writing about the history of the Internet. This emphasis amplifies the contributions of both successful and failed entrepreneurial firms, the lessons learned during competition between incumbent firms and innovative entrants, and the consequences of regulatory decisions for entrepreneurial activities. This focus is challenging to execute because of the genuine technical and regulatory complexity of events, which resist simplified narratives. It also is challenging due to the ensemble cast of participants in the Internet access market. Because they were drawn from disparate backgrounds, there was too much variety to define the innovation coming from a "typical" Internet access firm at any point in time, or for that matter, over time.

The Organization of the Chapter

I examine three time frames, or snapshots, of the market structure during the first decade—namely, 1993, 1998, and 2003. These snapshots capture the salient features of each year, while allowing readers to perceive changes over time.

The first snapshot, in 1993, is loosely speaking just after the commercialization of Internet access in the United States. While later observers recognized the significance of this event, it hardly received notice at the time. The first advertised prices for commercial access appear in this year, mostly from descendants of the NSFNET and not from the many other commercial firms that soon would have interests in the commercial network. Entrepreneurial firms explored the opportunity, but only a few incumbent firms did. These investigative actions eventually demonstrated the viability of the commercial market.

The second snapshot, in 1998, captures the urgent investment in infrastructure markets to meet the growing demand for Internet services. A clash of strategic visions and commercialization approaches reached its apex this year and the next. This clash took place during the emergence of contracting norms and business practices that helped a large-scale network reach mass-market users with a wide variety of needs.

The third snapshot, in 2003, comes in the midst of a unique combination of economic triumph and hardship, stability and change. A second wave of investment in broadband began to cannibalize the services from firms that developed dial-up access. Investments in wireless access augmented the existing network, but a diversity of approaches clashed. These investments coincided with changes in the identities of leading firms.

Market Structure, Innovation, and Variety

I will focus on a particular feature of competitive behavior: the variety of innovative commercial actions. Here *variety* means the following: firms use different commercial assets in different locations, different personnel with distinct sets of skills, different financial support structures with different milestones for measuring progress, and even different conceptual beliefs about the technical possibilities. As a result, one firm's assessment of the returns from innovating does not need to be the same as another's. Different assessments result in different methods for achieving the same commercial goals, which may lead to different costs or different commercial goals altogether, such as targeting different customers.

Variety receives attention because it increases the possibilities for learning externalities across firms. Even in failure, one firm's innovation may fail, but in failing, it may teach lessons to others about what would succeed. In this sense, when a variety of firms flourish, the accumulated experience from all the firms can be greater than any individual firm could or would have had on its own.

This chapter will show that entrepreneurs played a key role in shaping the amount of variety, both because they supported many innovative commercial actions and, more important, because entrepreneurial firms pursued market opportunities *not* chased by incumbent firms. Most remarkable, the variety continued to flourish through most of the first decade, even after some technical and commercial uncertainty declined.

How did this happen? A crucial part of the explanation highlights the commercial origins of the Internet. The commercial leaders of computing and communications did not forecast the growth of Internet access into a mass-market service. Only a few entrepreneurial firms and descendants from the NSFNET offered service at the outset. In addition, a number of other Internet participants took foresighted actions to support the entrepreneurial firms that entered the commercial Internet access market. Many of these participants lacked profit motives or did not have the status of an established firm. As a result, this market lacked the typical business markers of the computing or communications markets, yet many entrepreneurs (and their financial backers) concluded that they had a reasonable chance to profit from developing a new business. That surprised most computing and communications incumbents, many of whom scrambled to react to a flood of entrepreneurial entrants.

Another part of the explanation emphasizes the resources devoted by firms to exploring a variety of potential contracting and operational routines in order to discover those that worked efficiently. These searches succeeded in developing low-cost and reliable operations for exchanging data and operating the Internet on a vast national scale. That raises questions about whether the emergence of routines enabled or deterred a variety of experiments. In this chapter I will argue that the emergence of routines largely enhanced experimentation in this market for much of the decade, especially in the second snapshot, in 1998. This remained so even later in the decade, in 2003, though with significant qualifications. By then firms had incentives to pursue innovations that would be capital deepening, reinforcing preexisting routines. In spite of that incentive, substitutes for existing services continued to emerge.

These conclusions will beg another question: What role did policy play in the Internet access market? Regulatory policies could have shaped many facets of Internet access markets because these markets involved telephone service and equipment, which are heavily regulated. A thorough answer, therefore, requires a close review of regulatory policy.

The chapter stresses that while this market structure arose in part from several forward-looking decisions at the NSF, the NSF's managers built on a propitious combination of inherited market structures and regulations from communications and computing markets. These regulations came into existence for historical reasons only loosely connected to their eventual impact on exploratory behavior in Internet access markets. I underscore the importance of regulatory rules for pricing local telephone calls and changes to regulations supporting business-line restrictions at local telephone companies. Ongoing regulatory decisions also shaped the market's evolution, sometimes nurturing innovation and sometimes not. I highlight those regulatory decisions fostering or hindering diverse ownership over key assets such as backbone infrastructure as well as those regarding the implementation of the 1996 Telecommunications Act. These decisions gave the U.S. Internet access industry several unique features in comparison to the same industry in other countries.

While the chapter largely focuses on how the market structure shapes innovative action, I will also consider how innovative actions altered the market structure. The relationship is not straightforward because innovative behavior by itself tends not to alter the market structure in direct ways. Rather, innovative action leads to the emergence of new ideas about services and business processes. In turn, this might induce new entry from entrepreneurial firms, which either contributes to altering the identities of leading firms or causes incumbent firms to respond to newly perceived opportunities or threats, inducing actions that alter the range of their services. These might or might not alter the market structure.

More to the point, these changes are challenging to observe. New assessments about a market's potential as well as the changes rendered can be observed only with close study. To keep the discussion focused, I will limit the scope here to changes in pricing, operational practices, and expanded geographic scope. I will show how these changed in response to innovative actions that led to a new understanding about the sources of market value.

These conclusions differ from prior work mostly in emphasis. To date, scholars of Internet infrastructure have concentrated on factors shaping the digital divide, the dot-com boom, or the policies for broadband investment.[3] While there is considerable analysis of specific regulatory rules, no research has traced the patterns of interplay between regulation, market structure, and exploratory behavior over the first decade. This omission should be rectified. Exploratory behavior and entrepreneurship are key ingredients in economic development and growth, and do not happen by accident. Some settings encourage them, while others discourage them, and the difference fades with time as observers forget what actually happened and what might have happened under different circumstances.

I also will touch on themes of interest to those studying Internet use across countries. While the chapter does not discuss the origins of the market structure in other countries, it does inform conjectures about why U.S. market structure and firm behavior was unique. For example, contemporary news accounts noted several unique features of the U.S. access market. The United States contained several thousand retail access providers, while no country other than Canada had more than a few. The U.S. firms also were the first to introduce flat-rate pricing for Internet access, while firms in many other countries never did.[4]

1993: A Nascent Internet Access Market

The administrators at the NSF did not privatize the Internet with the intent of starting a boom in exploratory investment in Internet infrastructure markets. They privatized it in response to concerns that seem parochial to later observers—concerns such as whether or not a federal agency could operate and fund the NSFNET as it grew larger. Privatization also allowed nonacademic vendors to sell Internet access to private users without violating government "acceptable use" policies, thereby ending a simmering dispute that would have eventually boiled over.[5]

To ensure that the Internet would continue to support researchers and universities, the NSF implemented a privatization plan with many farsighted features designed to enable growth. This included the removal of government funding and the management of Internet backbone assets, designing institutions for the regional (nongovernmental) funding of data-exchange points, and bidding out the allocation of domain names to a private firm. Yet most information technology users outside of universities did not react to the NSF's actions. Most information technology users and managers of commercial vendors acted as if nothing of any commercial significance had occurred.

That "nonreaction" would persist for a couple years. In that setting, entrepreneurial firms with views outside the mainstream had incentives to make risky and exploratory investments, which as it turned out, were sufficient to induce many to take action. Their initial commercial success catalyzed a confrontation between different approaches to the commercial opportunity in Internet access markets.

Internet Connectivity in and outside the Academy

The collection of firms that operated the Internet in 1993 together formed a propitious combination. While pursuing their own interests, most of these firms simultaneously cooperated with one another. The cooperation at this stage established precedents for operating a commercial network. Such operations were crucial to enabling firms to explore the commercial Internet over the next few years.

Internet service providers (ISPs) require a physical connection because of the architecture of the Internet. Under both the academic and commercial network (as shown in figure 3.2), the structure of the Internet is organized as a hierarchical tree. Each layer of connectivity is dependent on the following one. The lowest level of the Internet is the customer's computer, which is connected to the Internet through a local ISP. An ISP maintains its own subnetwork, connecting its points of presence and servers with Internet protocol networks. These local access providers, or ISPs, get their connectivity to the wider Internet from other providers upstream, either regional or national ISPs. Regional networks connect directly to the national backbone providers.[6] Prior to 1992, private backbone providers connected to public backbones at network access points. This arrangement raised awkward questions about growing a commercial service, which could have (and under any plausible scenario would have) involved using public assets to carry private commercial traffic. That is, commercial traffic might go over the public backbone, violating the NSF's policies forbidding the use of public assets for commercial purposes.

The academic ISP was an informal operation at many universities, but it resembled what would soon become a commercial ISP at both the technical and operational level. Academic ISPs maintained modem banks and servers, and administered passwords for

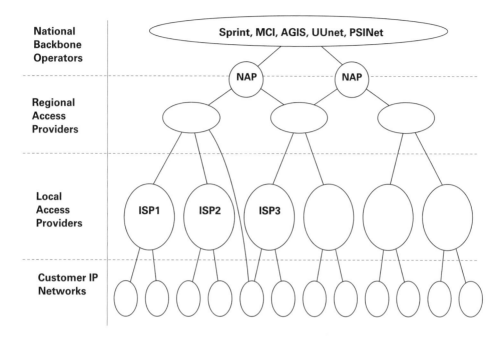

Figure 3.2
Consumers and business customers

log-ins; some also maintained a local area network at the university. Unlike a commercial ISP, however, the academic one did not charge its users for access.[7] Many also did not monitor user time online—a practice that some commercial firms would adopt for billing purposes.

The NSF had subsidized the establishment of academic ISPs throughout the United States, but rarely their operations. At most locations, university ISPs operated with a mix of professional managers and students. Many of the complementary network operations—such as the management of routing tables, data-exchange facilities, and the Internet backbone—were set up by the NSF or, more precisely, the firms with whom the NSF subcontracted, such as the Michigan Education Research Information Triad, IBM, MCI, and Advanced Network and Services, Inc. (ANS).[8] These firms and groups continued to operate after privatization, funded within a regional structure. A few intrepid ISPs and data carriers then took actions that the NSF's managers had anticipated—namely, planning to offer commercial Internet access to users not affiliated with a university.[9] Most of these vendors were offspring of the NSFNET, such as IBM, MCI, ANS, and Performance Systems International, among others.

Overall, descendants of the NSFNET began 1993 with a set of clients and customers, and wanted to attract more, fostered by the NSF's plans to privatize the network. It is hard to believe in retrospect, but this collection of firms did not receive much attention from its contemporaries, except a few professional insiders. No prominent business magazine hailed these firms as leaders of an Internet revolution (as was done to the founders of Netscape only two years later). Their business prospects were not deemed unusually strong. That would all change once they demonstrated the commercial viability of the Internet access market to later market participants. Relatedly, many of their key managerial decisions stayed within a fairly small professional community. Only later did observers realize that they had established precedents for cooperation for data exchange, defining the role of a private firm and the operations for the entire network.

The NSF's managers encouraged participation from a variety of firms during the privatization of the Internet, setting up a structure that had multiple firms in positions of leadership—for example, more than MCI and IBM were involved from the outset. Unlike many other countries, which handed the management of the Internet over to their telephone companies, nothing in the United States prevented many firms from participating in the operations as the network grew. That outcome had two economic underpinnings that were obvious to contemporaries but easy to overlook in retrospect. At the time the communication industry's structure was in flux, still unsettled due to ongoing court hearings over Judge Harold Greene's administration of the modified final judgment. It would have been foolish for the NSF to rely on a single firm for all services, such as AT&T, which was not the only firm offering data communications services. At the time, there were at least three firms with national fiber networks—AT&T, MCI, and Sprint—but many others appeared to have the ability and resources to plausibly develop national networks too—though to be sure, that is not equivalent to saying they had a business case for doing so. Second, the NSF's managers and researchers were important users of the Internet, and anticipated continuing. When users design their own industries, they have incentives to invite multiple suppliers and encourage innovative behavior from multiple firms. Implementing this in practice is easier said than done, but in this case, the academic Internet already had multiple participants. It had established technical and procedural precedents for moving data between the servers in different organizations.

The Imminent Surprise

To a knowledgeable insider in 1993, the Internet was still progressing, but it received no attention outside a small technically oriented community. The earliest advertisements for ISPs in *Boardwatch Magazine* appeared in late 1993, as the magazine attempted to expand from its role as the primary news publication for the bulletin board marketplace. As it turned out, advertisements grew slowly until mid-1995, at which point the operation became so large that *Boardwatch* began to organize its presentation in table format.[10]

Although the Internet access market was considered a good growth opportunity for the descendants of the NSFNET, for a number of reasons discussed below it was not seen as a huge immediate market opportunity, as it ultimately turned out to be. E-mail was popular among experienced users, so it was easy to forecast that growth in such activity could generate steady revenues for the firms providing the infrastructure.

This potential should be understood in context. Many market participants had anticipated some form of mass-market electronic commerce for households. Online service providers, such as CompuServe, Prodigy, Genie, and America Online (AOL), continued to try to grow the home market for online information in a bulletin board format. These services had difficulty widening their appeal beyond technically sophisticated home personal computers users, however, though they were trying. In addition, Microsoft Network, or MSN, made ambitious plans for a proprietary network with features similar to AOL's, but the plans called for a gradual development cycle over several years, anticipating that the mass-market opportunity for MSN would emerge slowly, giving it enough time to learn from experience and improve.[11]

In 1993, the conversion of large-scale computing to client-server architecture or any other form of networking was the revolution du jour among information technology consultants and others in the market of enterprise computing. According to the standard mantra, the Internet contributed to that movement in some useful ways, since exchanging data between computing systems was cumbersome in other formats, such as EDI. Most of the consulting practices in this area did not forecast the large impact the Internet would have on enterprise computing.

A couple foresighted contrarian investors made moves in 1993 and 1994, but no major change in investing occurred until 1995—and even then, the boom did not occur until a few months prior to (and then especially after) the Netscape initial public offering in August 1995.[12] The first high-profile analysis of the Internet access market on Wall Street did not arrive until Mary Meeker at Morgan Stanley organized one, resulting in a publication for general audiences in 1996, a bit later after her team had done related work for clients.[13]

In other words, with few exceptions, firms with strong interests in Internet infrastructure markets did not take actions in 1993 that would have served their interests a few years later. Like the abundant clues about impending disaster in a B movie, later observers could easily see many symptoms of the coming surprise that few of the contemporaries noticed or prepared for.

Phone Companies, Bulletin Boards, and Policy

In 1993, local telephone firms and bulletin board operators already had an uneasy and asymmetrical relationship with one another, defined by years of lawsuits, regulatory disputes, and distinct attitudes about technology and operations. This relationship arose for reasons idiosyncratic to the United States (and Canada), and for a short time, these two types of firms formed an unlikely partnership in providing Internet access to the United States. Hence, this relationship defines many facets of the unique features of the U.S. experience in comparison to other countries.[14]

Local telephone firms were compelled by regulators to do two things that helped bulletin board operators. First, state regulators in virtually every state required unmeasured pricing for local telephone calls over short distances, such as a ten- to fifteen-mile radius (or more in rural areas). Local calls involved extremely low costs per minute, if any. This regulation arose out of policies designed to encourage universal service in telephone use. More to the point, such rules were not motivated by the effect on bulletin board business, though they had huge consequences for them.

Second, the Federal Communications Commission (FCC) compelled local telephone companies not to "discriminate" against bulletin boards as well as any others that were classified as "enhanced service providers." These complex regulations grew out of years of antitrust and regulatory lawsuits and inquiries at the FCC. They resulted in a series of regulatory rulings for communications equipment and services provided by the phone company and those who interacted with it. The set of regulations became known as Computers I, II, and III. By 1993, only Computers II and III were in effect.[15] Computer II was issued in 1980, and was viewed as an improvement to Computer I, an inquiry that began in 1966 and resulted in an order in 1971. The first Computer III order was issued in 1986, and it underwent subsequent revision and court challenges. These would eventually shape Internet use, access, and business in profound ways, though as one can see from the timing of their adoption, they had motivations other than their impact on the Internet.

Computer II offered protection for entrants into new information service markets from postentry discriminatory behavior from telephone companies. It permitted telephone firms to enter these same markets only if the telephone firm contained a structurally separate division offering services that competed with new entrants. Telephone firms were compelled by this order to offer any service to a competitor that it would offer to its own division. By the time the Internet was commercializing, the FCC was trying to write (and rewrite) Computer III, which was designed to be less burdensome yet accomplish similar goals as Computer II and was in the midst of a court challenge. Under Computer III, a telephone company could have an integrated division as long as it also had a detailed and approved plan for providing interconnection to others. More to the point, by the early 1990s, every local telephone firm lived with rules that required it to treat a competitor the same as its own division, and these applied to bulletin board operators.

These rules arose from several tendencies in U.S. law and regulation. First, U.S. antitrust law enforcement agencies have a long history of attempting to carry out antitrust rulings. These prevent a dominant incumbent from using its dominance in one market (i.e., local telephone service) for commercial gain in another where competitive entry and innovative activity might emerge (e.g., selling data). A second set of concerns arose out of the tradition of common carrier law, which requires a monopoly carrier not to use its monopoly to the benefit of one business partner over another—that is, not to discriminate against any potential business partner.[16]

These rules were regarded as a nuisance for local telephone firm operations in the details, but by 1993, the broad principles were widely appreciated. As a technical matter, any such engineering details behind interconnection had long ago been worked out.[17] More concretely, the timing and features of calls to bulletin boards looked distinct from voice calls, placing peculiar demands on capacity during peak load times. They were also a small source of revenue since they led users to demand second lines. Overall, because the bulletin board volumes were comparatively small in relation to voice telephony, most local telephone firms and local regulators thought the burdens were a manageable nuisance.

Why did this matter? Because ISPs would absorb the regulatory norms from the bulletin board industry. Unlike the experience in much of the rest of the world, when the Internet privatized, neither U.S. telephone company executives nor regulators assumed that the telephone companies should be the only provider of Internet service. Nor, for that matter, did users, who did not resist the idea of calling another party to make their Internet service operable. In brief, ISPs were free to explore the commercial possibilities without much constraint on their actions, protected by the ample precedents, especially as they were embodied in Computers II and III.

To be sure, these rules were *the most* important business parameters for bulletin board operators and academic ISPs. A bulletin board operator attempted to locate its modem banks at phone numbers close to where its customer base could make a local phone call. Once these investments were made, a bulletin board operator could not alter its networks or change advertising about its coverage without accruing cost. Many bulletin board operators also believed the telephone companies would not be cooperative without being compelled to do so—whether this belief was true or not is another question. Many also believed they would not have a viable business unless telephone companies were precluded from being in their market space, though there was no legal basis for excluding telephone firms that complied with either Computers II or III.

Three types of bulletin board operators emerged in the 1980s and early 1990s. Most would transition into the Internet market as they learned about its potential. The first were the online service providers, such as AOL, Prodigy, and Compuserve, as already mentioned. These firms all had ambitious management and would attempt to become national ISPs a few years later as Internet demand boomed, with varying degrees of success. Of the three types of bulletin board operators, only the online service providers tended to have any experience in political or regulatory arenas.

The second set of bulletin board operators resembled the first group in technical operations, but tended to be smaller because the organization attempted to accomplish a special function, such as product support or sales (e.g., information sales or product manuals), club support (e.g., airplane hobbyists), or games (e.g., groups devoted to Dungeons and Dragons). Some provided limited user-group functions, such as daily updates of news, organized Usenet group postings, or periodic e-mail. Some of these did become ISPs later.[18] Unlike the academic ISPs, many of these firms were at home with commercial behavior, and as the Internet diffused, quietly transformed themselves into content providers or hosting sites. This group also produced some leaders of amateur groups and industry associations. Importantly, the descendants from this group had perspectives that differed from the NSF descendants, online service providers, and because they were present in many cities nationwide, technology business leaders in places like Silicon Valley, New York City, and Boston.

The third type of operator, and arguably the most numerous, was the one that supported pornography. Though any individual firm tended to be small, these operators were present in every U.S. city, even those with small populations.[19] Once again, many of these operators were comfortable with commercial practices that the NSFNET descendants did not engage in. The pornographers were social pariahs in many cities, protected under the First Amendment and defended by such groups as the American Civil Liberties Union in light of the perceived larger principles. Many of them would become ISPs, and most would become content providers and hosting sites when the Internet developed.

Overall, these firms seeded the newly privatized Internet with a myriad of commercial experiments. After the Internet privatized, there was no single typical response from these firms. They had widely different backgrounds. They did not speak with a single voice on any of the key questions regarding forecasting, strategy, or public policy. The backgrounds and experiences of these firms differed from those that descended from NSFNET, giving them distinct viewpoints about the commercial potential of the Internet. Some, especially the smaller firms, were most comfortable casting themselves as outsiders, adopting the attitude that they were "Netheads," in contrast to the "Bellheads" that did not see the Internet in the same terms.[20] Many sought to satisfy local needs or niche customers they perceived the large national firms ignored. Others, such as AOL or MSN, viewed the phone companies as cooperative partners in some respects and intended to use the federal regulatory apparatus to settle disputes.

Commercial Platforms in Computing

Did many firms in the computing and communications markets know what would be coming in the next few years? By elementary reasoning the answer is no. If they had seen it coming, they would have acted differently in 1993. They would have invested heavily in anticipation of a coming boom.

There is more to that observation than just private regret. This nonaction later shaped the industry's development in two ways. First, entrepreneurs with so-called contrarian views had incentives to look for opportunities in Internet access markets because no established firm had an insurmountable competitive advantage founded on early-mover actions. Second, because entrepreneurs succeeded in exploiting the lack of prior investment by incumbents, existing platforms for communications and computing did not exert much early influence over the direction of technical change.

One early mover was Cisco, a supplier of routers and hubs for academic ISPs and operations. In 1993, the company went through a restructuring and change in management, guided by its venture capitalists and board of directors. At this point, the new managers for the company adopted a strategy to expand beyond just hubs and routers, positioning Cisco as the leading Internet equipment supplier. As perception about the growth potential of the Internet market changed, so did Cisco's tactics, such as accelerating its acquisitions; but the company's ambition to "become the lead architect and provider for TCP/IP voice, data, and video," as all its company material states, remained unchanged from 1993 onward. In retrospect, it is apparent that Cisco's early repositioning was rare. Most equipment firms took partial steps toward strong positions in server markets, such as Nortel, Novell, Alcatel, or 3Com, but did not make such overt changes in strategy as early as Cisco. The equipment division of AT&T, which would become Lucent in a few years, also did not make any such declarations about the Internet.

IBM and MCI both offer interesting contrasts. Both firms held unique positions as descendants of the NSFNET. Both took actions to support their provision of data-carrier services. In IBM's case, this was enhanced by the change in strategic direction when Louis Gerstner took over as CEO. In 1993 he and many others regarded the organization as in the midst of a severe business crisis, so it is hard to know just how much of this was a defensive reorganization as opposed to forward-looking strategies to take advantage of prior experiments with the Internet.[21] In retrospect, it appears that the changes initiated in 1993 made it easier for IBM's later transition to serve the booming commercial Internet markets as an ISP for business and a service integrator for enterprises using the Internet.

In MCI's case, its experience with NSFNET left it well positioned for offering services in two rather distinct markets. First, MCI had developed skills it could translate into a business for national data transport over its own fiber network. Second, it could offer ISP service for both home and business in the coming commercial data market, even though it did not have an identity as an old-line computer firm. As it turned out, as the demand for Internet services boomed, MCI did gain a large amount of business in both, especially the former.

It has also been widely noted that most firms in the software industry did not anticipate the Internet. Few existing software firms initiated new projects in 1993 related to the Internet, and almost none were founded this early.[22] Specifically, Microsoft's inattention to the Internet later received a spotlight in its federal antitrust trial. The publicly available documents consequently reveal the foundations for Microsoft's inattention: because Microsoft's strategists studied the actions of many other firms before settling on a plan, Microsoft's actions reflected the consensus of views elsewhere across the country, thus illustrating the types of forecasting errors made at the time.

In 1993, Microsoft was developing a product that later would be called Windows 95. Windows 95 was supposed to replace the disk operating system (DOS) as the standard operating system in personal computers. Up to that point, Windows was an application built on top of DOS. The architects of Windows 95 were confident that the main competitive issues affiliated with the Internet *had* been addressed. Reading Transmission-Control Protocol/Internet Protocol (TCP/IP) compatible files had been a standard feature of Unix systems for some time—an operating system with which Microsoft anticipated competing in server markets. Making Windows 95 TCP/IP compatible, both in the personal computer and server versions, Microsoft anticipated that Internet applications would be built on top of the operating system by others.[23] Microsoft did not, however, foresee that any other application would be pervasive—except e-mail, for which it had designed an application that in fact would become popular in a few years.

Predicting that Microsoft would need a platform to engage in the widely anticipated developments in electronic commerce, Microsoft's strategists concluded that only a proprietary network could be profitable and the applications for it *could not* be built on nonproprietary technical standards, such as those that supported the Internet. Hence, Microsoft foresaw that its new MSN division would use proprietary standards and other Microsoft-specific designs, and anticipated borrowing the lessons learned from the actions of firms such as CompuServe and AOL.

There was never any *technical* error in these assessments—a statement that some naive contemporaries made about Microsoft's decisions. The strategy team had up-to-date technical information and sufficient technical skill to understand how these technologies functioned. Rather, the error was one of misinterpreting the commercial prospects. This mistake would take four interrelated forms: anticipating much less commercial activity around the Internet than what occurred, mis-underestimating the Internet's value to users, underestimating the ability of Internet firms to support applications that threatened Microsoft's profitability in the marketplace, and the late recognition of the first three errors. The first three were common to many firms, and only the fourth error of timing was unique to Microsoft—the top strategists did not recognize the first three errors until spring 1995.[24] That was later than many other firms that had begun to investigate these issues in late fall 1994, around Netscape's founding.

By spring 1995, when Microsoft eventually decided to change its commercial priorities about the Internet, it had already made a number of decisions that would limit its short-term ability to follow through on these new aims. For example, while the designers of Windows 95 could change some aspects of it quickly, business partners could not. The first browsers in Windows 95 were not designed to support hypertext markup language (HTML), so as a result, software application firms in 1995 and 1996 had to use tools and solutions from available shareware—such as those available from the World Wide Web Consortium—and other firms such as Netscape that had correctly anticipated the need to support HTML. Such behavior ran counter to Microsoft's strategic goal to be the indispensable supporter of all major tools for application development on the personal computer operating system.[25]

Nurturing Institutions

If established firms did not support application development by entrepreneurs in the early years, what institutions arose to aid exploration? The research community acted partly as the breeding ground for organizations that would later foster standard protocols for many applications. Their actions would jump-start private development by making it much easier for a small firm to begin commercial life without an obligation to the commercial interests of an established firm. Because this institutional setting looked so different from any precedent, most observers did not forecast its importance at the outset.

Unlike many new industries, the Internet did not start its early commercial life beholden to the technical or commercial vision of a single dominant firm that (perhaps) held a key patent or dominant market position. In addition, the NSF received no explicit congressional directives on crucial issues (such as which suppliers to favor), so the NSF's managers had discretion to do what they thought would work best.[26]

The National Center for Supercomputing Applications, a well-funded operation subsidized by the NSF, at the University of Illinois would play an unexpected role in the history of all commercial Internet firms. Joining the large community of programmers employed there, a few computer science undergraduates were hired to create a browser to make use of Tim Berners-Lee's HTML language, which was a 1989 creation. That project resulted in the Mosaic browser, which became available in November 1993, just *after* the Internet was privatized. Over the next eighteen months Mosaic was downloaded by millions, primarily users in research labs and universities, the main users of the Internet at the time. More important, its invention and diffusion were not anticipated by the NSF's administrators. While the University of Illinois arranged to license the Mosaic browser to others, the same students went on to found Netscape Corporation with Jim Clark in late 1994.

The NSF managers also did not foresee the dilemma concerning the management of the Domain Name System (DNS)—one of the Internet's earliest and most visible policy dilemmas, which had informal institutional arrangements until the early 1990s. In 1992, the NSF put the registry function out for bid. It was apparent even then that there was no sense in frequently altering the DNS management after its privatization, a feature that implied registries required some sort of regulatory oversight as a quasi-natural monopoly; however, the NSF had not put in place an oversight system for registries. To be fair, the unanticipated growth in the Internet would have made institutional design extremely challenging; even if a regulatory framework had been put in place early on, the mission would have had to change dramatically and quickly.

In response to the perceived need, and after much growth in the Internet, in 1997–1998 the Clinton administration set up the Internet Corporation for Assigned Names and Numbers (ICANN) under the domain of the Department of Commerce. This birth was contentious, and the difficulties made for good copy among Internet insiders.[27] By this point, though, most of the valuable domain names already were allocated (e.g., CNN owned CNN.com and so on), and an active secondary market had emerged for many of the other domain names, some of which were still quite valuable. It was not a blindingly efficient system, but it allowed Internet access firms to get many things done in due time.

Again, unlike most new industries, the Internet access market was founded with several of the institutions normally seen in a *mature* market, such as an industry association and a set of standing active committees to consider upgrading the interoperability standards. For example, many of the other important, though perhaps less well-known, technological building blocks for commercial ISPs came from the Internet research communities that had long used and refined them. Unix compatibility with TCP/IP had long been a Department of Defense procurement requirement, and as a result, all vendors for commercial Unix operating systems had built in compatibility as a matter of routine, obviating the need to alter most private server software as the Internet diffused. Other building blocks included e-mail (i.e., the send-mail shareware), file transfer protocol, Kerberos (security), and the basic protocols to make TCP/IP data work in Unix-based systems. As it turned out, many of these building blocks became a part of the suite of protocols and standards endorsed by the World Wide Web Consortium, founded in October 1994 by Berners-Lee, who located his office at the Massachusetts Institute of Technology.[28]

This was not the only such organization to play a related role. Several helped coordinate developments of the Internet when it was under Defense Advanced Research Projects Agency (DARPA) and NSF management. The Internet Engineering Task Force (IETF), among others, had played a part at DARPA as part of the Internet Activities Board (IAB). The IETF had become the focal organization for protocol development and technical concerns as privatization approached, folding in efforts that had been spread among several groups until then. With privatization, the Internet Society, a nonprofit organization, took formal control over the IETF and the IAB, complementing both groups with additional information-dissemination tasks. The IETF continued to accept proposals for national standards and protocol development, operating like an industry consortium for standard setting.[29]

Another example of a vendor group supporting the young industry was the Commercial Internet eXchange (CIX). It was founded in 1991 by several early ISPs to facilitate the operation of data exchange, both inside and eventually outside the public access points.[30] CIX went on to become an organization that all the ISPs joined in the early years. Playing the role of an access industry trade association during the mid-1990s, CIX facilitated conversations among its members, and policy platforms for Washington, DC, lawmakers, agencies, and their staff.

Open source projects also made a difference. These were building on another precursor institution that also continued to play a role: shareware distributed over bulletin board systems, with floppy disk libraries, and increasingly over the Internet. The Linux movement had already begun prior to 1993, but it was not an important player in commercial Internet markets for a few years, except insofar as it was a model of how to successfully organize such a community of developers. Perhaps more directly essential to the growth of the commercial Internet was a project begun by a set of programmers in 1995: Apache. It would become the typical server software for ISPs; again, it was another outgrowth of the experience in the shareware communities.

Overall, most of the leading incumbent firms in computing had not taken action in anticipation of the Internet. As a result, none of the usual technical support activity existed. Yet that did not hinder private development of the Internet, as would happen in most young industries. A number of descendants from the research communities attempted to serve a similar function in the area of interoperability standards, languages, and basic tools for managing access at the access firms. Much of this was an outgrowth of the cooperative behavior supporting the decentralized Internet in the research communities. In 1993, there was still the larger open question of whether this would be sufficient to support the growth of the commercial Internet.

Overview

In retrospect, the handoff of this technology from public management to privatization had several features whose significance was hard to forecast. First, there were the Internet's unusual origins and the equally unusual institutions that accompanied that history. Second, the NSF's managers planned for large-scale operations in e-mail. As it turned out, many of these plans could accommodate a different set of applications that the NSF had not anticipated, built on the World Wide Web. Third, the timing of key discoveries fostered surprise, reducing the possibility for anticipatory investment by established firms and leaving ample commercial room for multiple entrants.

The structure of the marketplace and the regulatory rules that governed the Internet access market at this point played to the advantages of venture-funded entrepreneurial firms. The commercial setting contained a rare set of unprecedented features, and most incumbent firms were not investing in anticipation of large growth. An eclectic mix of economic actors made investments while exploring the possibilities for building viable businesses. They generated experiences that in turn, would provide lessons that others would notice soon. Soon the Internet access business would grow unexpectedly fast, initiating a competitive contest among entrepreneurial firms as well as with established ones.

1998: An Era of Impatience

By 1998, the U.S. Internet access market had experienced the benefits and frenzy of commercial entry unconstrained by a single commercial vision. Firms with distinct views about the commercial prospects for the Internet tried to build businesses to support their perspectives about the source of value. Those stances reflected different local knowledge bases about customer needs, different visions about how to commercialize a national service, and distinct strategies for taking advantage of existing assets. New retail pricing practices also changed under competitive pressure. In addition, there was no contemporary consensus about where or how the competition would end.

The growth in demand for the Internet justified enormous investments by both users and vendors. To accommodate this change in the scale, national operators instituted a set of pricing policies for governing commercial transactions in the backbone for Internet access. The ISPs with ambitions to sell mass-market services also instituted regular policies. Both actions were part of a general movement toward the (almost) inexorable use of a small number of standardized technical processes for routine tasks to operate the network.

The Emergence of Routine Network Operations

A large-scale national network involving so many participants could not operate efficiently without standard modes of contracting. Almost from the outset of commercialization, many of the largest firms began to insist on them at both the wholesale (e.g., backbone) and retail (e.g., ISP) level. After some changes coincident with the diffusion of the browser, regular and predictable patterns of contracting emerged, and in tandem, a sense of routine to this collective activity also began to appear. Any individual firm could take for granted this routine when it made its own parochial decisions. In this sense, the routines affiliated with the market in 1998 shaped the character of exploratory activity by all firms.

Regular operations had to accommodate an extensive variety of demands. Most households physically connected through dial-up service, though both cable and digital subscriber line (DSL) technologies gained some use among households near the end of the millennium.[31] Broadband connections were more typical for businesses in major cities, if they were anywhere, but ISPs in some areas tried to sell T-1 lines or the integrated services digital network (ISDN), as did some local phone companies. Many large business users made the physical connection through leased lines or other direct connections, while smaller users often tended to have dial-in connections, if they had any at all.

There were several different types of ISPs by 1998. The national private backbone providers (i.e., MCI, Sprint, etc.) were the largest carriers of data, and many of these firms also provided retail ISP services to consumers or other ISPs that rented rights to resell the use of their modem banks. The remaining ISPs ranged in size and scale from wholesale regional firms down to the local ISP handling a small number of dial-in customers. Many of the large firms were familiar incumbents, such as Earthlink, Sprint, AT&T, IBM Global Network, AOL, and Mindspring. Other large firms included entrants or Internet insiders from the NSF, such as PSINet, Netcom, ANS, GTE (which acquired assets from Bolt Beranek and Newman Planet in 1997), and others. Some of these firms owned their own fiber (e.g., MCI), and some of them ran their backbones on fiber rented from others (e.g., UUNet). Still others offered ISP services to consumers (e.g., AOL and MSN), but did not own any facilities. They rented it from others (e.g., ANS), although users did not know this and often did not care as long as everything worked.

The market share at the retail level was skewed. A couple dozen of the largest firms accounted for 75 percent of the market share nationally, and a couple hundred made up 90 percent of the market share. In other words, the majority of these ISPs were small dial-ups covering a small regional area, but the majority of users employed national providers.[32]

The so-called mom-and-pop ISPs reflected their small size and informal origins—in contrast with national firms such as AT&T WorldNet. In August 1996, *Boardwatch* listed prices for 2,934 ISPs; in February 1997 for 3,535; in January 1998 for 4,167; and in January 1999 for 4,511. In each case, though, the magazine listed many more ISPs for which it did not have price information. The highest reported number in *Boardwatch* was just over 7,000 in March 2000.[33] As another illustration of the variety of ISPs available, one estimate found over 65,000 phone numbers used by just over 6,000 ISPs in fall 1998.[34] A national firm would support anywhere from four hundred to six hundred phone numbers. Half of these ISPs supported only one phone number, which means there were about 3,000 ISPs supporting one phone number, and the rest supporting anywhere from two to six hundred numbers.

Many small ISPs ran their own network point of presences and provided limited geographic coverage. Many also leased such lines or externally managed point of presences through a third party, such as AT&T or MCI, in locations where they did not have coverage. This allowed them to offer national phone numbers to their local customers, competing with commercial online services with national coverage.

How did anyone get a sense of order out of such heterogeneity? To the delight of some market participants and the dismay of others, by 1998 a commercial structure had emerged that some called *tiered*.[35] Tiers indicated which firms carried data over long distances and collected charges from others for transit service. The size of its footprint, the volume of traffic, and the propitious location of one firm's lines determined the *direction* of charges from one firm to another. More important, this was a commercial mechanism that involved the private interchange of data and revenues, funding operations in a way other than what the NSF envisioned. It grew up next to the regional data exchange points set up at the beginning of commercialization.

The largest national backbone firms all became tier-1 providers. This included AT&T, IBM (before being sold to AT&T), MCI (whose backbone was sold to Cable and Wireless as a condition for the merger with WorldCom), ANS (which received many of its facilities from AOL), UUNet (eventually sold to WorldCom), and Sprint, among others.[36] Most regional and local ISPs became lower-tier ISPs, purchasing interconnection from either one of several national providers or a larger regional provider, which then passed on the traffic to the tier-1 firms.

Nobody doubted the existence of the tiers, but other observers were skeptical of the rigidity of them. They called this system a *mesh*, where participants faced many options for interconnection.[37] For example, many ISPs arranged to use multiple backbone providers (known as multihoming), thereby diminishing the market power of any one backbone provider in any single location. Similarly, dial-up ISPs with large networks could receive calls in one location, but backhaul them to another location to be connected to the Internet. This network design gave ISPs multiple options for connecting to the Internet, thereby limiting the discriminatory power of any single backbone firm. Finally, many of the largest national firms acted in ways that contributed to the mesh. Using a contract form known as the indefeasible right of use, some firms rented their facilities for twenty years to other firms that needed connections along a particular path (e.g., from one city to another).[38] Some firms also rented lists of phone numbers at different locations to retail ISPs that wanted to offer their traveling customers local phone numbers to call in different cities.

The organization of transit services in the United States overlapped with the tier designation. Backbone firms that exchanged traffic of roughly equal size adopted practices that facilitated trade—that is, they did not charge each other. This practice was equivalent to volume discounting for the sake of saving costs on monitoring traffic and billing one another; however, it had other consequences as well. The practice raised the incentives of the major backbone firms to exchange traffic bilaterally, bypassing the public exchanges. Private interchange raised issues about the quality of data services for those who used the public exchange. While both practices generated inquiries to the FCC, nothing substantively was done about either one.[39]

By 1998, a number of other routine technical and business processes had begun to characterize the operations of ISPs.[40] There was no consensus about the set of services or operations that led to the greatest profitability in the long run. In light of such uncertainty, many ISPs sought to differentiate themselves from each other by offering services such as hosting, Web development, network services, and high-speed access.[41] These services were in addition to e-mail, newsgroup services, or easy access to portal content, online account management, customer service, technical support, Internet training, and file space.

Overall, this situation had both routine and variety. Routines fostered efficient operations, especially in the exchange of data between firms. Their emergence facilitated low-cost commercial experimentation by Internet access firms, which were growing rapidly at the time and attempting to serve a perceived (and anticipated) growth in demand.

Growing the New Mass-Market Internet

In 1998, an Internet user could surf thousands of sites from an extraordinary variety of sources, and participate in the Web equivalent of news, magazine, and hobby clubs. Shopping also changed for both mass-market and niche items. Every mass-market item was available, such as books, CDs, tickets, computing equipment, mattresses, or pet food. Pornography was widely available and accounted for a portion of the traffic. Users could also purchase specialty items, such as goat cheese from a remote farm in Wisconsin, a monarch chrysalis from a Caribbean butterfly farm, and hundreds of thousands of other products that specialized vendors were trying to sell. It was unbelievable, exciting, and overwhelming.[42]

Helping new users navigate these possibilities became a facet of the Internet access business. Some firms specialized in profiting from selling to new users. For example, building on its business philosophy—which it developed prior to the Internet's diffusion—AOL devised simple instructions and online help as well as services specialized for new users, such as community sites and easy-to-use e-mail. By the end of 1998 (and for a variety of reasons discussed below), AOL had grown into the dominant national provider to the home, with a market share of between 40 and 50 percent, depending on who was counting and when.[43]

Most of the online service providers—Prodigy, Genie, CompuServe, MSN, and AOL— had also begun converting to Internet service in 1995, each with different results. Except for AOL, all failed to gain much additional market share from this move. Several savvy decisions contributed to AOL's success. For instance, in 1996, AOL sold off its physical facilities, relying on a long-term contract with another firm that operated the modem bank. It chose to concentrate its investments on content software development and marketing. It successfully made the transition to unlimited pricing at about the same time.

In addition, AOL's version of instant messaging became the dominant provider by far, especially after it bought ICQ (an acronym for "I seek you"), an entrepreneurial firm that filed for several patents and, after entering in 1996, had established the largest instant messaging network on the Internet.[44] After establishing dominance in this application, AOL went to great lengths to prevent others from interoperating with its system without paying licensing fees. In fact, some smaller ISPs did pay licensing revenues to AOL to allow their users to interoperate. Yet that was not uniformly true. In 1998, AOL, Yahoo!, and MSN all supported their own instant messaging applications, and these could not interoperate with one another.

One of the most important business decisions occurred in early spring 1996, when AOL agreed to make Internet Explorer its default browser in exchange for two things: several hundred million dollars in cash, and Microsoft lifting its contract restrictions that prohibited AOL from putting its logo on a personal computer's first screen. This deal went into effect in summer and fall 1996, when Internet Explorer 3.0 was rolled out, and had huge consequences throughout 1997 for the browser wars with Netscape, especially after Explorer 4.0 was released. Along with other factors, it allowed Microsoft to win the browser wars definitively. It also gave AOL a marketing tool that no firm other than MSN had. That, and large investments in marketing to new users, allowed AOL to grow quite rapidly.[45]

Finally, in February 1998, AOL bought CompuServe, a merger that many observers criticized, since it seemed to combine a firm that focused on technical users with one that did not. In retrospect, it solidified AOL's leadership of dial-up service for the next several years.[46]

The loyalty of AOL's user base earned AOL users contempt from technically oriented and experienced Internet users, who were comfortable using online resources without anyone's aid. AOL's approach became known as a *walled garden* for the way in which AOL protected or (in the words of critics) spoon-fed content to its users. Yet AOL was not the only firm pursuing such actions. MSN also became known for behavior that catered to new users. So too did @home as well as Juno, Earthlink, Mindspring, Netzero, and a few national firms.

As of 1998, AOL's success suggested that a large number of new users preferred this approach, but the success of others suggested that a large number preferred something else, such as the many mom-and-pop ISPs that accommodated local community needs. There were predictions that the Internet access business would bifurcate around these two approaches. There also were contrasting predictions that new users would become more sophisticated and depart AOL in time, while others forecasted that AOL would add new services to keep such customers.

Accordingly, the scope of service continued to differ among ISPs, with no emergence of a norm for what constituted minimal or maximal service in an ISP that was not AOL. Some ISPs offered simple service for low prices, depending on users to make do with Yahoo!, Excite, Lycos, or other portals. Other ISPs offered many additional services, charging for some and bundling other services into standard contracts. Private-label ISPs emerged when associations and affiliation groups offered rebranded Internet access to their members. These groups did not operate an ISP; instead, their access was being repackaged from the provider that supplied the connection. For business users there were many options. By 1998, many local telephone firms had begun basic service. For instance, the long-distance and data carrier AT&T emerged as the largest retail Internet provider to businesses; AT&T had already had a data carrier business with firms, but it grew larger when IBM sold the company its operation in 1997. Likewise, WorldCom was not far behind AT&T, having acquired UUNet and Metropolitan Fiber Systems.

When MCI and UUNet became part of WorldCom in 1998, WorldCom became the largest backbone provider and a large reseller of national point of presences to other firms. The so-called fringe backbone players were not trivial, however. Sprint, AT&T, and GTE also had large networks, collocated with their existing networks. In addition, Level3 had raised considerable funds and announced plans to enter on a national scale with a newly configured architecture, so the long-term structure of supply did not appear set or stable.

The retail market also still appeared to be open to recent entrants, such as Erols, Earthlink, Mindspring, Main One, Verio, and many others. All these firms gained large market positions, but plenty of smaller firms also had acquired enough of a market share to sustain themselves. The so-called free ISP model also emerged in late 1998 and grew rapidly in 1999, offering free Internet access in exchange for advertisements placed on the users' screen. These firms eventually signed up several million households, generally for use of a second account.

The growth in ISPs led to an increase in Internet calls, which in turn increased the demand for second lines, a source of revenue for local telephone firms.[47] This market had become too large for any telephone firm to ignore. By 1998, most local telephone companies also had entered the dial-up ISP business.

In summary, growing the mass market placed new demands on Internet access firms. Many ISPs took actions to serve this growing market; many did not, choosing instead to limit their focus to technically sophisticated users, niche users in a small locale, or users learning at their own expense. This variety reflected a lack of consensus about which approach was superior for offering ISP service.

The Geographic Coverage of Access Providers

The massive Internet investments after 1995 gave rise to two geographic features: the widespread geographic availability of access, and "overbuilding" in some places.[48] These outcomes did not have to be connected, since the former referred to the spread of retail access while the latter generally referred to the investment in fiber for the backbone to provide transport services. Yet these were linked in one sense: the spread of backbone fiber helped retail ISPs flourish in some places.

Internet access became widespread for a several reasons. First, it lent itself to small-scale customization at the user's location. It was easy to adapt to personal computer use or available local area network technology.[49] Second and related to this, economies of scale did not arise, since ISPs could survive on a small scale. Third, the standard technical software for supporting an ISP, Apache, was widely available, and the necessary technical know-how for getting started did not differ greatly from routine knowledge found at a firm performing computing services prior to commercialization. Fourth, the growth of standardized contracting practices for renting facilities from other firms in other locations added to the geographic reach of individual ISPs, even in areas where they did not operate facilities.

By 1998, many ISPs had located in all the major population centers, but there were also some providers in sparsely populated rural areas. One estimate showed that more than 92 percent of the U.S. population had access by a local phone call to seven or more ISPs by 1998. Less than 5 percent did not have any access.[50] Almost certainly, these estimates are conservative. The true percentage of the population without access to a competitive dial-up market is much lower than 5 percent.

The spread of firms, such as those operated by Akamai or Digital Island, contributed to the geographic spread of Internet access. These firms operated cache servers in multiple locations, often colocating with data carriers. They would offer updated "mirrors" of the most frequently accessed sites, so users could receive data quickly, rather than from a central server operated by, say, AOL or Yahoo! Popular Web sites such as Yahoo! made deals with these firms, which then eliminated much of the competitive differences between locations of backbones and thus removed differences in qualities of performance for users.

The only locations lacking Internet access were the poorest of urban areas, or the smallest and most remote rural populations. Some of this small minority of the country were in areas that bore signs of permanently slowed development.[51] The rural areas often lacked a competitive supply of providers, and even when suppliers existed, they sometimes provided limited services or focused on specific segments, such as business users.[52] There was never any issue about getting some service, even in the worst situation. A user could make a long-distance phone call and get service, but the effective cost was higher.

In addition, there was a dichotomy between the growth patterns of entrepreneurial firms that became national and those that became regional. National firms grew by starting with major cities across the country and then progressively moving to cities of smaller populations. Firms with a regional focus grew into geographically contiguous areas, seemingly irrespective of urban or rural features. By 1998, many rural telephone cooperatives were opening Internet services—following long-standing traditions in small and rural cities to use collective quasi-public organizations to provide utility services that other private firms do not find profitable.

The vast majority of the coverage in rural areas came from local firms. In 1996, the providers in rural counties with a population under fifty thousand were overwhelmingly local or regional. Only in populations of fifty thousand or above did national firms begin to appear. In fall 1998, the equivalent figures were thirty thousand or lower, which indicates that some national firms had moved into slightly smaller areas and less dense geographic locations. Figures 3.3 and 3.4 offer a visual sense of those patterns in 1996 and 1998.[53]

Whereas the massive geographic spread of Internet access availability evolved from both national and regional firms, concerns about overbuilding focused on the concentration of national backbone fiber in major urban areas. Some of these firms owned facilities—that is, national fiber-optic networks. Other firms built their backbone services, using indefeasible rights of use that they arranged with the owners of the fiber. Some of these firms only offered transit services to other ISPs, while others also offered sets of phone numbers in some locations and rented them to other ISPs. In these two senses, the overbuilding of backbone in major urban areas contributed to the availability of Internet access at the retail level, either by making transit services more available (and presumably cheaper) or by making point of presences available to others to rent.

The interest in this situation must be put in context. In many industries, observers *would not* find it surprising that different ambitious firms would attempt to serve the same customer in the same capacity. Yet no such behavior had ever been seen within a large-scale communications service market. And this break with precedent was the source of astonishment.

Several factors simultaneously led to the overbuilding. For one, the U.S. backbone was built in increments by many firms, not by a single firm with monopoly ownership. In the late 1990s, notably AT&T, Sprint, WorldCom, GTE, Level3, Qwest, Global Crossing, Cable and Wireless, and Williams all had plans to build networks with national geographic footprints—that is, presence in all or most major cities.[54] These plans were not coordinated.

Also, the impatient financial environment of the late 1990s provided four related incentives to grow. The first was due to Wall Street's exuberance, as stock prices for publicly traded firms responded to announcements of building plans. Initial public offerings for new Internet ventures displayed a similar pattern too. This generated backbone investments by many of the participants as well as investments to support retail ISP service in many locations, even when it resulted in significant redundancies between rival suppliers.

Second, there was a common perception among ISPs that demand for Internet traffic would grow quickly for many years. In retrospect, it is hard to distinguish between the

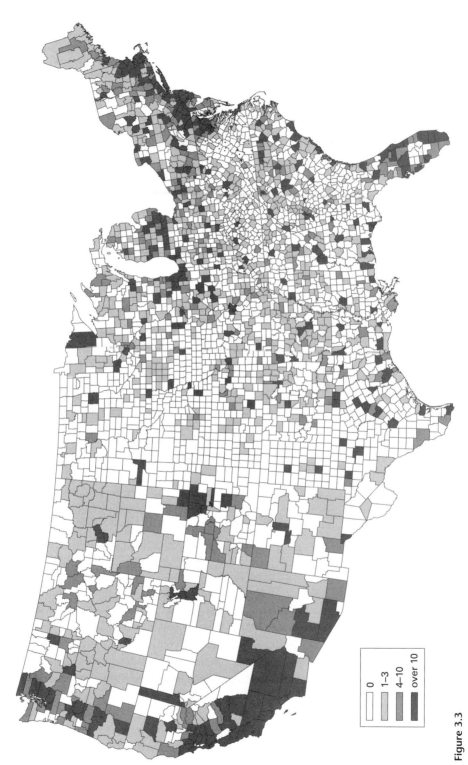

Figure 3.3

Distribution of ISPs, September 1996. Copyright © 1998 Shane Greenstein.

0
1–3
4–10
over 10

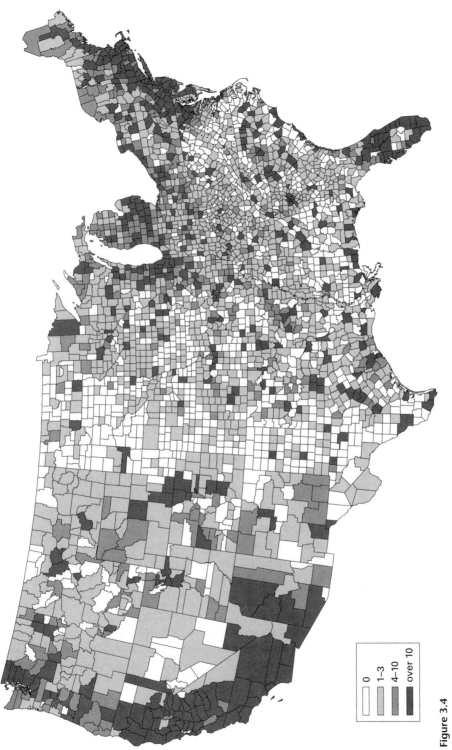

□	0
▨	1–3
▦	4–10
■	over 10

Figure 3.4

Distribution of ISPs, October 1998. Copyright © 1998 Tom Downes and Shane Greenstein.

reality, hyperbole, and dreams that underpinned this perception about demand. This also fostered the next factor: aggressive supplier behavior.

Third, there was a common understanding that savvy business practices required laying backbone capacity (i.e., sinking fiber optics into the ground) ahead of buyers' demand. Backbone suppliers saw themselves racing with each other to sign (what at the time was forecast to be) lucrative contracts for services with low-variable operating expenses. This view led firms to sign up retail customers at a financial loss at that time, with the belief that because users would not want to switch later, the firms were supporting reliable and well-paying revenue streams in the future.

Fourth, many backbone firms anticipated reselling capacity. Such resale also generated incentives to build capacity early for resale later (as long as demand continued to grow). Indeed, firms such as UUNet were ready buyers of such capacity for a time, running their network over such fiber.

This incautious behavior fostered uncoordinated build-outs among rival firms in the same locations, which in turn fostered overlapping footprints and other redundancies in supply. By 1998, it was becoming apparent that commercial firms replicated potential capacity along similar paths. Locations such as San Francisco, Chicago, Los Angeles, Washington, DC, and New York City, among others, benefited from this redundancy, because every firm with a national footprint thought it had to be in such centrally located major cities. At the time much of this fiber along these routes remained "unlit," which was industry parlance for unused.

The situation might be summarized simply: at the time there was plenty of demand for data services and well-publicized predictions about the growth in demand. Along with other factors, these predictions motivated investment, though these forecasts would prove to be too optimistic. There were also predictions about anticipated increases in the efficiencies of transmission technology. These notions played little role in investment behavior, and even if these were expected, they came to realization faster than most investors and their advisers expected. So as it would turn out, not enough growth in demand and too much growth in supply would put pressure on prices, and only a couple years later, make it challenging for investors to realize high returns on their investments.[55] This is discussed more below.

Pricing Practices

By 1998, the dominant price point for all dial-up ISPs was twenty dollars a month. Survey data showed that it was truly modal among actual purchases. That said, there was considerable variance around this price point, with roughly one-third of households paying less and one-third paying more.[56] How did twenty dollars a month arise as a focal contract?

To begin with, this price point represented a dramatic change from the pricing norms governing bulletin boards, where the pricing structure of the majority of services involved a subscription charge (on a monthly or yearly basis) *and* an hourly fee for usage. For many applications, users could go online for "bursts" of time, which would reduce the total size of usage fees. The emergence of faster, cheaper modems and large-scale modem banks with lower per port costs opened the possibility for a different pricing norm—one that did not minimize the time users spend on the telephone communicating with a server. The emergence of low-cost routines for accessing a massive number of phone lines was complementary because it enabled many ISPs to set up modem banks at a scale only rarely seen during the bulletin board era.

As the ISP industry began to develop, users demonstrated preferences for unmonitored browsing behavior, which translated into resistance to contractual limits on their time spent online. In response, some vendors began offering unlimited usage for a fixed monthly price. These plans are commonly referred to as *flat-rate* or *unlimited*.

As with any proposal changing existing practices, this proposal was not initially obvious to many vendors. Even if some users liked flat-rate pricing, there were questions about whether new users could adapt to old norms. Moreover, ISPs easily could see the difficulties of managing the modem line loads over the day or week, especially during peak times. Peak load issues would cause ISPs to invest more in modem equipment, so this type of contract was perceived to be costly to support.

A key event for the development of this focal price was the entry of AT&T's Worldnet service, which was first aimed at business in late 1995 and then explicitly marketed at households in early 1996. Far from being the first ISP to offer home service (PSINet and Netcom, among many others, could legitimately claim to have developed large-scale customer bases sooner), AT&T was among the first to offer service from a branded and established national company. The sole exception, IBM, supported a worldwide business service but not the mass-market home user.

Initially AT&T's entry went well. Choosing twenty dollars a month with the intent of becoming the dominant national provider of household Internet service, AT&T acquired over one million users within a year. That choice also imposed pricing pressure on other ISPs throughout the country.

But a funny thing happened on the way to dominance. Many small ISPs reacted, and AT&T Worldnet did not achieve dominance. More precisely, the number of new users grew so quickly that plenty of other new firms also could compete for their business. Simply said, AT&T grew large and fast, but so did others.

When AOL converted fully to flat-rate pricing in 1996, it experienced a difficult transition. Although AOL also adopted a price at near twenty dollars a month, its management had not anticipated its users' enthusiastic response to flat-rate pricing: there were insufficient modem banks nationwide to handle the increasing traffic, and many users experienced busy signals. The bad publicity induced further entry by other ISPs looking to acquire customers fleeing the busy phone lines. Ultimately, AOL survived the bad press through a series of new investments in facilities, content, and intense marketing, as well as several savvy deals, such as those with ICQ, Microsoft, and CompuServe.[57]

Eventually, many ISPs in 1996 and 1997 introduced plans that looked like a twenty-dollar-per-month fee, but many were not (actually). The fine print in these contracts included hourly limits and high marginal pricing above the limit. Most such limits were not particularly binding (i.e., involving monthly limits ranging from sixty to one hundred hours) unless the user remained online for hours at a time most days of the month.[58]

One pricing pattern had emerged by 1998, and it would continue for the next few years—namely, ISPs were largely unable to raise prices as the Internet improved. Many ISPs improved their own service in myriad ways, as did many complementors. For example, most ISPs had adopted technologies to enable Web pages to upload more quickly, either by improving dynamic Web page allocation through caching or by making arrangements with Akamai for its service. In addition, browsers got better (as a result of the browser wars and learning), so every user's experience improved. Those providing help services were also experienced and knew how to resolve issues. Yet such improvements were part of the standard package rather than vehicles for premium pricing. Take, for instance, the upgrade from 28K to 56K modems. There were two price levels for a short time—with a higher premium for faster modem service—but by late 1998, the twenty-dollar price umbrella prevailed once again.[59] In other words, the primary gatekeepers for access did not capture a higher fraction of the value from improvements in the network's many components.

Remarkably, the opposite trend for pricing also did not emerge: prices did not fall to the floor in spite of some tremendous drops in transmission costs. From 1996 to 1999, prices for non-AOL dial-up services showed a small and steady decline downward, while AOL's did not decline at all. General estimates are hard to come by, but the declines each year (on average) were no more than a reduction in a dollar for a monthly contract.[60] Several factors contributed to price buoyancy, but the key ones seemed to be user inertia and loyalty. Many ISPs saw no reason to reduce prices if users were reluctant to give up e-mail addresses or other services to which they had grown accustomed. Accordingly, many ISPs adopted policies refusing to forward e-mail for former customers as a way to make them reluctant to switch ISPs. Indeed AOL, the largest ISP in 1998, gave many of the same reasons for not lowering its prices.

Public Policies

The NSF's privatization did not come with any quid pro quo about how ISPs implemented technologies. Aside from the loosely coordinated use of a few de facto standards—coming from groups such as the World Wide Web Consortium, the IETF, or the Institute of Electrical and Electronics Engineers—mandates after commercialization were fairly minimal.[61] The ISPs were able to tailor their offerings to local market conditions.

As with other investments, impatience was a key motivation for the decision to adopt new technology.[62] Policy could play a useful role in such an environment, up to a point. For example, to avoid interference with existing equipment, the FCC imposed a number of technical restrictions on the design of 56K modems, as it had with prior transmission technologies, such as ISDN. When the International Telecommunications Union intervened in the 56K modem war and designed a new standard, adoption was voluntary; but participants decided to adopt the standard because they foresaw a huge potential loss if they delayed the upgrade further. In other words, policy helped design a better standard from a technical standpoint, and publicly supported institutions helped firms not let competitive rivalry interfere with an upgrade that was in everyone's interest.

The most important policy issues arose from the implementation of the recently passed 1996 Telecom Act. This was the first major piece of federal legislation for telecommunications since the 1934 act that established the FCC. While the 1996 Telecom Act contained many complex features, several deserve attention here because of their effect on exploratory behavior.

Some parts of the 1996 act attempted to establish regulatory continuity. For example, the act reaffirmed FCC policies to define ISPs as providers of enhanced services, following precedents set in Computers I, II, and III. Courts reaffirmed such an interpretation. For all the reasons discussed above, this had the short-term effect of encouraging the entry of ISPs.

Relatedly, ISPs did not face obligations to pay the universal service fees that telephone companies had to. The act also exempted cable companies, which because of the asymmetrical burden placed on telephone companies, became especially important to regulators when cable companies began converting their lines for carrying Internet traffic to homes. This situation festered for several years, however. Its resolution could have involved one of two actions: imposing universal service fees on Internet access provided by cable firms, or removing these fees for the suppliers of DSL (principally provided by telephone companies and third parties). Through a rather circuitous route and after several years of debate, the United States tended toward the latter choice. To understand how, I need to describe other features of the act.

The act formalized legal definitions for a Competitive Local Exchange Carriers (CLEC). Though CLECs bore some resemblance to the Competitive Access Providers of the recent past, this definition did represent a legal discontinuity.[63] The new definition was embedded in a broad set of provisions governing the access of CLECs to facilities from an Incumbent Local Exchange Carrier (ILEC). These provisions were intended to further competitive local telephony. As it turned out, the activities of CLECs presented an acute and immediate issue for the intersection of U.S. telephone and Internet policy.

Specifically, although some CLECs built their own facilities, some rented facilities from ILECs. As directed by the act, state and federal regulators had set prices for renting elements of the ILEC's network, such as the loops that carried DSL. A related set of policies concerned the billing and compensation of CLECs for exchanging traffic with ILECs. In brief, the national billing system for telephony assumed that interconnecting firms made as many calls as they received. In the U.S. compensation system, a firm paid for only one of these: the calls made, not those received.

In other words, if a CLEC received about as many calls as it sent, then no issue would have arisen. While some CLECs did just that, not all did, which gave rise to a particularly urgent question for regulators. Taking advantage of the intercarrier rules, a number of CLECs set up businesses to *receive* ISP calls from households but *send* very few, which effectively billed other telephone companies for "reciprocal compensation."

Initially this strategy went unnoticed by regulators, and consequently, regulatory decisions for the reciprocal compensation of CLECs encouraged CLEC entry, which partly encouraged ISP entry through interconnection with CLECs. The practice received attention as it grew, resulting in hearings at the FCC in 1998. The FCC passed a set of rules to do away with the practice in February 1999.[64]

Although these billing strategies and subsequent policy decisions had impacts, the effects should not be exaggerated. Their scale grew between 1997 and 1998, but ISP entry started well before then and continued afterward (until the dot-com bust in spring 2000). Moreover, most of the effect was felt in urban areas; such locations would have had a great deal of ISP entry even without this implicit subsidy to CLECs.

Though this was the first attempt by FCC staff to interpret the act's principles and ambiguities, it ended the de facto assumption of forbearance from intervening in entrepreneurial events in enhanced service markets—a precedent that Computers II and III had attempted to establish. To be fair, later some intervention was not a choice for FCC commissioners, as court decisions and congressional pressure also compelled the FCC to revisit specific actions.

The act also contained provisions for the "E-rate program"—with both real and symbolic significance. Among other goals, the E-rate program was aimed at alleviating inequities in the provision of the Internet and was proposed as a funding scheme for bringing the Internet to disadvantaged users, particularly in schools and libraries. Closely identified with the ambitions of Vice President Al Gore, who had made fostering next-generation information technology a special interest, the E-rate program was labeled the "Gore Tax" by opponents. It survived several court challenges and regulatory lobbying efforts after its passage. Although delayed by the legal challenges, the E-rate program eventually raised over two billion dollars a year from long-distance telephone bills. This money was administered by the FCC. In 1998, this program was just getting under way. It would eventually have an impact, especially on isolated locations.

In one other sense, the Internet access market got an additional implicit and explicit subsidy. The Internet Tax Freedom Act, passed in October 1998, placed a federal moratorium on taxing the provision of Internet access. Unlike several other communications technologies, such as cellular or land-line telephony, Internet access was free from local attempts to tax the service (except those that were grandfathered in prior to October 1, 1998). The market's young status justified the law, according to supporters, who worried that excessive local taxation could deter growth for the new nationwide commercial applications of electronic commerce.[65]

There was some confusion about the scope of the Internet Tax Freedom Act. Some observers incorrectly thought it outlawed sales taxes on electronic commerce. Actually, other laws already determined that retailing e-commerce had to be treated as equivalent to a catalog or mail-order seller. Online entities were not subject to local sales taxes as long as the transactions crossed state lines, which they did—for all intents and purposes—if the firms that sold the goods maintained no active physical organization in the state.

Finally, 1998 saw the beginning of a debate about merger policy for telecommunications. The 1996 Telecommunication Act did not include any overt guidance about merger policy in the telephone industry. Yet merger policy had a large effect on the restructuring of access markets. Domain for merger policy in the United States rests with the Department of Justice and the Federal Trade Commission as well as the FCC when the merger involves a national telephone firm.

During the Clinton administration, several high-profile mergers met with opposition. For example, the divestiture of some Internet backbone became a condition for government approval of the MCI-WorldCom merger—a condition that appears to have been a significant action in retrospect, as it helped to partially deconcentrate the ownership of those assets. Similarly, opposition to the proposed merger between Sprint and WorldCom, which the European regulators were the first to oppose and which the Department of Justice almost certainly would have opposed too, at least in part (the merger was called off before official Department of Justice action), looks wise in retrospect in light of WorldCom's later troubles. Finally, in late 1998, Bell Atlantic proposed a merger with GTE (forming Verizon). As a condition for approval, GTE spun off the backbone as a separate entity (forming Genuity). These actions encouraged less concentrated ownership in a setting where the number of decision makers was small.

In the local telephone market, in contrast, merger policy was comparatively focused on fostering competition in voice telephony—specifically, encouraging one telephone company to open its facilities to entrants and enter the geographic territory of another, as envisioned by Section 271 of the 1996 act.[66] The consequences for the growing Internet were secondary. In this sense, the Clinton administration did not place restrictions on several mergers involving local telephone companies, but used merger as a quid pro quo for conditions that shaped entry by one voice firm into another's territory or facilitated entry by CLECs that the ILEC opposed.

Overall, policies during this period encouraged exploratory behavior among young firms. This can be seen in both the fiscal rules that minimized taxation and fostered technical advance among young firms and the FCC's forbearance in enhanced service markets, which preserved the regulatory umbrella for ISPs. It also can be seen in the intervention in merger cases—actions that partially fostered less concentration in backbone markets, and partly not. Policy, however, did not speak with one voice, and there were limits to forbearance. By 1998, the FCC began to intervene in intercarrier compensation issues, taking actions to protect incumbent compensation.

Overview

In just half a decade, firms built a functioning network that made the information of the online world available to many participants at a low cost. Indeed, the whole of the Internet access industry was greater than the sum of individual firm's efforts, both on an operational level and in terms of its exploratory activity. A set of routines began to emerge, and so did a variety of approaches to the delivery of service. The Internet access business grew quickly and became pervasively available in every major location. Different firms perceived different customer needs, employed different visions about the commercial possibilities, and invested in assets at an unprecedented level, even for the computer market.

Broadly, such a combination of expansion in sales, variety in approaches, and increasing standardization in operations was not unusual for a young entrepreneurial market. Yet to contemporaries, the specific accomplishments appeared remarkable. No operating large-scale communication network had ever been so entrepreneurial and expanded so fast, except, arguably, the new and competitive U.S. telephone industry a century earlier. The situation in 1998 is all the more remarkable when compared to 2003, when a patina of stability informed a great deal of activity.

2003: Reliability and New Transitions

By 2003, the quality of the average Internet user experience had increased visibly, with some of that due to the increased reliability of ISPs and backbone firms, and some of that due to improvements in complementary services, such as cache sites. Accordingly, the average number of hours online and the amount of data traffic had increased too. In addition, the access market became swept up in the broad forces altering the commercial Internet, such as the dot-com bust (i.e., the financial bankruptcy of many new Internet businesses), which diminished investor enthusiasm for this sector. The widespread availability of backbone diminished long-haul prices and created further financial hardship at any firm vested in the national backbone. These factors and others altered the identities of commercial leaders who had shaped prior growth in the access market.

Though many standard operations for many large-scale activities emerged, innovative activity had not ceased. Two additional modes for access had evolved, here labeled *broadband* and *wireless*. Although the arrival of broadband and wireless access had been expected for some time, the perceived slowness of its build-out shaped a national debate about broadband policies.

The Emergence of Broadband

After the Internet demonstrated its potential to become a mass-market service, the value of upgrading to broadband—where users experienced higher access speeds—was anticipated by nearly every market vendor, user, and policymaker. In an earlier era, broadband was associated with upgrading the telephone network to support ISDN, which achieved speeds of 128K. These initiatives had not generated widespread adoption. The broadband of the new era, in contrast, had a few key features. First, it was always on, which meant a user could gain access to the Internet nearly instantaneously once a session began, in contrast to dial-up. Second, it had a larger bandwidth than dial-up or ISDN. Third, and probably most important, a whole slew of software applications had already been written for slower lines and would improve on faster lines. Many users thus had the motivation to pay for an upgrade, which in turn motivated vendors to invest to meet the anticipated demand. To put it simply, in the absence of that content, it was not obvious that the user had such motives. Hence, the presence of so much content reduced the commercial uncertainty affiliated with developing broadband services—in comparison to a decade or even a half decade earlier.

Broadband Internet access firms took three forms. First, many business firms could acquire direct access in the form of T-1 lines from telephone firms or CLECS. This had been true well before 2003, though this was normally observed only in business districts. By 2003, T-1 lines had a low market share due to their costs, and the demand for this form was in decline. Local telephone companies largely dominated this market in their home areas. A competitive provision had also arisen in some major cities in the form of metropolitan rings of fiber, which appeared to be the likely form for providing future direct access to business.[67]

Second, cable television firms retrofitted their lines and switches to carry Internet data, usually with much faster bandwidth to the user than from the user. This kind of broadband access had the earliest build-out, partly due to the regulatory advantages that will be discussed subsequently. Its market share was highest in 2003.[68] This involved a small number of (mostly national) firms, with Comcast (having acquired the cable assets of AT&T in late 2002), Time Warner, Charter, and Cox serving the largest number of households. By 2003, these firms had taken complete control of their Internet service, ending five years of outsourcing software development and other aspects of ISP service to @home.[69]

Third, local telephone firms upgraded their lines and switches to carry DSL, usually with an asymmetric digital subscriber line implementation, which once again was faster at sending data to the user than from the user. The actual bandwidth of these modes varied with implementation and location, but in general peak DSL had higher bandwidth than dial-up, although not as high as cable.[70] By 2003, few CLECs remained in this market, and those that did so, only did so with the protection of regulator-enforced agreements with ILECs. As a result, local telephone firms largely dominated the supply of DSL.[71]

In the earliest years of diffusion to households—that is, prior to 2003—supply-side issues were the main determinants of Internet availability. Cable and telephone firms needed to retrofit existing plants, and that constrained availability in many places. Cable and telephone companies found highly dense areas less costly due to economies of scale in distribution and lower expenses in build-out.

DSL access was inhibited for some consumers due to the infrastructure and distance requirements. The maximum coverage radius for DSL is approximately eighteen thousand feet from a central switching office, which is a large, expensive building.[72] Furthermore, the radius is closer to twelve thousand feet for high-quality, low-interruption service. Therefore, those living outside this radius from the central switching offices already built before DSL was available were more likely to suffer from a lack of service.

The crucial factors that affected the decision to offer DSL or cable were similar: the cost of supplying the service across different densities, the potential size of the market, the cost of reaching the Internet backbone, and telephone company regulations.[73]

As of October 2003, 37.2 percent of Internet users (i.e., just under 20 percent of U.S. households) possessed a high-speed connection. Broadband penetration was uneven, however: 41.2 percent of urban and 41.6 percent of central-city households with Internet access used broadband, whereas only 25.3 percent of rural households did. Consistent with the supply-side issues, the FCC estimated that high-speed subscribers were present in 97 percent of the most densely populated zip code areas by the end of 2000, whereas they were present in only 45 percent of the zip code areas with the lowest population density.[74]

Prices for broadband were ostensibly higher than the twenty-dollar-per-month norm for dial-up, so users faced a price/quality trade-off. Different preferences over that trade-off shaped the margin between an adopter and a nonadopter. For example, if the user was already paying twenty dollars a month for ISP service plus an additional charge for a second line, moving to broadband at forty dollars a month while retiring the second line would not seem like a large trade-off in price.[75] Indeed, because so much demand moved from dial-up to broadband, official U.S. price indexes did not show any appreciable decline or rise in the price of monthly contracts for Internet access during this time period.[76] While this shift in demand hurt pricing at dial-up firms, it supported prices at broadband firms.

The availability of broadband motivated some new Internet use in households, but not much among the heretofore nonadopters. Hence, broadband demand largely cannibalized existing dial-up demand, resulting in one-quarter of U.S. households using broadband in 2003. Access revenues for dial-up were just over $10.5 billion, while cable modem and DSL were over $12.9 billion.[77] Overall, broadband diffused within the context of reduced demand uncertainty—a factor that made it different from the diffusion of dial-up access. There were still many questions about the details of these business operations, such as their cost and pricing, but there was a consensus about the broad direction of change.

Wireless Access in the United States

The value of upgrading to wireless data services was anticipated by nearly every market vendor, user, and policymaker, but the form of its delivery was undetermined. As with dial-up, the American experience differed from the patterns that emerged in other countries, which also suggests that innovation in Internet access markets still could lead to unexpected outcomes. The United States saw three major modes of wireless access.

The first was a significant and large group of users who retrofitted a connection to their laptop computer—a mode that was labeled *Wi-Fi*. Wi-Fi required a hot spot in a public space, or a special server at home or work. A hot spot in a public space could either be free, paid for by the café or restaurant trying to support its local user base or attract a new one, or be subscription based, with users signing contracts. The latter was common at Starbucks, for example, which subcontracted with T-Mobile to provide the service throughout its cafés. The specific details behind the growth of Wi-Fi had a "bottom-up" quality reminiscent of the first wave of enthusiasm for the Internet. It also lacked regulatory guidance (except for restrictions on unlicensed spectrum, which the FCC guided). Both perceptions appealed to technically sophisticated users, who experimented with Wi-Fi.

The development of Wi-Fi occurred as follows: Wi-Fi involved a technical standard from the Institute of Electrical and Electronics Engineers subcommittee for Committee 802. Committee 802 was well-known among computing and electronics engineers because it had helped diffuse the Ethernet standard that Bob Metcalfe designed decades earlier. Subcommittee 802.11 concerned itself with wireless traffic for local access networks using Ethernet protocol. In 1999, it published Standard 802.11b, which altered some features of an earlier attempt at a standard for local Ethernet protocol (increasing the speed, among other things). Because many vendors had experimented with earlier variations of this standard, the publication of 802.11b generated a vendor response from those already making equipment.

Committee 802.11 did not initially intend to design a wireless standard for generating Internet access in coffee shops or other public spaces such as libraries. The designers focused on traditional big users (e.g., FedEx, UPS, Wal-Mart, Sears, and Boeing) that would find uses for short-range Ethernet. In this sense, their original charter was quite narrow. The publication spurred more commercial experiments.

Around the same time, pioneers of the standard—including 3Com, Aironet (now a division of Cisco), Harris Semiconductor (now Intersil), Lucent (now Agere), Nokia, and Symbol Technologies—formed the Wireless Ethernet Compatibility Alliance (WECA). As a marketing ploy for the mass market, WECA branded the new technology Wi-Fi. Related to these goals, the group also performed testing, certified the interoperability of products, and promoted the technology. In 2003, in recognition of its marketing success, WECA renamed itself the Wi-Fi Alliance.

In the midst of 2003, Intel announced a large program to install wireless capability in its notebooks, branding it *Centrino*. This action was regarded as an unusual strategic move by many Wi-Fi participants because embedding a Wi-Fi connection in all notebooks did not involve redesigning the microprocessor, which Intel made. It involved redesigning the motherboard, eliminating the need for an external card. Intel made prototypes of these motherboards and branded them. It hoped that its endorsement would increase the demand for wireless capabilities within notebooks. Nontrivially, it also anticipated that the branding would help sell notebooks using Intel chips and designs, much as the "Intel Inside" campaign had.

Intel ran into several snafus at first, such as insufficient parts for the preferred design and a trademark dispute over the use of the butterfly, its preferred symbol for the program. Also, and significantly, motherboard suppliers, card makers, and original equipment manufacturers did not like Intel's action, as it removed some of their discretion over the design of notebooks. Yet by embedding the standards in its products, Intel made Wi-Fi, or rather Centrino, easy to use, which proved popular with many users. Only Dell was able to put up any substantial resistance, insisting on selling its own branded Wi-Fi products right next to Intel's, thereby supporting some of the card makers.

Despite Dell's resistance, the cooperation from antenna makers and (importantly) users helped Intel reach its goals. Centrino became widely diffused. Intel's management viewed this outcome as such a success that it invested in further-related activities, such as upgrades (to 802.11n) and a whole new wireless standard (to 802.16, aka Wi-Max).

The second significant set of wireless users were Blackberry enthusiasts, who numbered several million. Research in Motion sold these and had explored these applications for many years. The Blackberry was a small, lightweight device designed solely to send and receive e-mail text messages, using digital cellular infrastructure to carry the signal. Research in Motion focused on corporate e-mail users, to whom it sold a software process that easily and securely forwarded business e-mail. Blackberry was thought to have over two-thirds of the market for mobile e-mail use. No other firm's implementation was as popular, either from Microsoft or a partner, or from Palm or a partner.

The third and least popular mode for wireless data services in the United States was a form of text messaging using cellular handsets. Most of these used second-generation global system for mobile communication designs, imitating practices in other developed countries where these designs were much more widespread. Despite their limited availability by 2003, these applications had not captured the popular imagination in the United States. The alternatives were more functional and less expensive.

Vendor and user perceptions in the wireless area were in flux at this time, because no sensible observer would have forecast that this situation would persist. The cellular carriers and their equipment providers were open about their ambitions to develop applications. There were well-known plans to upgrade Wi-Fi. Nor was Microsoft shy about its ambitions to foster alternatives on handheld devices using Windows CE, nor was Palm about its desire to add wireless capabilities to its popular organizers. More experimentation with hardware form factors and software applications was anticipated. No observer could have reliably forecasted which of these would most appeal to users, and generate the most revenue or profit.

From New Adoption to Capital Deepening with Business Users

Stocks of information technology capital grew at a 20 percent annual rate from the end of 1995 to the end of 2000.[78] By 2000, computer hardware and software stocks had reached $622.2 billion.[79] The majority of this investment was affiliated with enabling business applications. In 2000, the total business investment in information technology goods and services was almost triple the level for the personal consumption of similar goods.[80] The level and growth of investment dropped off considerably after 2000, flattening in 2001, 2002, and 2003.

There seemed to be several reasons for the flattening of investment in information technology. To begin with, there was a saturation of certain types of Internet adoption. In some businesses, the Internet had been adopted across all facets of economic activity, while in others adoption was not widespread. What explains this variance? There were many purposes for the Internet in business.[81] The simple first purpose, *participation*, relates to activities such as e-mail and Web browsing. This represents minimal use of the Internet for basic communications. By 2003, most businesses had made the investments necessary to participate. Indeed, adoption of the Internet for the purpose of participation was near saturation in most industries as early as 2000.

A second purpose, *enhancement*, relates to investment in frontier Internet technologies linked to computing facilities. These latter applications are often known as e-commerce, and involve complementary changes to internal business computing processes. Hence, most of the investment by 2003 was affiliated with refining the more complex applications (i.e., enhancement), and those business segments were more specific rather than widespread. Heavy Internet technology users tended to come from the historically heavy information technology users, such as banking and finance, utilities, electronic equipment, insurance, motor vehicles, petroleum refining, petroleum pipeline transport, printing and publishing, pulp and paper, railroads, steel, telephone communications, and tires.

Why did this pattern between participation and enhancement investment emerge? First, the applications with the most demand were e-mail and browsing. Investments that went beyond that were complex and costly, but they were potentially valuable to some firms. Specifically, those that had invested in advanced information technology in the past had the staff, equipment, and need to invest in complex applications.

Second, most firms are incremental in their approach to complex investment in information technology—compromising the benefits of frontier technology and the costs of keeping an existing process, they pick and choose among those new possibilities that make the most sense for their business. Hence, few industries with little experience in advanced information technology suddenly chose to become a heavy investor when the Internet commercialized.

Third, investment in innovative information technology is directed toward automating functional activity or business processes within an organization, such as accounts receivable, inventory replenishment, or point-of-sale tracking.[82] Such processes only change slowly, if at all. Even if firms wanted to invest, it was difficult to do so without affecting current operations.

Finally, in addition to the types of investment available, the dot-com bust impacted the Internet access market, as a large secondary market for used equipment from recently bankrupt dot-com and CLECs depressed prices for new equipment. Firms with major businesses in selling equipment into Internet access markets suffered from the drop in sales in 2003. These demand conditions affected every leading equipment firm, such as Cisco, Lucent, Nortel, JD Uniphase, Corning, and many others. Moreover, a large number of start-up firms with projects developing frontier products found themselves without any potential buyer and without a realistic possibility for an initial public offering. Many of the venture capital funds for these communications equipment firms had not reached a profitable state by 2003.

Enterprise computing stood at a crossroads, and therefore so stood a large segment of the Internet access market. By 2003, many business operations were increasingly dependent on the reliable operation of the Internet. That heightened questions about the ability of the Internet to support secure transactions, withstand terrorist attacks to infrastructure, or survive malicious virus attacks to the operating software. It also heightened questions about scaling the Internet to many devices. For example, IPv6 allowed for a large expansion in the number of Internet addresses, alleviating a potential problem that most technical insiders forecast. Though defined in 1994 by the IETF, the slow diffusion of IPv6 did not inspire confidence about the ability of the Internet to scale in the next decade when it involved the uncoordinated actions of so many firms.

In 2003, a number of potential future high-value applications were being discussed, but none had yet diffused widely. Voice-over Internet protocol and applications of the wireless Internet were some of the most popular among futurists. Most examples of high-impact applications were still confined to frontier users in businesses and homes, except for a few mobile applications. That was no more than a small percentage of the total Internet usage.

Market Leadership

It is inherent in exploratory activity that vendors have discretion to take risks, assemble information, and generate assessments about future prospects. It is inefficient for stockholders, regulators, or auditors to question every aspect of a firm's decision at every moment in time. Yet publicly traded companies, and even most privately held ones, do not retain such discretion indefinitely. Periodically managers' decisions will be reviewed by someone, such as a corporate board, an internal or external audit team, skeptical stockholders, or financial reporters from newspapers.

Such reviews at some key firms in the Internet access business generated a number of scandals. These scandals raised questions about the long-run economic viability of the U.S. Internet networks, and these questions led observers to wonder in retrospect whether some Internet infrastructure investment had been excessive.

From 2000 to 2002, Internet investment took a downturn. When quite a few of the CLEC firms did not realize their commercial promises, losing significant financial value in a short period, their losses became known as the Telecom Meltdown. Financial support for dot-coms declined in spring 2000 and was popularly labeled "the dot-com bubble burst." Then the September 11 terrorist attack in 2001 shook business confidence in long-term investments. This low continued as the WorldCom financial scandal was publicized in spring 2002.

Many observers believed that the United States had extraordinarily high levels of "dark fiber"—that is, capacity for carrying data that went unused (or unlit). Qwest, Level3, Sprint, Global Crossing, MCI-WorldCom, Genuity (formerly the backbone for GTE), Williams, PSINet, AT&T, and others came under this suspicion because all of them invested heavily in redundant transmission capacity during the boom. In addition, technical advances in multiplexing allowed owners of fiber to use existing capacity more efficiently, increasing the efficiencies from (or reducing the costs of) using existing conduit by orders of magnitude. By 2003, industry insiders forecast that there would not be sufficient growth in demand to use up the capacity in existing national fiber for anytime into the indefinite future. As expected in a market with overcapacity, plenty of evidence suggested that buyers of large amounts of bandwidth experienced drops in prices for carrying data over long distances from one major city to another.[83]

Financial scandal also hit the Internet carrier business. A division of WorldCom, UUNet was the largest backbone data carrier in the United States. Although UUNet did not have any accounting problems, accounting scandals at its corporate parent led to the bankruptcy of WorldCom. PSINet and Global Crossing also overextended themselves, and had to declare bankruptcy. Genuity did the same to facilitate merging with Level3. And Qwest and AT&T overextended themselves financially, and went through dramatic management changes and restructuring. So too did AOL after a merger with Time Warner.

The crisis in the financial conditions of many leading firms did not generate a coordinated response from the agencies in charge of communications policy, such as the FCC. Following long-standing norms, regulators would not act unless U.S. communications were interrupted by bankruptcy and restructuring. Thus, improving the financial solvency of companies was left to the discretion of managers, their corporate boards, and debt holders.

An objective observer in 2003 would have found it difficult to have full confidence in the leaders of the Internet access industry from a similar list made a half decade earlier. The market leaders from five years ago had all made errors of strategy or accounting. AOL, WorldCom, AT&T, Enron, Qwest, Global Crossing, Genuity, PSINet, and MCI had all lost their status as trustworthy leading firms in many aspects of their businesses. Many had lost their identities altogether as distinct firms, while others tried to hire managers to work through the crisis and help the organizations emerge anew with sound operations in the remaining businesses. To be sure, despite financial crises, the largest of them—AT&T, Qwest, and WorldCom—all continued to offer services, and none of them suddenly lost many customers for their data services.

Only a few of the entrants of the prior decade could claim a leadership position, such as Earthlink/Mindspring, Juno/Netzero, or Level 3. Other firms—such as IBM, Accenture, Microsoft, Intel, or Cisco—that spoke with authority about the Internet access market were involved in the broad Internet infrastructure business in many respects, not just Internet access as the only market service.

A new set of market leaders also emerged from among the firms that had had more conservative behavior—especially broadband carriers. Cable companies, due to their investments, established the largest market share—such as Comcast, which had bought the cable plant for home provision from AT&T. The three gigantic and financially sound local telephone firms (e.g., Verizon, SBC, and Bell South) had the next largest market share. These firms were increasingly providing the wireline broadband access to homes across the country. Mergers of local telephone companies further resulted in a consolidation of managerial decision making over the assets affiliated with deploying DSL in the United States as well as partially over the major backbones.

To the enthusiasts who helped initiate and catalyze the growth of the Internet, these outcomes in 2003 seemed like a cruel cosmic joke. The scandals, strategic missteps, and mergers helped bring about a decline in once high-flying entrepreneurial firms, which naturally led to a decline in the founding of new firms after 2000. With the aid of regulatory rulings favoring incumbents that owned facilities, a set of firms that were among the least entrepreneurial at the outset of the new technology market in 1993 began to acquire a position of commercial leadership in 2003. These firms also stood for the centralized management of technological opportunities, the cultural antithesis of what the more rebellious parts of the Internet community supported.

These firms took a different view, arguing that their prudence allowed them to capitalize on the business excesses of others. In addition, they pointed to their distinct assets and competitive advantages, maintaining that the new technological opportunities favored their organizations' comparative advantages under almost any regulatory setting.

By the end of 2003, a new debate began about whether any regulatory intervention was needed to curb the actions of the largest data carriers. Given the label "net neutrality" by its proponents, the regulation proposed curbs on retail and whole discriminatory practices by broadband carriers. That is, it proposed bans on blocking access to content and some forms of tiered pricing for different services. At the end of 2003 this debate was far from resolved.

Merger and Regulatory Policy

Legal challenges to the 1996 act continued well into 2003, altering all the important rules for interconnection and access—what could be unbundled, prices for renting unbundled elements, and eventually, the definitions for the boundary between an information service and a telecommunications service. In general, seven years of rule changes did not give participants a clear prediction about the market conditions that governed their investments. Even seven years after the passage of the 1996 act, one key provision after another received a new interpretation or implementation, either due to changing court rulings or regulatory interpretations from new FCC commissioners. Prior investments either became more or less valuable. All firms learned their lessons quickly: the value of exploratory investments depended critically on decision making in Washington, DC, and the rulings at the courts hearing one lawsuit or another.

For example, the FCC gradually reduced the ease with which competitors could make use of the unbundled elements of incumbent networks. Mandated by court rulings, in 2002 the FCC began to adopt a set of rules designed to discourage others from using new investments in broadband, thus encouraging local telephone companies to invest in broadband. By 2005, ILECs were not obligated to make most DSL service available on a wholesale basis to resellers.

Interpreting and implementing this area of regulation became one of the most contentious chapters in U.S. telecommunications policy. The policy debates did not take place in the face of calm deliberations of historical facts but became opportunities for strident expressions of belief and viewpoints, further amplified by lobbyists and the concerns of key politicians. Needless to say, the role of concentrated provision in fostering exploration was not the most salient factor in either side's public stance. More to the point, the summary below is but a brief oversimplification of a complex and extensive set of arguments.[84]

To some, the FCC's decisions to favor "facilities-based competition" were seen as aiding U.S. competitiveness and productivity. For example, the FCC made it easier for telephone and cable firms to deny access to any third-party ISP or CLEC that had taken advantage of "regulatory expropriation" rather than invest in its own physical facilities. In addition, telephone company executives and shareholders had long considered themselves as unfairly disadvantaged by the unbundling requirements in the 1996 Telecommunications Act as well as by the asymmetries in the applications of enhanced service rules. Hence, these rule changes also were interpreted as righting a prior wrong.

Views of the opposing camp were equally as strident. The rule changes were considered as an outrageous attempt to prevent CLECs from competing and a means to close channels to firms other than owners of bottleneck assets in the national communication infrastructure. Many of the bankruptcies of CLECs in the last few years were portrayed as the result of anticompetitive actions by cable and telephone companies. In this light, the changes were viewed as a capitulation to political and regulatory lobbying by phone and cable companies, not fixing the correct problem, nor using any measured consideration of policy. Cynics pointed to the explicit political pressure coming from the House of Representatives, which passed the Tauzin-Dingle Bill in February 2002. Though the bill did not pass the Senate that year, it called for some of the same features that the FCC adopted in 2003. Cynics argued that policy seemed to help well-connected telephone and cable companies at the expense of the less-experienced firms. To Internet enthusiasts, who viewed the growth of the Internet as a rebellion of outsiders, these rule changes also were seen as Goliath's conspiracy to support the old establishment—that is, cable firms and telephone firms—in the face of a technical entrant—that is, CLECs, ISPs, and many related infrastructure firms.

Merger policy also took on increased importance. The Bush administration continued with earlier trends of allowing mergers among local telephone firms. Eventually, Verizon combined the assets from what used to be Bell Atlantic, Nynex, GTE, and in 2005, the financially weakened MCI/WorldCom. SBC combined the assets of Southwest Bell Corporation, Southern New England Telecommunication, Ameritech, the Pacific Bell Company, and in 2005, what was left of AT&T's long-distance and local service, including its Internet business. Also, SBC took AT&T as the corporate name thereafter.[85]

Concerns about the ownership of bottleneck facilities played little role in these mergers. No divestitures were required as a condition for the 2005 mergers—a striking feature since the market coverage of the firms did geographically overlap, and the acquiring firms in this case, local telephone firms, managed many points of access to telephone switches. One condition did emerge from the negotiations. The telephone companies were required to sell "naked DSL" services—that is, purchase of DSL service without a dial tone. In plain language, a customer did not have to purchase a voice service to get DSL services. This was not an onerous requirement for telephone companies in comparison to the many things that could have been imposed.

Overall, the most notable policy decisions in 2003 facilitated a shift in policy favoring the consolidation of ownership of assets in the operation of telephone networks. This came from two fronts: the change in rules for the resale of unbundled elements, and merger policy. More consolidation from merger would come within the next few years. This consolidation opened questions about whether the public policy directive to achieve financial solvency in both access and upstream backbone markets led to too much reduction in redundant investment. Had the United States retained sufficient redundancy for firms to competitively discipline one another for unwarranted price increases or undesirable nonprice vendor behavior? This reduction also raised questions about whether owners of bottleneck facilities could expropriate returns from innovation conducted by firms offering complementary services. Only events in the next decade could answer such questions.

Overview

The Internet of 2003 would shock a market participant from a decade earlier. The novel had become routine. Access markets became a reliable functioning part of the Internet, invisible to most users, yet still changing in reaction to new opportunities and new regulatory rulings.

By 2003, the set of identities of the market leaders expanded dramatically. Financial scandals, and events known popularly as the Telecom Meltdown and the dot-com bubble burst, reduced the flow of financial resources to this sector. This and the absence of growth in demand led to financial hardship. In addition, financial and accounting scandals came to light, involving some of the industry's highest-profile participants.

Summary of the First Decade

There is a cliché from canonical narratives of technical new industries: in a setting where cautious or unaware market leaders have no reason to alter their business, the intrepid entrepreneurial firms may be the first to take risks and seek customers. The absence of concentration enables those so-called contrarians to reach the marketplace sooner and initiate innovative responses from incumbents. The incumbent response then determines whether the commercial efforts of the entrants lead to changes in market leadership or not.

The Internet access experience does reflect this cliché in some respects. In 1993, few observers recognized that the setting would change so dramatically. A variety of firms saw a market opportunity and developed their commercial services before incumbent firms, ultimately demonstrating the viability of commercial Internet access service. After that, competitive pressures accelerated the adoption of innovative practices by a variety of providers. Their actions collectively initiated a commercial revolution. Any reasonable reading of this history has to conclude that this revolution would have been delayed had the entrepreneurial firms not acted.

Yet the cliché also simplifies the role for variety. The commercial Internet access market did not involve only one key difference between the views of entrants and incumbents. Rather, an ensemble of firms approached the new opportunity in 1993. At first a few firms invested, while most did not. They did not speak with one voice or even necessarily share the same vision about the source of commercial value. By 1998, all participants could reliably forecast growth in demand. In that setting a variety of approaches flourished. Yet experts still disagreed about the most valuable form of service and business operations, or for that matter, the likely identities of future commercial leaders. Even by 2003, after a number of routines emerged for commercial behavior and consolidation reduced the range of differences between some leading firms, the access market had not lost all its variety. Firms were developing new access modes in broadband and wireless markets. Innovation was creating new value yet again.

The canonical cliché also overlooks the nurturing institutional setting in which the commercial Internet diffused. The commercial dial-up access market was the unanticipated result of the partnership between telephone companies and ISPs. At the outset, this relationship was mutually beneficial but fragile. Legal precedent held it together at first. The unexpected and massive commercial opportunity turned it into a key component of almost every firm's activity.

As the first decade of the commercial Internet came to a close, this relationship had come under severe strain, affected by two powerful and distinct roles for regulation in spurring innovation. On the one hand, entrepreneurial firms were unlikely to undertake risky innovative activity if they experienced changes in the regulatory and legal environment that protected them. On the other hand, publicly traded firms, such as a telephone company or cable firm, were unlikely to undertake large-scale multimillion dollar investment in the face of regulatory and legal uncertainty about the financial returns on their investments. From the start, policy inherited rules that emphasized the former principle, nurturing innovative activity out of ISPs. At the end of the decade, policy tried to improve incentives for large telephone companies, and only later events would tell whether this came at the expense of the first principle.

The entire decade is also astonishing for the number of unplanned but ultimately complementary circumstances that transformed the Internet from a communications tool for researchers into a strategic investment priority for every firm in commercial computing and communications. The NSF's managers gave the Internet access market a nurturing beginning, but that did not determine the outcome, nor could it. The NSF had the good fortune to inherit competitive norms from computing and regulatory rules from communications that fostered commercial experiments from a variety of firms. Other ongoing regulatory decisions could have done a great deal of harm by discouraging the development of new operational practices and the discovery of the sources of value in nascent demand. Yet a fair reading has to conclude that such massively poor judgment did not occur. Though not all regulatory actions in the first decade facilitated innovation, a balanced characterization of the first decade can acknowledge both the challenges faced by and the impressive achievements of the participants in the young Internet.

Acknowledgments

I would like to thank Bill Aspray, Pablo Boczkowski, Tim Bresnshan, Paul Ceruzzi, Barb Dooley, Chris Forman, Avi Goldfarb, Rebecca Henderson, Christiaan Hogendorn, Scott Marcus, Kristina Steffenson McElheran, Roger Noll, Greg Rosston, Jeff Prince, Alicia Shems, Scott Stern, Tom Vest, and seminar audiences for suggestions. All errors are mine.

Notes

1. To be sure, this is a statement about the size of activity, not the private or social rate of return on investment. In the 2004 Service Annual Survey (released by the U.S. Census Bureau on December 29, 2005) NAICS 514191, Internet service providers made $10.5 billion in access fees in 2004; in NAICS 5132, cable network and program distribution made $8.6 billion in Internet access service; and in NAICS 5133, telecommunications carriers made $4.3 billion in Internet access services. This does not count the revenue for other online services. See U.S. Department of Commerce, Bureau of Census, *Statistical Abstract of the United States*, 125th ed. (2005).

2. Bruce Owen, "Broadband Mysteries," in *Broadband: Should We Regulate High-Speed Internet Access?* ed. Robert W. Crandall and James H. Alleman (Washington, DC: AEI–Brookings Center for Regulatory Studies, 2002), 9–38.

3. On the dot-com boom, for example, see the series of studies on use by the National Telecommunications Information Administration (NTIA): "Falling through the Net: A Survey of the

'Have-Nots' in Rural and Urban America'' (1995); ''Falling through the Net: Defining the Digital Divide'' (1997); ''Falling through the Net II: New Data on the Digital Divide'' (1998); ''A Nation Online: How Americans Are Expanding Their Use of the Internet'' (2002); ''A Nation Online: Entering the Broadband Age'' (2004). All are available at ⟨http://www.ntia.doc.gov/reports.html⟩. See also the surveys on use done by the Pew Internet and American Life Project, available at ⟨http://www.pewinternet.org⟩. For a summary of the writing on the geography of the digital divide, see Shane Greenstein and Jeff Prince, ''The Diffusion of the Internet and the Geography of the Digital Divide,'' in *Oxford Handbook on ICTs*, ed. Robin Mansell, Danny Quah, and Roger Silverstone (Oxford: Oxford University Press, 2006). There is considerable writing on both the dot-com boom and broadband investment policies. See, for example, Lawrence Lessig, *Code and Other Laws of Cyberspace* (New York: Basic Books, 1999); Gregory Sidak, ''The Failure of Good Intentions: The WorldCom Fraud and the Collapse of American Telecommunications after Deregulation,'' *Yale Journal of Regulation* 20 (2003): 207–267; Robert Crandall, *Ten Years after the 1996 Telecommunications Act* (Washington, DC: Brookings, 2005); J. E. Nuechterlein and P. J. Weiser, *Digital Crossroads: American Telecommunications Policy in the Internet Age* (Cambridge, MA: MIT Press, 2005); Marjory S. Blumenthal and David D. Clark, ''Rethinking the Design of the Internet: The End-to-End Arguments vs. the Brave New World,'' in *Communications Policy in Transition: The Internet and Beyond*, ed. Benjamin Compaine and Shane Greenstein (Cambridge, MA: MIT Press, 2001), 91–139. This list is hardly exhaustive.

4. See, for example, Organization for Economic Cooperation and Development, *Local Access Pricing and E-Commerce* (July 2000). Other practices also have received attention. For instance, the U.S. market attracted more than half a dozen backbone providers, while most countries had one or only a few. See Nicholas Economides, ''The Economics of Internet Backbone,'' in vol. 2, *Handbook of Telecommunications Economics*, ed. Martin Cave, Sumit Majumdar, and Ingo Vogelsang (Amsterdam: Elsevier Publishing, 2005); Christiaan Hogendorn, ''Excessive (?) Entry of National Telecom Networks, 1990–2001,'' mimeo, Wesleyan University, available at ⟨http://chogendorn.web.wesleyan.edu⟩. In addition, for many years U.S. access firms sold portfolios of new services, taking organizational forms unlike that of any other country's firms. See Shane Greenstein, ''Commercialization of the Internet: The Interaction of Public Policy and Private Actions,'' in *Innovation, Policy, and the Economy*, ed. Adam Jaffe, Josh Lerner, and Scott Stern (Cambridge, MA: MIT Press, 2001). The causes behind the dot-com boom and bust have also attracted attention. See, for example, Martin Kenney, ''The Growth and Development of the Internet in the United States,'' in *The Global Internet Economy*, ed. Bruce Kogut (Cambridge, MA: MIT Press, 2003); Matthew Zook, *The Geography of the Internet Industry* (Malden, MA: Blackwell Publishing, 2005); Brent Goldfarb, David Kirsch, and David Miller, ''Was There Too Little Entry during the Dot-com Era?'' Robert H. Smith School research paper, April 24, 2006, available at ⟨http://ssrn.com/abstract=899100⟩. This chapter will address overlapping determinants in access markets.

5. This simplifies a complex history. See, for instance, Brian Kahin, *Building Information Infrastructure: Issues in the Development of the National Research and Education Network* (Cambridge, MA: McGraw-Hill Primis, 1992); Karen Frazer, *Building the NSFNET: A Partnership in High-Speed Networking*, available at ⟨ftp://nic.merit.edu/nsfnet/final.report/.index.html⟩; Janet Abbate, *Inventing the*

Internet (Cambridge, MA: MIT Press, 1999); David C. Mowery and Tim S. Simcoe, "The Origins and Evolution of the Internet," in *Technological Innovation and Economic Performance*, ed. R. Nelson, B. Steil, and D. Victor (Princeton, NJ: Princeton University Press, 2002); Paul E. Ceruzzi, "The Internet Before Commercialization," herein; Thomas Haigh, "Software Infrastructure of the Commercializing Internet," herein.

6. In an intriguing footnote, Abbate (*Inventing the Internet*, 239) states that this structure was probably inspired by the telephone industry at the time, with a mix of regional networks and national interconnection. It is unclear how much of this inspiration was "conceptual" and how much was simply "following the path of least resistance," because the national telephone carriers—Sprint, MCI, and AT&T—could use their existing assets to provide backbone services.

7. This statement necessarily oversimplifies a long institutional history. There were many earlier attempts to establish internal pricing within universities for the incremental use of computing services. Most of these had died out many years earlier, instead replaced by indirect funding mechanisms supporting Internet department budgets that provided services for students and faculty without usage charges. This principle extended to the use of modem banks. For the early history of this practice, see William Aspray, and Bernard O. Williams, "Arming American scientists: NSF and the provision of scientific computing facilities for Universities, 1950–1973," *IEEE Annals of the History of Computing* 16, no. 4 (Winter 1994): 60–74.

8. Once again, this is a simplification. See Kahin, *Building Information Infrastructure*; Frazer, *Building the NSFNET*; Abbate, *Inventing the Internet*; Mowery and Simcoe, "The Origins and Evolution of the Internet"; Ceruzzi, "The Internet Before Commercialization"; Haigh, "Software Infrastructure."

9. Note that the period for the existence of the first commercial ISPs is 1991–1992, *prior* to when the NSF began to lift the acceptable use policy for commercial activity over the Internet. These firms were already anticipating operating a private network that would bypass the publicly funded Internet backbone. For example, three of these networks—PSINet, CERFNet, and Alternet— formed the Commercial Internet eXchanges in July 1991 for the purpose of exchange traffic outside the publicly funded data-exchange points.

10. Other sources show a similar trend. In one of the earliest Internet "handbooks," *The Whole Internet User's Guide and Catalog* (Sebastopol, CA: O'Reilly, 1992), Ed Krol lists 45 North American providers (8 have a multicity national presence). In the second edition (1994), Krol lists 86 North American providers (10 have a national presence). April Marine, Susan Kilpatrick, Vivian Neou, and Carol Ward, in *Internet: Getting Started* (Englewood Cliffs, NJ: Prentice Hall, 1993), list twenty-eight North American ISPs and six foreign ones. Karen G. Schneider, in *The Internet Access Cookbook* (New York: Neal-Schuman Publishers, 1996), lists 882 ISPs in the United States and 149 foreign ones.

11. For an accessible comparison of the AOL/MSN differences and similarities—especially their market positions and strategic approaches—during the mid-1990s, see, for example, Kara Swisher, *aol.com: How Steve Case Beat Bill Gates, Nailed the Netheads, and Made Millions in the War for the Web* (New York: Random House, 1998).

12. Krol (*The Whole Internet*, 1994) gives a good sense of movement prior to 1995. For a description of the type of resistance encountered by foresighted entrepreneurs during this time, see, for instance, Charles Ferguson, *High Stakes, No Prisoners: A Winner's Tale of Greed and Glory in the Internet Wars* (New York: Crown Business, 1999).

13. See, for example, Mary Meeker and Chris Dupuy, *The Internet Report* (New York: HarperCollins Publishers, 1996). This team of analysts was not the first to organize a systematic analysis of the vendors in the market. The first-stage venture capital firms certainly were earlier. To my knowledge, however, it is among the earliest publications for general investors from an established Wall Street organization.

14. Note, however, that a full comparison of the regulatory/industry relationship in other countries will be left for future work; this chapter concentrates on only the U.S. experience.

15. There is a long history behind these events, and this chapter reviews only a part. See, for example, G. Roger Noll and Bruce M. Owen, "The Anticompetitive Uses of Regulation: United States v. AT&T (1982)," in *The Antitrust Revolution*, ed. John E. Kwoka and Lawrence J. White (Glenview, IL: Scott, Foresmann and Company, 1989); Kevin Werbach, *A Digital Tornado: The Internet and Telecommunications Policy*, FCC, Office of Planning and Policy working paper 29, March 1997; Jason Oxman, "The FCC and the Unregulation of the Internet," FCC, Office of Planning and Policy working paper 31, 1999; Robert Cannon, "Where Internet Service Providers and Telephone Companies Compete: A Guide to the Computer Inquiries, Enhanced Service Providers, and Information Service Providers," in *Communications Policy in Transition: The Internet and Beyond*, ed. Benjamin Compaine and Shane Greenstein (Cambridge, MA: MIT Press, 2001); Bruce Owen, "Broadband Mysteries," in *Broadband: Should We Regulate High-Speed Internet Access?* ed. Robert W. Crandall and James H. Alleman (Washington, DC: AEI-Brookings Center for Regulatory Studies, 2002), 9–38; and Christiaan Hogendorn, "Regulating Vertical Integration in Broadband: Open Access versus Common Carriage," *Review of Network Economics* 4, no. 1 (March 2005).

16. In addition, in Judge Greene's administration of the modified final judgment of the divestiture of AT&T, there were bright lines regulating local telephone firm involvement in enhanced service markets. See Noll and Owen, "The Anticompetitive Uses of Regulation"; Nuechterlein and Weiser, *Digital Crossroads*.

17. This is a simplification for the sake of brevity. Vendors did learn what was possible, but the engineering challenges turned out to be manageable. See Werbach, *A Digital Tornado*; Oxman, "The FCC."

18. This transition is apparent by the 1997 and 1998 editions of *Boardwatch Magazine's Directory of Internet Service Providers* (Littleton, CO.)

19. See the discussions in William Aspray, "Internet Use," in *Chasing Moore's Law: Information Technology Policy in the United States*, ed. William Aspray (Raleigh, NC: SciTech Publishing, 2004), 85–118; Blaise Cronin, "Eros Unbounded: Pornography and the Internet," herein. As further evidence, the 1996 edition of *Boardwatch Magazine's* directory makes it apparent that these bulletin board firms were geographically dispersed across the country.

20. See, for example, Robert Friedan, "Revenge of the Bellheads: How the Netheads Lost Control of the Internet," *Telecommunications Policy* 26, no. 6 (September–October 2002): 125–144.

21. See, for example, Gerstner's account of the resistance he met trying to move the firm to support technologies, such as Unix and Microsoft-client software, from outside those found in mainframes. He argued that in retrospect, these changes were among those that were most beneficial for the long-term health of the firm. Louis V. Gerstner, *Who Says Elephants Can't Dance?* (New York: HarperCollins, 2002).

22. Ferguson (*High Stakes, No Prisoners*) contains a detailed account of the 1994 founding of Vermeer, a company that aimed to provide software for Internet applications and servers. The book offers a variety of reasons why so few new firms were founded this early. See also the discussion in Krol (*The Whole Internet*, 1994).

23. Under the original design plans for Windows, there were two target markets, one aimed at personal computer clients and one at servers. TCP/IP compatibility had value for server software as a direct competitor to Unix systems. It also had value because it eased data exchange between a server and a client.

24. Design decisions for the operating system were receiving attention from the highest level of the company, including Bill Gates, Steve Balmer, Nathan Myrvold, and all the other corporate officers who made up the "brain trust" for the firm. The acknowledgment of the error and change in strategy came in April–May 1995. See William Gates, "The Internet Tidal Wave," internal Microsoft memo, May 20, 1995, Redmond, Washington, available at ⟨http://www.usdoj.gov/atr/cases/ms_exhibits.htm⟩, government exhibit 20.

25. This is a necessary summary of a long set of complex events. See, for example, Haigh, "Software Infrastructure"; Timothy Bresnahan and Pai-ling Yin, "Setting Standards in Markets: Browser Wars," in *Standards and Public Policy*, ed. Shane Greenstein and Victor Stango (Cambridge: Cambridge University Press, 2006); Timothy Bresnahan, Shane Greenstein, and Rebecca Henderson, "Making Waves: The Interplay between Market Incentives and Organizational Capabilities in the Evolution of Industries," working paper, 2006. ⟨http://www.Keffgs.Northwestern.edu/faculty/greenstein⟩.

26. To my knowledge, no research has traced fully the pork barrel in the NSFNET, which favored IBM, MCI, and Bolt Beranek and Newman, the local economies built around the supercomputer centers that the NSF subsidized, and arguably, some of the young entrepreneurial firms that descended from the NSFNET. It also favored the many universities that received subsidies to establish Internet connections as well as the computer science departments whose faculty and students worked in the supercomputer and broader computing centers that provided Internet access. The contemporary public justification for these subsidies emphasizes the traditional rationale—that is, the technical goals or research accomplishments of the projects and other publicly spirited aims affiliated with subsidizing research and development (see, for example, the discussion in Frazer, *Building the NSFNET*; Abbate, *Inventing the Internet*).

27. This is a much longer story. See the accounts in Ceruzzi, "The Internet before Commercialization"; William Aspray, "Internet Governance," in *Chasing Moore's Law: Information Technology*

Policy in the United States, ed. William Aspray (Raleigh, NC: SciTech Publishing, 2004), 55–84; Milton Mueller, *Ruling the Root: Internet Governance and the Taming of Cyberspace* (Cambridge, MA: MIT Press, 2004).

28. See, for instance, Tim Berners-Lee with Mark Fischetti, *Weaving the Web: The Original Design and Ultimate Destiny of the World Wide Web By Its Inventor* (New York: HarperCollins, 1999).

29. Once again, this simplifies a complex story. See, for example, Abbate, *Inventing the Internet*; Timothy Simcoe, "Delay and *De Jure* Standardization: Exploring the Slowdown in Internet Standards Development," in *Standards and Public Policy*, ed. Shane Greenstein and Victor Stango (Cambridge: Cambridge University Press, 2006).

30. For more about these events, see Ceruzzi, "The Internet before Commercialization."

31. Approximately 26 percent of U.S. households had Internet access in 1998, up from 18.6 percent the year before. Cable companies accounted for only 2 percent of subscriptions. Local telephone companies accounted for only 10 percent, so the number of broadband connections was relatively small to homes.

32. NTIA ("Falling through the Net II") shows that national service providers, telephone companies, and cable companies accounted for just under 90 percent of the household subscriptions. The NTIA surveys do not begin to track broadband users until August 2000, when the survey finds that 4.4 percent of U.S. households are broadband users (with 41.5 percent of households being Internet users).

33. See, for example, Thomas Downes and Shane Greenstein, "Universal Access and Local Commercial Internet Markets," *Research Policy* 31 (2002): 1035–1052, and "Understanding Why Universal Service Obligations May Be Unnecessary: The Private Development of Local Internet Access Markets," *Journal of Urban Economics* (forthcoming). See also Gregory Stranger and Shane Greenstein, "Pricing at the On-Ramp for the Internet: Price Indices for ISPs in the 1990s," in *Essays in Memory of Zvi Griliches*, ed. Ernst Berndt, Chuck Hulten, and Manuel Trajtenberg (Chicago: University of Chicago Press, 2006).

34. See Tom Downes and Shane Greenstein, "Do Commercial ISPs Provide Universal Access?" in *Competition, Regulation, and Convergence: Current Trends in Telecommunications Policy Research*, ed. Sharon Gillett and Ingo Vogelsang (Mahwah, NJ: Lawrence Erlbaum Associates, 1998), 195–212.

35. See, for example, Friedan, "Revenge of the Bellheads"; Brian Kahin, "The U.S. National Information Infrastructure Initiative: The Market, the Web, and the Virtual Project," in *National Information Infrastructure Initiatives, Vision, and Policy Design*, ed. Brian Kahin and Ernest Wilson (Cambridge, MA: MIT Press, 1997), 150–189.

36. Among those sometimes counted as tier-1 providers are Genuity, Qwest, IXC, Williams, and Level3. For different discussions, see Michael Kende, "The Digital Handshake: Connecting Internet Backbones," FCC, Office of Planning and Policy working paper 32, 2000; Hogendorn "Excessive (?) Entry of National Telecom Networks"; Economides, "The Economics of Internet Backbone."

37. See, for instance, Stanley Besen, Paul Milgrom, Bridger Mitchell, and Padmanabhan Sringagesh, "Advances in Routing Technologies and Internet Peering Agreements," *American Economic Review* (May 2001): 292–296; Stanley Besen, Jeffrey S. Spigel, and Padmanabhan Srinagesh, "Evaluating the Competitive Effects of Mergers of Internet Backbone Providers," *ACM Transactions on Internet Technology* (2002): 187–204.

38. For a description of this institution, see Hogendorn, "Excessive (?) Entry of National Telecom Networks."

39. For contemporary summaries of these debates, see Oxman, "The FCC"; Kende, "The Digital Handshake."

40. Technical practices improved. In 1993, the connections offered tended to be Unix-to-Unix CoPy (UUCP) connections that were capable of exchanging files, newsgroups, and e-mail, but they had no interactive features. By 1998 (and probably as early as 1995), all of the ISPs supported serial line Internet protocol access, a more highly interactive connection that has all the capabilities of UUCP plus additional features (including multimedia capabilities). This became obsolete comparatively soon, as most ISPs began to use point-to-point protocol. For a broader discussion of these and other technical issues, see Haigh, "Software Infrastructure."

41. See, for example, Shane Greenstein, "Building and Delivering the Virtual World: Commercializing Services for Internet Access," *Journal of Industrial Economics* 48, no. 4 (2000): 391–411; Shawn O'Donnell, "Broadband Architectures, ISP Business Plans, and Open Access," in *Communications Policy in Transition: The Internet and Beyond*, ed. Benjamin Compaine and Shane Greenstein (Cambridge, MA: MIT Press, 2001).

42. Many observers remarked about the wide variety of content at this time. For a statistical sample of use, see, for example, Peter C. Clemente, *The State of the Net: The New Frontier* (New York: McGraw-Hill, 1998).

43. AOL had a sizable market share in 1994–1995, but not this large. This dominating share emerged in 1996–1997 and especially after the CompuServe acquisition. For a recounting of the philosophical and marketing strategies that led to this outcome, see, for example, Swisher, *aol.com.*

44. Most services offer a feature known as *presence*, indicating whether people on one's list of contacts are currently available to chat. This may be called a *buddy list*. In some instant messaging programs, each letter appears as it was typed or deleted. In others, the other party views each line of text after another is started.

45. Microsoft was willing to pay cash, but AOL drove a hard bargain, perceiving that Microsoft's program of compelling promotion of Internet Explorer 3.0 among all ISPs and original equipment manufacturers could not succeed without signing a large vendor such as AOL. After considerable negotiation back and forth, AOL signed Netscape up as a browser, but not the exclusive default browser. Under the pressure that this contract would go unanswered, Microsoft relented to AOL's terms for making Internet Explorer the default browser. Once Microsoft's manager capitulated to

AOL's terms, many MSN employees left. Altogether, the deal helped Microsoft's browser business, but set back MSN's ambitions—MSN had been one of AOL's biggest competitors until that point. AOL lost the network of firms supported by Netscape, but gained dominance in the U.S. Internet access business. Microsoft did many other things to support Internet Explorer. See, for example, Michael Cusumano and David Yoffie, *Competing on Internet Time: Lessons from Netscape and Its Battle with Microsoft* (New York: Free Press, 1998); Timothy Bresnahan, "The Economics of the Microsoft Antitrust Case," available at ⟨http://www.stanford.edu/~tbres/research.htm⟩ (accessed July 2006); Swisher, *aol.com*.

46. In 1999, AOL bought Netscape—particularly its portal—well after the browser wars ended. The merger with Time Warner was proposed in January 2000.

47. See, for instance, James Eisner and Tracy Waldon, "The Demand for Bandwidth: Second Telephone Lines and On-line Services," *Information Economics and Policy* 13, no. 3 (2001): 301–309; Greg Rosston, "The Evolution of High-Speed Internet Access, 1995–2001" Stanford Institute for Economic Policy Research No. 5-19, esp. 10–12.

48. This section provides a summary of Shane Greenstein, "The Economic Geography of Internet Infrastructure in the United States," in vol. 2, *Handbook of Telecommunications Economics*, ed. Martin Cave, Sumit Majumdar, and Ingo Vogelsang (Amsterdam: Elsevier Publishing, 2005). See also Sean P. Gorman and Edward J. Malecki, "The Networks of the Internet: An Analysis of Provider Networks in the USA," *Telecommunications Policy* 24, no. 2 (2000): 113–134.

49. This is a simplification of a wide variety of circumstances. See, for example, Chris Forman, "The Corporate Digital Divide: Determinants of Internet Adoption," *Management Science* 51, no. 4 (2005): 641–654.

50. See Thomas Downes and Shane Greenstein, "Universal Access and Local Commercial Internet Markets," *Research Policy* 31 (2002): 1035–1052.

51. In a study of the Appalachians and some areas with histories of poor communications service ("Telecommunications and Rural Economies: Findings from the Appalachian Region," in *Communication Policy and Information Technology: Promises, Problems, Prospects*, ed. Lorrie Faith Cranor and Shane Greenstein [Cambridge, MA: MIT Press, 2002]), Sharon Strover, Michael Oden, and Nobuya Inagaki examine ISP presence in the states of Iowa, Texas, Louisiana, and West Virginia, and determine the availability and nature of Internet services from ISPs for each county. See also Sharon Strover, "Rural Internet Connectivity," *Telecommunications Policy* 25, no. 5 (2001): 331–347.

52. See, for example, Kyle H. Nicholas, "Stronger Than Barbed Wire: How Geo-Policy Barriers Coinstruct Rural Internet Access," in *Communications Policy and Information Technology: Promises, Problems, Prospects*, ed. Lorrie Faith Cranor and Shane Greenstein (Cambridge, MA: MIT Press, 2000), 299–316. This is a study of the multiple attempts to provide access to rural Texas communities. Nicholas shows how the construction of calling-area geographic boundaries shapes the entry patterns of ISPs. His study shows both the strengths and pitfalls of this policy approach.

53. See Downes and Greenstein, "Universal Access."

54. For a sense of these plans, see *Boardwatch* directories, various years.

55. For more on this, see Andrew Odlyzko, "Internet Growth: Myth and Reality, Use and Abuse," *Journal of Computer Resource Management* 102 (Spring 2001): 23–27; K. G. Coffman and A. M. Odlyzko, "Internet Growth: Is there a Moore's Law for Data Traffic?" in *Handbook of Massive Data Sets*, ed. J. Abello, Panos M. Pardalos, and Mauricio G. Resende (New York: Kluwer, 2006), 47–93.

56. Stranger and Greenstein, "Pricing at the On-Ramp."

57. This summarized an extended crisis at the company. See the full chronicle in Swisher, *aol.com.*

58. Some ISPs also instituted automatic session termination when an online user remained inactive, eliminating problems arising from users who forgot to log off. Customers, however, perceived this as poor service; consequently, many small ISPs hesitated to employ the scheme.

59. Stranger and Greenstein, "Pricing at the On-Ramp."

60. See the estimates in ibid.

61. For a history of standards in the Internet, see Simcoe, "Delay and *De Jure* Standardization."

62. For a discussion of the factors shaping this outcome, see Shane Greenstein and Mark Rysman, "Coordination Costs and Standard Setting: Lessons from 56K," in *Standards and Public Policy*, ed. Shane Greenstein and Victor Stango (Cambridge: Cambridge University Press, 2006).

63. There is a long history of regulating interconnection to the telephone network. For a discussion of many of the antecedents to and rationale for the approach taken in the Telecommunications Act of 1996, see Eli Noam, *Interconnecting the Network of Networks* (Cambridge, MA: MIT Press, 2001); Nuechterlein and Weiser, *Digital Crossroads*. For a discussion about CLECs, see, for example, Glenn Woroch, "Local Network Competition," in *Handbook of Telecommunications Economics*, ed. Martin Cave, Sumit Majumdar, and Ingo Vogelsang (Amsterdam: Elsevier Publishing, 2001), 642–715; Shane Greenstein and Michael Mazzeo, "Differentiated Entry into Competitive Telephony," *Journal of Industrial Economics* (2006): 323–350; New Paradigm Resources Group, *CLEC Report*, 2000, Chicago.

64. Even though the behavior was legal from a literal perspective, it was *not* the behavior Congress had intended to induce. FCC docket no. 99–38, Implementation of the Local Competition Provisions in the Telecommunications Act of 1996, Inter-Carrier Compensation for ISP-Bound Traffic, released February 26, 1999.

65. The act also barred multiple and discriminatory laws on e-commerce. It contained sunset provisions, but was renewed several times subsequent to its initial passage. It did sunset briefly in 2003, only to become renewed in early 2004.

66. Daniel Shiman and Jessica Rosenworcel, "Assessing the Effectiveness of Section 271 Five Years after the Telecommunications Act of 1996," in *Communications Policy and Information Technology: Promises, Problems, Prospects*, ed. Lorrie Faith Cranor and Shane Greenstein (Cambridge, MA: MIT Press, 2002).

67. See, for example, Woroch, "Local Network Competition."

68. NTIA, "A Nation Online: Entering the Broadband Age," shows a survey of households in October 2003, in which 20.6 percent had cable modem access, while 15.2 percent had DSL access, and 62.8 percent had dial-up access.

69. For a complete history, see Rosston, "The Evolution of High-Speed Internet Access."

70. For some statistics on this, see J. M. Bauer, "Broadband in the United States," in *Global Broadband Battles: Why the US and Europe Lag and Asia Leads*, ed. Martin Fransman (Stanford, CA: Stanford University Press, 2006), 133–163.

71. This is a long and complex story. For different views, see, for example, Crandall, *Ten Years after the 1996 Telecommunications Act*; Bauer, "Broadband in the United States"; Rosston, "The Evolution of High-Speed Internet Access."

72. This is the nonamplified radius; the signal could reach further with amplifiers. The eighteen thousand feet only applies to service of up to 1.5 Mbps; for higher speeds, the radius is more limited. For example, for speeds of 8 Mbps, the limit is nine thousand feet, and for very high bit-rate DSL, which could supply up to 55 Mbps, the limit is one thousand feet.

73. See Tony H. Grubesic and Alan T. Murray, "Constructing the Divide: Spatial Disparities in Broadband Access," *Papers in Regional Science* 81, no. 2 (2002): 197–221; David Gabel and Florence Kwan, "Accessibility of Broadband Communication Services by Various Segments of the American Population," in *Communications Policy in Transition: The Internet and Beyond*, ed. Benjamin Compaine and Shane Greenstein (Cambridge, MA: MIT Press, 2001).

74. NTIA, "A Nation Online: Entering the Broadband Age."

75. The cost of a second line varied, depending on the state. See Eisner and Waldon, "The Demand for Bandwidth."

76. See Bureau of Labor Statistics price series for Internet Services and Electronic Information Providers. This price series begins in December 1997 at 100. By December 2003, it had fallen to 97.6—in other words, measured transaction prices for Internet service declined by 2.4 percent in six years.

77. According to the 2004 Service Annual Survey, released by the U.S. Census Bureau on December 29, 2005, in NAICS 514191, Internet service providers had $10.5 billion in access fees in 2004; in NAICS 5133, telecommunications carriers had $4.3 billion in Internet access services; and in NAICS 5132, cable network and program distribution had $8.6 billion in Internet access service.

78. This includes computer hardware, computer software, and communications hardware and instruments. See David Henry and Donald Dalton, "Information Technology Industries in the New Economy," in *Digital Economy 2002*, U.S. Department of Commerce, available at ⟨http://www.esa.doc.gov/reports.cfm⟩; Lee Price with George McKittrick, "Setting the Stage: The New Economy Endures Despite Reduced IT Investment," *Digital Economy 2002*, U.S. Department of Commerce, available at ⟨http://www.esa.doc.gov/reports.cfm⟩. The growth rates are even higher if communications hardware and instruments are excluded.

79. These are constant (1996) dollars. See Henry and Dalton, "Information Technology Industries."

80. For 2000, the estimated personal consumption of information technology goods and services was $165 billion. For business it was $466 billion. See Henry and Dalton, "Information Technology Industries."

81. This discussion is based on Chris Forman, Avi Goldfarb, and Shane Greenstein, "The Geographic Dispersion of Commercial Internet Use," in *Rethinking Rights and Regulations: Institutional Responses to New Communication Technologies*, ed. Steve Wildman and Lorrie Cranor (Cambridge, MA: MIT Press, 2003), 113–145, "Which Industries Use the Internet?" in *Organizing the New Industrial Economy*, ed. Michael Baye (Amsterdam: Elsevier, 2003), 47–72, and "How Did Location Affect Adoption of the Internet by Commercial Establishments? Urban Density versus Global Village," *Journal of Urban Economics* (2005).

82. This is an oversimplification of Cortada's thesis for the sake of brevity. See James W. Cortada, *Information Technology as Business History: Issues in the History and Management of Computers* (Westport, CT: Greenwood Press, 1996).

83. For example, see the discussion in Rosston, "The Evolution of High-Speed Internet Access," Sam Paltridge, "Internet Traffic Exchange: Market Developments and Measurement of Growth," working paper on telecommunication and information service policies, OECD, 2006, Paris.

84. This complex topic strays into matters outside the scope of this chapter. For extensive description and analysis, see, for example, Steve Mosier, "Telecommunications and Computers: A Tale of Convergence," in *Chasing Moore's Law: Information Technology Policy in the United States*, ed. William Aspray (Raleigh, NC: SciTech Publishing, 2004), 29–54; Johannes M. Bauer, "Unbundling Policy in the United States: Players, Outcomes, and Effects," *Communications and Strategies* 57 (2005): 59–82; Nuechterlein and Weiser, *Digital Crossroads*. See also the court ruling a year later at the Washington, DC, Circuit Court, which restricted the FCC's prior decisions regarding CLEC entry and unbundled network elements cost-based rates; *United States Telecom Association v. FCC*, 359 F. 3rd 554 (DC Circuit, 2004).

85. For a list of the mergers among telecommunications carriers and the conditions placed on the final deals, see ⟨http://www.cybertelecom.org/broadband/Merger.htm⟩.

4 Protocols for Profit: Web and E-mail Technologies as Product and Infrastructure

Thomas Haigh

The Internet evolved with breathtaking speed during the 1990s, from an obscure, academic system to a mainstay of the developed world's daily routines of communication, shopping, travel, entertainment, and business. As millions of ordinary people rushed to connect their computers to the network, most were driven by a desire to use two particular kinds of programs: e-mail systems, and the World Wide Web. These were the so-called killer applications of the Internet during its commercialization: specific application programs so compelling that people were prepared to purchase an entire computer system in order to use them.[1] Both played two roles at once: as products in their own right for a small number of companies such as Netscape and Microsoft, and as crucial parts of the communal software infrastructure on which countless other companies tried to build their own online business empires.

In this chapter, I explore the evolution of Internet e-mail and the Web since the late 1980s, looking at the markets that developed for both kinds of programs, efforts to remake both technologies in more commercial ways, and the ways in which the culture and technology of the precommercial Internet of the 1980s shaped the development of its commercialized successor. Although space does not permit me to delve deeply into the experience of Internet users during this era, I do attempt to make connections between the fundamental architecture of these Internet technologies and the opportunities available to the businesses and individuals using them.

The Internet and Its Protocols

One of the things that makes it hard to write about the history of the Internet is uncertainty over exactly what "the Internet" is and was. Internet histories generally go back to the academic ARPANET of the 1970s, even though the concept of an "internet" dates from a later generation of networks built during the 1980s. The ARPANET first went into operation in 1969, and was funded by the U.S. Department of Defense to interconnect the computers it had purchased for researchers in different universities. No single piece of hardware, network infrastructure, or application software survived the transition from the ARPANET to the academic Internet of the early 1990s. During the 1990s the Internet changed still more rapidly, as its dominant users, applications, network operators, and traffic patterns changed beyond recognition. The situation recalls the apocryphal tale of an item said to be displayed in the Tower of London with a sign "Axe, eleventh century. Head replaced in early thirteenth century, shaft replaced in mid-fifteenth century." If every piece of something has been changed, then what are we really writing about when we tell its story?

In fact, the essence of the Internet lies not in hardware or software but in protocols: the agreed-on rules by which computer programs communicate with each other. When computers communicate on the Internet, it is the protocols that determine what is automatic and what is impossible; what is easy to accomplish and what requires complex or cumbersome special procedures.[2] But the protocols that define the Internet were designed by and for a group of users very different from that of the commercial Internet. The fundamental data Transmission-Control Protocol/Internet Protocol (TCP/IP) was developed during the late 1970s. The main protocol for Internet e-mail transmission, Simple Mail Transfer Protocol (SMTP), is from the early 1980s. And although the World Wide Web was not created until the early 1990s, its protocols were modeled on and designed to work with the existing standards of the Internet.

Events of the 1970s and 1980s have played a profound role in shaping today's commercialized Internet. As Lawrence Lessig has argued in his book *Code*, design decisions built into computer code can play a part in shaping the development of online life just as important as the role of legal codes in shaping the development of societies off-line.[3] Decisions built into the design of protocols perform a similar role. This insight parallels ideas long familiar to scholars researching the history of technology. Following the work of Thomas P. Hughes, these scholars have come to see technologies not as individual inventions or machines but as part of broader sociotechnical systems binding together elements such as standards, professional communities, institutions, patents, businesses, and users. The system outlives all its constituent parts. Successful technological systems such as the Internet develop what Hughes called "technological momentum," making them hard to displace or redesign.[4] Many historians and sociologists of technology have explored the ways in which the function and design of technologies were influenced by the choices and assumptions of their designers, which in turn reflect the values and cultures of the communities in which they worked. These scholars talk of studying the "social shaping" of technology, and of "opening the black box" of apparently inscrutable technical systems to discover the social choices hidden inside.[5] The technologies of the Internet have been clearly and powerfully formed by the circumstances of their creation and early use, in ways that explain both their rapid spread in the early 1990s and many of the problems faced by today's network users.

As Janet Abbate has shown in her book *Inventing the Internet*, the protocols of the Internet reflect its roots in the 1970s as a closed system, used by researchers, government research administrators, and military bureaucrats. The Internet was created during the early 1980s by the interconnection of different networks using TCP/IP. TCP/IP was designed to work with networks of all kinds, including satellite and radio systems as well as telephone lines and dedicated high-speed cables. It was so simple and flexible that it could work with any kind of transmission medium as long as suitable "link layer" software had been written. TCP/IP separated the transmission of data from the work of specific application programs. Likewise, computers of any kind could communicate with each other, provided that each was equipped with software faithfully implementing TCP/IP. This meant that someone writing a program to transmit a file or perform any other network task did not need to worry about what kind of computer was at the other end, or about the characteristics of the networks that data might be sent over.

During the 1980s, the most important constituent networks of the Internet included the ARPANET created for the Defense Advanced Research Projects Agency (ARPA) administrators and sponsored researchers, the National Science Foundation's NSFNET, and MILNET for military use. Connections to academic networks in Europe and Asia were gradually added during the 1980s and early 1990s. TCP/IP, SMTP, and the other Internet protocols were designed to support a wide range of applications, transmission mechanisms, and computers. They were not, however, designed for commercial use on public networks. As a result, they acquired a characteristic set of strengths and weaknesses that together go a long way to explain not only the rapid success of the commercial Internet but also its persistent problems. The special characteristics of the 1980s' Internet, and their influence on its protocols and architecture, may be summarized as follows.

• *The Internet was designed for a homogeneous population of academically oriented, scientifically trained users, granted access through their employers or universities.* As a result, it relied more on social mechanisms rather than technical ones to provide security and eliminate troublemakers who engaged in antisocial behavior (public insults, resource hogging, or commercial solicitations). Breaches of decorum were punished with warnings and eventually the termination of a user's access privileges by their home institution. This was impossible to do once commercial access brought a more diverse population online.

• *The Internet and ARPANET were designed as practical, working networks.* The early history of the Internet shows a constant interplay between theory and practice, with protocols and designs evolving rapidly on the basis of experience. From the beginning, the ARPANET was designed to do useful work, and its users found new and unanticipated applications such as e-mail. On the other hand, the Internet was never intended to be used commercially or set a global standard. This contrasted with long-running international efforts to set official standards for global public networks, which suffered from the need to satisfy every possible requirement and constituency.

• *The Internet was entirely noncommercial, and so provided no way to charge users according to the network resources they consumed or compensate the providers of network services.* The Internet provided no way of knowing exactly who was connecting to the services offered from your computer, still less of automatically receiving money from them as a result. This was a sharp contrast with existing commercial networks, including international telephone service and commercial packet-switched networks, for which the idea of precise billing for resources used was fundamental. Of course, universities paid a certain amount each month to network providers and received a connection of a certain bandwidth in return. But because TCP/IP was designed as a pragmatic and flexible technology for a noncommercial environment, it gave networks within the Internet no way of charging users to relay packets of data, even to recoup the costs of expensive resources such as intercontinental satellite links.

• *The Internet was designed to serve a research community, rather than to perform one specific task.* From the earliest days of the ARPANET, network designers wanted to make it easy to experiment with new network applications and make the network as flexible as possible. They initially did this by putting the network logic for each network site into a separate minicomputer known as the Interface Message Processor, but with the transition to TCP/IP the same goal was accomplished in software by separating the application layer protocols for things like e-mail from the transport and network layer code of TCP/IP itself.

• *The Internet was designed to support many different machine types.* Proprietary networks such as those offered by IBM and DEC often sought compatibility by requiring all computers involved to use the same hardware and/or software. In contrast, the Internet was built around publicly available protocols and could network many different kinds of computer.

• *Any computer connected to the Internet could send and receive data of any kind.* Many commercial networks were built around the assumption that a large number of cheap terminals would be connected to a handful of powerful computers, on which all information would be centralized. The Internet was designed from the beginning as a flexible mechanism to connect computers together, and any computer connected to it could offer services to any other computer. TCP/IP worked on a peer-to-peer basis. In practical terms, that meant that a computer could turn into a file server, an e-mail server, or (later) a Web server simply by running a new program. Users were expected to publish online information and provide services to each other, rather than to rely on a single central collection of resources.

• *The Internet integrated many different communications media.* The initial impetus behind TCP/IP came from a need to integrate data transmission over satellite and radio networks. The separation of media-specific aspects of communication from TCP/IP itself has made it possible to extend the Internet over many new kinds of links, such as dial-up telephone connections, cellular telephones, Wi-Fi networks, and Ethernet local area networks, which were unknown when it was originally designed. This flexibility was vital to the Internet's rapid commercialization, because once an Internet connection was established it could use exactly the same software (such as a Web browser) over a dial-up connection from home as an Ethernet connection in a university. On the other hand, because TCP/IP was designed for flexibility rather than performance, the Internet does not offer a guaranteed data delivery time for critical applications or the ability to avoid network congestion by paying a premium. This proved a huge challenge in the development of systems to provide audio and video over the Internet.

People often imagine that the success of the Internet was completely unexpected, but this is true only in a narrow sense. In fact, the breakthrough of computer networking, electronic publishing, and e-mail into the mainstream of personal communication was confidently predicted since the late 1970s. The two main surprises were that this took so long to happen, and that in the end it was the Internet rather than a specially designed commercial network that accomplished the feat. Millions of home computers flooded into American homes in the early 1980s, together with a wave of interest in the idea that the newly powerful and affordable electronic technologies based around the silicon chip would trigger a fundamental social upheaval. This was known variously as the "information society," the "microelectronic revolution," and the "home computer revolution."[6] Futurist Alvin Toffler wrote in *The Third Wave* (1980) of a new society in which electronic networks had broken apart large organizations and allowed most people to work from home.[7] E-mail's triumphant emergence as a replacement for most personal letters and many telephone calls took place more than a decade later than once predicted. In *The Network Nation* (1978), academics Star Roxanne Hiltz and Murray Turoff salted their scholarly examination of the online communication systems of the day with predications that by 1990, online polls would replace Congress and electronic mail would bring the postal service to its knees.[8]

For a few years in the early 1980s, telecommunications firms throughout the developed world were convinced that millions of ordinary people were about to start using their home computers or a cheap terminal connected to their television screens to shop, read news, book travel plans, search databases, and trade stocks online. The same idea returned a decade later; as enthusiasm for something to be called "the information superhighway" gripped America's politicians and businesspeople, cable television and computer companies were investing hundreds of millions of dollars to create high-bandwidth home networks accessed via computerized cable boxes plugged into television sets.[9]

Clearly, nothing like the Internet would ever have been designed or launched by a commercial telecommunications company. With no easy way to charge users for the resources consumed, and no obvious way to make money by publishing information online, the network would seem fundamentally incompatible with any plausible business plan. The Internet was decentralized and unplanned, relying on its users to fill it with content and services. It offered largely free access to a huge variety of amateur and academic content along with a growing mass of information that companies were giving away because there was no obvious way to charge for it. Yet these very characteristics appealed to early commercial Internet users. Bizarrely, it was the Internet's fundamental disregard for the essential features of a commercial network that made it such a commercial success during this period.

E-mail and Person-to-Person Communication

While Internet e-mail was practical, simple, well-proven, and universally compatible, its academic origins meant that it suffered from a number of limitations for non-academic use. I suggest that the same characteristics that in the short term explain the initial success of Internet e-mail for mainstream use, have also in the longer term made it hard to combat spam or use Internet e-mail as a reliable mechanism for important messages. It is perhaps in the case of e-mail that the limitations of Internet technologies for commercial use are the most apparent, and where the technological advantages of competing systems are best developed. With hindsight, historians and economists may come to view Internet e-mail as one of those cases, like the much-discussed Qwerty keyboard and the VHS videocassette recorder standard, in which an allegedly inferior technology gained a permanently entrenched position because of a temporary initial advantage.[10]

People have been using computers to send messages to each other even before computers were networked to each other. By the early 1960s, computers were employed in corporate communication hubs to process sales data or information requests sent via teletype, and to automatically process and retransmit telegram-style messages sent between the offices and factories of large corporations.[11] A few years later, the development of time-sharing operating systems made it possible for several users to simultaneously work interactively on the same computer. It was natural and easy to allow users to chat with each other, and to send messages to be stored in the user accounts of other system users and read when convenient.[12] Today we tend to think of e-mail, instant messaging, and discussion forums as distinct methods of communication, but there are no obvious or inherent boundaries between them. Systems have often blended aspects of these approaches—for example, mixing public discussion areas with private messaging, blending word processing with e-mail, or offering real-time chat features as well as file exchange.

When and where networks were built to link computers together, this created an obvious opportunity for the exchange of messages between the users of the different computers on the network. Exactly when this was first done is not clear. As Abbate has shown, e-mail unexpectedly emerged as the main application of the ARPANET in the early 1970s. E-mail soon accounted for most of the network traffic, becoming the communications medium of choice within ARPA itself. Although e-mail had not been planned for when the network was being designed, its flexible nature made it easy for users to evolve their own applications. Experimental e-mail systems were used to transfer messages between sites from 1971 onward, and in 1973 the existing standard for ARPANET file transfers was modified to support e-mail transfer, greatly simplifying the task of transferring messages between computers of different types. Users quickly created programs to view, sort, and delete mail as well as to maintain e-mail discussion lists. E-mail was effectively free to its users, because the ARPANET infrastructure was bankrolled by the Department of Defense to interconnect the academic computer scientists it was supporting. Neither universities nor individuals were billed when a message was sent.[13]

E-mail remained a key network application as the ARPANET was replaced by the Internet, and from the early 1980s onward, specially created protocols for Internet e-mail ensured that people using different e-mail programs and different types of computers would be able to exchange messages freely. Mail was relayed to the recipient's machine using SMTP, designed in the early 1980s.[14] As an application protocol this ran over the Internet's native TCP/IP, meaning that all the complexity of routing messages and verifying their successful transmission was taken care of by TCP/IP. SMTP had to do little more than use the Internet's existing infrastructure to contact the recipient's mail server, and then transmit the name of the sender, name of the recipient, and the text of the message itself. E-mail servers were connected to the Internet, and were registered with the Internet's domain name servers as the place to which all e-mail addressed to recipients in a particular domain (e.g., ibm.com) or subdomain (e.g., Wharton.upenn.edu) should be delivered. The first widely used SMTP message delivery software was Sendmail, and today SMTP is supported by virtually all e-mail clients to send outgoing messages. While SMTP could originally transmit only plain (seven-bit encoded) text, it was later supplemented with the Multipurpose Internet Mail Extensions standard for encoding file attachments and non-Western alphabets.[15]

During the 1980s and the early 1990s, Internet users generally accessed e-mail and other Internet resources by logging in with a video terminal or terminal emulator program. They would access their e-mail by running a text-based mail reader, such as the basic "mail" command built into Unix or the more sophisticated Elm, a public domain product. These worked by directly reading the contents of the mail file into which the user's e-mail had been deposited. These programs ran on the Internet service provider's (ISP) or university's computer, meaning that e-mail could only be sent or received when logged into the computer.

The Internet as an E-mail Gateway

But although the Internet provided the key technologies behind today's mass-market e-mail systems, during the 1980s it was not open to the public or amateur computer enthusiasts. Collectively various other corporate, commercial, and academic e-mail networks boasted an estimated six million American users by the end of the 1980s— far more than the number of people directly connected to the Internet.[16] But except in a few specialized communities such as computer science research, e-mail had conspicuously failed to establish itself as a medium for communication between organizations, unlike the fax machine, which had spread far more rapidly during the decade to become a fixture of businesses large and small. Many problems conspired to hold back e-mail services: they were rarely easy to use, they sometimes suffered from reliability problems, modems and computers often required elaborate configuration process to work with them, message delivery was sometimes slow, and their complex billing systems gave users an incentive to minimize use. The biggest problem, however, was the lack of interconnection between networks. As one reporter noted, although e-mail was "growing fast and becoming a standard business tool" within many organizations, it had "developed as a lot of small, closed systems...rather than one big network that everyone can use."[17] When the Internet first reached a reasonably broad academic audience, at the end of the 1980s, it was not through a direct connection but via the interconnection of the Internet e-mail network with other amateur and commercial networks. Internet e-mail became a lingua franca between other e-mail systems.

The amateur enthusiasts of the 1980s found refuge on local bulletin boards, which formed an archipelago of thousands of little information islands across America, dotted most closely together in major urban areas and centers of high technology. At its simplest, a bulletin board was a personal computer left running a special program, and connected to a single telephone line and modem. Users could wait their turn to dial in and leave messages for each other, mimicking the asynchronous communication methods of a physical bulletin board. More ambitious boards offered multiple telephone lines for simultaneous access by several people, together with other services, such as online chat. By the mid-1980s, people had started connecting them together, most commonly through a piece of bulletin board software known as FidoNet. This worked as a poor person's version of the Internet e-mail and Usenet discussion systems, batching together packets of mail and newsletter updates intended for the users of other FidoNet nodes, and passing them on in cheap, late-night phone calls. At its peak in the mid-1990s, FidoNet had more than thirty thousand nodes worldwide (many with hundreds of users each). The FidoNet software was free, and many of the system's constituent nodes were run by volunteers (known as "sysops"), who made no charge to their users. This made FidoNet a popular system for hobbyists in the United States, smaller businesses in Europe, and institutional users in poorer regions with expensive and unreliable communication links such as Africa and the former Soviet Union.[18]

Another ad hoc system, BITNET, was used by academic users to pass messages between IBM mainframes. It was the original mechanism used to create "Listservs," or automated e-mail discussion lists.[19] A third system, UUCPNET, was an informal and decentralized network of university and research lab servers running free software. UUCPNET maintained its own e-mail system, with messages passed from machine to machine via cheap, late-night phone calls. It also distributed Usenet, a thriving system of thousands of hierarchically arranged discussion forms running the gamut from comp.os.mac to alt.sex.swingers. Usenet discussions were synchronized between servers, with large sites such as Bell Labs, DEC, and Apple shouldering most of the long-distance communications costs.[20]

E-mail services were also available to individuals and small businesses through commercial online services. By the start of the 1990s, four main firms were selling online services to consumers. Each offered e-mail as one of a bundle of services, available for a monthly fee and hourly usage charges. The largest and most successful of these services, CompuServe, had been founded in 1969 as a time-sharing business, launched a new service for microcomputer owners during the 1980s, and finished up with a solid business selling affordable e-mail and online database systems to smaller businesses. Genie and Prodigy were newer and smaller services, created in the 1980s for home users. The smallest of the four, America Online (AOL) had just 110,000 subscribers, far behind the more than 500,000 each boasted by CompuServe and Progidy.[21] More specialized online services, offering only e-mail, were run by major American telecommunications firms such as AT&T and MCI. These mail networks imposed a monthly subscription fee, plus charges for messages sent, received, and stored. One glowing 1987 profile of MCI Mail noted that the service was "cheap" because its 100,000 subscribers would pay only a dollar to send a 7,500-character e-mail within the United States. For an extra fee, MCI would print out the e-mail, and then deliver it by post or courier.[22]

Large corporations built their own e-mail networks based around servers and networks supplied by firms such as IBM, Wang, and DEC. E-mail was a key component of the office automation systems promoted during the early 1980s as a gateway to the paperless office of the future.[23] IBM's mainframe-based PROFS system, for example, was widely used by large organizations such as the U.S. government. DEC's ALL-IN-1 ran on its popular VAX minicomputers. By the early 1990s, the vendor-specific corporate e-mail systems sold by mainframe and minicomputer manufacturers were joined by a new breed of e-mail systems able to run on standard microprocessor-based servers using OS/2, Unix, or Novell Netware. Among the most successful of these was Lotus Notes, first sold in 1989, which used a novel and flexible system of replicated databases to provide e-mail services, shared calendar functions, and discussion groups, allowing corporate customers to build their own custom applications around these communication and textual database capabilities. Notes required users to install and configure special client software. A rival system, Microsoft Exchange, emerged as a strong competitor during the mid-1990s in conjunction with Microsoft's Outlook client software.

There was certainly no shortage of e-mail networks and technologies. But this was actually a problem, because during the 1980s, users of commercial online services and corporate e-mail networks were generally able to send mail only to users of their own network. Someone wanting to communicate with users of Genie and CompuServe would need to purchase subscriptions to both systems, learn two completely different user interfaces, and make separate phone calls every day to each system to retrieve new mail. This dramatically limited the spread of e-mail for ordinary home or small business users. Likewise, users of corporate e-mail systems could easily reach their boss or colleagues by e-mail, but not their customers and suppliers.

Online services and corporations thus had an incentive to allow for the exchange of e-mails with other systems, even as they kept their other network content proprietary. The need for this kind of e-mail exchange between networks was recognized early, and addressed through two standards developed by CCITT, the international federation of telecommunications carriers, and adopted by the computer communication standards group Open Systems Interconnection (OSI). These were the X.400 standard for message delivery and the X.500 standard for directory access. In contrast, the Internet's SMTP was little known in the business world. During the 1980s and the early 1990s, the OSI standards effort was supported by mandates from many national governments (including the United States) and was almost universally viewed as the future of computer networking. Plans were even made for the Internet itself to abandon TCP/IP as its foundation, shifting instead to a related OSI standard protocol, TP/4.[24] Given this backing and the clear need for e-mail interchange, X.400 was taken very seriously. As well as the obvious base of support among national telecommunications carriers, hardware and software vendors including DEC, Novell, Hewlett-Packard, and Microsoft added X.400 capabilities to their e-mail systems.

Yet by the end of the 1980s it was Internet e-mail, rather than X.400, that was beginning to emerge as the leading method for e-mail transmission between other networks. The number of computers connected directly to the Internet was still small, in part because of the restrictions placed on commercial use. But other networks were already beginning to build gateways to forward messages initiated by their users on to the Internet and deliver messages received from the Internet. This did not require the networks to adopt either the SMTP Internet e-mail standard or TCP/IP for their internal use, so it was quite easy to accomplish. Messages were simply reformatted and retransmitted, perhaps hours later. E-mail exchange took a big step forward in 1989 when MCI Mail and CompuServe began to exchange e-mail with the Internet and each other.[25] FidoNet and UUCPNET were also connected by gateways to the Internet. AOL and Genie followed suit in 1992, though these still had some limitations, such as a cap on the message size at eight kilobytes for the personal computer version of AOL.[26] Until well into the 1990s, gateways of this kind accounted for most of the e-mail sent between users of different services.

The core technologies of the Internet were created as practical, short-term solutions to real needs, often those of the teams or organizations creating them. People did not try to design standards that would endure for decades or meet the needs of every possible user. While this led to shortsighted decisions in some areas, in many cases the simplicity and flexibility of the Internet has made it possible to evolve its technologies along with their changing uses. A comparison of SMTP, the simple and effective Internet e-mail system, with the agreed-on world standard X.400 makes the difference clear. In the late 1980s, Internet e-mail was capable of little more than the delivery from one server to another of textual messages (in which, using separate tools, graphics and attachments could be encoded). X.400 offered a long list of features, including security, notification when a message had been received or read, different levels of priority, protection against the faking of addresses, and the automatic conversion of messages and attachments between different formats. It was designed to support centralized directories, via the X.500 protocol, so that there was a standard way to find the e-mail address of the person you wanted to contact. Internet e-mail systems lacked all these features, and although various additional standards have been proposed to add them, they are still missing from today's typical e-mail experience. In 2000, Microsoft Exchange switched its default protocol from X.400 to SMTP, recognizing the triumph of Internet e-mail technologies. Despite the superior functionality promised by the X.400 standard, perhaps even because of it, X.400-based systems were large, expensive products that were harder to use and never included all the features specified by the committee.[27]

One of the clearest examples of this comes in a comparison of the e-mail address formats used by the two standards. An X.400 e-mail address took the form of a series of attributes, with the shortest-possible version looking something like "G = Harald; S = Alvestrand; O = sintef; OU = delab; PRMD = uninett; ADMD = uninett; C = no."[28] That was hardly an attractive thing to put on a business card or print in an advertisement. X.400 e-mail addresses could get even more baroque. The committee stuffed every possible attribute into the address to satisfy the demands of corporate users as well as international postal and telecommunications bodies, meaning that e-mail addresses could include up to four different "organizational units" (companies, divisions, laboratories, etc.) and even the user's street address. Every eventuality was catered for. In contrast, an Internet e-mail address was something in the format ⟨johndoe@xytech.edu⟩. As we shall see later, the simplicity and lack of security in Internet e-mail left it highly vulnerable to spam and other abuses once its technologies were transplanted into a more hostile environment.

Internet E-mail for the Masses

As Internet access spread during the early 1990s, direct access to Internet e-mail became practical for millions. Most ISPs included at least one Internet mailbox with each personal subscription, and multiple e-mail addresses with small business subscriptions. Extra mail storage space and further addresses were available for additional charges. Since the e-mail address would include the ISP's own domain name (for example, ⟨Jane.Smith@earthlink.com⟩), ISPs had no incentive to break out e-mail service from their basic subscriptions and charge for it separately. Each message sent advertised the customer's chosen ISP, while users would continue to receive e-mail at the address only while they maintained their subscriptions. Changing an e-mail address was a chore, and so many people continued to stick with their original ISP even after discovering the availability of cheaper, faster, or more reliable service elsewhere. (Organizations and individuals with their own Internet domains usually either set up their own mail server or contracted this service out together with their Web hosting needs.)

Unlike the commercial online services and specialist e-mail businesses of the 1980s and the early 1990s, ISPs did not charge users additional fees for each message sent, or even (in most cases) for the time spent online. And an Internet e-mail could easily be sent to anyone with an Internet e-mail address or gateway, which by the mid-1990s included pretty much everyone with an e-mail account on any service. As a result, e-mail use increased dramatically. E-mail was truly a killer application for the Internet, and indeed for the personal computer itself. Businesses bought computers, modems, and service plans to communicate with their customers; parents with their college student offspring; emigrants with their friends and families across the globe; and enthusiasts in every field with others sharing their interests. Between November 1992 and November 1994, the volume of Internet e-mail almost quadrupled, to exceed more than a billion messages a month.[29] By the end of the decade, an estimated three billion e-mail messages were being sent every single day, outstripping the volume of physical mail sent in the United States.[30]

Internet E-mail Software: The Business That Wasn't
Early ISPs followed the same patterns established by university computer centers to provide e-mail addresses to their campus users. They would dial in, and then use a text terminal window to access an e-mail client program such as Pine or Elm running on the ISP's central computer. But as personal computers were increasingly connected directly to the Internet itself, demand grew for a new kind of e-mail client able to download messages from the central server to read, archive, organize, and compose replies directly on a personal computer.

This was achieved by coupling SMTP with a new protocol, POP (Post Office Protocol). POP was particularly useful for dial-up service users, allowing them to connect briefly and retrieve all accumulated messages without having to stay online to read them. ISPs were quick to add POP capabilities to their mail servers. Because the mail client was running on the local machine, it used a convenient graphical user interface rather than the primitive text interface of earlier mail readers. Shifting work to the cus-tomers' computers also lightened the load on the ISPs' servers.

This created a new market for desktop e-mail client software. Eudora, the most widely used early POP client, was first released as a freeware package for the Apple Mac-intosh in 1990 by its author Steve Dorner of the University of Illinois. Its name was a humorous tribute to the writer Eudora Welty, one of whose best-known stories was "Why I Live at the P.O."[31] Eudora was produced in Macintosh and Windows versions, with a freely downloadable basic version joined by a commercially distributed "Pro" one. Millions of copies of the free version were given away by universities and ISPs to their customers. Qualcomm, better known for its modems than its software expertise, sold Eudora commercially, though the program evolved slowly and gradually lost its once-dominant position. In 2006 Qualcomm gave up, dropping Eudora as a commer-cial product.[32]

Most users relied either on free software (including the free version of Eudora) or used one of the e-mail clients bundled with other packages. Netscape included a clunky but functional Internet e-mail client with its Navigator browser from version 2.0 (1995) onward. Microsoft offered an e-mail client as a standard part of Windows 95, and bundled e-mail and news clients with its own Internet Explorer browser from version 3.0 (1997) onward. From 1997 this was known as Outlook Express, and today it is the world's most widely used e-mail client, though it has not received a significant update since 1999 or any new features at all since 2001.[33] In 2005 Microsoft assembled a new development team to work on the program, and has announced plans to include an improved version, renamed again to Windows Mail, in its forthcoming Windows Vista operating system.[34]

Microsoft also offered a powerful "Internet mail only" mode in its Outlook package (which despite the name was entirely unrelated to Outlook Express). Outlook was mar-keted as a personal information manager, a software category created during the 1980s to combine address book, calendar, and note capabilities. While many of Outlook's most powerful features worked only with Microsoft's own Exchange mail server, it was a popular package among Internet e-mail users, if only because of its bundling with Microsoft's ubiquitous Office suite. Thus by the 1990s anyone using a recent ver-sion of Windows, downloading either of the two leading browsers or purchasing Microsoft Office, would have acquired a perfectly usable e-mail client. Rather than sup-plying Eudora or another package to their users, ISPs could simply include some instructions on how to configure Outlook Express or Netscape to access their user's e-mail. It is little wonder that only a small minority of users chose to spend time and money on commercial competitors such as Eudora Pro.

Internet e-mail was the first hit application of the personal computer to break the pattern established with the early firms of the 1970s and 1980s. Each new class of pop-ular application spawned at least one major company: Microsoft and Borland from programming languages, WordPerfect and MicroPro from word processors, Lotus and VisiCorp from spreadsheets, and Aldus from desktop publishing. More people spent more time using e-mail than any of these earlier kinds of programs. Yet nobody got rich selling Internet e-mail programs. In part this was a result of Microsoft killing the market for e-mail client programs by giving away its own e-mail software. It also reflects the preexisting norms of the Internet, however, where software was usually written by academics or enthusiasts with no need to support themselves from royalties, and people were used to downloading programs rather than going to buy them in a store.

The sale of Internet mail server software also failed to become a big business, or really any kind of business. Internet e-mail transmission has usually accomplished either using open-source products (such as sendmail and the increasingly popular qmail), or tools bundled as a standard part of all recent releases of proprietary server operating systems such as Windows Server and Novell Netware. In contrast, the market for the more capable proprietary e-mail systems remains healthy. Even today, large organizations continue to use Microsoft Exchange and IBM's Notes to handle internal e-mail, relying on gateway capabilities built into these systems to translate and retransmit mail intended for external recipients. Some estimates showed that Exchange overtook Notes in 2001 to become the most widely used corporate e-mail system. In 2006, it had an estimated 118 million users worldwide.[35] According to computer industry analysts at Gartner, in 2005 Microsoft and IBM collectively controlled more than 86 percent of the market for "enterprise e-mail and calendaring" systems measured by revenue volume.[36] While many times smaller than the market for Internet e-mail in terms of the number of users, this market remains much bigger in dollar terms, as most Internet e-mail software and e-mail accounts are given away free.

Web Mail

With the rise of the Web, however, an entrepreneurial opportunity did present itself in the shape of Webmail systems. The first of these, hotmail.com, was launched in 1996 by Sabeer Bhatia, a young Indian immigrant working as an engineer in Silicon Valley, and his coworker Jack Smith. Hotmail offered users free hotmail.com e-mail accounts, allowing them to read and compose their e-mail using standard Web browsers. Free Web mail services had two main advantages from the point of view of users: they were independent of their ISPs and so would remain valid when graduating from school, changing jobs, or moving; and they could be accessed from any Web browser. This meant that new and stored e-mails were accessible when traveling, and from office or public locations where other access methods might be blocked by firewalls or restrictions. Webmail also removed the need to install and configure an e-mail client program.

Webmail was quite simple to accomplish technically: essentially, the e-mail application ran on the server like a traditional e-mail reader such as Elm, but directed its output to a Web page rather than a terminal window. Yet Hotmail was the first such service and received sufficient venture capital to cement its advantage by giving away eight million e-mail accounts without having to worry about the costs. At the end of 1997 Microsoft acquired Hotmail for $400 million, making Hotmail one of the emblematic success stories of the early years of the Internet boom: a fable of overnight wealth based on the confident and timely execution of a good idea.[37] Unlike most of the other Internet businesses created to give away things that usually cost money, Webmail made considerable sense. Providing the service required only modest technical and network resources, but because users would visit the site frequently to check and reply to e-mail, they would spend a great deal of time exposed to advertising messages. Advertising, initially for Hotmail itself, could also be included at the bottom of each outgoing e-mail message. Finally, as users grew more reliant on the service, some could be induced to pay monthly fees for premium services such as greater mail storage or the ability to download e-mail to a desktop e-mail client such as Eudora.[38]

The Hotmail idea was not hard to duplicate, and other firms such as the start-up RocketMail and the Yahoo, Excite, and Lycos portals entered the free Webmail business. In 1997 Yahoo acquired RocketMail, whose technology it had previously been licensing, and built up e-mail as a key feature of its popular Web portal. Yahoo!'s e-mail service expanded to include calendar capabilities, discussion groups, and instant messaging. Yahoo Mail eventually overtook not just Hotmail (to become the most widely used free e-mail service) but also AOL (to become the most widely used e-mail service of any kind).[39] In 2004, Google shook up the Webmail business with its announcement of Gmail.com, a new service with a powerful and uncluttered user interface, an unheard of gigabyte of online message storage (250 times the amount offered with free accounts from market-leader Yahoo), and the ability to download mail to standard clients and forward incoming mail to other accounts. Google's free e-mail service was far superior to the premium services offered by its competitors in exchange for paid subscriptions. Initially Gmail was available by invitation only, with each user given a number of invitations to pass on to their friends and colleagues. By 2005 this policy had been relaxed. Though other services quickly moved to match Gmail's generous storage limits, Gmail has recently added calendar, voice communication, and text chat capabilities.[40] Still, people are reluctant to switch e-mail services without a good reason, and by mid-2006 Gmail's 8.6 million active users only earned it fourth place in the U.S. market behind Yahoo (with 78 million users), AOL, and Hotmail.[41]

Such was the popularity of Web mail that leading mail servers such as Microsoft Exchange and Lotus Notes (or Domino as the server has been renamed) now offer it as a standard feature, meaning that ISPs and organizations generally provide Web browser access e-mail. It still has the disadvantage that users cannot read or compose e-mail while off-line, but in today's world of high-speed broadband connections and Wi-Fi hot spots, this is less of a drawback than it once was. Similarly, new kinds of discussion and collaboration systems based on Web technologies—such as blogs, wikis, and on-line community sites—have largely usurped the once-separate technologies used to disseminate and access Usenet newsgroups. The Web browser's role as a universal inter-face to online resources grows ever broader.

Spam

The other big economic opportunity opened up by e-mail has been an unexpected one: the sending and blocking of spam (unsolicited commercial bulk e-mail). SMTP was designed for use on a small, closed network where commercial activities were expressly forbidden, and whose user population shared an academic culture where trust and the free exchange of information were assumed. This was in sharp contrast with the ill-fated X.400 protocol, designed from the beginning for commercial use by a diverse and potentially untrustworthy user population. Internet e-mail provided no easy way to charge users for each e-mail they sent (which would have destroyed the economics of spam), no easy way to determine the source of a message (and hence to block or prosecute spammers), and no protection against address faking. Because Inter-net e-mail systems treated composing, sending, relaying, receiving, delivering, and reading e-mail as six separate tasks, accomplished with different programs, most SMTP relays of the early 1990s were happy to relay messages across the network without demanding authentication from the sender. Once Internet e-mail became the standard for hundreds of millions of users, its own strengths became almost fatal weaknesses: the rapid, anonymous, and free transmission of an unlimited quantity of spam.

Internet lore identifies a message transmitted in 1978 as the first bulk unsolicited commercial e-mail sent on the ARPANET. The message, sent by a DEC sales representa-tive, invited every known network user on the West Coast of the United States to a sales presentation. This faux pas solicited rapid and widespread criticism from network users, and it was not repeated.[42] By 1994, however, the Internet was used by millions rather than thousands. On March 5, the small and rather seedy husband-and-wife law firm Canter and Siegel set aside all the rules of "netiquette" (as good online manners were then called) by posting advertisements for its immigration services in every one of the thousands of Usenet newsgroups.[43] Canter and Siegel then founded a firm called Cybersell to offer spamming services to others. Aggrieved Internet users fought back with large e-mails designed to overwhelm the firm's e-mail boxes and unsolicited mag-azine subscriptions to fill its physical mailboxes, but it was already clear that the tradi-tional social norms of the Internet would be insufficient to battle this new surge of unashamedly antisocial commercial activity.

The term spam itself, which gained currency to describe this kind of bulk solicitation, was a classic example of the type of geek whimsy central to the Net's precommercial culture. It referred to a comedy sketch from the surreal 1970s' British television show *Monty Python's Flying Circus*, in which a couple entered a cheap café only to discover that every choice on the menu held greater or lesser amounts of Spam, a processed meat product emblematic of the poor state of British cuisine during the immediate postwar decades. No substitutions were permitted, so while Spam lovers could choose delicious items such as "spam spam spam spam spam spam spam baked beans spam spam spam and spam," even the most Spam-averse customer was forced to order "egg bacon spam and sausage" as the dish with the least Spam in it.[44] In homage to this, the term spam appears to have established itself during the 1980s as an obscure geek idiom meaning "something unpleasant forced on people indiscriminately and in large quantities," before springing into widespread usage to describe bulk e-mail and newsgroup messages.

Spamming e-mail accounts was a little harder than spamming newsgroups, largely because no central directories of e-mail addresses existed. Spammers solved the problem by using software to roam the Web and newsgroups in search of e-mail addresses, or simply by sending messages to randomly created addresses in popular domains such as AOL and CompuServe. Spammers used free e-mail accounts, specialized bulk e-mailing software tools, and "zombie" computers (stealthily taken over by worm attacks) to send out billions of e-mail messages. Because spam annoyed its recipients, the return addresses supplied were usually bogus. The advertisements promoted fraudulent or disreputable offers: invitations to embezzle funds from Nigerian bank accounts, cut-price impotence drugs, bad investment advice, and pirated software. One particularly dangerous kind of spam, so-called phishing messages, requested that users visit fake but convincing replicas of leading Web businesses such as eBay and Citibank to enter their passwords. The fake Web site logged the account information, allowing the criminals behind the scam full control of the user's online account.[45] By the start of 2005, spam accounted for 83 percent of all e-mails sent.[46] According to one widely quoted estimate, the effort required to deal with all these messages wasted an annual average of $1,934 worth of productivity for each and every corporate employee in the United States.[47]

The rising tide of spam floated the fortunes of a whole new industry providing antispam products and services. The most effective of these filtered incoming e-mail to detect spam based on a number of characteristics, including its format, the presence of certain words such as "Viagra," and its statistical similarity to known spam messages. Spam could be blocked by software running on the personal computers of e-mail recipients (in 2003, a simple spam filter was built into Microsoft Outlook), or detected on mail servers and deleted before ever being downloaded by users. Many antispam systems have been produced, both commercial and open source. Some spam blockers have even been packaged into self-contained hardware units to minimize the effort required to add them to a network. As of 2006 the antispam market remains fragmented, though filters appear to be working increasingly effectively. My own Gmail spam filter catches around fifty messages a day, letting only a couple through. In a generally unsuccessful attempt to circumvent such filters, spams have been growing ever more surreal, often consisting of meaningless sequences of blank verse, interspersed with occasional Web addresses, mangled names of products such as "SOFT CIAzLIS" or "V/AGRA," and the opportunity to "reffnance" or to buy "s0phtw_aRe."

Instant Messaging

The Internet is widely used for another kind of personal communication: instant messaging. E-mails can be sent at any time and, like traditional mail, accumulate in the recipient's mailbox. Instant messaging, in contrast, is more like a phone call. The people communicating are online at the same time, and send a conversational stream of short messages back and forth. Instant messaging software allows users to see which of their contacts are online and available for chat at any given time.

The basic concept was established long ago. Time-sharing systems had allowed users to "chat" with each other since the 1960s. Internet users could "finger" their friends to see whether they were currently logged in, and use the "talk" command to print short messages on their screens. A popular free protocol, Internet Relay Chat, was widely used to create Internet chat rooms from the late 1980s onward.[48] In the world of commercial online services, a large part of AOL's success has been attributed to its hugely popular public chat rooms. The term instant messaging was popularized by AOL as part of its proprietary online service. Users could establish a "buddy list" and would then see which of their acquaintances were online. Using AOL's Instant Messenger (AIM) feature, they could send messages to pop up on their friends' screens. There was, however, no established standard for person-to-person private instant messaging over the Internet prior to its commercialization. Internet e-mail packages such as Eudora and Outlook Express had merely to implement proven protocols for users to interface with their e-mail accounts. Internet-based instant messaging systems, in contrast, had to establish their own protocols. They were competing not merely to offer better software but to build their own proprietary networks.

Instant messaging over the Internet was pioneered by Mirabilis, an Israeli start-up firm founded in 1996. Word of its new ICQ service (pronounced "I seek you") spread rapidly. According to one January 1997 report, the firm gained sixty-five thousand users within seven weeks of the creation of its first software, before it had advertised or even put up a home page.[49] By the end of 1997, a number of firms were competing in the fast-growing market, including Excite, Yahoo, and a variety of specialist start-ups. But ICQ was by far the most popular, and ten months after its release it claimed more than three million users.[50] Fighting back, AOL released instant messenger software for users of Netscape's Navigator browser, allowing them to send messages to each other and customers of its online service. The companies competing in this market aimed to make money by selling advertisements to be displayed to their users, or by licensing the software to corporate users or ISPs. In June 1998, AOL announced that it was purchasing Mirabilis and merging its own instant messenger network with ICQ's network of twelve million users.[51]

Instant messaging programs gradually acquired additional features, including voice conversation over the Internet or with conventional telephones, file exchange, and integration with e-mail. As a result the markets for instant messaging systems, Internet telephony systems (such as Skype), and Webmail systems (such as Gmail) have largely converged. The instant messaging field has moved slowly toward standardization, but remains divided into several incompatible networks. Although numerous standards have been created, the most popular services prefer to keep their customers locked into their proprietary software. AIM is now the most popular, and AOL has extended its reach to cell phones and other wireless devices. Apple's iChat software is able to communicate with AIM users. AOL took legal and technical measures to prevent other companies from producing software compatible with this service, however, blocking attempts by Microsoft to couple its MSN Messenger (released in 1999) with AOL's existing user network. To compete with AIM, Yahoo and Microsoft agreed in 2005 to merge their instant messaging networks. According to Internet traffic measurement firm ComScore, AIM, MSN Messenger, and Yahoo Messenger were the most widely used systems in North America during February 2006. The study showed that instant messaging was widespread in the United States with 37 percent of the online population sending a message that month. Interestingly, this lagged far behind the proportion in other regions, with 49 percent of European Internet users and 64 percent of those in Latin America relying on instant messaging.[52] This mirrors the exceptionally slow adoption of cell phone text messaging by North Americans.

All the major instant messaging networks remain free to users, are associated with proprietary access software, and are supported by showing advertisements. The most successful open instant messaging standard, Jabber, claims a relatively modest ten million users working with a variety of software (most notably Google's Google Talk service). Several small companies and open-source teams have produced programs able to communicate with users of all the popular networks, but programs such as Trilian, though popular, account for only a small fraction of all instant messaging users.

Instant messaging appears to have found its niche among younger people using the Internet from home. Unlike e-mail, instant messaging has been slow to emerge as a business tool. This is perhaps because e-mail initially presented itself to business users as an extension of a known quantity: the interoffice memo. Instant messages, in contrast, are more personal, more informal, and harder to archive. As well as being a significant Internet communication technology in its own right, as an application that came of age during the commercialized era of the Internet, instant messaging provides an important contrast with e-mail. Because no dominant standard was established during the precommercial era, Internet instant messaging remains balkanized between the closed standards established by powerful companies. This has hurt the ability of users to communicate, and slowed the mainstream adoption of instant messaging.

The World Wide Web

The ambitiously named World Wide Web was created by Tim Berners-Lee, a British computer specialist with a PhD in physics working at the European particle physics lab CERN.[53] The first Web site was launched on August 6, 1991. Berners-Lee's prototype browser and server software worked only on NeXT workstations, an obscure and commercially unsuccessful type of computer based on a heavily customized version of the Unix operating system.[54] Unlike most later browsers, it also allowed users to edit existing pages and create new ones. On the other hand, it could not display graphic elements within a Web page and used a cumbersome navigation system. Only fifty thousand NeXT computers were ever built, so the initial market for the Web was not large.[55]

The Web was assembled from existing building blocks in a matter of months. Berners-Lee had neither the resources to develop Web software for more widely used systems (in fact, CERN denied his request for assistance in creating a browser for more common versions of Unix) nor the inclination to create a company to market his invention.[56] CERN devoted a total of about twenty person years to the Web during the entire course of its involvement with the project, not all of it authorized and much of it from interns.[57] This unavoidable reliance on proven, widely used technologies was a key factor in the Web's success.

Three crucial standards defined the Web, each of which survived essentially intact from Berners-Lee's 1991 prototype into the Netscape and Mosaic Web browsers used by millions a few years later. Each built extensively on, and was only possible because of, existing Internet infrastructure.

1. The hypertext transfer protocol (HTTP) used by Web browsers to request pages and by Web servers to transmit them. This runs on top of TCP/IP, which handled all the hard work of sending data between the two. In this it followed the model pioneered by earlier Internet application standards, such as SMTP for mail transmission.
2. The hypertext markup language (HTML). After Web pages have been transmitted to the browser via http, the browser decodes the HTML instructions and uses them to dis-

play the Web page on the screen. In the early days of the Web, pages were hand-coded directly in HTML using a simple text editor. HTML was an application of the standard generalized markup language, designed as a universal and extendable way of embedding information on document structure into text files. HTML included the ability to link to Web pages and other Internet resources, and "form" capabilities so that information entered into the browser could be relayed back to the server.

3. The uniform resource locator (URL; originally the universal resource identifier). This extended the existing Domain Name System of using meaningful names (such as public.physics.upenn.edu) rather than numbers to identify computers connected to the Internet. The URL added prefixes such as nttp://, FTP://, and telnet:// to identify the particular kind of request to make to the machine in question. The prefix http:// was used to identify Web pages, in conjunction with an optional filename such as /pub/reports/newreport.html to identify the particular page required, and a TCP/IP port number such as :80 to identify how to communicate with the Web server. A full URL might therefore look like ⟨http://public.physics.upenn.edu/pub/reports/newreport.html:80⟩.

The Web might have been a primitive hypertext system, but it was a first-rate interface for the disparate resources scattered across the Internet. Berners-Lee's great contribution was to produce a simple and workable method by which it was as easy for the author of a document to link to a page on the other side of the world as to another part of the same document. Given the existing capabilities of the Internet this was technically trivial, but Berners-Lee's invention of the URL and its use to specify links to resources on other computers made it possible for the first time to access resources on the Web without needing to know what systems they were housed on.

The Web was by no means the only attempt to built a browsable interface to the ever-growing mass of resources scattered across the Internet. A system named Gopher had been publicly released by the University of Minnesota in 1991. It spread much more rapidly than the Web at first, largely because Gopher software was offered for widely used computers well before fully featured Web browsers. Like the Web, Gopher displayed text documents, could index to resources held on other servers as well as indexes on the same server, and worked as a convenient means of cataloging Internet resources including telnet and file transfer protocol (FTP) servers. Gopher was less flexible than the Web because it worked on a more formal menu system, rather than allowing for the insertion of hypertext links within documents. But Gopher's rapid eclipse by the Web from 1993 onward has often been attributed more to attempts by the University of Minnesota to control its development and make money from it than by any technical shortcomings. CERN made it clear that anyone was free to create a Web browser or server based on the published specifications of HTTP, HTML, and the URL system. In contrast, after the University of Minnesota announced in 1993 that it planned to charge fees for its own Gopher server software, many feared that it would seek to assert its intellectual property rights against rival Gopher servers.[58]

At first, the great appeal of the Web was that it provided an easy and consistent way to catalog and browse the existing resources of the Internet. Any new communication technology faces what most people would term a chicken-and-egg issue, and what economists would call a "lack of network externalities." Why build a Web site when nobody has a Web browser? Why download and configure a Web browser when there are no Web sites to visit? The challenge is to persuade a group of enthusiastic users to invest time and money in the new technology, so as to build up a sufficiently large and visible community to attract others. Early browsers cleverly solved this problem by making themselves into a universal interface for existing Internet resources. In 1992, early in the Web's development, Ed Krol wrote in his seminal *Whole Internet Catalog and User's Guide* that "the World-Wide Web really hasn't been exploited fully yet.... You can look at a lot of 'normal' resources (FTP archives, WAIS libraries, and so on), some of which have been massaged into Hypertext by a clever server.... Hypertext is used primarily as a way of organizing resources that already exist."[59] Even the line-mode browser made available at CERN in 1991 integrated support for Gopher, Usenet news browsing, and file downloading and browsing from FTP servers. Browsers running on personal computers could automatically open other programs as needed for kinds of resources such as telnet sessions, Postscript documents, and video files. Thus, the browser was far more useful than if its powers had been limited to viewing pages on the handful of Web sites then available.

In the meantime, the Web was gaining popularity following the release of browsers for other, more widely used types of Unix workstations. Berners-Lee's own browser vanished rapidly (though not without a trace; his "libwww" code was freely available and gave an easy starting point for later efforts). The important thing, however, was not the software but the free-to-use protocols and standards on which it was based. Because the Web relied on open, published, and simple standards, anyone with an Internet connection and a modicum of programming ability was free to create a Web browser or server, and then hook it up to the ever-growing Web. After all, the Internet had been created specifically to connect computers using different kinds of hardware and running different kinds of software. Within months, amateur teams across the world were at work on improved browsers.

The most influential of the 1992 crop of releases was the Viola browser for Unix workstations, created by a student at the University of California at Berkeley. Viola introduced several key features of later browsers: forward and back buttons, the history function, and bookmarks for favorite pages.[60] Unlike Berners-Lee's browser, Viola worked with the standard X Windows system. Because most personal computers with Internet connections were running Unix in the early 1990s it was the most important platform for both browsers and servers during the first few years of the Web.

The first browser to achieve widespread use, though, was Mosaic. Constructed by a small team at the National Science Foundation–funded National Center for Supercomputer Applications (NCSA) of the University of Illinois, Mosaic was the first browser to be released for multiple platforms and the first to extend HTML to permit the display of images within a Web page. The public release of the first X Windows version of Mosaic in early 1993 was followed by Mac and Windows versions. In January 1994, a survey of Web users suggested that 97 percent of them used Mosaic as their primary browser, and 88 percent used Unix (Windows and the Mac had yet to make serious inroads into the Internet).[61] By October 1994, Mosaic had an estimated two million users.[62]

Mosaic's support for Gopher servers was particularly important because it opened the way for the rapid shift of Gopher users to the Web; they could now access the existing mass of Gopher pages and the fast-growing population of Web pages with a single stylish, graphic client. Users could wander from Web pages to Gopher indexes to downloading files via FTP without even being fully aware of the transition between protocols. Even the name Mosaic reflected the idea that the browser was pulling together shards of information already present on the Internet, assembling them for the first time into a coherent picture. While Gopher was text based, and so could be used on simple terminals and computers without direct connections to the Internet, even this advantage was countered by the increasing popularity of the text-based Web browser Lynx, launched in 1992. By 1994, Web traffic had overtaken Gopher traffic, and over the next few years most of the leading Gopher sites shifted their content over to the Web.

The Web quickly won favor as a universal interface mechanism for online applications, such as library catalogs and institutional telephone directories. Before this, using an online database had meant either working with a plain-text interface and terminal-emulation software, which was ugly and hard to use, or installing a piece of software specially written to work with the database in question. Special software usually required a lot of work to configure and tweak, and would have to be updated whenever a new version of the system was released. But if the system was hooked up to a Web server, then anyone with an up-to-date Web browser could work with it, regardless of what kind of computer it was running on. The complexity was all hidden away behind the Web server at the other end.

These interactive Web systems relied on dynamically generated pages. Whereas a standard or "static" Web page is stored on the server exactly as it will be transmitted to a browser, a dynamic page is tailored to each user and filled with customized information. HTML included the capability to add "form fields" to Web pages, such as boxes to type information into or buttons to click. This information was passed back to the Web server and used to customize the page sent back to the user. For example, if someone entered the name of an author on a library page and then clicked the "submit" button, then the system might perform a database search, format the results into a Web page, and transmit this back to the searcher.

The challenge was in hooking Web servers up to databases and programming tools. The Mosaic team solved the problem by adding a simple but flexible feature called the Common Gateway Interface (CGI) to their Web server. CGI was just a way for a Web server to take information entered into a user's browser, execute a specified program to process this data, and then receive back the results for transmission to the user. But because a CGI call could trigger any kind of software, this kept the Web server itself very simple, and let system builders use their preferred programming language, database system, or scripting tool to handle the task. It could be inefficient with busy systems, but made it quick and easy to hook existing applications up to the Web. CGI followed the classic Unix and Internet philosophy of building a number of flexible, specialized software tools and connecting them together as needed to solve particular problems. Here, as elsewhere, the Web succeeded by building on the mass of established technologies.

Mosaic seemed to offer the University of Illinois a huge financial opportunity. By the 1990s, major American universities had built well-funded technology transfer offices, staffed with teams of lawyers to scrutinize research contracts, patent inventions made by university staff, and license the rights to develop those inventions commercially. A handful of spectacular successes, most notably Stanford's patent on gene splicing and the University of Florida's rights to the Gatorade sports drink, had convinced university officials that savvy exploitation of the right invention might bring in hundreds of millions of dollars. But the successful licensing of software proved particularly difficult because of the gulf between a lab prototype and a salable system. Computing research had given a few universities such as MIT and Stanford equity in some reasonably successful start-up firms, but it had produced no blockbusters.

Here, however, was a university with rights to the most promising new computer application since VisiCalc. While Mosaic itself remained free to use, university authorities decided to offer rights to develop its code commercially to firms interested in marketing their own browsers. The initial terms were $100,000 per license, plus a royalty of $5 for each browser sold.[63] The university licensed the Mosaic code to several firms. The most important licensee was Spyglass, Inc., a firm created to commercialize technologies developed by the NCSA. Spyglass produced its own improved Mosaic browser, and in turn licensed this code to many other companies. By 1994 a rash of commercial browsers had appeared, almost all of them based on either the original NCSA Mosaic or the enhanced Spyglass version code. One 1995 review of Web browsers included no less than twenty offerings.[64] But unfortunately for Spyglass, which had negotiated royalty agreements with its own licensees, the market for browsers would be both short-lived.[65]

Netscape and AOL Bring Browsing to the Masses

By 1995, Mosaic and its commercial descendants had themselves been marginalized with astonishing rapidity by a new browser designed from the beginning as a commercial product: Netscape Navigator. When first released in its unfinished "beta" version, in October 1994, it was already more usable, reliable, and powerful than its many competitors. Netscape was a well-funded Silicon Valley start-up firm that hired many key members of the original Mosaic team, including its cofounder Marc Andreessen, a twenty-three-year-old programmer who served as the chief technology officer. Navigator was much more efficient than Mosaic, and so gave acceptable performance over dial-up modem links and Internet connections slower than the high-speed links enjoyed by computer science labs and research centers. From the beginning, it was available in ready-to-use versions for Windows, Macs, and several different versions of Unix. Netscape eventually claimed more than 80 percent of the browser market, crushing Mosaic and most of its licensed derivatives.[66]

The unprecedented success of Navigator made Netscape into a symbol of the so-called Internet Revolution. Andreessen himself became one of the public faces of the Web, a technology business prodigy apparently cast in the mold of the young Bill Gates.[67] By early 1995, discussion of the Web had spread far beyond campus labs and corporate research departments, into the mainstream of the computer industry, the financial press, and the international media. Overnight, the rhetoric and expectations lavished on the information highway during the early 1990s were shifted to Web browsers and the Internet. As *Wired* magazine wrote in 1994, "Prodigy, AOL, and CompuServe are all suddenly obsolete—and Mosaic is well on its way to becoming the world's standard interface.... The global network of hypertext is no longer just a very cool idea."[68]

The spread of the Web also provided the first really compelling reason for ordinary people to connect their computers directly to the Internet via its TCP/IP. The leading commercial online services of 1993 already allowed users to exchange e-mails with the Internet, and savvy users could even use this capability to request files to be sent from Internet servers via e-mail. For users who needed more, a few specialized services offered text-based Internet accounts. Customers used terminal-emulation software to dial into a central server, on which they could run text-based applications such as e-mail, Internet news, or programming tools. Only the server was connected to the Internet or could execute Internet applications, but the service was affordable, and everything worked fine with a slow modem and a low-end personal computer. Krol's *Whole Internet User Guide and Catalog* said a "dial-up Internet connection" of this kind was available for as little as twenty dollars a month.[69]

But accessing the Web required a different kind of setup: a computer with a fast processor, ample memory, and high-resolution display connected directly to the Internet.

In 1993, when Mosaic first appeared, only powerful Unix workstations costing tens of thousands of dollars and hooked to high-bandwidth campus networks could offer a vaguely satisfying Web experience. As one 1995 report noted, "You can't pick up a newspaper or magazine without reading about the World-Wide Web.... Despite the hype, only a small percentage of the huge numbers that are touted to be 'on the Internet' are really there, except by e-mail. Right now, relatively few have the full, direct access that allows full-screen GUI interfaces, graphics and sound—and access to the Web."[70]

This was changing rapidly, as advances in browser software coincided with the inclusion of newly capable processors (Pentium chips for Windows computers and PowerPC chips in Macintoshes) in mass-market computers, improvements in operating systems, and a glut of inexpensive and relatively high-speed (up to 33.6 Kbit/s) modems. Krol dealt only briefly with the exotic and expensive idea of using the new SLIPP or PPP protocols to make a direct TCP/IP connection to the Internet over a telephone line, which he suggested would cost around $250 a month.[71] But by 1995 hundreds of companies, large and small, were competing to sell true Internet access over telephone lines and so turn household computers into full-fledged Internet nodes for the duration of the connection. Suddenly it was practical, even fun, to browse the Web over an ordinary telephone line on a personal computer costing three thousand dollars.

In late summer 1995, Netscape had around ten million users and was about to ship version 2.0 of its browser. On August 9, just seventeen months after its founding, the firm made an initial public offering of stock. Its share price doubled on the first day of trading, valuing it at more than two billion dollars.[72] Netscape was hailed as the most successful start-up firm in history, and suddenly venture capitalists and stock investors were rushing to invest their money in anything Internet related. While other software companies benefited, most of the so-called Internet stocks were companies trying to make money by providing some kind of product or service over the Internet.

Despite *Wired*'s warning that they were obsolete, the established online services quickly realized the importance of the Web. The most critical of these was now AOL. Building on its user-friendly software, AOL had expanded frantically. From 1993 until the end of the decade, the company relied on a "carpet-bombing" program, mailing out copies of the software with a free trial offer by the millions, attaching them to the covers of computer magazines, burying them in cereal boxes, and even handing them out on planes. By 1996 it had become the largest of the online services, with more than six million subscribers.

AOL gradually shifted itself toward the Internet, evolving piecemeal from a closed network responsible for finding its own content to a hybrid model where it eventually offered full Internet access. It dealt with the threat by buying small Internet software firms to acquire skills and code, shifting its own internal networks toward standard technologies and gradually incorporating Internet capabilities into its existing client software. In 1994, AOL incorporated access to Usenet newsgroups into its service, unleashing what many newsgroup veterans regarded as a flood of ill-mannered plebeians into their private club. In late 1994, after Prodigy became the first of the traditional online services to offer Web access, AOL purchased an obscure Web browser company called BookLine and built its technology into a new version of the AOL client software. It became the world's biggest distributor of Internet software, bundling, modifying, and rewriting utilities into its smoothly integrated client program.[73]

Even after Web access capabilities were added, AOL continued to steer all but the most determined of its users to its own content rather than the broader Web. With version 3.0 of AOL, released in 1995, it became possible to use standard Internet tools and browsers with the service. AOL's own tools were sometimes limited, earning it the dismissive description of being the "training wheels for the Internet." But for new users its friendly, integrated, and nicely sorted interface remained a selling point. This was a real advantage, because in 1995 setting up a typical one-year-old computer running Windows 3.1 to use standard Internet software might require obtaining separate e-mail, news, dialer, TCP/IP, chat, and Web browser programs, and configuring each of these with various obscure settings. AOL installed everything automatically, and hid the cracks so well that many users were not fully aware of the difference between AOL and the Internet. By end of the 1990s, AOL captured about half of the entire American market for home Internet users and claimed more that twenty-five million users.[74]

Remaking the Web as a Business Platform

The early Internet had been an explicitly noncommercial space. With the success of Netscape and the anointing of the Web as the ubiquitous online network of the future, businesses rushed to make money buying and selling things over the Internet. While the use of the Internet by business is the topic of several other chapters within this volume, it is appropriate here to consider the changes made to the Internet infrastructure to support this commercial activity.

The most obvious piece of software a business needed to get online was a Web server. Traditionally, the code for server software for Internet applications such as file transfer and e-mail exchange had been given away free of charge by its authors. Despite the commercialization of the Internet, this remained largely true. The most successful early server software was produced by the NCSA Mosaic team and distributed without charge. Plenty of companies tried to produce commercial Web server software, and both Sun and Netscape enjoyed considerable success in this market during the late 1990s. Other popular Web servers were bundled with commercial operating systems, most notably Windows. But the most widely used Web server software from March 1996 to the present has been Apache, an open-source package used to host around 70 percent of all Web sites in recent years.[75] Apache is produced by the nonprofit Apache

Foundation, originating in 1995 as "a patchy" bundling of fixes for the NCSA server. Since then it has become a crucial part of the Web infrastructure, incorporated into commercial products from IBM, Oracle, Apple, and Novell, and used as a platform for many widely used Web applications. Because the Web relies on clearly defined standards to govern the interaction of browsers and servers, few people have any idea that their Microsoft Web browser spends most of its time fetching information from open-source Web servers.

Unlike commercial online services such as AOL, neither the Web nor the underlying technologies of the Internet had been designed to support financial transactions. There was no easy way to bill users for the information they viewed, or for other goods and services ordered on the Web. This had profound consequences for the development of the Web publishing and Web navigation industries. When someone ordered a book from Amazon.com (launched in 1995) or a used computer via an eBay auction (also launched in 1995), he or she generally used a credit card to pay for the purchase. This was a simple extension of practices pioneered for telephone sales, with which consumers, merchants, and banks were already comfortable. But even this simple task was beyond the capabilities of Mosaic and other early Web browsers. They did not provide any secure method of transmitting credit card information—and the Internet did not help authors by offering a standard method of encrypting or authorizing messages. This had to be fixed before the Web could be a platform for commerce, and Netscape did so when it added a then-unique feature to its first browser and server systems: the optional encryption of data entered into Web pages as they were transmitted back to the server. (Netscape displayed a lock icon on the screen to let users know that the page was secured.) This advance, known as a Secure Sockets Layer (SSL), made the Web a practical platform for financial transactions and other sensitive data. SSL built on and extended the existing Internet traditions of layered protocols and open standards, to create a new protocol layer between TCP/IP and application-specific protocols such as the Web's HTTP.

To support commercial applications, Web servers needed an easy way of tracking who the user was, at least for the duration of a particular usage session. This capability was not built into the Internet itself. Its TCP/IP foundation treated each individual packet of data dispatched from one computer to another as a self-contained event, in contrast with a traditional telephone call where a connection is opened from one party to another and maintained until the call is over. Many online applications, however, naturally follow a sequence in which contact is established, a series of interactions take place, and the user finally disconnects (for example, searching and then ordering from an online catalog, or logging into an online banking system and paying a bill). Some existing Internet application protocols, such as telnet, had been built around the concept of a session, but Berners-Lee did not build any such capability into HTTP.[76] Each request for a Web page was treated as a separate and unrelated event.

Yet the Web server needed some way of tracing which user was which from one screen to another, if only to know whose bank balance or shopping cart to display. Various crude work-arounds were devised for this problem by early online application developers, until Netscape introduced the idea of "HTTP cookies" with its first browser release. Cookies were essentially tracking numbers assigned by a server and then passed back by the browser with subsequent requests so that the server could more easily identify which user it was dealing with. While the Web still forced programmers to do a lot of work to create the illusion of an ongoing connection between browser and server, cookies removed a good deal of the messiness.

Although Netscape did much of the work to make the Web into a viable platform for online business, it generally made its enhancements public and proposed them as standards for adoption by others. Cookies, SSL, the JavaScript scripting language (discussed later), and a number of enhancements to HTML were all released first by Netscape but later codified by external groups. This ensured that Web applications would be accessible regardless of the browser being used, as long as both the creators of the pages concerned and the designers of the Web browsers adhered to the relevant standards. The most important Web standards body was the World Wide Web Consortium, headed by Berners-Lee since its founding in 1994 as the guardian of HTML. In the tradition of the Internet its specifications were royalty free, with the work of the consortium supported by its member companies, research grants, and host institutions. While Microsoft's interest in adherence to HTML standards has decreased dramatically since its Internet Explorer achieved market dominance, the consortium has continued its work in the new area of extensible markup language for Web-based data exchange.

Selling things over the Web demanded much more complex server facilities than simply publishing a Web site full of documents, including code to verify and bill credit cards, update inventory, and maintain user accounts. As Web sites grew in scale and became more commercial, dynamic page generation was used in many different ways. Among the most obvious were up-to-date catalogs showing the current availability of goods, and the various registration and checkout pages accompanying most online retail systems. Some sites went further. Amazon.com, for example, used dynamic page generation to personalize recommendations so that each user would see a unique main page when connecting to the system. Newspaper sites used it to show stories of particular personal interest, and smoothly and constantly update their sites to display breaking news and link to related stories. By the end of the 1990s, a database and set of "fill-in-the-gaps" templates sat behind almost every commercial Web site of any complexity.

Although they never enjoyed a majority for the market for Web servers, commercial software producers had more success in the market for other aspects of Web software infrastructure. A host of new Web-related software niches bloomed in the mid-1990s, too numerous and specialized to deal with fully here. One of the most important was the so-called middleware server and development tools designed to sit between Web servers and databases, and provide a specialized environment for the easy development of Web-based business applications. The flexible and easy-to-use ColdFusion, produced by a start-up firm called Allaire, enjoyed considerable success in this market. Other firms produced tools to search corporate Web sites, tools to integrate Web servers with existing corporate applications such as enterprise resource planning systems, development aids for Web programmers, bundled systems intended to simplify the process of creating a particular kind of online application, and other pieces of software infrastructure aimed at easing the burdens imposed on Web application programmers. In many of these areas, the initial advantage held by proprietary systems has gradually been eroded by open-source competitors. For example, the open-source package PHP is now a standard part of the Internet infrastructure and is used by an estimate twenty million Web domains.[77]

Microsoft Discovers the Web

Even as Mosaic and Netscape took the Internet by storm, and AOL opened it to millions, the top-level managers of Microsoft remained strangely oblivious. By mid-1995, Microsoft had established itself as the world's largest software company, and was crushing its last serious opposition in the markets for desktop operating systems (with Windows) and office productivity applications (with its Microsoft Office suite). Throughout 1994 and early 1995, the firm was embroiled in the Herculean task of finalizing and launching its repeatedly delayed Windows 95 operating system. This distracted its leaders from noticing the Web's new prominence.

In August 1995, tens of millions of Windows 95 cartons were finally piled high in shops around the world, accompanied by the most expensive promotional campaign in the history of the computer industry. The disks inside these cartons did not include a Web browser. Windows 95 did include greatly improved support for networking, and recognized the growing importance of Internet technologies by providing TCP/IP support, a provision for dial-up connections to ISPs, and basic utilities for Internet file transfer and remote log-ins. This made it much easier and more reliable to configure a computer for Internet use, as ISPs no longer needed to provide or support many of the utilities demanded by earlier Windows versions.[78] The firm had even licensed the Mosaic browser code from Spyglass, after which a handful of programmers had been allowed to slap on the Microsoft logo, rename it Internet Explorer, and tinker with it to run smoothly on Windows 95.[79] But this browser was not included on the main Windows 95 disk (which somehow managed to squeeze in such essentials as multiple versions of a video of the geek rock group Wheezer singing their hit song "Buddy Holly"). Instead, Internet Explorer was relegated to the optional "Microsoft Plus!" pack, a thirty dollar impulse buy holding a lucky dip of Space Cadet Pinball, screen savers, and other trivia. The big networking push at the Windows 95 launch was for

MSN, a proprietary online service modeled on the pre-Internet AOL.[80] While Internet Explorer languished in obscurity, MSN programs were built into Windows 95, with a prominent icon displayed on the Windows Desktop and an invitation to sign up as part of the Windows installation procedure.

Behind the scenes, Microsoft had begun to work with increasing urgency to incorporate the Web into its product line. The origin of this shift is conventionally attributed to May 1995 when Bill Gates sent a now-famous memo titled "The Internet Tidal Wave" to his fellow executives. It warned that "the Internet is crucial to every part of our business...the most important single development...since the IBM PC" on which Microsoft had built its business. As a first step, Gates ordered the shift of "all Internet value added from the Plus pack into Windows 95 as soon as we possibly can."[81] The memo, and a series of other events and reports through the rest of 1995, spread the message that integration of the Internet technologies into all Microsoft products was now the most pressing task facing the firm. Gates even delayed publication of *The Road Ahead*, ghostwritten book delivering the shocking news that computer networks were about to transform the world, so that references to the Internet could be inserted alongside its lengthy discussion of the information highway.[82]

In his memo, Gates had ordered that improved Web capabilities be "the most important element" of all new releases of Microsoft application software, demanding that "every product plan go overboard on Internet features." Microsoft quickly created a download to allow Word, its word processor, to create Web pages without the need to manually code HTML tags. In 1996, it released a free Web server for use with its Windows NT server, and acquired a promising Web site creation tool called Front Page from Vermeer Technologies, a Massachusetts-based start-up firm.[83] The same year it launched a new tool, NetMeeting, which was soon bundled with Windows to provide online audio and videoconferencing features. When it shipped the next version of Microsoft Office in 1997, the company went to great lengths to promote the newfound ability of Word, Excel, and PowerPoint to output documents as Web pages. Since then, each new version of Office has added capabilities to make it easier to share documents over the Internet, edit Web documents, collaborate with remote users, and so on. Many of these features, like a number of the special features offered by later versions of Front Page, required the use of Microsoft Web browsers and servers to function properly. In this way, Microsoft tried to use the popularity of its desktop software to build a stronger position in the booming market for corporate intranets (internal networks based around internet technologies).

Microsoft also rushed to get into the business of providing online information and entertainment. In some ways, this was a natural evolution of three of its main strategic thrusts of the early 1990s. One of these was the publication of reference and educational materials on multimedia CD-ROMs. The arrival around 1993 of compact disc players, audio capabilities, and high-resolution graphics as standard items on high-end personal computers had opened what seemed at the time to be a huge and growing market for interactive books. Microsoft created the widely acclaimed Encarta encyclopedia, Cinemania film review archive, and dozens of other CD-ROM titles covering topics from ancient civilizations to scary wildlife. The second thrust, to overthrow AOL and the other online services with its new MSN, brought it into the online content business, because online services licensed existing content for their subscribers (such as material from the *New York Times*), and then produced their own proprietary materials and services. Microsoft was also making a major effort to enter the cable television business, guided by the belief (held widely for much of the 1990s) that the forthcoming technology of digital cable television would mean that the personal computer and television industries would inevitably merge. The rise of the Web redirected all of these efforts toward the Internet, as Microsoft launched a host of news and entertainment services. These included the MSNBC cable news channel and Web site (launched in collaboration with NBC), the online culture and politics magazine *Slate*, the Sidewalk series of online city guides, and the travel agent Expedia. Synergies have generally proved elusive in the media business, and this was no exception. Just as Sony had discovered that there were no real advantages to be gained by owning the movie studios and record labels that produced the films and records played on its equipment, so Microsoft found few benefits from producing the Web pages viewed with its browsers. While at the time these businesses were seen as an integral part of Microsoft's Internet push, with the benefit of hindsight it became apparent that they were strategically irrelevant and all have since been disposed of.

Java and the Browser as Application Platform

The most visible aspect of Microsoft's Internet campaign was its promotion of its Internet Explorer browser as a substitute for the hugely popular Netscape Navigator. Why was Microsoft so determined to make its own browser the standard? The answer lay in the ability of the Web browser to serve as a universal access mechanism for different online systems. Microsoft's domination of the personal computer operating system market had become self-reinforcing, because the strength of the Windows platform ensured that application software developers would write their exciting new programs to run on Windows, which in turn would attract new users. The Web and other Internet systems worked on open standards, however. When browsing a Web site or sending an e-mail, a Windows computer had no inherent advantage over a Macintosh or Unix computer using the same Web site or composing the same e-mail. They all used the same protocols. If the exciting new applications of the future were to be delivered over the Internet to a Web browser, rather than installed and run on a personal computer, then Microsoft was in danger of losing its hard-won strategic position within the computer industry. Gates called this a "scary possibility" in his famous memo.[84]

As the dominant producer of browser software, Netscape seemed liable to seize the ground lost by Microsoft.

As Netscape captured the attention of the technology world in 1995 it began to promote itself as the next Microsoft. To make its product into a ubiquitous standard, Netscape developed browsers for no less than sixteen different operating systems. In 1995, Andreessen is reported to have announced that new advances in Web technologies would make Windows 95 into nothing more than a "mundane collection of not entirely debugged device drivers" sitting between Netscape and the underlying hardware.[85] This claim might appear rather fanciful given the rather slow, limited, and awkward nature of Web-based systems during the mid-1990s when compared with Windows applications. At the time, Web browsers could do no more than display simple pages and accept input data. A program written for Windows could display moving images, make sounds, perform calculations, and validate or query data as the user entered them. It was hardly a fair fight, and in retrospect it is one Netscape would have done better to avoid picking when its own technologies were so immature.

But fight it did, and Netscape rushed to add feature after feature to Navigator to boost its credibility as an application platform. Netscape created a simple programming language, JavaScript, to make Web pages more interactive by handling tasks such as calculating a total or verifying a date without reloading the entire page. It added support for exploration of "virtual reality spaces," which for a few months in 1995 was widely discussed as the future of the Web.

The big shock came when version 2.0 of Navigator, released in September 1995, included a then-obscure programming technology from Sun Microsystems known as Java (JavaScript and Java, confusingly, are entirely different things). Over the next few years Java received a degree of avowed support, public discussion, investor hype, and general enthusiasm never granted to a programming language before or since. Java was not only a programming language but also an application platform in its own right. Java programs were written to run one specific computer. The novel thing was that this computer was never designed to be built in hardware. Instead, an additional layer of software, known as a "virtual machine," mimicked its internal functioning. Java was a kind of Esperanto: nobody was a native speaker, but it was designed as a universal second language. Java's marketing slogan was "write once, run anywhere," because any properly written Java program was supposed to run on any properly implemented Java virtual machine, and hence to be usable without modification on any computer.

The original idea behind Java was that this would avoid having to rewrite complex interactive television software for every different kind of cable box. But by the time Java was released attention had shifted to Web browsers. At this point Sun was the leading supplier of Web server hardware. By including a Java virtual machine in every

Web browser, Netscape aimed to make Java a denominator even more common than Windows. If a company needed its customers to do something they couldn't do easily on a regular Web page, such as editing formatted text or working interactively with a technical model, it would write a little Java program known as an "applet" for inclusion on the page. When the page was loaded, this code would run in any Netscape browser whether the customer used a Windows computer, a Macintosh, a Unix workstation, or a special Internet device.

Java enthusiasts hoped that within a few years it would have spread beyond these simple tasks to become the standard platform for computer programs of all kinds. Corel, the firm that acquired the faltering WordPerfect word processor, wanted to revive its fortunes by rewriting the entire package in Java to beat Microsoft into what it thought would be a huge market for Java office applications. Opponents of Microsoft, most vocally Larry Ellison of Oracle, claimed that the spread of Java would make the traditional personal computer altogether irrelevant. Rather than a complex, crash-prone personal computer running Windows, Ellison believed that the standard device for home or office computing would be a stripped-down "network computer" running Java software from a server. This would save on hardware costs, but more important, would be much easier to manage and configure than a traditional personal computer.

The Browser Wars

When Windows 95 was launched, Netscape had an estimated 75 percent of the browser market, and its share was still rising rapidly as its early competitors faltered.[86] The first release of Internet Explorer sank without trace, weighed down by the twin burdens of running only on the newly released Windows 95 (and NT) operating systems and not being very good. As the entrenched leader in the browser market, Netscape had significant advantages over Microsoft, including the goodwill and awareness of users, a steady stream of visitors to its Web site looking for new downloads, and the fact that many Web sites had been designed to take advantage of the nonstandard features present in Navigator but not in official HTML standards. This meant that some Web sites would not appear correctly viewed in any other browser. As Michael Cusumano and David Yoffie observed in their book *Competing on Internet Time*, Netscape had a strategy of "open but not open" with respect to Internet standards, using and supporting public standards only to a point while trying to maintain a competitive edge through its de facto control of the browser market.[87]

Yet these advantages proved little match for the might, discipline, and aggression of Microsoft. Seldom in the history of the computer industry has a large, entrenched company responded so effectively to the threat posed to its established business by a radical new technology. Microsoft had three main advantages: thousands of skilled programmers and experienced managers, enormous and assured income flows from its established businesses, and control over the technical architecture of Windows itself. It exploited all of these advantages mercilessly in its struggle against Netscape.

Like most other personal computer software companies with popular products, Netscape had hoped to make money by selling its programs. Internet users expected to be able to download software freely, and this was the only practical way for a small firm like Netscape to get its software out quickly to millions of people. So Netscape adopted a version of the classic shareware software business model: Navigator could be downloaded without payment, but commercial users were required to register and pay for the software after a trial period. The basic price was ninety-nine dollars a copy, with discounts for corporate licensing, and shrink-wrapped copies could be purchased in computer stores.[88] But Explorer could be used by anyone for any purpose without paying a penny. By giving away its own product, Microsoft cut off Netscape's main source of income and crushed the emerging market for Web browsers. Only in 1998 did Netscape finally follow suit and make its own browser officially free for all users.

Microsoft assigned large numbers of skilled programmers to a crash program designed to boost the quality of Explorer to match Netscape. The second version of Internet Explorer, released just a few months after the first, was a big improvement and ran faster than Netscape, though it still lacked some features. Abandoning its once-exclusive focus on Windows 95, Microsoft produced versions of Internet Explorer for Macintosh computers, Unix systems, and computers running the older Windows 3.1 operating system, challenging Netscape on all the major personal computer operating systems of the era. By 1996, its Internet Platform and Tools Division comprised about twenty-five hundred people, more than twice as large as the whole of Netscape.[89] By the time the third version of Explorer was released, in mid-1996, Internet Explorer was rated by most independent reviewers as evenly matched against Netscape in technical terms.[90] Microsoft even licensed Java from Sun, matching what was perceived as a huge advantage for Netscape. (Sun later sued, after Microsoft breached its terms by promoting an improved Windows-specific version of Java.)[91] Navigator retained its market dominance, but its technological edge had eroded. By the time both firms released their version 4 browsers in mid-1997 it was gone entirely. Explorer was widely judged to be faster and more stable than Navigator.[92]

Netscape's other main potential source of revenue was the sale of its extensive family of Web server products. In 1996 it shifted its strategic focus from browsers to intranet products, including Web servers, e-mail servers, and collaboration support tools for internal corporate use.[93] But in the Web server and intranet markets too the firm faced competition. Microsoft also bundled a fully functional Web server into its Windows NT business-oriented operating system (later renamed Windows 2000), though as Windows had a much less dominant position in the server market than the desktop market, the effects of this on Netscape were less pronounced.

Meanwhile, estimates of the threat to Microsoft from Java proved massively overblown. Developing large applications in Java was much harder than had been expected, leading Corel to abandon its efforts without ever getting WordPerfect to work properly. Java didn't even live up to expectations as a replacement for HTML in building interactive Web site interfaces. Including a Java program as part of a Web site caused a long download for dial-up Internet users, and it would not function correctly except with a specific browser version and exactly the right Java release. Differences between different Java implementations, and the frequent release of new and incompatible versions, made the promise of universal compatibility a cruel joke. The cost of personal computer hardware dropped so rapidly that few customers ever felt tempted by the specialized "Internet appliances" sold for e-mail and Web browsing, when for a few more dollars they could have a general-purpose computer with much more flexibility. Sony, 3Com, Compaq, and other major firms all suffered abject failure when they attempted to sell Internet appliances. Network computers were a little more successful in the corporate world, but remained a niche item for the kinds of jobs formerly performed on less-powerful video terminals rather than a general-purpose replacement for the personal computer. This may not have been entirely inevitable. Had Microsoft not deliberately targeted Netscape's control of browser design and taken steps to weaken Java's cross-platform compatibility by introducing its own extensions, then perhaps things would have gone differently.

The third of Microsoft's advantages—control of the desktop operating system market—was perhaps the most important. It seized on its control of the operating system installed on desktop personal computers as a chance to push its own browser to the exclusion of Netscape. Its aggressive action here led to complaints from Netscape, and in turn to a major antitrust case filed against Microsoft by the U.S. Department of Justice in 1998. One of the major issues in this case concerned Microsoft's domination of software distribution channels. From 1996 onward, it was impossible to purchase a Windows computer without Internet Explorer preloaded on to it. At the same time, Microsoft threatened major manufacturers of personal computers with the withdrawal of their licenses to sell Windows if they did not remove Netscape from the software bundles supplied with new computers. Microsoft also persuaded AOL, CompuServe, and Prodigy to make Internet Explorer the only browser distributed on their installation disks. So keen was Microsoft to capture the browser market that it was willing to sacrifice MSN's special place on the Windows desktop to do so. It agreed to supply software from these rival services with each copy of Windows, and even to promote them with icons placed on the desktop of each new computer.[94] People could continue to download and install Navigator themselves, but these moves ensured that Netscape was denied the two most important channels for the distribution of browser software: preinstallation on new computers, and inclusion in the promotional mailings with which AOL and its competitors saturated American households.[95]

The other main accusation against Microsoft in the antitrust case was that it had illegally bundled its Explorer browser, with which Netscape was competing, with the Windows operating system. U.S. law draws a sharp distinction between the competitive practices tolerated from monopolists and those acceptable for ordinary firms. Bundling two products together is normally legal, but a firm with a monopoly in one market is forbidden from using bundling to wipe out competition in another market. In this sense, Microsoft's browser was not really free to its users. The thousands of employees in its Internet division still had to be paid, after all. Microsoft supported this work through its monopoly profits from the desktop operating system market, effectively making the cost of Internet Explorer a tax paid on every Windows computer sold whether or not its buyer wanted Explorer. This appeared to put Netscape at a disadvantage.

Microsoft's push to tie Explorer to Windows went far beyond joint distribution. In 1996, Microsoft had been encouraging Web developers to include small Windows-only programs, dubbed "ActiveX controls," on their Web pages as a way of adding interactive features. Internal battles within Microsoft during 1997 destroyed the autonomy of the Internet team, subordinating its single-minded push to create strong Internet tools to a new strategy based on merging the browser into the heart of the Windows user interface.[96] With the launch of Windows 98, a modestly improved version of Windows 95, Microsoft made it technically impossible to remove Internet Explorer without disabling Windows itself. Its most hyped feature was Active Desktop, a buggy, cumbersome, and little-used method to display constantly updated Internet content on one's desktop.[97] Microsoft insisted that there were obvious, overwhelming technological reasons for this integration, but few outside the company found this claim credible. One technology reporter wrote that he had "been avoiding the subject of the Active Desktop because frankly, I don't really get it. I don't use it, I don't know anyone who does use it and I don't think it works very well for those who do use it."[98] Another remarked that it "sticks its tendrils into every part of your operating systems and will wreak havoc on your computer" with a "nasty habit of crashing every thirty minutes." A third wrote of the new features, "Don't like 'em, don't want 'em, don't need 'em and I say, to heck with 'em."[99]

Microsoft's actions were motivated by a desire to turn the disruptive power of the Web into nothing more than an extension of its proprietary world of Windows. Its insistence that Web browsers and operating systems were conceptually indistinguishable became the cornerstone of its legal defense in the case as well as an expression of Gates's personal determination to maintain Windows as the centerpiece of Microsoft's business. If the browser was merely a legitimate and necessary feature of Windows, rather than a separate and bundled product, then nothing illegal had taken place. This was particularly important for Microsoft to prove, since earlier instances of monopoly abuse had led it to sign a 1994 consent decree promising not to tie any of its other products to the sale of Windows. The case progressed badly for Microsoft, as its executives struggled to explain away the powerful evidence mustered against them. Gates refused all compromise with the government, but tarnished his reputation as a technical and business genius with testimony during which he quibbled over linguistic nuances while denying all knowledge of his own e-mails and memos.[100] So disastrous was his performance during the trial that in 2000, he was quickly eased out as CEO in favor of Steve Ballmer, stepping back from business management to a new position as chief software architect, in which his day-to-day involvement in the business of the firm was gradually diminished in favor of charitable work.[101]

In April 2000 Microsoft lost the case. It was ruled to have used anticompetitive means to maintain a monopoly in the market for personal computer operating systems, illegally tied Internet Explorer to its Windows monopoly, and destroyed the browser market by adopting predatory pricing to crush Netscape and stifle competition.[102] The trial judge, Thomas Penfield Jackson, wrote that "Microsoft paid vast sums of money, and renounced many millions more in lost revenue every year, in order to induce firms to take actions that would help enhance Internet Explorer's share of browser usage at Navigator's expense." Given that Microsoft had promised never to charge for its browser, Jackson continued, this could "only represent a rational investment" if its purpose was to protect the Windows monopoly from the emergence of a rival applications platform.[103] After Microsoft failed to reach an agreement with the government during mediation, Jackson granted the Department of Justice's request that the firm be broken into two separate businesses—one confined to operating systems, and the other to applications.[104]

Jackson's ruling on monopolization was upheld on appeal, but the breakup was deferred for consideration of alternative remedies. Microsoft was then able to negotiate a much less dramatic settlement with the newly installed Bush administration by promising to fully document the techniques needed to write applications for Windows, make it easier for customers to remove desktop icons for Microsoft products or set competing products as defaults, and stop blackmailing computer makers into dropping products written by its competitors.[105] These measures amounted to a reprieve for the firm, after which Ballmer moved to settle the other major lawsuits against Microsoft. A private suit brought by Netscape on the basis of the antitrust findings was settled for $750 million in 2003, and a group of suits brought by Sun for almost $2 billion in 2004.[106] A separate antitrust case launched against Microsoft by the European Union proved harder for the company to contain. Again the issue was anticompetitive bundling. After failing to comply with the terms of a 2003 order to fully document the behavior of its server products (to make it easier for competitors to achieve compatibility) and offer a version of Windows without Windows Media Player, the company has been hit with a series of fines. As of October 2006, the firm had still not satisfied the European Union and had accumulated more than $1 billion in fines while it continues to appeal.[107]

But even as Microsoft's legal problems mounted during the late 1990s it was winning its commercial struggle against Netscape. By 1998, market research reports were suggesting that the usage of Internet Explorer had almost pulled level with that of Navigator.[108] Many large companies were standardizing on Explorer.[109] Meanwhile, Netscape's software developers found it increasingly difficult to keep up as they tried to match Microsoft feature for feature, support numerous versions of Navigator for different operating systems, tidy up Navigator's increasingly unwieldy code base, and supplement it with e-mail, news, Web page editing, and calendar functions. Netscape's efforts eventually foundered entirely, as programmers working on the release 5.0 of Navigator labored for several years before eventually giving up. In late 1998 a diminished Netscape agreed to be acquired by AOL, consolidating two of the best-known Internet firms of the era.

Throughout the late 1990s, AOL invested heavily in Internet technologies. The acquisition of Netscape, for stock worth nearly ten billion dollars when the deal was consummated, seemed at the time to make AOL into a major force in the world of Internet software.[110] A joint venture with Sun, the developer of Java and a leading supplier of Web servers, was quickly announced with a promise that the alliance would become "the dot-com software company."[111] Contemporary reports suggested that the new alliance would focus on developing the Internet as a retail environment, pushing the spread of Internet commerce to small businesses as well as on to wireless devices and into cable television boxes.[112]

Whatever plan drove the acquisition was quickly rendered irrelevant by one of the great, all-time disasters in the history of business. Despite its push into the Internet software business, AOL ultimately decided that its future was as an entertainment company. In January 2000, it acquired Time Warner to form what the deal's architects hoped would be the perfect marriage: a virile young firm controlling tens of millions of home Internet connections spliced to a creaking but powerful media empire churning out vast quantities of "content" just waiting to be digitally delivered to eager consumers. Though AOL was a much smaller company, AOL shareholders received a majority stake in the new firm, dubbed AOL Time Warner, because theirs was perceived as a fast-growing business.

Before the new firm could even finish building its stylish New York City headquarters, AOL's advertising base collapsed and its users began to defect to high-speed alternatives. Within two years of the merger all the senior AOL executives had been purged, and in 2003 the combined firm had dropped AOL from its name. Meanwhile, the collapse of the formerly booming market for Internet stocks and services crippled the prospects of AOL's alliance with Sun.[113] Starved of resources, support, and leadership

Netscape's browser withered with astonishing speed. The eventual arrival of Navigator version 6.0 (in 2000) and 7.0 (in 2002) merely accelerated the collapse, as Netscape die-hards discovered their long-awaited upgrades were sluggish, prone to crashing, and labored under a clumsy user interface. Navigator's estimated market share, close to 50 percent at the time that the acquisition was announced in late 1998, plummeted to well under 5 percent over the next four years.[114] Time Warner never even shifted users of its own AOL client software to the Netscape browser. In 2003, it closed its Netscape division and eliminated the few remaining employees. It had, however, finally found a use for the Netscape brand: as the name of a cut-price dial-up Internet service. The name that just a few years earlier had been attached to one of the most spectacularly successful pieces of software ever created, appeared to have met a sad end.

The Web Grows Up

By the time Microsoft edged past Netscape in its share of the browser market it had a faster, more stable, and more stylish browser. Furthermore, Microsoft gave its browser away, whereas Navigator cost money. Were customers really hurt by Microsoft's triumph, whatever its legal transgressions? Although the question is hard to answer, a look at events since 1998 is highly suggestive. Version 5 of Internet Explorer was released in 1999, as the Netscape threat had begun to recede. Following this, work to enhance the browser almost stopped. Internet Explorer for Macintosh saw no significant improvements after 2000, and was officially discontinued three years later.[115] The Unix version of the browser was also terminated, and even the Windows version stagnated. Version 6, released along with Windows XP in 2001, added only minor improvements. In 2003, Microsoft announced that future browser versions would be made available only with the purchase of new versions of Windows (mirroring its claim during the antitrust trial that the Web browser was an operating system feature rather than a separate application).[116]

During its trial, Microsoft had argued that an enforced split between its browser and operating system businesses, as sought by the government's lawyers, would hurt its ability to delight consumers through rapid and coordinated innovation in both areas. Gates repeatedly demanded that Microsoft be given "freedom to innovate and improve our products," contending that "continuing to evolve the Internet technologies into Windows is great for consumers."[117] The firm even set up a self-proclaimed "grass-roots" organization, the Freedom to Innovate Network, complete with newsletters and press releases to help satisfied customers lobby their elected representatives on its behalf.[118] Yet with the case settled on terms favorable to Microsoft, the promised stream of delightful improvements has somehow failed to find its way into the hands of users. In August 2006, Internet Explorer marked a remarkable five years without an increment of its version number from 6.0. The next month, Windows celebrated the fifth anniversary of Windows XP (also known as version 5.1) as the company's flagship personal computer operating system. Every personal computer sale added a little more money to Microsoft's famously ample stash of cash and short-term investments (which has fluctuated between thirty and fifty billion dollars in recent years), but little in the way of new technology was coming out of Redmond, Washington, in return. Furthermore, Microsoft's rush to integrate the Web browser deep into the heart of the operating system and allow ActiveX programs to run within it—actions spurred at least in part by legal and business considerations—had the nasty side effect of opening many vulnerabilities to worms, viruses, and spyware.

The browser story has a twist. In 1998, as Netscape's browser business unraveled and its corporate strategy shifted toward the Web portal market, its management had gambled on the novel but fashionable strategy of open-source development as a way to leverage its limited resources. The full source code for its browser suite was released over the Internet in the hope that this might lead to a flood of variant and improved versions to counter the threat from Microsoft. After the initial excitement wore off, little seemed to be happening. Following its acquisition by AOL, Netscape continued to support the open-source project, dubbed Mozilla, but the results were disappointing. The doomed Navigator versions 6.0 and 7.0 were based on the Mozilla code. Incremental updates improved both somewhat, but Mozilla-based browsers won support only from the most dedicated of open-source enthusiasts.

Then in 2004, word suddenly began to appear in national publication such as the *Wall Street Journal* and *New York Times* of a new browser: Firefox.[119] Version 1.0 was officially released in November, but even before that the program won a large and loyal following. Firefox was based on the Mozilla code, which because of its open-source license was free for anyone to modify or improve. Although the Mozilla project had produced some excellent code, its own browser remained clumsy to use and was burdened with a mass of extraneous features. In 2002, a Netscape programmer and a teenager working as an intern had resolved to create a new version in which the browser was pared down to its essentials and given a stylish yet functional interface.[120] The project gained steam rapidly, recruiting experienced developers, and eventually became the flagship product of the Mozilla project. Its appearance coincided with a rash of security attacks exploiting Internet Explorer's tight integration with the Windows operating system, giving ordinary users a convincing reason to shift. Firefox boasted other simple but useful features missing from Explorer, particularly the ability to open several pages within a single browser window and shift rapidly between them. Its capabilities have been extended by dozens of add-on programs, customizing the browser to do things like block Internet advertisements.

Firefox became the first open-source desktop computer application to win widespread usage by Windows and Macintosh users. Within a year of its first official release, its share of the worldwide browser market had risen above 10 percent in many surveys.[121] Microsoft invested an enormous amount of time and money in redesigning Windows to eliminate security holes, hoping to match Firefox's reputation as a safer browser. Firefox even rekindled browser development at Microsoft, which quickly started promoting the new features planned for version 7.0, the first significant enhancement in five years. Reversing its previous announcement, Microsoft revealed that Internet Explorer 7.0 would be available as a download for users of its existing Windows XP operating system as well as being bundled with its the forthcoming Vista version of Windows. Early reviewers found the new release welcome but unspectacular, suggesting that Microsoft had successfully duplicated many of the features pioneered by its competitors over the past five years but included few new ideas of its own.[122]

Meanwhile, a number of successful and innovative Web-based services have grown up in the past few years. By 2005, these were being collectively referred to as "Web 2.0," a term that must owe some of its success to its conspicuous vagueness. As defined by technology publishing veteran Tim O'Reilly, whose company was initially responsible for spreading the phrase, Web 2.0 involves a new wave of Web services based around user communities and collective tagging efforts (such as Wikipedia and YouTube). But the term also signifies new Web technologies to provide users with rich, interactive user interfaces comparable to those found on traditional desktop applications. This rekindled the enthusiasm for the Web browser as an application-delivery platform seen by Java enthusiasts during the mid-1990s. Like earlier bursts of Web creativity this is an example of what Eric Von Hippel has called "user innovation," where technological change takes place not as the result of a strategic vision of a company building a new product but through the creative actions of its customers in reconfiguring and combining existing products.[123]

Google applications such as Gmail and Google Maps helped to demonstrate and popularize the ability of Web browsers to support attractive, powerful, and responsive user interfaces. This was not due to any single technological breakthrough but instead rested on the creative combination of a range of existing capabilities to snappily update the data shown on Web pages without reprocessing and reloading the entire page. One popular blend of technologies has been dubbed Ajax, for Asynchronous JavaScript and XML. Most of the underlying technologies (formerly called Dynamic HTML or DHTML) first appeared in browsers toward the end of the so-called browser wars as Microsoft and Netscape competed to add new features as rapidly as possible. Over the last few years, as high-bandwidth Web connections spread, browser bugs and incompatibilities were solved or worked around, programmers came to grips with the nuances of these techniques, Web developers found ways to achieve many of the objectives originally set for Java without having to deal with its weaknesses. Online applications have some real advantages in some areas, including access to large databases, the elimination of the need to configure or install software on the user's computer, and easy data sharing between users. Google now offers a reasonably powerful online spreadsheet and word processing programs accessed through a Web browser. While unlikely to dent the sales of Microsoft's Office suite, these applications do offer easy sharing of documents with other users, collaborative editing, secure backup, and easy accessibility of saved documents from any computer with a Web browser. Many users are choosing online applications for both personal tasks such as tax preparation and, with the success of Salesforce.com, corporate applications such as customer relationship management.

The push by Netscape, Sun, and Oracle to use Java as a direct and aggressive challenge to Windows failed miserably a decade ago. But even without major updates to Internet Explorer, the Web has gradually matured as an application platform. Few new companies have tried to enter the market for stand-alone desktop computer applications since the mid-1990s, and the markets for software such as word processors, personal finance software, and antivirus utilities are all dominated by a handful of well-entrenched companies. Excitement, innovation, and investment dollars have long since shifted to web-based applications, with community networking sites such as YouTube and Facebook providing the latest round of hits.

Within the software industry, Google is widely seen as the next challenger to Microsoft's domination of the market for operating systems and office applications. Given Google's refusal to publicly outline any such strategy this may just be wishful thinking on the part of Microsoft-hating bloggers, but analysts speculate that its push into online applications and the aggressive hiring of gifted programmers portends an eventual showdown. Microsoft's leaders have also described Google as their biggest threat, perhaps recognizing the usefulness of an external enemy in motivating their own employees. Gates, his own youthful swagger long gone, said of Google, "They are more like us than anyone else we have ever competed with."[124] The Web's growing maturity as an application delivery platform has not yet hurt Microsoft's profitability or its stranglehold on its core markets but may slowly be eroding the strategic value of that stranglehold.

Conclusions

The Web and e-mail technologies of the early 1990s were profoundly and directly shaped by the environment in which they had been produced. They were excellent solutions to the problem of a relatively small and homogeneous network of academic researchers. Nobody designing systems to be used by billions of people for commercial purposes would have designed anything remotely like them. Yet in a few short years, the Internet and its application protocols became the foundation for the first and only universal, commercial, and global computer network in human history. The funny thing was that their rapid success in the commercial environment had a lot to do with the very characteristics that no commercial designer, manager, or consortium would have produced: simplicity, flexibility, openness, decentralization, the omission of billing and accounting mechanisms, and so on. In particular, the Web offered itself to businesses and enthusiasts as a quick, easy, and widely accessible platform on which to establish their online publications and applications. Current battles to protect net neutrality (the principle that ISPs treat data from all Internet sites equally rather than favoring particular business partners) reflect the influences of principles built into TCP/IP long before the Internet was opened to business.

The Internet's strengths and weaknesses have fundamentally influenced the evolution of Internet businesses, making some niches (such as commercial publishing) hard to fill and creating others (such as search engines) that did not exist on other online systems. The Internet's basic approach of building new systems and standards on top of old ones made it easy to extend it to new applications, including many commercial ones, but hard to go back and change the underpinning infrastructure on which the new applications rested.

One of the most distinctive features of the precommercial Internet was its reliance on open standards and protocols, published and free for all to use. Its cultures, values, and practices shifted rapidly during the 1990s as the Internet commercialized. In some ways, however, they proved surprisingly resilient. Netscape remade the Web as a commercial platform, adding features such as credit card encryption and cookies, but it did so largely by building on existing standards and establishing its own innovations as new standards. Even Microsoft was initially forced to follow the same pattern to establish its own browser as a credible alternative. Since Microsoft's victory in the browser wars, the rate of changed has slowed, but because Web browsers must continue to work with a wide variety of servers, Microsoft has not been able to remake the Web as a closed system. The Web continues to be defined by open standards rather than proprietary code. Meanwhile, the focus of innovation has shifted from Web browsers themselves to Web applications, as the new approaches referred to as Web 2.0 evolved around new uses for clusters of existing technologies. The Web remains a platform for user-driven innovation, far beyond the control of any one company.

More than anything else, the Internet was a collection of protocols, and these protocols have continued to shape its evolution long after their designers lost institutional authority over the network. Over the past fifteen years, many changes have been made to the software technologies of the Internet as the Web has been remade as a secure platform for online commerce. But the fundamental characteristics of protocols such as TCP/IP, SMTP, and HTTP continue to shape our daily lives. Whole industries have emerged to tackle problems such as e-mail spam, yet without basic changes to the software architectures inherited from the precommercial Internet, these woes can be alleviated but not cured. In some ways today's Internet is crippled by the very design choices that provided its initial strengths. In any event, the software technologies whose evolution I have described here formed the sometimes shaky infrastructure on which companies of all kinds built the different types of Internet businesses discussed in the rest of this book.

Notes

1. The concept of the killer application was popularized in Robert X. Cringely, *Accidental Empires: How the Boys of Silicon Valley Make Their Millions, Battle Foreign Competition, and Still Can't Get a Date* (Reading, MA: Addison-Wesley, 1992).

2. Scholars have recently paid considerable attention to the role of protocols in regulating the Internet, particularly the governance of its Domain Name System of domain name allocation and resolution by the Internet Corporation for Assigned Names and Numbers. See Alexander R. Galloway, *Protocol: How Control Exists after Decentralization* (Cambridge, MA: MIT Press, 2004); Milton Mueller, *Ruling the Root: Internet Governance and the Taming of Cyberspace* (Cambridge, MA: MIT Press, 2002); Daniel Paré, *Internet Governance in Transition: Who Is the Master of This Domain?* (Lanham, MD: Rowan and Littlefield, 2003). My concern in this chapter is not with the overtly political work of assigning domain names but with the broader influence of the Internet's precommercial technologies, cultures, and practices on the development of its software infrastructure during the 1990s.

3. Lawrence Lessig, *Code, and Other Laws of Cyberspace* (New York: Basic Books, 1999).

4. Thomas P. Hughes's specific ideas on the development of "large technological systems" came as a generalization of his study of the evolution of electric power networks in *Networks of Power: Electrification in Western Society, 1880–1930* (Baltimore, MD: Johns Hopkins University Press, 1983). A variety of approaches to the topic were collected in the highly influential Wiebe Bijker, Thomas P. Hughes, and Trevor Pinch, *The Social Construction of Technological Systems* (Cambridge, MA: MIT Press, 1987).

5. The concept of opening the black box was popularized by Donald MacKenzie, who explored the social shaping of missile guidance systems in *Inventing Accuracy: A Historical Sociology of Nuclear Missile Guidance* (Cambridge, MA: MIT Press, 1990).

6. These ideas are critiqued in Langdon Winner, "Mythinformation in the High-tech Era," in *Computers in the Human Context*, ed. Tom Forester (Cambridge, MA: MIT Press, 1991), 82–132. A roundup of scholarly thinking on the topic is in Frank Webster, *Theories of the Information Society* (New York: Routledge, 1995).

7. Alvin Toffler, *The Third Wave* (New York: William Morrow, 1980).

8. Star Roxanne Hiltz and Murray Turoff, *The Network Nation: Human Communication via Computer* (Reading, MA: Addison-Wesley, 1978).

9. No detailed history or analysis of the early 1990s' push for home networking around cable boxes has yet been published, but an evocative sketch is in Michael Lewis, *The New New Thing* (New York: W. W. Norton, 2000), 70–79, 119–120.

10. An accessible introduction to the economic literature on path dependence is in Douglas Puffert, *Path Dependence* (EH.Net Encyclopedia, edited by Robert Whaples, 2003), available at ⟨http://eh.net/encyclopedia/article/puffert.path.dependence⟩ (accessed September 15, 2006).

11. Westinghouse boasted one of the most advanced such networks. By 1963, its automated hub in Pittsburgh already routed more than fifteen thousand messages a day between 265 corporate locations. See R. C. Cheek, "Establishing the Telecomputer Center," *Journal of Data Management* 2, no. 6 (June 1964): 28–33, and "A System for General Management Control," in *Control through Information: A Report on Management Information Systems (AMA Management Bulletin 24)*, ed. Alex W. Rathe (New York: American Management Association, 1963), 5–10.

12. The CTSS system developed at MIT from 1961 onward appears to have allowed users to send messages to each other, and is reported to have added an internal mailbox system in 1965. Tom Van Vleck, *The History of Electronic Mail* (Multicians.org, September 10, 2004) ⟨http://www.multicians.org/thvv/mail-history.html⟩ (accessed October 29, 2006).

13. Janet Abbate, *Inventing the Internet* (Cambridge, MA: MIT Press, 1999), chapter 3.

14. Jon Postel, *RFC 821: Simple Mail Transfer Protocol* (Network Working Group, August 1982), available at ⟨http://www.faqs.org/rfcs/rfc821.html⟩ (accessed October 22, 2006).

15. Glyn Moody, "E-Mail: Polishing Your Communications," *Guardian*, August 18, 1994, T18. A good overview of Internet e-mail technologies in the early commercial era is in David Wood, *Programming Internet Email* (Sebastapol, CA: O'Reilly, 1999).

16. John Burgess, "Electronic Mail's Delivery Problems" *Washington Post*, April 21, 1989, F1.

17. Ibid.

18. FidoNet has so far received almost no academic or journalistic discussion. Its creator has placed some historical material online at Tom Jennings, *Fido and FidoNet*, available at ⟨http://wps.com/FidoNet/index.html⟩ (accessed September 19, 2006).

19. BITNET is discussed in David Alan Grier and Mary Campbell, "A Social History of Bitnet and Listserv, 1985–1991," *IEEE Annals of the History of Computing* 22, no. 2 (April–June 2000): 32–41.

20. Usenet is discussed in Howard Rheingold, *Virtual Communities: Homesteading on the Electronic Frontier* (Reading, MA: Addison-Wesley, 1993), chapter 4.

21. Kara Swisher, *aol.com: How Steve Case Beat Bill Gates, Nailed the Netheads, and Made Millions in the War for the Web* (New York: Random House, 1998).

22. T. R. Reid, "MCI Mail Lets You Set Up a Post Office Operation on Your Desk-Top," *Washington Post*, December 7, 1987, F41.

23. Office automation is discussed in Thomas Haigh, "Remembering the Office of the Future: The Origins of Word Processing and Office Automation," *IEEE Annals of the History of Computing* 28, no. 4 (October–December 2006): 6–31.

24. The OSI effort, its X.25 and TP/4 protocols, and their relationship to the Internet are discussed in Abbate, *Inventing the Internet*, 147–149. The practices and cultures of OSI and the Internet are contrasted in Andrew L. Russell, "'Rough Consensus and Running Code' and the Internet-OSI Standards War," *IEEE Annals of the History of Computing* 28, no. 3 (July–September 2006): 48–61. As late as 1994, one Internet discussion document still insisted that "currently in the Internet, OSI protocols are being used more and more," and implied that it was no longer "predominantly a TCP/IP network." Susan K. Hares and Cathy J. Wittbrodt, *RFC 1574: Essential Tools for the OSI Internet* (Network Working Group, February 1994), available at ⟨http://www.rfc-archive.org/getrfc .php?rfc=1574⟩ (accessed September 19, 2006).

25. Various, *CompuServe* ⟨———⟩ *Internet Gateway (Thread in comp.sys.ibm.pc)* (Usenet, August 23, 1989), available at ⟨http://groups.google.com/group/comp.sys.ibm.pc/browse_thread/thread/ 4d0ece2be6c1cec0/ebe6db9230b3a7d4⟩.

26. Tom McKibben, *GeoReps Anyone? (Message in comp.os.msdos.pcgeos)* (Usenet, June 8, 1992), available at ⟨http://groups.google.com/group/comp.os.msdos.pcgeos/browse_thread/thread/ ce74d8bf02fc70eb/de27f35a78f675a5⟩.

27. Technical details on X.400 are given in Cemil Betanov, *Introduction to X.400* (Boston: Artech House, 1993). X.400's failure to gain widespread adoption as an e-mail gateway is chronicled in Dorian James Rutter, "From Diversity to Convergence: British Computer Networks and the Internet, 1970–1995" (PhD diss., University of Warwick, 2005), 172–200.

28. This example is taken from the discussion of X.400 addresses in Harald T. Alvestrand, *X.400 Addresses Are Ugly* (June 20, 1996), available at ⟨http://www.alvestrand.no/x400/debate/ addressing.html⟩ (accessed August 29, 2006). A simplified addressing mechanism was adopted in a 1988 revision to the original X.400 standard, but it introduced extra complexity because the system would need to access an X.500 directory and look up the full version of the address. Internet e-mail addresses, in contrast, were simple, but still included all the information necessary to deliver the message as long as the Domain Name System was working.

29. The Internet Society, *Growth of the Internet: Internet Messaging Traffic* (1995), available at ⟨http://www.chebucto.ns.ca/Government/IndustryCanada/tab2.html⟩ (accessed August 29, 2006).

30. Joan O'C. Hamilton, "Like It or Not, You've Got Mail," *Business Week*, October 4, 1999, 178–184.

31. For an early report on Eudora, see Ron Anderson, "Qualcomm's Eudora," *Network Computing*, May 1, 1994, 86. Dorner discusses the origins of the project in "Dorner Speaks: A Conversation with the Creator of Eudora," *Enterprise Times* (Qualcomm, 1994), available at ⟨http://www.ibiblio .org/wpercy/welty/html/dorner.html⟩ (accessed August 25, 2006).

32. Qualcomm, *QUALCOMM Launches Project in Collaboration with Mozilla Foundation to Develop Open Source Version of Eudora Email Program* (Eudora.com, October 11, 2006), available at ⟨http:// www.eudora.com/press/2006/eudora-mozilla_final_10.11.06.html⟩ (accessed October 20, 2006).

While work on the existing Eudora program has ceased, the Eudora name is to be attached to a new, free program based on a customized version of the open-source Thunderbird code managed by the Mozilla Foundation. Dorner is still involved in the project.

33. At one point, a Microsoft manager publicly announced that development work had ceased on Outlook Express because Microsoft preferred to steer customers toward its paid-for Outlook package. Angus Kidman, *Microsoft Kills off Outlook Express* (ZDNet, August 13, 2003), available at ⟨http://www.zdnet.com.au/0,139023166,120277192,00.htm⟩ (accessed September 20, 2006). This seemed to accurately reflect its actual policy from 2000 until 2005, though the announcement was later retracted. Angus Kidman, *Outlook Express Gets Last Minute Reprieve* (ZDNet Australia, August 15, 2003), available at ⟨http://www.zdnet.com.au/0,130061791,120277332,00.htm⟩ (accessed September 20, 2006).

34. John Clyman, *For a Better Internet Experience* (PC Magazine, April 16, 2006), available at ⟨http://www.pcmag.com/article2/0,1759,1950279,00.asp⟩ (accessed September 24, 2006). Microsoft has also announced Windows Live Desktop, another e-mail client program, designed to work with Microsoft's own Windows Live Web mail service (formerly Hotmail) and other e-mail services.

35. The Radicati Group, *Microsoft Exchange Market Share Statistics, 2006* (March 2006), available at ⟨http://www.radicati.com/brochure.asp?id=284⟩ (accessed October 29, 2006).

36. Tom Austin, David W. Cearley, and Matthew W. Cain, *Microsoft E-mail Momentum Growing at IBM's Expense* (Stamford, CT: Gartner, 2006).

37. Po Bronson, "HotMale," *Wired* 6, no. 12 (December 1998): 166–174.

38. Sharael Feist, *Microsoft to Charge for E-mail Retrieval* (News.com, June 5, 2002), available at ⟨http://news.com.com/2102-1023_3-933024.html?tag=st.util.print⟩ (accessed September 28, 2006).

39. Figures on usage are taken from Saul Hansell, "In the Race with Google, It's Consistency vs. the 'Wow' Factor," *New York Times*, July 24, 2006.

40. John Markoff, "Google to Offer Instant Messaging and Voice Communications on Web," *New York Times*, August 24, 2005, C4.

41. Hansell, "In the Race with Google." These figures represent estimates of the "audience size" during June 2006 for the different services rather than the numbers of registered users. ComScore Medix Metrix figures on global Web mail usage show the services ranked in the same order, but with much higher numbers of users (more than 250 million for Yahoo), according to Ina Fried, *Hotmail's New Address* (News.com, April 26, 2006), available at ⟨http://news.com.com/2009-1038_3-6064507.html⟩ (accessed September 28, 2006).

42. Brad Templeton, *Reaction to the DEC Spam of 1978*, available at ⟨http://www.templetons.com/brad/spamreact.html⟩ (September 20, 2006).

43. Glyn Moody, "Spam, Spam, Spam," *Computer Weekly*, October 3, 2002, 34.

44. Graham Chapman et al., *The Complete Monty Python's Flying Circus: All the Words, Volume 2* (New York: Pantheon, 1989), 27–29.

45. Mary Wisniewski, "Banks Work on Hooking 'Phishers': By Yearend: More Proof of Who You Are Online," *Chicago Sun-Times*, May 29, 2006, 49.

46. Anonymous, "Winning the War of Spam," *Economist*, August 20, 2005.

47. Nucleus Research, *Spam: The Serial ROI Killer* (Nucleus Research, 2004), available at ⟨http://www.nucleusresearch.com/research/e50.pdf⟩ (accessed May 20, 2006). While widely quoted, this number does not seem plausible. It assumed that employees received twenty-nine spam e-mails a day, which is plausible, and that each one took thirty seconds to deal with, which is not. Yet it did not include the costs of information technology staff, software, hardware, and network bandwidth devoted to transmitting, storing, and clocking spam.

48. Internet Relay Chat is specified in Jarkko Oikarinen and Darren Reed, *RFC 1459: Internet Relay Chat Protocol* (Network Working Group, 1993), available at ⟨http://www.irchelp.org/irchelp/text/rfc1459.txt⟩ (accessed September 5, 2006).

49. Virginia Baldwin Hick, "Snowbound? Log On and Keep On Talkin,'" *St. Louis Post-Dispatch*, January 20, 1997, 1.

50. Elizabeth Waserman, "Desktop to Desktop in a Flash," *Pittsburgh Post-Gazette*, October 19, 1997, E-3.

51. Jon Swartz, "AOL Buys Net Chat Company," *San Francisco Chronicle*, June 9, 1998, E1.

52. ComScore Networks, *Europe Surpasses North America in Instant Messenger Users, ComScore Study Reveals* (2006). www.comscore.com/press/release.asp?press=800.

53. For the story of the Web's original design, see Tim Berners-Lee and Mark Fischetti, *Weaving the Web: The Original Design and Ultimate Destiny of the World Wide Web by Its Inventor* (San Francisco: Harper, 1999); James Gillies and Robert Cailliau, *How the Web Was Born: The Story of the World Wide Web* (Oxford: Oxford University Press, 2000).

54. In addition to the NeXT browser, a student intern working with Berners-Lee at CERN created a simple "line-mode" text-only browser that could be used from anywhere on the Internet via telnet to CERN. Gillies and Cailliau, *How the Web Was Born*, 203–205. This allowed people to experiment with the Web, but it had many limitations, including a lack of support for different fonts and the need to type numbers in order to activate hyperlinks. Its use is described in Ed Krol, *The Whole Internet User's Guide and Catalog* (Sebastapol, CA: O'Reilly and Associates, 1992), 227–229.

55. Ken Siegmann, "Canon Drops Deal to Buy Next Inc.'s Hardware Unit," *San Francisco Chronicle*, April 3, 1993, D1.

56. Gillies and Cailliau, *How the Web Was Born*, 200.

57. Ibid., 234.

58. Philip L. Frana, "Before the Web There Was Gopher," *IEEE Annals of the History of Computing* 26, no. 1 (January–March 2004): 20–41.

59. Krol, *The Whole Internet User's Guide and Catalog*, 232.

60. Viola is discussed in Gillies and Cailliau, *How the Web Was Born*, 213–217.

61. James Pitkow and Mimi Recker, *Results from the First World-Wide Web User Survey* (1994), available at ⟨http://www.static.cc.gatech.edu/gvu/user_surveys/survey-01-1994/survey-paper.html⟩ (accessed June 20, 2006).

62. Robert H. Reid, *Architects of the Web: 1,000 Days That Built the Future of Business* (New York: John Wiley and Sons, 1997), 38.

63. Gary Wolfe, "The (Second Phase of the) Revolution Has Begun," *Wired*, October 1994.

64. Clive Parker, "The Reckoning," *Net: The Internet Magazine*, September 1995, 76–81.

65. Although Spyglass conspicuously failed to dominate the Web, it still proved an exciting investment for the university. The firm remade itself as a provider of embedded Internet software for cell phones and other electronic devices, went public, and was acquired at the height of the Internet boom by cable television box firm OpenTV, in exchange for stock briefly valued at more than two billion dollars. Claudia H. Deutsch, "OpenTV, a Software Provider, to Buy Spyglass for $2.5 Billion," *New York Times*, March 27, 2000, C2.

66. Michael A. Cusumano and David B. Yoffie, *Competing on Internet Time* (New York: Free Press, 1998), 11, includes a composite chart of Internet browser market share over time.

67. For an example of the glowing press given to Andreessen, see Elizabeth Corcoran, "Software's Surf King; Marc Andreessen, 24, Is Making Waves on the Internet's World Wide Web," *Washington Post*, July 23, 1995, H01.

68. Wolfe, "The (Second Phase of the) Revolution Has Begun."

69. Krol, *The Whole Internet User's Guide and Catalog*, 336.

70. Nancy Garman, "A New Online World," *Online* 19, no. 2 (March–April 1995): 6–7.

71. Krol, *The Whole Internet User's Guide and Catalog*, 335.

72. Reid, *Architects of the Web*, 44.

73. AOL distributed hundreds of millions, probably billions, of packages holding its software during the 1990s. As well as being mailed out and glued to the front of computer magazines, according to a history of the company, the software was "handed out by flight attendants on American Airlines, packaged with flash-frozen Omaha Steaks," and "dropped onto stadium seats at football games." Swisher, *aol.com*, 175. AOL's willingness to spend money to attract new subscribers threatened to ruin the company—by 1996, it was spending $270 to acquire each new customer. Amy Barrett and Paul Eng, "AOL Downloads a New Growth Plan," *Business Week*, October 16, 1996, 85. The disks themselves are now prized by collectors looking to acquire all of the thousands of variations used.

74. AOL, *Who We Are: History* (2006), available at ⟨http://www.corp.aol.com/whoweare/history.shtml⟩ (accessed September 30, 2006).

75. Netcraft, *July 2006 Web Server Survey*, available at ⟨http://news.netcraft.com/archives/web_server_survey.html⟩.

76. Jon Postel and Joyce K. Reynolds, *RFC 854: Telnet Protocol Specification* (Network Working Group, May 1983), available at ⟨www.faqs.org/rfcs/rfc854.html⟩ (accessed October 31, 2006).

77. PHP Group, *Usage Stats for June 2006*, available at ⟨http://www.php.net/usage.php⟩ (accessed June 29, 2006).

78. Windows 95 components replaced the separate dialer software and TCP/IP stack (usually called winsock, for Windows Sockets) needed by Windows 3.1 users. These components were hard to configure without expert knowledge. Windows 95 also included telnet, FTP, and Internet e-mail clients.

79. David Bank, *Breaking Windows: How Bill Gates Fumbled the Future of Microsoft* (New York: Free Press, 2001), 59–69. The figure of four employees working on the first version of Internet Explorer comes from Kathy Rebello, "Inside Microsoft," *Business Week*, July 15, 1996, 56.

80. Kathy Rebello, "Microsoft's Online Timing May Be Off," *Business Week*, no. 3433, July 17, 1995, 41.

81. Bill Gates, "The Internet Tidal Wave" (U.S. Department of Justice, May 26, 1995), available at ⟨http://www.usdoj.gov/atr/cases/exhibits/20.pdf⟩ (accessed May 20, 2006).

82. Bill Gates, Nathan Myhrvold, and Peter Rinearson, *The Road Ahead* (New York: Viking, 1995). The book appeared at the end of 1995, more than a year behind schedule.

83. The founder of Vermeer tells its story in Charles H. Ferguson, *High Stakes, No Prisoners: A Winner's Tale of Greed and Glory in the Internet Wars* (New York: Times Business, 1999).

84. Gates, "The Internet Tidal Wave."

85. Bob Metcalfe, "Without Case of Vapors, Netscape's Tools Will Give Blackbird Reason to Squawk," *Infoworld*, September 1995, 111. In their book on Netscape, *Competing on Internet Time*, Cusumano and Yoffie write that Andreessen claimed that "the combination of Java and a Netscape browser would effectively kill" Windows, and that Andreessen's famous remark was made "time and again" (40). Netscape's early determination to attack Microsoft directly is discussed in ibid., 114–120.

86. Maggie Urry, "Surfers Catch the Wave of a Rising Tide," *Financial Times*, August 12, 1995.

87. Cusumano and Yoffie, *Competing on Internet Time*, 133–138.

88. Netscape Communications, *Netscape Communications Offers New Network Navigator Free on the Internet* (October 13, 1994), available at ⟨http://news.com.com/2030-132_3-5406484.html⟩ (accessed May 30, 2006); Richard Karpinski, "Netscape Sets Retail Rollout," *Interactive Age* 2, no. 16 (June 5, 1995): 1.

89. Bank, *Breaking Windows*, 68.

90. See, for example, Anonymous, "Best Web Browser," *PC World*, August 1996, 136–139.

91. Jeff Pelline, Alex Lash, and Janet Kornblum, *Sun Suit Says IE Failed Java Test* (News.com, 1997), available at ⟨http://news.com.com/2100-1001_3-203989.html⟩ (accessed September 20, 2006). Microsoft eventually agreed to settle this and a range of other complaints by Sun with a payment of almost two billion dollars and an agreement to work cooperatively. Stephen Shankland, *Sun Settles with Microsoft, Announces Layoffs* (News.com, April 2, 2004), available at ⟨http://news.com .com/2100-1014_3-5183848.html⟩ (accessed September 20, 2006).

92. For examples of head-to-head comparisons, see Mike Bielen, "Battle of the Web Browsers: Part Two," *Chemical Market Reporter*, October 13, 1997, 16; Yael Li-Ron, "Communicator vs. Internet Explorer. And the Winner Is," *PC World*, December 15, 1997, 206–209.

93. Cusumano and Yoffie, *Competing on Internet Time*, 28–34.

94. The deals are reported in Rose Aguilar, *CS, Prodigy Cozy up to MS Too* (News.com, March 13, 1996), available at ⟨http://news.com.com/2100-1023-207369.html⟩ (accessed September 19, 2006). The implications of the AOL deal for Netscape are explored in Cusumano and Yoffie, *Competing on Internet Time*, 111–118.

95. Microsoft's behavior here was eventually judged not to breach antitrust law, on the grounds that Netscape had still been able to distribute million of copies of its browser. Ken Auletta, *World War 3.0: Microsoft and Its Enemies* (New York: Random House, 2001), 200–202, 363.

96. Bank, *Breaking Windows*, 78.

97. Ibid., 88–89.

98. Paul Smith, "Bringing the World Wide Web Right to your Desktop," *Scotsman*, August 12, 1998, 12.

99. Myles White, "'Explorer' Closes Gap," *Toronto Star*, September 4, 1997, J3.

100. A great deal was written at the time on the Microsoft case, as newspapers and magazines followed every twist and turn. A closely observed account is given in Auletta, *World War 3.0*.

101. Bank, *Breaking Windows*, 159. Bank's account is well sourced, and agrees in substance, if not tone, with the Microsoft-authorized version of the Gates-Ballmer transition and division of labor in Robert Slater, *Microsoft Rebooted: How Bill Gates and Steve Ballmer Reinvented Their Company* (New York: Portfolio, 2004), 51–71. Slater writes that a key motivation behind the transition was to remove the polarizing figure of Gates after the antitrust debacle, and that during attempts to settle the case "better a clever, less emotional deal maker such as Ballmer in charge at this crucial time than someone who blindly believed his company has done no wrong" (61). According to plans made public in 2006, Gates will complete his retirement from Microsoft management by 2008, handing over his remaining technical responsibilities to Ray Ozzie.

102. Joel Brinkley, "U.S. Judge Says Microsoft Violated Antitrust Laws with Predatory Behavior," *New York Times*, April 4, 2000, A1.

103. Thomas Penfield Jackson, "Excerpts from the Ruling That Microsoft Violated Antitrust Law," *New York Times*, April 4, 2000, C14.

104. Joel Brinkley, "Microsoft Breakup Is Ordered for Antitrust Law Violations," *New York Times*, June 8, 2000, A1.

105. The Bush administration's proposed settlement is in United States of America, *Revised Proposed Final Judgement* (November 6, 2001), available at ⟨http://www.usdoj.gov/atr/cases/f9400/9495.htm⟩ (accessed May 5, 2006). Reaction to the settlement by those who had been following the trial was generally negative. See, for example, Lawrence Lessig, "It's Still a Safe World for Microsoft," *New York Times*, November 9, 2001, A27. Nine of the eighteen states involved continued to press their case separately, but with little success.

106. Ian Fried and Jim Hu, *Microsoft to Pay AOL $750 Million* (News.com, May 29, 2003), available at ⟨http://news.com.com/2100-1032_3-1011296.html⟩ (accessed August 28, 2006); John Markoff, "Silicon Valley Seeks Peace in Long War against Microsoft," *New York Times*, April 4, 2004, 1.

107. Paul Meller and Steve Lohr, "Regulators Penalize Microsoft in Europe," *New York Times*, July 13, 2006, C1.

108. Paul Festa, *Study: Netscape Share below 50%* (News.com, September 29, 1998), available at ⟨http://news.com.com/2102-1023_3-216043.html?tag=st.util.print⟩ (accessed September 15, 2006).

109. Erich Luening and Janet Kornblum, *Study: Navigator Ahead in Office* (October 15, 1998), available at ⟨http://news.com.com/2100-1023_3-216743.html⟩ (accessed September 16, 2006).

110. The figure of $4.2 billion was widely quoted as Netscape's price at the time of the merger. AOL's stock rose rapidly during the next few months, however, and the stock it exchanged for Netscape was worth $9.6 billion on March 17 when the deal finally closed after approval by the Department of Justice and Netscape shareholders. Shannon Henry, "AOL-Netscape Merger Official; $9.6 Billion Deal Completed after Last Hurdles Are Cleared," *Washington Post*, March 18, 1999, E03.

111. Kim Girard and Tim Clark, *AOL, Sun Detail Software Tie-Up* (News.com, March 30, 1999), available at ⟨http://news.com.com/2100-1023-223687.html⟩ (accessed September 15, 2006).

112. Louise Kehoe and Roger Taylor, "AOL to Power up E-commerce with $4.2bn Buy," *Financial Times*, November 25, 1998, 31.

113. A nice treatment of the AOL Time Warner merger is in Nina Munk, *Fools Rush In: Steve Case, Jerry Levin, and the Unmaking of AOL Time Warner* (New York: HarperCollins, 2004).

114. Browser estimates for the late 1990s are consolidated in Maryann Jones Thompson, *Behind the Numbers: Browser Market Share* (CNN.com, October 8, 1998), available at ⟨http://www.cnn.com/TECH/computing/9810/08/browser.idg/⟩ (accessed September 18, 2006). A survey released in late 2002 gave Netscape and its open-source derivatives just 3.4 percent of the browser market. Matthew Broersma, *Tech Doesn't Buoy Netscape Browser* (News.com, August 28, 2002), available at ⟨http://news.com.com/2100-1023_3-955734.html⟩ (accessed September 19, 2006).

115. Anonymous, "Microsoft to End Mac Browser Versions," *Chicago Sun Times*, June 14, 2003, 30.

116. Paul Festa, *Microsoft's Browser Play* (News.com, June 4, 2003), available at ⟨http://news.com.com/2100-1032_3-1012943.htm⟩ (September 29, 2006).

117. Bill Gates, *Bill Gates' Antitrust Statement* (ZDNet News, December 6, 1998), available at ⟨http://news.zdnet.com/2100-9595_22-513040.html⟩ (accessed September 12, 2006).

118. The Freedom to Innovate Network's claims to be nonpartisan and grassroots are reported in Adam Cohen, "Microsoft Enjoys Monopoly Power," *Time*, November 15, 1999, 60. The group still exists as a hub for Microsoft's lobbying efforts, though it now bills itself as "Microsoft's Freedom to Innovate Network."

119. Byron Acohido, "Firefox Ignites Demand for Alternative Browser," *New York Times*, November 10, 2004, 1B; Walter S. Mossberg, "Security, Cool Features of Firefox Web Browser Beat Microsoft's IE," *Wall Street Journal*, December 30, 2004, B1.

120. Josh McHugh, "The Firefox Explosion," *Wired* 13, no. 2 (February 2005).

121. Ibid.

122. See, for example, Paul Taylor, "The Leading Windows on the Web Receive a Fresh Polish," *Financial Times*, July 7, 2006, 9; Walter S. Mossberg, "Microsoft Upgrades Internet Explorer—But Not Much Is New," *Wall Street Journal*, October 19, 2006, B7.

123. Eric Von Hippel, *Democratizing Innovation* (Cambridge, MA: MIT Press, 2005).

124. Quoted in Fred Vogelstein, "Search and Destroy," *Fortune*, May 2, 2005, 72–82.

5 The Web's Missing Links: Search Engines and Portals

Thomas Haigh

As a dense but disorganized jungle of information, the Web has always relied on automated search engines, human-compiled directory services, and what came to be known as Web portals to steer its users toward the material they seek. Unlike other electronic publishing systems the Web had no central directory or integrated search function, so these services played an integral role in establishing the Web as a useful publishing medium. Search engine firms such as Excite and Lycos were among the first Internet companies to make initial public offerings, fueling the boom for dot-com stocks during the late 1990s. Although the Web has changed greatly since the early 1990s, throughout its evolution the most visited Web sites have been navigation services. Today the world's four most visited Web sites are Yahoo, Microsoft's MSN portal site, Google, and the Chinese search engine Baidu. Search sites and portals command the lion's share of the world's Internet advertising revenue, making them the most successful of all Internet information businesses. Yahoo and Google together receive more advertising dollars than the combined prime-time offerings of the traditional big-three U.S. television networks.[1]

Yet contrary to this brief and happy summary, the development of the Web navigation industry has been anything but straightforward. This chapter tells its tempestuous history, beginning with the special characteristics of the Web and the Internet when compared with earlier electronic publishing systems. The very features of the Web that brought it instant success also created an urgent need for navigation services. I contrast the hypertext approach to electronic publishing, used by the Web, with the information retrieval approach taken by earlier commercial online publishing systems. Early firms took one of two approaches. Web directories, such as Yahoo, exploited the hypertext nature of the Web and the power of human labor to create online guides combining features of traditional Yellow Pages business directories and library catalog files. Search engines—such as Lycos, Excite, and AltaVista—adapted traditional information retrieval techniques to the new and fragmented world of the Web to create huge, automatically generated, searchable indexes of all the text held on Web pages.

By the late 1990s both groups of firms had evolved into Web portals, competing with each other as well as new competitors such as AOL, Microsoft, Netscape, and Disney to create one-stop Web sites stuffed with so many different kinds of attractions (among them news, weather, e-mail, music, and shopping) that users would browse there for hours at a time. The rush to create these "full-service" portals, their apparent success in the late 1990s, and their subsequent collapse can be explained only by reference to the exceptional economic conditions that warped normal business logic during the dot-com boom. Finally, I turn to the success of Google, which rose to dominate the Internet search field by focusing on excellence in search just as its established rivals began to downplay search in their bid to become portals. Google's focus on improving the user's search experience, and refusal to emulate the crass and intrusive advertising practices of its rivals, has paradoxically made it the most successful seller of online advertising in the history of the Internet.

The Web as an Electronic Publishing System

Within five years of its 1991 introduction the World Wide Web was already the biggest, most heavily invested in, most publicized, most widely used, and most geographically distributed online publishing system in world history. It was also the first major online publishing system to be built with no centralized index or built-in method to search its contents for items of interest. These two accomplishments are not unrelated. The appealing characteristics of the early Web were achieved only because the thorny issue of search was initially ignored: its simplicity, its flexibility, and the ease with which new Web sites could be created.

The Web was by no means the first online publishing system. In fact, by the time it made its debut the industry was around twenty years old. The foundations for these online systems were laid in the 1950s, when "information retrieval" first surfaced as a field of study among academic and corporate researchers. Information retrieval combined the use of new technologies (including punched cards, microfilm, and specialist electronic devices as well as early computers) with information theory and the application of techniques of classifying and structuring information drawn from library science.[2] During the 1950s and 1960s, information retrieval was a central concern of the nascent community of "information scientists," who were concerned particularly with managing academic journal articles and other scientific information they believed was increasing exponentially in an "information explosion."[3] As the name information retrieval suggests, specialists in this believed that *retrieving* information was the key challenge, working on methods of organizing, selecting, and displaying information to maximize the effectiveness of this process. As the field developed, its focus was on methods of selecting the relevant results from large collections of electronic records, such as tagging documents with metadata (keywords, date, author, and so on), indexing them, abstracting them, finding the most effective search algorithms, and analyzing users' searching patterns.

Companies first began to use computers to provide online searching and retrieval from textual databases during the 1960s. In 1969 the *New York Times* announced its Information Bank service, a publicly accessible online database including abstracts from many other magazines, newspapers, and magazines, although the system only became fully operational in 1973.[4] By the start of the 1970s, Lexis, one of the first successful online information services, was created to provide access to legal text databases. Dialog, one of the first online search packages, was developed by Lockheed in the mid-1960s, with NASA as its first customer.[5] From the early 1970s onward, Lockheed used Dialog as the basis for a publicly accessible online service.[6] Now owned by specialist publishing giant Thomson, Dialog is still a hugely successful service providing access to the databases of journals, newspapers, magazines, analyst notes, patents, government regulations, and other sources.

But Dialog, and its competitors such as LexisNexis, evolved quite separately from the Internet and have remained distinct from the Web even though most subscribers today use a Web browser to access them. They are sometimes called the "deep Web," and provide their subscribers with a much more orderly world of professional and technical information inaccessible through Web search engines. Unlike the Web, online information services are centralized, with all documents indexed and held in a central repository. Because documents are tagged with metadata, users can choose to search only for results written during a particular month, published in a certain source, or written by a specific reporter.

The Web took a fundamentally different approach to publishing and finding electronic documents. Previous electronic publishing systems involved a number of users logged into a centralized server, on which documents were stored and indexed. The Web had no central server, and thus no central directory or index. The peer-to-peer structure of the Internet meant that any computer on the network could publish Web pages. If one already had access to a computer with a direct and reasonably fast Internet connection, which in the early 1990s was a common occurrence at universities and in computing research labs and a rather uncommon one elsewhere, then all one needed to do was install the Web server program and create a directory holding a Web page or two. There were no forms to fill out, no licenses to apply for, no royalties to pay, and no special fees to negotiate. The Web made online publishing almost ludicrously easy when compared with the enormous amount of work involved in setting up a traditional online information retrieval service or arranging to publish material through a commercial online service such as AOL or CompuServe.

The very ease with which a Web page could be published created a new issue: how would anyone ever find it? If one had the address of a Web page, one could jump directly to it. But Tim Berners-Lee expected users to move effortlessly around the Web as they followed links from one document to another. The connections between Berners-Lee's design for the World Wide Web and earlier work on hypertext is well-known. By 1991, hypertext had already been incorporated into some widely used computer systems. Since 1987, the Hypercard system bundled with each Macintosh system had allowed people with no programming knowledge to create complex "stacks" incorporating sounds and animations as well as text and links. In today's terms, Hypercard was halfway between a Web browser and PowerPoint. Hypertext was also an integral part of the standard help system used by Microsoft Windows itself and Windows application programs.

Hypertext represented a fundamentally different paradigm for electronic publishing from traditional information retrieval. Information retrieval systems searched through a huge body of independent documents to find the best matches for a users' query. Hypertext allowed readers to browse from one page to another, following cross-references, jumping around tables of contents, and meandering into footnotes and away along digressions into other works. Long before the Web existed, hypertext pioneer Ted Nelson was popularizing the idea of a worldwide hypertext network knitting together independent documents within what he called the "docuverse."[7] Even Vannevar Bush, whose 1945 speculative essay "As We May Think" is often claimed to have invented the hyperlink, imagined a miniaturized library in which users could create and share their own "trails" linking material found in many different books and papers.[8]

The idea of a hypertext system or full-text database spanning material held on multiple, independently administered servers was nothing new. Indeed, hypertext specialists initially felt that Berners-Lee had done nothing to advance their research, rejecting his paper describing the Web when he submitted it for consideration at the prestigious Hypertext'91 conference.[9] But while the concept of a worldwide web of hypertext was nothing new, the reality most certainly was. The actual hypertext systems of the late 1980s were limited in scope. Users could browse about within a single hypertext document, such as a book or a reference manual, but there was no obvious way to create a link from one author's work to another's. Establishing a distributed, public hypertext network posed some challenging problems. How could a central database be established so that all links to a particular document would automatically be preserved when its physical location on the network shifted or the document itself was revised? How could every document be seen that had been linked to the document one was currently reading? Nelson himself was never able to fully answer these riddles in practice despite decades of work toward a system he called Xanadu, and these questions also preoccupied an established community of hypertext researchers during the late 1980s and the early 1990s.[10]

Berners-Lee, who was a practicing programmer rather than an academic researcher, dealt with these thorny, fundamental problems by ignoring them. His concerns were more practical: he had to struggle to win permission from his managers at CERN to spend time on his pet project and was obliged to justify the early Web as a helpful document-sharing system for use within the lab rather than as a long-term research project to create fundamental advances in hypertext.

The Web offered neither an information retrieval search capability nor the full richness of Nelson's original conception of hypertext. Published material was often deleted or changed, meaning that links pointed to missing or irrelevant pages. Links between documents could be followed forward but not backward. Nelson was outraged: "The World Wide Web was not what we were working toward, it was what we were trying to *prevent*," he wrote a few years later, complaining that it had "displaced our principled model with something far more raw, chaotic and short-sighted."[11] Famous computing researcher Alan Kay likened the Web's hypertext markup language (HTML) to Microsoft's crude but ubiquitous MS-DOS operating system, which in computer science circles is one of the nastiest things you can say about a system.[12]

But the Web's spectacular success during 1993 and 1994 shows that a distributed hypertext system did not actually have to deal with these underlying problems to find an enthusiastic group of users. The Web grew fast, and the ambitions of its early users grew with it, until they began to write of it as a new Alexandrian library, tying together the sum of human knowledge. The problem was that the Web was a massive and ever-growing library without reference librarians, quality control, or a card catalog. Any foray on to the Web was likely to bombard the explorer with trivia on any number of esoteric topics. But finding the answer to a specific question could be hard indeed. As one well-known science writer reported in mid-1995, "I have briefly signed up with a number of Internet providers, only to become exasperated by the maddening randomness of the Net.... [T]he complicated searches... feel like a waste of time."[13]

As the commercialization of the Web began in earnest in 1994, it was already obvious that creating fast and easy ways of searching or navigating its contents was a necessary precondition for the success of Web-based businesses. One early user reported that "the Web is anarchy right now, and the librarians among us will quickly learn there is no easy way to search the unruly masses of home pages and Internet information." Yet as she correctly suggested, "That's the next step and it will happen quickly, just as rudimentary tools for searching Internet gophers have developed."[14]

The problem was particularly pressing for online businesses. Putting up a Web site was easy. But how would a customer ever find it? A host of new Web sites and businesses sprang up to address the problem. Many of them had off-line parallels. In the physical world, for example, indexing and searching happens manually. A great deal of work goes into gathering information to create telephone books, encyclopedias, and Yellow Pages business directories. Businesses advertise their existence in specialist publications and trade directories. But the need for such mechanisms was much greater on the Internet. There was no such thing as foot traffic, and the Web was so new that its users lacked the purchasing habits, established routines, or brand recognition that most businesses rely on. Yet the payoff has also been huge. Web searching is, in retrospect, one of the few truly great business opportunities created by the Internet: a pure information business to supply a missing but vital function of the Web itself.

Clearly, some new kind of company was going to get rich bringing Web sites and their potential visitors together to the benefit of both. But because the Web was so new there was no established way of doing this. In 1994 the commercial viability of Web publishing, Web searching, and Web indexing was an article of faith rather than an observed fact. While it was certainly possible to spend a lot of money to create a Web directory or search engine, or to fill a Web site with witty and entertaining writing, nobody knew what kinds of things ordinary people would use the Internet for, what they might be willing to pay for, and what new kinds of online publishing businesses or genres would prove viable. Publishing genres such as newspapers, trade magazines, and directories had evolved over many decades to create well-defined business models. In the virtual world of the Internet, a catalog could also be a store or a trade publication could run an auction. A newspaper could be sold one article at a time, and an encyclopedia could be supported by advertising revenue.

One might expect this to give rise to a period of cautious experimentation on a limited scale. Instead a frenzy ensued, as corporate giants and well-funded start-up companies rushed to colonize every conceivable niche within the emerging ecosystem of Internet business. This was era of day-trading, stock splits, investment clubs, financial news channels in bars, companies hoping to make money by giving away computers, and the best-selling book *Dow 30,000*. Much has been written about the follies of the dot-com era, so there is no need here to recount its sorry history in full.[15] But to understand the story that follows a few aspects of this era must be remembered.

First, politicians, corporate leaders, and investment advisers all endorsed the idea of the Internet as a technological force destined to sweep away every aspect of the existing business world. Any idea or business, no matter how silly or staid, could instantly legitimate itself through association with the Internet.

Second, although the huge importance of the Internet was universally recognized, the precise means by which it was to translate into great business success was not. Fads swept the world of Internet business every few months, each one bringing a new rash of start-up companies and promoting existing businesses to make wrenching strategic shifts. As conventional wisdom changed again and again the early search companies grappled with advertising, rushed to provide original content, explored "push" technologies, tried to build personalization capabilities, and started to call themselves portals.[16]

Third, from 1994 to 2000 the ordinary laws of nature governing management and corporate finance were suspended. Excite, Lycos, and other search firms were generously funded by venture capitalists, went public, saw their stock prices soar, and thus were able to issue more stock, hire thousands of employees, make expensive acquisitions, and branch out into many new areas of business. Through all of this, their losses mounted ceaselessly. Indeed, the rate at which they lost money only increased as their businesses grew. But investors believed that growth, market share, and the number of users were more important gauges of an Internet company's worth than whether it made a profit. In the final years of the boom, this idea was cynically promoted by a powerful alliance of brokerage houses, investment banks, and venture capitalists. Because Internet-related firms could make initial public offerings of their stock without having a clear prospect of profitability or more than a token number of customers, venture capitalists funded start-up companies secure in the knowledge that a bank could reap huge fees by taking them public long before their viability had been established.[17] Dot-com firms may have been no more likely to fail than the average business, but they were able to fail much more expensively and spectacularly. When the bubble finally burst in 2000, all the major search and portal firms faced an abrupt and usually fatal encounter with reality. Search firms were a crucial part of the dot-com business world, and despite the ultimate success of Yahoo and Google, the story of their industry cannot be separated from the broader bubble.

Web Directories

Because it was made of hypertext the Web could, with a little work, become its own index. As in a paper book, there is nothing magic about an index page; it is just a page full of references to other pages. Anyone who wanted to could set up a directory page full of links to other Web sites addressing a particular topic. For the first few years of the Web, this was not particularly challenging—when CERN stopped updating its first master list of Web servers in late 1992, it held just twenty-six links.[18] In the early days of the Web, thousands of visitors were drawn daily to now-forgotten directory sites such as EINet Galaxy, GNN, and CERN's own World Wide Web Virtual Library Subject Catalog, founded by Berners-Lee himself in 1991.

Some early directory sites depicted the Web visually. Flaunting the geographic reach of the Web, each known Web server in a country or region was shown as a labeled point on the map. Clicking on the point opened up the Web site in question. But the rapid spread of the Web made it impossible to track the location of every server, still less squeeze their names on to a map. According to an MIT survey, there were 130 Web servers on the Internet in mid-1993, 623 by the end of the year, and more than 10,000 by the end of 1994.[19] Even the most determined part-time indexer would probably have succumbed when the number of Web servers doubled in the next six months, reaching 23,5000 in June 1995. The server came online around March 1997, and the 10 millionth in early 2000.

As Web directories struggled to deal with the proliferation of sites, they adopted a hierarchical form, adding new levels whenever individual pages threatened to become unmanageably large. For example, a page of links to music Web sites might be replaced with multiple pages dealing with different musical genres, which in turn might eventually consist of little more than a set of links to pages cataloging Web sites dealing with particular bands. This made it possible to organize a list of hundreds of thousands of Web sites, but creating and maintaining such a list required a large and well-organized team of indexers.

Running a popular index could also be expensive because it consumed a lot of server capacity and network bandwidth. By far the most successful of the general-purpose Web directories, and the first to switch to a commercial mode of operation, was Yahoo. Yahoo began in early 1994 as "Jerry and David's Guide to the World Wide Web," a typical amateur directory, created by two electrical engineering students at Stanford University. The service quickly won a following, but keeping it up-to-date proved ever more time-consuming. In early 1995 its founders, David Filo and Jerry Yang, incorporated the business, and following a path well beaten by Stanford spin-off companies before them, won start-up funds from the venture capital companies lining Sand Hill Road on the edge of campus.[20]

Yahoo benefited from a great deal of free publicity. During 1994, publications such as *Business Week*, *Time*, *Newsweek*, and hundreds of more specialized titles began to run stories about the exciting new World Wide Web. Over the next few years untold tutorials were published telling readers what the Web was, how to access it, and what to do with it. Web directories were the obvious place to send new Web surfers. These tutorials often included the addresses of Yahoo and other directory sites. For example, Bill Gates's notorious 1995 "Internet Tidal Wave" memo alerting Microsoft executives to the strategic power of the Internet included a link to Yahoo! under heading "Cool, Cool, Cool."[21] Yahoo was also linked to home pages created by universities, Internet service providers (ISPs), and other institutions.

Yahoo displayed its first advertisement in August 1994, which at least proved that a source of revenue existed to support the new business. It began to add other kinds of information such as news, weather, and stock quotes, and affixed its name to a print magazine, *Yahoo Internet Life*. In April 1996, Yahoo made an initial public offering of stock, providing a rapid payback for its first investors and cementing its position as a leading Internet firm. Like Netscape the previous year, its shares more than doubled on the first day of trading. At this point Yahoo employed fifty full-time people to surf the Web, evaluating Web sites for inclusion in the directory.

The ever-increasing size of the Web made it ever more expensive for competitors to enter the directory business, because of the amount of human effort necessary to match Yahoo's success. This was an expensive business to do well. Yahoo's biggest early competition in the Web directory business came from McKinley with its Magellan directory, but the start-up faltered after running out of cash before it was able to sell its stock to the public.[22] LookSmart, launched in 1996 by *Reader's Digest*, survived but never enjoyed a huge success. Some search engines and portal sites continued to offer their own directories, but Yahoo dominated the market. Likewise, the human-generated directory listings remained at the heart of Yahoo's business throughout the 1990s.

The commercial importance of Web directories has dwindled steadily since the mid-1990s. In recent years, Yahoo's only main competition in the Web directory field came from the Open Directory Project. This was founded in 1998 to compete with Yahoo using volunteer labor to produce a directory.[23] In 1999 the effort was acquired by Netscape, and at some point in 2000 the size of its Web directory (around 1.6 million entries) was estimated to have overtaken Yahoo's.[24] While the project itself is not a household name, its directory listings were made freely available, and have been used as the basis of Web directories offered by many popular and once-popular Web navigation services, including AltaVista, Netscape, and Lycos. By incorporating the Open Directory Project's results, these firms could offer a similar capability to Yahoo's directory without having to spend a fortune to create and maintain it.

Web Search

One way to find something on the Web was to use a Web directory and click down through the subject headings to find a selection of Web sites dealing with the topic of interest. The other way, of course, was to search for Web pages that contained a particular word or phrase. Yahoo offered a search facility, but until October 2002 this defaulted to searching the keywords and headings in the Yahoo directory, rather than the contents of the Web sites themselves.

The advantage of this approach was that the necessary indexing could be done automatically by a computer, eliminating the need to hire huge teams of Web surfers and allowing more thorough coverage of the Web. The disadvantage was that creating an automatic index of this kind was a much knottier programming challenge, and required an exceptionally fast network connection and a powerful collection of servers.

In the early days of the Web these two approaches were expected to coexist. When thousands of Web pages contained a popular phrase it was unrealistic to expect a user to visit more than the first few dozen in search of enlightenment. While the human surfing of the Web for Yahoo could exercise a measure of quality control and editorial judgment, it was hard for automated systems to make an informed judgment as to which of the many Web pages about a popular subject, such as Britney Spears, deserved to be highlighted. So search engines appeared to have an edge in looking for every instance of an obscure phrase, whereas Web directories were better suited to recommending the best material on popular subjects.

The very ease with which material could be published on the Web meant that indexing and searching it posed a huge challenge. Searching the Web was actually a harder problem than designing the Web in the first place. Neither Berners-Lee nor any of his handful of volunteer collaborators and student interns at CERN were specialist hypertext or database researchers, but they managed to design the Web as well as create the first servers and browser programs in a matter of months. Neither did they have access to powerful hardware: the only special equipment used for the project were two NeXT desktop computers, whose purchase was authorized for other reasons.[25] By necessity, the Web was designed around the existing infrastructure and capabilities of the Internet, and so inherited both the strengths and weaknesses of the underlying network. Search, on the other hand, went against the grain of the disorganized, decentralized, and unregulated Internet. Creating a search engine needed specialist technical expertise and esoteric algorithms, large powerful servers, fast network connections, and teams of people ready to constantly tweak its hardware and software configuration.

The concept of an Internet search engine was already established before the creation of the first Web search engines. File transfer between computers was one of the original applications of the Internet, inherited from the early ARPANET.[26] All it took to share files with others was to run a piece of software called a file transfer protocol (FTP) server, allowing any computer on the Internet to connect, procure a list of available files, and download any of interest.[27] The facility was used in many ways during the 1980s and the early 1990s: by computer science departments to publish technical reports electronically, by public domain software archives to disseminate programs, and by the administrators of the Internet itself to share new technical documents.

But the ease of setting up an FTP server created a new problem. Because it required nothing more than downloading and installing a piece of software, there was no central index of files shared. Someone had to know the Internet address of the server they were looking for and the path to the directory holding the file they needed (perhaps by reading it in a newsletter or via an online newsgroup). FTP servers hosting large or popular files consumed large amounts of network bandwidth, forcing the universities and research institutes hosting them to limit the number of simultaneous downloads allowed. This encouraged others to share the burden by establishing "mirror sites." But spreading multiple copies of files around the network just made things worse if nobody knew where they could be found. It was to address this problem that a team of McGill University students introduced the popular utility Archie in 1990. Users could enter all or part of the name of the file they were interested in acquiring, and Archie would scour the Net in search of it. It relied on a "spider" program to automatically query popular servers, and then create a centralized, searchable database of all known files and locations. Archie did not search the contents of text files, however, merely the names of files and the directories in which they were held.[28]

In 1992 the success of Gopher inspired the creation of Veronica, a searchable database of the contents of thousands of Gopher servers.[29] Veronica searched Gopher index headings, but did not access text within the files that these headings pointed to. During the same era, Wide Area Information Servers (WAIS) gave users a simple and consistent user interface with which to search the actual contents of text files held on servers all over the Internet. WAIS was promoted by supercomputer company Thinking Machines, which advertised its own computers as WAIS. Yet WAIS was limited compared with later, Web-oriented search systems. Rather than hosting a single massive central index, it relied on the people and organizations publishing text documents on the Internet taking the trouble to create and share their own indexes covering this material. Searchers then had to specify which of the hundreds of indexes they would like their search to include. This built on a long-discussed but seldom-implemented concept known as the "federated database." The search capabilities of the widely used public domain versions were crude, always providing all the documents including a particular set of words regardless of whether they occurred as a phrase or were scattered around different sections of the text.[30]

The sudden proliferation of Web sites from 1993 onward posed essentially the same problem, but on a bigger scale and with a more difficult target. The challenge was to find and index these pockets of information scattered randomly across the Internet. The basic technology behind automated Web indexing systems is known as a "Web crawler" or spider. In the absence of any official directory of Web servers, the Web crawler starts with a small set of Web pages and analyzes them to extract the addresses of all the other Web sites they link to. The program thus crawls from one site to another, discovering new destinations as it goes. Each page discovered by the crawler is saved to a special database, and the words occurring on it are indexed. Then when a user submits a query to the search engine, the Web crawler checks the index to find pages in which all the specified terms occur. Techniques to search large bodies of text were already well developed prior to the creation of the Web for online technical and legal databases. The Web crawler, however, was a new invention.

The first Web crawlers appeared in 1993, as the new Mosaic browser inspired the creation of thousands of Web sites. Washington University's WebCrawler was the best known of the early crawlers, and the first to allow users to search the full contents of the pages it indexed. Over the next few years, dozens of automated Web search services were launched. Running a successful service posed two main challenges: mustering sufficiently powerful hardware and efficient software to crawl a reasonable proportion of the Web on a regular basis, and creating a search system powerful enough to hunt through the resulting database of millions of saved pages in a reasonable time to present the most relevant results.

The most successful search engines built on existing expertise in these areas. AltaVista, launched in December 1995, was created by DEC and promoted as a tool to demonstrate the power of servers based on its new Alpha processor architecture.[31] AltaVista won an enthusiastic following by delivering rapid searches of a comprehensive database via a powerful but uncluttered user interface, which included some of the advanced search features common in commercial information retrieval systems.

The Excite search engine emerged from Architext, a firm created by several Stanford University students with an interest in automatic textual analysis. After receiving funding from Kleiner Perkins Caulfield and Byers, one of Silicon Valley's leading venture capital firms, it quickly became one of the leading Web search sites.[32] Lycos, another popular search engine, was developed from a Carnegie Mellon research project and commercialized after being sold to a venture capital fund based (for a change) in Delaware.[33] Lycos boasted the biggest of the early Web databases, with a claimed 11.5 million pages indexed by November 1995.[34] Both Excite and Lycos supplemented their automatically generated search results with Web site reviews. Colorful banners signifying inclusion in Lycos's "top 5 percent" directory were proudly displayed on many early Web sites. Inktomi, a 1996 start-up, was based on technology developed at the University of California at Berkeley. Inktomi's technology was licensed to other sites and powered the popular HotBot search service offered by Web publisher HotWired. Ask Jeeves joined the competition late, in April 1997. Its gimmick was a friendly butler displayed on the home page to whom users were encouraged to type questions in full English sentences rather than brusque keywords.

Web Advertising

None of the major Web directories or search engines had any income during their first months in operation. This was perfectly normal for early Internet businesses, but obviously could not go on forever. Once an online business had a functional Web site and a stream of visitors, its next challenge was to make some money.

Nelson's concept for the Xanadu global hypertext network involved the use of so-called micropayments, where the creator of a document would automatically be credited with a tiny payment every time somebody read it.[35] Something similar had been realized in traditional online services such as AOL, where it had been easy to charge customers for online transactions: additional fees were just added to the user's monthly statement. These were split between the online service and the creator of the content or service—such as the popular Motley Fool personal finance guide. This was similar to the arrangements used to charge users of premium telephone services. Only a small number of companies published information on AOL, but one could imagine a similar mechanism where small businesses and individuals could publish documents and be paid for them.

The economics of providing information on the Internet were different and rather discouraging. The distinctive technological choices made during the design of the Internet and the Web had profound implications for the commercial development of the Internet. The early Web was primarily an electronic publishing system, used to browse static pages. Because the Internet had not been designed for commercial use, there was no way for a Web site or network owner to collect money from an ISP in return for services consumed by the user. That seemed to pose an economic problem, since Web sites could not easily charge their readers for access. The economics of offline publishing were fairly straightforward: publishers made money on each book, musical score, or record sold. Selling more copies meant more money, and one hit could underwrite the cost of several flops. In contrast, a popular Web site would run up huge bills for network bandwidth and servers, without receiving any income to cover this expense. Grabbing more readers meant bigger losses, not bigger profits.

During 1994 and 1995, many people believed that the missing tools to process micropayments could easily be retrofitted to the Web to provide a sound foundation for commercial Web publishing. Numerous companies were launched to create "electronic cash" systems for online use, using advanced encryption techniques to create secure and efficient billing systems to support very small transactions. In reality they were not widely adopted by Web sites or users, and were never integrated into Web browsers. By the end of the 1990s, a host of well-funded, high-profile micropayment start-up firms such as FirstVirtual, CyberCash, and Pay2See had thrown hundreds of millions of dollars of investor's money at the problem with no success. The problem of simultaneously signing up a critical mass of users, enhancing browser and server software to make payments easy, and enlisting Web publishers proved intractable.[36]

The one major commercial success among Internet payment firms was Silicon Valley start-up PayPal, founded in 1998 and acquired by eBay for around $1.5 billion in 2002. PayPal's eventual business model was a hybrid between a credit card processor and an unlicensed online bank rather than a true micropayment company, but it found a profitable niche among small-time online auction sellers for whom the monthly fees demanded by banks to provide credit card processing services would be uneconomic.[37]

Instead, the economics of Web publishing developed around advertising. Rather than harvesting micropayments from users, it was easier to leave browser technology untouched, and then deploy new "ad server" programs on Web sites to incorporate banner and pop-up advertisements into each Web page delivered. The publisher was still receiving a tiny sum for each page displayed, but the sum was coming from an advertiser rather than the reader. Clicking on the advertisement took viewers straight to the advertiser's own Web site—a crucial advantage over traditional advertising media. HotWired, one of the first commercial Web publishers, pioneered the large-scale sale of Web advertising in 1994.[38] The largest Web publishing firms employed their own advertising sales teams, on the model of traditional magazine and newspaper publishers (both of which relied primarily on advertising to cover their costs).

The shift to advertising favored large, commercial Web sites able to attract the interest of advertisers and deploy ad server technologies. In the absence of micropayments, amateur Web publishers had no easy way to profit from their work, however popular their sites. This shifted the balance of power toward popular Web directories and search engines, and away from small publishers. By the late 1990s, just three firms were estimated to be receiving 43 percent of all online advertising revenues: AOL, Yahoo and Microsoft.[39]

Early Web advertising was fairly unsophisticated. In theory, the Web allowed companies to track their customers and profile not only their purchasing history but also their browsing patterns—which topics they searched for, which products they lingered over, and what kinds of advertisements they had proved likely to click on in the past. This, according to Internet advertising specialists such as Doubleclick.com, would allow fundamentally new kinds of interactions with customers, showing them exactly the right advertisement or offer at the right time.[40] But in practice, most Web advertising brokers just placed ads on to Web sites indiscriminately and charged the advertiser ten or twenty dollars per thousand views. Just as with print magazine advertising, this relied on a neutral auditor to certify the actual number of readers exposed to the advertisement. Advertisements might be targeted at a general level (advertisements for video games on Web sites about video games), but were rarely turned to the content of a particular article, still less to the profile of a particular viewer.

Internet navigation companies quickly realized that they had a significant commercial advantage: they knew what their visitors were thinking about at any given moment. This did not require any elaborate data-mining techniques. All they had to do was to see what words the user entered into the search box. Clearly, somebody who has just searched for "BMW 3 series" is a much more valuable viewer of an advertisement for a luxury car than someone who happens to be browsing political news on the *New York Times* Web site. In 1995, Infoseek had already begun to sell advertising with each search keyword, with popular words such as "Music" sold for forty dollars per thousand searches.[41] These premium rates made advertising an appealing option for search businesses.

Several alternative sources of revenue were explored by search sites. One was imposing paid subscriptions or usage charges. Before the Web, this had been the dominant business model for online information retrieval. Services such as LexisNexis and Dialog charged both subscription fees and usage charges to search databases of newspaper articles, scientific journals, patents, and so on. These services made a successful transition to the Web, but remained almost entirely separate from Web search services. In contrast, subscription-based Web search services had little appeal, because the same service was always available free from other search engines. Before shifting to its ad-supported business model, Infoseek briefly tried to justify subscription and usage fees for its search engine by offering extra features such as the ability to save searches and an index of Usenet newsgroups.[42] Another company, Northern Light, tried to bridge the gap between the separate worlds of premium online information services and Web search engines by combining Internet search with paid access to proprietary content from academic journals and government sources.[43]

Another other potential source of revenue was payment from companies in exchange for preferential treatment. This model had long been used by the Yellow Pages and other business directories. Yahoo and other Web directory services began to charge businesses for rapid or premium listings in the directory, supplementing their advertising income. Preferential treatment in search results was more complicated and controversial. It was pioneered by Overture, formerly known as GoTo.com. Overture focused on providing search services to other companies rather than building traffic to its own Web sites. During the early 2000s, Overture ran the search functions on both Yahoo and Microsoft's MSN site, making it one of the biggest search specialists on the Web. Overture's technology, introduced in 1998, allowed firms to bid for specific keywords. Selling advertising on keywords wasn't new, but Overture wasn't just showing an advert together with the results. It was selling top place in the search results themselves. Its other innovation was to use an automated auction system, so that prices fluctuated constantly with supply and demand. This prompted protests from some Internet users, who felt that paid search results were a breach of the trust they placed in search engines to provide accurate answers.[44] Overture's creator, Bill Gross, defended its honor by arguing that a company willing to pay to be listed was bound to be relevant to the query. This, he believed, made Overture's results more useful than those of traditional search engines.[45]

Overture's other key innovation was what has been called the "pay-per-click" model for Web advertising. By the end of the 1990s, users had largely stopped clicking on traditional banner advertisements, thus deterring advertisers from paying merely to have their advertisement displayed. With pay-per-click advertising, the advertiser paid only when somebody clicked on the link to visit their Web site. This innovation revitalized the economics of Internet advertising in general and search advertising in particular.

The third and final possible source of revenue for search technology companies was to sell their technology to other companies. This could be offered as a service (the model of Inktomi and Overture) or by selling programs as software packages. Big Web sites generally offered search capabilities to help users navigate through their depths. For this market, Excite provided Excite for Web servers and Alta Vista offered an intranet version of its search engine, with licenses starting at sixteen thousand dollars.[46] Indeed, Excite was originally created as a search product for Web site owners and was launched as an online service only after a change of business strategy.[47] Microsoft bundled Microsoft Index Server with its Web server to handle Web site searches.[48]

But as Internet use spread rapidly during the mid-1990s, many companies began to use the same technologies to create intranets full of private information.[49] These used Web servers and browsers to publish documents internally, creating a market for the use of Internet search technology. For obvious reasons, companies could not rely on public search engines to index their confidential information. A new wave of text-searching companies appeared, focused on providing search engine capabilities for intranets. One leader in this field, Open Text, was founded in 1991 to commercialize search technology developed by Waterloo University during the creation of the second edition of the *Oxford English Dictionary*.[50] Early in the development of the Web, it was the operator of a popular public search engine (used at one point by Yahoo to supplement its own directory). Indeed, a 1995 roundup of search tools reported that Open Text had "shot to the forefront of WWW databases with powerful search and display features," and the largest of all Web indexes.[51] Despite this strong start, Open Text soon withdrew from the Web search field to focus on systems for searching internal corporate documents.

Web Portals

During the late 1990s, the major Web search companies were all seeking to achieve rapid growth by broadening their operations. As one 1997 report put it, "Analysts have been hounding search companies to break away from the advertising-only revenue model and into a more diversified business."[52] As public companies they were vulnerable to such pressure. Despite a conspicuous lack of revenues, Excite, Infoseek, and Lycos had all staged initial public offerings in 1996, bringing rapid rewards to their first investors. Ask Jeeves, Overture (or GoTo.com as it was then known), and Inktomi followed suit later in the dot-com boom. Excite used some of the money to acquire other navigation services, including WebCrawler and the once-popular Magellan directory. AltaVista was the odd one out. The business was bought and sold several times, but never managed to stage the initial public offering for which it was being groomed. Along with the rest of DEC it was purchased by Compaq in 1998, and eventually disposed of to Overture.[53]

During the late 1990s the leading Internet search services Excite, Lycos, and AltaVista followed the same business strategy, which in retrospect was clearly flawed. Like other advertising-supported sites they were trying to achieve what was then called "stickiness," meaning that customers would visit often and remain within the site for a long time, clicking from one item to another. The problem with building a business around a search service was that a good search engine would come up results very quickly. Users would arrive, enter a search term, and click on one of the results. At that point they would leave the site and might not return until the next time they needed to look something up. During this process they would read little advertising. Search company strategists also worried that it was hard to differentiate search engines or build brand loyalty, since their results were likely to be pretty interchangeable. Even if search results could be improved, doing so would just send visitors on their way even faster. The solution, it seemed, was to make search into just one feature among many and remake Web sites into Web portals.

The portal concept, which first appeared around 1997, suddenly became ubiquitous in 1998. The first use of the term Web portal in a major U.S. newspaper was by the *New*

York Times in an October 1997 article, which stated that "Yahoo has been the most successful so far in establishing an identity as a hip, if quirky, Web portal."[54] The term fitted with Yahoo's ostentatiously wacky corporate culture, complementing irreverent job titles such as the "Chief Yahoo" tag given to one of its cofounders. Portal, literally just another word for an entrance gate, suggested a device through which one might travel to distant lands in an instant. Magic portals were a cliché of fantasy fiction, and featured prominently in the plotline of the *Dungeons and Dragons* children's cartoon of the 1980s as well as in various spells and magical items within the game it was based on.

Portals aimed to provide something akin to AOL's well-integrated user experience for people confused by the variety and complexity of the Web. They combined search capabilities with real-time chat functions, online games, weather, news, stock information, television listings, shopping opportunities, discussion areas, maps, and other services. All of them frantically licensed technologies and acquired businesses to merge new features into their portals. Portals offered customization capabilities, so that users could adjust the topics displayed on the main page to suit their own interests. The idea was that with all these different services available in one place, users would set the portal as their home page, visit frequently, and spend a lot of time skipping happily from one area to another while reading advertisement after advertisement.[55]

The conventional wisdom was that portals represented the most valuable part of the entire Internet and would be the launching point for every exploration of the Web. And it must be admitted that early Web users really did rely on search and portal sites. According to a list published in June 1996 of the most visited Web sites, the twelve most popular included five search specialists: Yahoo, WebCrawler, AltaVista, Lycos, and Excite. The other six sites (AOL, Netscape, Prodigy, GNN, CompuServe, and MSN) were all Internet access or technology companies whose Web sites functioned as portals. So-called destination sites such as Amazon, eBay, the *New York Times*, and CNN had yet to break into the top twenty-five.[56]

The portal field was increasingly crowded, as online services and browser companies realized that they could turn their popular Web sites into Web portals. In 1997 Netscape, which then enjoyed a steady stream of visitors to its Web site, decided to remake its home page as a portal called Netcenter. Netscape.com had long been one of the most popular sites on the Internet. As a supplier of browser software Netscape had a particular advantage. Each Web browser opens a particular Web page automatically when it is started, known as the home page.[57] It is not obvious how to change this, and people do so slowly or not at all (according to one 2001 report, 59 percent of Web users still had the initial setting).[58] Realizing that all this Web traffic was more attractive to investors than the company's struggling attempts to sell its Web browser, the firm created a new division to build up Netcenter into a full-fledged portal.[59] When Netscape Navigator slipped below a 50 percent market share for the first time since its launch, a company spokesperson shrugged off the news, saying, "It's not the browser wars anymore, it's the portal wars. . . . We are getting huge growth in Netcenter users, and that's what counts."[60]

Both AOL and its new rival MSN realized that they could complement the proprietary material offered to their online service providers with public Web sites providing the typical portal content such as news and e-mail. Microsoft launched msn.com as a public portal in 1998, consolidating material created in its existing Web sites such as Hotmail, MSNBC, and Expedia.[61] Microsoft made msn.com the default home page for its increasingly dominant Internet Explorer Web browser and later unleashed an additional piece of software, MSN Explorer, intended to give Web surfers a nearly packaged client similar to that offered by AOL to its users.[62]

The line between Web published and portal began to blur, as search engines, browser firms, and online services morphed into portals, and began to fill their site with information and services as well as links to other Web sites. Specialist Web publishing companies such as CNET began to call themselves portals as they added new Web sites (which were sometimes, on the model of television, called channels) to their networks. This followed the model attempted by media giant Time Warner early in the history of the Web. In 1994, it became the first major media company to invest heavily in the Web when it created Pathfinder.com, a Web site full of electronic content from all Time Warner's subsidiaries.[63] This single domain was shared between operations as diverse as *Time*, *Fortune*, special topic Web publications such as O. J. Central, Warner Brothers' movie *Batman Forever*, and Warner Books' publication *The Bridges of Madison County*. The site eventually included discussion and chat areas, free e-mail accounts, and a personalized news service. Individual Time Warner publications were forbidden from using their own domain names to build independent Web sites. The attempt to rival AOL by creating a huge mass of professionally produced media content had an obvious appeal to the managers of an unwieldy conglomerate whose entire existence was predicated on a faith in the existence of "synergies" between unrelated businesses. But the practical benefits of this integration were limited. And burying popular brands such as *Fortune* within Pathfinder was perverse. Although much-visited domains such as ⟨http://www.time.com⟩ merely forwarded users to the appropriate area of Pathfinder, users continued to identify with the individual publication brands rather than their corporate parent. A company spokesperson reportedly said that 98 percent of visitors to the site were forwarded from other domains rather than going directly to Pathfinder.[64] As Web users proved unwilling to pay subscriptions for online publications, Pathfinder never came close to covering its costs through advertising. In early 1999, the Pathfinder strategy was officially abandoned and Time Warner's surviving Web sites were free to go their own separate ways.

Despite Pathfinder's problems, the idea of building a self-contained Web world around the different media brands owned by one huge corporation still appealed in late 1998 when Disney and Infoseek (then the eighth-busiest site on the Web) announced the merger of their Web efforts to create the Go.com Web portal.[65] Disney's media holding included the ABC television network and ESPN as well as its own brand of sugary family entertainment. Disney launched a heavy campaign to plaster Go.com promotions across its television channels, books, theme parks, and other outposts of its corporate empire. While intended to compete with Yahoo and the other major portals, Go.com would give preferential treatment to search results and links featuring the firm's own offerings. Disney aimed to spin off Go as a separate company, offering a special class of stock representing its operations to take advantage of the enormous sums of money that investors were funneling into Internet stocks. Unfortunately, Web surfers continued to think of, and visit, Disney's online sites such as ESPN.com and ABC.com as separate entities. As an unknown brand, Go.com struggled to make an impression. It lost more than $1 billion in 1999, most of it spent to acquire Infoseek.[66]

In the late 1990s, the portal companies looked up to AOL as a model not only for its click integration of different services but also for its clout with advertisers. AOL came late to advertising. In 1996, it was facing stiff competition from Internet services offering unlimited usage for a flat monthly fee of around $20, while it continued to charge several dollars per hour. When AOL followed suit, this irreversibly changed the economics of its business. That same year it also acquired CompuServe, cementing its dominance of the industry. Previously it had made money out of keeping users online longer, but now it faced ruin if they spent the whole day online. Its modem pools were massively overloaded, and for a while it was jokingly known as "America On Hold." In October 1996 the firm reported a $353 million quarterly loss, reflecting previous customer acquisition costs it could no longer justify as an investment to produce future usage charges. Instead, AOL increasingly relied on advertising to make up for the costs incurred as its users spent more time online. AOL users spent a great deal of time in AOL's own proprietary areas (particularly its chat rooms) and on its Web site, providing plenty of opportunities to display advertisements to them.

By 1999, 16 percent of AOL's revenue was coming from advertising.[67] It was selling more online advertising than anyone, but it was also extracting huge payments from "partners" to feature their products prominently on its system. AOL had a dominant position as the largest ISP. During the late 1990s, thousands of well-funded start-up firms were desperate to attract visitors to their sites. The conventional wisdom was that the first popular site in a particular market, such as online pet food, would enjoy a huge advantage. Companies could therefore go public or sell themselves for hundreds of millions of dollars purely on the basis of a healthy stream of visitors to their Web sites, regardless of their financial position. As a result, start-ups were taking in millions of dollars from venture capitalists, and many were throwing this money at AOL with the idea that a prominent spot would be a quick way of bringing in visitors and boosting their profile in the media. The trend started in 1996 when short-lived telephone firm Tel-Save offered a cash payment of $100 million to be the exclusive long-distance telephone provider advertising on AOL. Many other deals followed. Such a deal, usually talked up as a "strategic partnership," might give the company enough credibility to make its own initial public offering.[68]

Following the model of AOL, the Web portals were able to bring in money by signing partnership deals with other Internet companies. How much sense this made is unclear. Even after AOL connected its service to the Internet, its users continued to rely on special AOL software that gave pride of place to AOL's own offerings and sponsors. In contrast, portals such as Excite and Lycos had no special grip on their users, meaning that being the "exclusive" provider of online books on Lycos (as BarnesandNoble.com did) or the exclusive CD retailer on Yahoo (as CDNOW did) was unlikely to do much to justify the millions of dollars it cost.[69]

During the boom years, deals of this kind appeared to confirm the strategic power of portals as the hubs of the Internet. They continued at a dizzying rate. Between August and October 1999, Lycos (now billing itself as "the world's largest online community") claimed to have created "the Internet's first true full service e-commerce portal with the launch of LYCOShop, the Web's most complete integrated shopping destination," to have launched more than a dozen new "localized" versions of the portal for different countries, to have formed or extended "strategic alliances" with IBM, Fidelity, American Greeting Cards, and AOL, and to have acquired or invested in a half-dozen smaller firms including a maker of MP3 audio players. None of these deals amounted to much in the end, but they certainly provided the impression of a dynamic and ambitious firm.[70]

The major portals were joined by a wealth of specialist portals. Portals sprang up in China, India, Europe, and elsewhere. Although the Web itself was worldwide, details such as shipping costs and customs regulations meant that most Internet businesses functioned within national boundaries. Even purely information sites, such as online magazines, were of interest only to people who could read the language involved and had some familiarity with the culture of the country involved. National portals made a lot of sense. By the same logic, some Web sites (particularly those operated by local newspapers) began to sell themselves as regional portals, integrating news, weather, business listings, cultural information, restaurant reviews, and the like for a particular area. The *Boston Globe*'s site, Boston.com, was one of the first and most successful of these, including material from local television and radio stations and magazines as well as the paper's own stories.[71] Newspapers saw the regional portal model as a means of safeguarding their local stranglehold on classified advertising, traditionally a major source of revenue, from predation by specialist online marketplaces.

Meanwhile, so-called industry portals were set up as online hybrids of trade publications and business-to-business marketplaces. In the mid-1990s, many assumed that the Web would replace traditional relationships between industrial customers and suppliers with specialist commodity markets, in which spot prices for all kinds of goods and services fluctuated constantly as electronic bids were placed and met. Few companies were rushing to adopt this new model for their purchasing needs, and so the more ambitious industry portals collapsed rapidly. For example, three big metal industry portals—Aluminium.com, MetalSpectrum, and MetalSite—all folded within a two-week period in June 2001.[72] Another flurry of doomed firms tried to set themselves up as "eGovernment Portals" to bring citizens together with government institutions. Among these was govWorks, featured in the memorable documentary film *Startup.com.*

Large organizations set up portals for their employees and customers. Portals were intended to provide a single, easy-to-use entry point for all the different offices, services, and sources of information spread out among the constituent parts of large bureaucratic organizations such as city governments, universities, and major corporations. Perhaps the most ambitious in scope were the national and local government portals announced by politicians eager to show that their governments were accessible, forward-looking, and in touch with their citizens. The United States offered not just "FirstGov.gov, the U.S. Government's Official Web Portal" but also the cartoonlike "FirstGov for Kids," complete with a Web "treasure hunt" to reward children for finding the correct answers to questions such as "What federal agency provides citizens with information about the Air Quality Index?"[73] So popular were portals that a new industry grew up to supply portal software for use by corporate customers. Portal packages were sold by major corporate information technology providers such as Sun Microsystems, IBM, and Oracle, specialists such as Plumtree and Hummingbird Communications, and Yahoo itself. Over time, many portal packages evolved to incorporate the capabilities of content-management systems, generating Web content dynamically from databases and templates.

Enthusiasm for portals grew along with the bubble in dot-com stocks. Excite, Lycos, Netscape, and Yahoo were much better known as stocks than as companies, and so their success was gauged on their soaring share price rather than their actual prospects. Stock tips were everywhere in the late 1990s, as financial television channels grew, investment clubs thrived, and newspapers profiled the sensational initial public offerings of the day. YHOO was perhaps the greatest of the pure Internet stocks. Adjusted for splits, it rose from an offering price of $24.50 in 1996 to a peak of $2,850 at its all-time high in January 2000. The other portal firms enjoyed similar success. In May 2000, Terra Networks offered stock then worth a staggering $12.5 billion in a successful bid for Lycos, then the third most visited portal site. But by then the dot-com crash had already begun.

The Crash

Between March 2000 and October 2002 the NASDAQ composite, a measure of share prices on the high-technology oriented NASDAQ exchange, fell from 5,047 to 1,114. Internet stocks fared far worse. Portals saw their advertising revenues dry up, as most online advertising came from other online businesses. The plunge in the value of existing Internet stocks meant that no more Internet initial public offerings were possible, which meant that venture capitalists were no longer handing millions to doomed or speculative start-ups, which meant that there was no flow of easy money to portals from start-up companies rushing to buy visitors to their Web sites. The sudden drop in online advertising was a particular challenge to the portals, whose existence was premised on the idea that it would rise rapidly for many years to come. When the flow of new money ceased they faced a crisis.

Often this was terminal. Excite, for example, had merged with high-speed Internet pioneer @Home and squandered its cash on dubious acquisitions, such as the payment of $780 million for an online greeting card company.[74] It ran out of money soon afterward and was liquidated. Go.com, in contrast, was clearly doomed even before the crash, having failed to attract advertisers or visitors. Disney tried to salvage the portal by refocusing it on entertainment and leisure, but it continued to hemorrhage money and slip down in the Internet rankings.[75] In early 2001 Disney closed the division, shut down the portal, and disposed of its assets.[76] The once-mighty Lycos withered away more slowly. In 2004 what remained of Lycos was sold again, this time for less than 1 percent of the price that Terra Networks had paid four years earlier. Its original business was essentially destroyed, though efforts are now under way to revive the brand.[77]

AOL began to fall apart in 2001, just a few months after using its stock to buy Time Warner to form AOL Time Warner. Time Warner was the world's biggest media company, but the failure of Pathfinder and a number of other Internet initiatives had left its stock stagnant. AOL, on the other hand, had an enormously high share price because (inasmuch as any rational reason can be provided) it appeared to have been growing rapidly for a long time and was expected to continue to do so indefinitely. Facing pressure to show continued growth, AOL began to claim revenues based on exchanges of advertisements with other companies in which no money really changed hands.[78] (It eventually agreed to pay $510 million to settle an investigation by the Securities and Exchange Commission into these practices.)[79] But such tricks could work only for a short time, and within eighteen months of the merger almost all the senior AOL executives had been purged from the firm. Far from being the engine of growth for the merged company, AOL was dragging down the performance of successful parts of the conglomerate, such as the HBO subscription television channel. Its base of loyal dial-up subscribers saved AOL from total annihilation, but its diminishment was clearly signaled in 2003 when the AOL Time Warner board decreed that henceforth, plain old Time Warner would be a better name for the firm.

Its main competitor, MSN, never lived up to expectations. By 2001, it had become the second most visited portal in the United States (after Yahoo), though by that point the portal business was collapsing.[80] It did become the second-biggest provider of dial-up Internet access in the United States by the end of the 1990s, and according to Web traffic metering specialist Alex.com the MSN Web site is currently the fourth most visited in the country. For any other company, those results would have reflected a triumph. But there was really little to cheer about, given all the money that Microsoft had pumped into MSN, the $400 rebates it had handed out liberally to win three-year subscriptions from new personal computer purchasers, the special placement it received within Windows, and its huge advantage as the default home and search pages of Internet Explorer. Fortunately for Microsoft, its Windows and Office monopolies continued to generate money much faster than the rest of the company could spend it.

Beyond the immediate crisis affecting all firms reliant on online advertising, the portal industry faced a particular problem: as the Internet matured, most people did not really want or need portals. As users grew accustomed to the Web they created their own lists of links and bookmarks, visiting different sites for different kinds of information. People no longer needed a portal to tell them that they could buy a book from Amazon, read the *New York Times* online, or purchase a plane ticket from Expedia. While a search engine was still a good place to go to investigate a new topic, the rush to remake search engines as portals had drawn resources away from the development of their search technologies, and filled their front pages with slow-loading and distracting clutter. Furthermore, as portals sought to provide every possible service, they had no real way to differentiate themselves from their competitors. As *Washington Post* reporter Rob Pegorano noted in 2000, "These sites are all unique in pretty much the same way. Most of their content even comes from the same third-party sources—news from the Associated Press and Reuters, forecasts from the Weather Channel, shortcuts to buy books through Amazon.com, and so on."[81] Personalization features were supposed to be the compelling benefit of portals, but few users even bothered to configure their preferences, and still fewer kept them updated.[82]

Of the independent would-be portal sites, only Yahoo survived the crash. Its luster as a stock vanished quickly, as YHOO lost 97 percent of its early 2000 peak value in less than eighteen months.[83] But Yahoo the business retrenched, downsized, brought in a new CEO, and redoubled its efforts to drum up advertising revenues and create premium services that users were willing to pay for. Since 2002 Yahoo has been consistently profitable, dominating what was left of the portal field and enjoying some success with features such as online video sharing, social networking, auctions, and job listings.

Google and the Resurgence of Search

By the end of the 1990s the Internet search pioneers Lycos, Infoseek, Excite, and Alta-Vista were no longer very interested in the search business. Their technical, financial, and managerial resources were aimed squarely at the Web portal business. Their Web search capabilities were just one of many items, packed into home pages full of eye-grabbing attractions. In 2002, a Yahoo spokesperson explained that his firm was "first and foremost a media company," and that "search and directory is an increasingly small part of what we do."[84]

Starved of resources, the search engines were actually becoming less useful. As the Web got bigger and bigger, the issue was no longer finding a sufficient number of pages containing the search term but ranking the results so that useful pages came first. Studies found that only about 20 percent of users would go beyond the first page of search results in search of relevant pages.[85] But search engines might find many thousands of matching pages, and used quite simple methods to rank results, looking particularly at the number of times the keyword appeared. As Web site operators became more sophisticated, they found it easy to manipulate these results. For example, some search engines could be manipulated into displaying a page in response to a query on a popular term such as "antivirus software" just by inserting that term into the raw code for the page thousands of times in such a manner that it would never actually be displayed. The top results for many queries were hijacked in this manner by operators of pornographic Web sites. A small industry grew up promoting these techniques to Web site operators as a cost-effective alternative to paid advertising. Its apologists dubbed the practice "search engine optimization," but critics preferred the less flattering "search engine spamming."[86]

It seemed that search engines had reached the limits of their capabilities. In 1999 Danny Sullivan, whose bloglike site Search Engine Watch is a major source of search engine news and comments, boldly announced that "this was the year that humans won. [Earlier] you had one major search service, Yahoo, that used human beings to categorize sites while the others were trying to use technology to do the same thing. But now with six out of the top 10 services, the main results you get will be by people."[87] He was apparently referring to the incorporation of Open Directory results by the major portals. Sullivan was not alone in this belief. The next year Chris Sherman, a search engine consultant, insisted that the project was "leading a resurgence of human-compiled Web directories, toppling spider-complied search engines from their dominant positions as principal gateway to the Internet."[88]

Sullivan and Sherman were quite wrong. As the established search companies marched in unison over the cliff in pursuit of portal status they had left behind a niche for a company focused on providing the best search experience to users. A new search company, Google, seized this opportunity and proved that Web search was one of the Internet's most profitable business opportunities. Like Excite and Yahoo a few years before, Google began as the personal project of some Stanford University computer science students. Larry Page and Sergey Brin launched their service as google.stanford .edu. They made the most of Stanford's cultural and technical resources, using huge quantities to network bandwidth and storage space to build a fully functional version of the service with a massive database before seeking funding for their idea. In 1998, they founded Google, Inc. to turn the service into a business. Drawing on Stanford's well-established connections with Silicon Valley businesses, the pair won initial investments from various technology industry entrepreneurs, followed in 1999 by a generous injection of $25 million, of which most came from the valley's two most storied venture capital firms: Kleiner Perkin and Sequoia Capital.

But while Google enjoyed privileged access to the sea of easy money floating around Silicon Valley toward the end of the boom years, its founders resisted many of the ideas imposed on other search firms by the experienced managers brought in to steer their evolution. In contrast with the ever-more-crowded front pages of the portals, Google's Web page consisted of nothing more than a white space in which floated a simple logo, a box to type search terms into, and a button labeled "Google Search." The results pages were similarly spartan. With no advertisements, Google pages loaded almost instantly even over a dial-up connection. Indeed, its founders were initially quite dismissive of the idea of advertising:

The goals of the advertising business model do not always correspond to providing quality search to users. . . . [T]he better the search engine is, the fewer advertisements will be needed for the consumer to find what they want. This of course erodes the advertising supported business model of the existing search engines. . . . [T]he issue of advertising causes enough mixed incentives that it is crucial to have a competitive search engine that is transparent and in the academic realm.[89]

Behind this simple interface lay a search engine of exceptional power. Google's coverage of the Web was soon unsurpassed, as its crawler (dubbed "the googlebot") inched its way into the mustier fringes of the Web. But its biggest attraction was the consistently high relevance of its results.

Without human intervention, Google somehow pushed high-quality, relevant sites toward the top of its search results. This has been attributed to its much-discussed "PageRank" algorithm, patented by Stanford. Unlike the crude techniques used by early search engines, this method consistently put the most important and widely linked-to sites on a particular topic within the first page or two of the search results. PageRank looked not just at the page itself but also scoured its database to assign the page a rank according to the links created to it from other sites. This idea was inspired by the long-established practice of ranking the importance of scientific papers according to the number of citations they received. (Citations were indexed by the Institute for Scientific Information, giving raw data for researchers in the field of bibliometrics and the institute's own annual rankings of science's greatest hits.) Google extended the idea so that links from highly ranked sites were themselves accorded more weight than links from poorly ranked ones. Google also incorporated some established information retrieval principles, considering the print size of terms, their closeness to each other, and their position on the page in determining the ranking. While this method proved to have its own vulnerabilities to exploitation, it continues to produce high-quality results.

Google's founders tried to license their invention to AltaVista, Excite, Infoseek, and Yahoo before making their commitment to commercialize it themselves. All these firms turned it down—search was no longer a priority for them. Google's advantage came not just from its algorithm but also from its unwavering focus on providing an effective search service. Its staff constantly tweaked the internals and user interface of its search engine, making little adjustments and adding features. As the service became more popular, they worked hard to eliminate search spam and keep results relevant.

Google's success hinged on technical feats in operating systems and parallel computing as well as the original cleverness of its ranking algorithm. Google searched more of the Web than its competitors, and gave more useful answers, faster, and to more users. This was possible only by applying unprecedented volumes of computer power to the problem. Google accomplished this without running out of money only by finding innovative ways to combine many thousands (and eventually hundreds of thousands) of cheap computers based on commodity personal computer hardware running the free Linux operating system. Its competitors often relied on small numbers of expensive, multiprocessor systems using proprietary hardware along with operating systems from firms like IBM and Sun.[90]

Having created the Web's best and fastest-growing search service, Google still needed a way to make money. In 2000 it began selling advertisements, with an approach it called AdWords. Like Overture, Google accepted bids from potential advertisers to have their ads appear when users searched on particular terms. It also copied Overture's pay-per-click model. Google realized that users disliked large, distracting advertisements, however, and would not trust a search service in which the top results appeared only because users had paid for them. Instead, Google presented a single paid result, clearly labeled as a "sponsored link," and a handful of simple advertisements, each consisting of three short lines of text, grouped on a separate part of the screen (initially at the top of the screen, and later in a column at the right). Google added a twist of its own to Overture's model by factoring in how often advertisements were clicked on as well as the amount bid by the advertiser in deciding how to order them. Even the highest bids brought little revenue to Google unless people were prepared to click on the advertisements once they were shown, so this tweak simultaneously improved the relevance of the adverts displayed and boosted profits.[91] The year 2000 was not good for the online advertising industry, but Google's increasing popularity and ability to present users with advertisements directly related to whatever they were researching soon allowed it to charge a premium. It turned out that sticking to search could be lucrative indeed. Google's advertising is among the world's least obtrusive but most profitable. In fact, the very sparseness of the ads raised their value by decreasing the supply.[92]

As well as its own popular search service, google.com, Google also won contracts to provide search services to many of the Web's most popular sites. It shared its advertising revenues with the sites concerned. By 2002 these included the MSN, AOL, and Yahoo Web portals. Google's superior information retrieval capabilities and more efficient advertising technologies allowed it to edge out Overture and Inktomi for these crucial accounts. The portals continued to think of search as a peripheral feature best outsourced to a specialist firm. Google was happy to support this misapprehension. A 2001 article stated that "Google vehemently denies that it has designs on its portal customers' turf," and quoted a company spokesperson as saying that "we have 130 customers ... they don't feel we're competing with them, and we're comfortable with that model."[93]

Google's success even revived some of the old motifs of the .com era. Google's leaders pride themselves on its distinctive corporate culture, including its aggressively uncomplicated motto "Don't be evil," the prominent display of pianos, video games, and gourmet food in its offices, and a corporate mandate to make work fun. In 2004 it went public, and while the stock broke with tradition by failing to rise on the first day of trading, its value quadrupled over the next eighteen months. Page and Brin achieved prominent spots on the *Fortune* list of the world's most wealthy people while still in their early thirties, and stock options made its early employees millionaires overnight. The determination of Google's founders to impose their culture of technical tinkering on the world of business extended even to the initial public offering, which was conducted using a unique auction process to set the share price and structured in such a way as to leave control of the company in their hands via a special class of stock with superior voting rights.

The Web Navigation Business Today

After the success of Google became apparent, Yahoo changed course to concentrate heavily on search technology (as did Microsoft, Amazon, and many other firms). Yahoo bought Inktomi in 2002 and Overture in 2003. As Inktomi had already purchased what was left of AltaVista, that left Yahoo with a powerful array of search technologies. In 2004, it switched its main Web search over to its own technologies, marking the first time that Yahoo had operated its own public search engine rather than outsourcing the work to others.[94] This investment has failed to stem Google's rise. According to the comScore Media Metrix rankings of Internet search traffic, Google held a 44.1 percent share of the 6.5 billion Web searches in the United States during August 2006, versus 28.7 percent for Yahoo (now the owner of Overture) and 12.5 percent for MSN.[95] Other estimates give Google more than 60 percent of the search market.[96]

Driven in large part by the success of search advertising, Internet advertising on Yahoo and other leading sites rebounded, and has shown steady growth.[97] In 2005, online advertising revenues reached a new high of $12.5 billion, according to the most widely accepted estimate. Of this money, 43 percent came from search advertisements.[98] By 2006, Google was selling far more advertising than any other Internet firm, and had overtaken traditional media powerhouses such as the broadcast television networks and Gannett Company (owner *USA Today* and more than a hundred other daily newspapers, twenty-three television stations and more than a thousand other periodicals).[99] It achieved this by displaying advertisements to the users most likely to click on them. According to one newspaper report, "For every page that Google shows, more than 100 computers evaluate more than a million variables to choose the advertisements in its database to display—and they do it in milliseconds."[100] These variables are said to include the time of day, the location of the user, and the type of Internet connection, but not personal information. It has finally delivered on the idea, much discussed during the .com era, that Web advertising can be much more efficient than off-line advertising because of the potential to target it more precisely.

Yahoo's purchase of Overture and its innovative search advertising technology underpinned its return to financial health. It so far remains less effective than Google in maximizing advertising revenues by displaying the adverts that users are most likely to click on, though by the end of 2006 it aimed to have pulled ahead of Google in this respect with a new system incorporating information on users' demographics, query history, and browsing habits.[101] Yahoo's recent progress illustrates the absurdity of stock valuations during the dot-com era. Yahoo reported earnings of close to two billion dollars for 2005, around thirty times higher than those for 1999. Yet its stock, which is by no means undervalued, has regained less than 45 percent of its peak value.[102]

As well as selling advertisements on their own Web sites, Google and Yahoo have also become brokers of advertising to be displayed on smaller Web sites. Again this was not entirely new. Since the mid-1990s, advertising firms such as DoubleClick had sold advertisement space in bulk. They controlled which advertisements were shown where, and hosted the ads on their own servers, so that all an Internet publisher had to do was insert a link to the ad server in the appropriate place on their pages. But Google's AdSense system refined the concept by offering the same bid system and pay-per-click model used on its own Web site.[103] Revenue is shared between Google and the operators of the Web sites where advertisements are displayed. This has helped to shift the economics of Web publishing back toward smaller amateur and semiprofessional ventures. (It has also created a new and hugely profitable industry of sites holding nothing but Google advertising links, placed on attractive yet vacant domains such as clothes.com or on sites such as nytomes.com, yagoo.com, or ebey.com reached by mistyping more popular domain names.)[104]

For all its success, widely syndicated pay-per-click advertising led inexorably to a new problem: click fraud. Unscrupulous Web site operators arrange for automated programs to repeatedly mimic the effect of Web surfers clicking on the advertisements displayed on their sites. Google then collects money from the advertisers and passes a cut on to the fraudsters.[105] While Google quickly recognized click fraud as a serious challenge to the viability of its business, this practice has been hard to eliminate. Estimates of the proportion of Google AdSense clicks that are fraudulent have ranged widely, but some are as high as 35 percent.[106] Google itself has refused to release detailed information on the scale of the problem.

So far, Google has succeeded in making itself the world's most successful Internet company without falling victim to the distractions and overreach that destroyed earlier search companies. Theoretically its position is vulnerable, because search users could switch to a new and better search service much more quickly and easily than, for example, a different operating system. Microsoft, Amazon, Yahoo, and Ask Jeeves have all spent large amounts of money to try to produce a Google-killing search service. In practice, Google continues to give better, faster, and more useful search results than its competitors. It has extended its search capabilities to include images and videos, to search inside files other than Web pages (such as Acrobat, Word, and PowerPoint documents), to search for the best prices on specified goods, to search inside the contents of digitized books, to search the Usenet newsgroup archives, to search inside the e-mail messages stored by users of its gmail service, to search files on its users' personal computers, and to incorporate results from closed-access services such as academic journals.

While maintaining its lead in search, Google has taken advantage of its human and financial resources to develop or acquire a large number of other services. Its list of options recalls the plethora of features crammed into the portal sites of the late 1990s. They include the Google Maps cartography and route-finding service along with the related Google Earth satellite image browser, the Blogger blog hosting service, the Google Groups discussion group service, an online spreadsheet application, a chat and Internet telephony service, and an online calendar system. Gmail and Google Maps both gained large and enthusiastic user bases by offering services far more elegant, powerful, and interactive than their established competitors such as Hotmail and MapQuest. But in contrast to the cluttered, ugly pages of the late 1990s' portal sites, Google's home page remains a pristine sea of white, holding a single simple graphic for the company logo, a single input box for the search, and just two buttons: "Google Search" and "I'm Feeling Lucky." The search box can be used to obtain weather information or stock quotes, perform calculations, track UPS packages, convert currencies, or look up telephone numbers, but its unassuming exterior does nothing to frighten the novice user. Meanwhile, Google's full and potentially overwhelming list of services is displayed only by clicking on an unobtrusive link labeled "more" and choosing the option "even more."

Furthermore, and in contrast to the old portals, Google is not determined to confine users to its own site. Instead, it has been exploiting the original philosophy of the

Internet by making its services easy to customize as part of other applications. Google Maps, for example, can be configured by another Web site, such as a real estate listings service, to plot the location of points of interest. Its search engine can likewise easily be configured to display only results from a particular Web site, removing the need for Web sites to install their own inferior search systems for the benefit of visitors. Google's willingness to make itself part of the Web's software infrastructure by unleashing the power of independent developers may make it a crucial part of the emerging market for location-sensitive Internet services.

Google's steady consolidation of power over Internet search and advertising seems destined to involve it in an ever-growing number of controversies. Google has been sued by online businesses whose sales slumped after they fell in its search rankings, by advertisers suspicious of click fraud, and by publishers seeking to prevent it from digitizing their copyright material for its Google Print book indexing project. It has been denounced by a congressional committee for cooperating with the Chinese government in censoring search results, and taken to court by the U.S. Department of Justice to force the release of query records. Privacy activists fear that Google search makes it too easy to invade the privacy of others, and that the company itself has built up an unprecedented database on the interests and activities of its users. Google appears to have acted at least as responsibly as its peers in all these areas, but given the limitless ambition and self-confidence of its leaders the firm seems unlikely to escape the kinds of resentments that built up against Microsoft in earlier decades.[107]

As Google grew, AOL continued to dwindle from its position as the Internet superpower of the late 1990s. By 2006, it was clear that AOL's growing reliance on Internet advertising posed a strategic challenge for the shrunken business. Should AOL's special attractions such as its instant messenger software, news feeds, and e-mail service be used exclusively as a means of enticing and retaining dial-up customers to its Internet access service? Or should it focus on making its public aol.com Web portal as attractive as possible to all Internet users in the hope of selling more advertising? The problem was that the more features it reserved for its dial-up customers, the less special its public Web portal could be. In recent years, as AOL's customers shifted in ever-greater numbers to broadband services (many of them to its sister company Time Warner Cable) this issue became harder to ignore. Opening up its full range of services to the public might revitalize its portal business, but hasten the departure of its remaining nineteen million or so dial-up customers, many of whom were tied to the service by the difficulties involved in changing their aol.com e-mail addresses. (The service was also notoriously difficult to cancel, as the process involved long and sometimes repeated pleading sessions with "customer retention" specialists.)[108] AOL packed its extras for users of high-speed Internet connections as AOL for Broadband. Then in August 2006, it broke with the past and announced that its premium features, including e-mail accounts, would now be available free of charge to all Internet users.[109] In other words, it is now AOL that is trying to be like Yahoo.

Microsoft, meanwhile, is trying to be more like Google. Only two of Microsoft's portal services won any real following: the Hotmail mail service and MSN's instant messaging software. MSN has struggled to create momentum; in the first six months of 2006, its revenues dropped from a year earlier and it reported losses even as its competitors declared record profits.[110] Google's success in making capabilities such as Google Maps into a building block for other Web site developers has inspired Microsoft to follow suit. Since 2005, Microsoft has been downplaying the MSN service and redesigning popular components such as its Hotmail service as parts of a new "Windows Live" initiative.

The market for corporate intranet search products (often called "enterprise search") has continued to grow but remains quite small. On a business level, these systems are now a crucial part of many firm's efforts to comply with legal requirements to retain documents, improve document work flow and business processes, and back up business data against possible disaster. By 2005 Open Text held the largest single share, an estimated 13 percent, of a market that remained heavily fragmented among more than a dozen companies.[111] Autonomy, one of the other main suppliers, claims to provide more useful results than traditional keyword search by automatically generating taxonomies and associations between terms. In 2005, Autonomy acquired Web site search pioneer Verity in one of a series of mergers that are consolidating the field.[112]

Corporate search products are evolving in a different direction from their Web-based cousins, toward ever-closer integration with existing corporate applications. Unlike public search engines, systems of this kind must often tag documents according to security and access levels, and show only the material that a given user is authorized to see. As corporate search, content management, and portal software companies are acquired by larger enterprise software firms such as Computer Associates, IBM, Oracle, and BEA, it appears that these formerly distinct kinds of software have merged with each other and, increasingly, the complex systems these major firms sell to help organizations run their core operations. Autonomy's products, for example, are designed to integrate corporate information from sources such as e-mail, Enterprise Resources Planning systems (such as SAP), voice, and video into a single searchable corpus.

Enterprise search systems are licensed as software packages, providing their producers with money from sales plus updates, support fees, and consulting charges to implement systems. Autonomy reported revenues of $117 million in the first half of 2006, most of which came from its acquisition of Verity. Open Text projected around $200 million in revenues for the same period. While far from insignificant, these figures are dwarfed by Google's revenues of close to $5 billion. Interestingly, Google has come up with a different model to sell its technology for enterprise use: the "search appliance," a bright yellow server loaded with Google indexing and search routines ready to index intranet data as well as (with suitable bridges) data from other sources.[113]

Conclusions

It is impossible to imagine the story of Internet navigation services having developed the way it did without the enormous flood of money into the field during the dot-com boom. Search and portal firms were funded by venture capitalists in large numbers, encouraged to grow rapidly, and became publicly traded companies long before they were profitable or indeed before anyone had a clear idea of how the market would develop. The portal concept was attractive to an immature industry composed of firms desperate to show the rapid revenue growth their investors demanded, but this ill-considered diversification led AltaVista, Lycos, Infoseek, and Excite to ruin. Doing one thing well proved a better strategy than doing many things indifferently. Why would users favor a portal offering second- or third-rate implementations of e-mail services or music sales when the best sites in each category were just a click away? In a different business environment it would not have been necessary for the portals to waste quite so many billions of dollars to illustrate this. Only Google, fortunate enough to be able to learn from the mistakes of its predecessors and led by headstrong founders skeptical of conventional business wisdom, stayed focused on providing users with the best possible search results.

In the end, the story of search functions is a kind of parable to explain the strengths and weaknesses of the Web itself. Search is an essential feature of any distributed information system, but one neglected entirely in the Web's original design in favor of simplicity and decentralization. The Web effectively relied on free market innovation, rather than central planning, to provide its navigation capabilities. Dozens of rival search services battled to work around this fundamental deficiency, with considerable success. When the early leaders adopted suicidal business strategies, this free competition between technologies and business models allowed Google to step in with a clearly superior product. But even Google's search results are still full of broken links and out-of-date results. And it remains impossible to see a complete list of pages linking to a particular site—a key part of Nelson's original vision for hypertext. Yet it seems unlikely that any system encumbered with a central database or registry of links could have grown as quickly and evolved as fast as the Web did.

Given that Web pages supply almost none of the metadata (information snippets such as the name of the author, date, publisher, and keywords) relied on by traditional information retrieval systems, the effectiveness of current search engines is most impressive. But the lack of metadata makes search engines work hard to achieve sometimes mediocre results. For all Google's cleverness, it can only go so far in mitigating the fundamental limitations of the Web. Various efforts are under way to address this issue, most ambitiously a project known as the Semantic Web proceeding under the direction of Berners-Lee.[114] It is not clear whether this complex model will ever be widely adopted by Web publishers, but more and more areas of the Web are adopting simpler mechanisms to support metadata. One of the most discussed Web trends of recent years, the folksonomy, describes systems that allow users to assign their own arbitrary tags to things.[115] These tags can then be used by others to search or browse through the material. Popular new sites such as Flickr and YouTube are full of pictures and videos tagged with descriptive terms by visitors to the site. Other services, such as furl .net and digg.com, allow users to assign tags to Web sites.

Meanwhile, the Web currently supports just one hugely successful portal (Yahoo) and just one hugely successful search engine (Google). As Yahoo has added strong search capabilities and Google has rounded out its range of services, their capabilities have begun to converge even as their personalities remain quite different. Between them, they dominate the market for Internet search advertising and syndicated advertising displayed on other Web sites. Whereas conventional wisdom in the late 1990s favored the full-service portal concept and held that Web publishing would be dominated by a few major companies, today people celebrate the Internet's ability to make niche markets profitable by reducing the distribution and production costs necessary to supply books, films, and other entertainments to small audiences (an idea known as the "long tail"). Amazon and eBay make it easy for buyers to find formerly obscure works or goods that fit their taste. Likewise, in the world of Web publishing Yahoo and Google have established themselves as the vital central points where searchers can find what they are looking for and advertisers can find buyers for their products. The profound centralization of search traffic and syndicated advertising revenue in the hands of these firms has supported an equally profound decentralization of Web publishing.

Notes

1. Kris Oser ("New Ad Kings: Yahoo, Google," *Advertising Age*, April 25, 2005, 1) forecast that this would take place during 2005, since which online advertising revenues have risen sharply.

2. For an early and enthusiastic report on the application of information retrieval to business, see Francis Bello, "How to Cope with Information," *Fortune* 62, no. 3 (September 1960): 162–167, 80–82, 87–89, 92.

3. The origins of the concept of information science are discussed, rather critically, in Hans Wellisch, "From Information Science to Informatics: A Terminological Investigation," *Journal of Librarianship* 4, no. 3 (July 1972): 157–187; Mark D. Bowles, "The Information Wars: Two Cultures and the Conflict in Information Retrieval, 1945–1999," in *Proceedings of the 1998 Conference on the History and Heritage of Science Information Systems*, ed. Mary Ellen Bowden, Trudi Bellardo Hahn, and Robert V. Williams (Medford, NJ: Information Today, Inc., 1999), 156–166.

4. Charles P. Bourne and Trudi Bellardo Hahn, *A History of Online Information Services: 1963–1976* (Cambridge, MA: MIT Press, 2003), 322–328.

5. Ibid., 141–184.

6. Ibid., 280–286.

7. Nelson's fullest explanation of his vision of a worldwide electronic hyptertext publishing network open to all is in Theodore H. Nelson, *Literary Machines* (Swarthmore, PA: Mindful Press, 1982).

8. Vannevar Bush, "As We May Think," *Atlantic Monthly* 176, no. 1 (July 1945): 101–108.

9. Tim Berners-Lee and Mark Fischetti, *Weaving the Web: The Original Design and Ultimate Destiny of the World Wide Web by Its Inventor* (San Francisco: Harper, 1999), 50.

10. A readable overview of the Xanadu project is in Gary Wolf, "The Curse of Xanadu," *Wired*, June 1995, 137–202.

11. Theodore Holm Nelson, "Xanalogical Structure, Needed Now More Than Ever: Parallel Documents, Deep Links to Content, Deep Versioning, and Deep Reuse," *ACM Computing Surveys* 31, no. 4es (December 1999).

12. Niklas Rudemo, "Beyond HTML: Web Issues Aired in Darmstadt," *Seybold Report on Desktop Publishing* 9, no. 9 (May 8, 1995): 10–11. An insider's discussion of the early Web's hypertext functions and their evolution is in Robert Cailllau and Helen Ashman, "Hypertext in the Web: A History," *ACM Computing Surveys* 31, no. 4es (December 1999). Other essays in the same electronic volume provide a valuable discussion of the relationship of the Web to hypertext research.

13. Charles C. Mann, "Is the Internet Doomed?" *Inc.* 17, no. 9 (June 13, 1995): 47–50, 52, 54.

14. Nancy Garman, "A New Online World," *Online* 19, no. 2 (March–April 1995): 6–7.

15. Readable overviews of the dot-com era can be found in John Cassidy, *Dot.Con: How America Lost Its Mind and Money in the Internet Era* (New York: HarperCollins, 2002).

16. The difficulties of running a company swept up in such fads is vividly captured in Michael Wolff, *Burn Rate: How I Survived the Gold Rush Years on the Internet* (New York: Simon and Schuster, 1998).

17. The operation of the initial public offering feeding chain is explored in Roger Lowenstein, *Origins of the Crash: The Great Bubble and Its Undoing* (New York: Penguin, 2004), 108–126.

18. Tim Berners-Lee, *W3 Servers*, CERN, 1992, available at ⟨http://www.w3.org/History/19921103-hypertext/hypertext/DataSources/WWW/Servers.html⟩ (accessed August 15, 2006).

19. Matthew Gray, *Web Growth Summary*, 1996, available at ⟨http://www.mit.edu:8001/people/mkgray/net/web-growth-summary.html⟩ (September 3, 2006).

20. The early history of Yahoo is recounted in Karen Angel, *Inside Yahoo Reinvention and the Road Ahead* (New York: John Wiley and Sons, 2002); Robert H. Reid, *Architects of the Web: 1,000 Days That Built the Future of Business* (New York: John Wiley and Sons, 1997), 241–279.

21. Bill Gates, "The Internet Tidal Wave," U.S. Department of Justice, May 26, 1995, available at ⟨http://www.usdoj.gov/atr/cases/exhibits/20.pdf⟩ (accessed May 20, 2006).

22. Julia Angwin, "Excite Will Buy Magellan Search Engine," *San Francisco Chronicle*, June 28, 1996, C1. Journalist and former Internet executive Michael Wolff gave a memorable portrayal of his experiences with Magellan's leaders in his *Burn Rate*, 69–104.

23. Chris Sherman, "Humans Do It Better: Inside the Open Directory Project," *Online* 24, no. 4 (July 2000): 43–44, 46, 48–50.

24. The date at which the Open Directory Project overtook Yahoo and its volume of pages is taken from the Wikipedia page on the project. I have been unable to verify this from a more stable source, though by 2001 the *New York Times* was referring to the project as "the largest directory of the Web." Pamela Licalzi O'Connell, "Mining the Minds of the Masses," *New York Times*, March 8, 2001, G1.

25. Berners-Lee discusses the creation of the first Web software in Berners-Lee and Fischetti, *Weaving the Web*, 12–50.

26. Abbay Bhushan, *RFC 114: A File Transfer Protocol*, Network Working Group, 1971, available at ⟨http://tools.ietf.org/html/rfc114⟩ (accessed October 20, 2006).

27. The standard for Internet file transfer was formalized in Jon Postel and J. Reynolds, *RFC 959: File Transfer Protocol*, Network Working Group, October 1985, which extended an earlier Transmission-Control Protocol/Internet Protocol–based FTP standard published in 1980.

28. Archie and its operation are profiled in Ed Krol, *The Whole Internet User's Guide and Catalog* (Sebastapol, CA: O'Reilly and Associates, 1992), 155–168.

29. Veronica is described and its creators interviewed in Billy Barron, "Tricks of the Internet Gurus," in *Tricks of the Internet Gurus*, ed. anonymous (Indianapolis: SAMS Press, 1994), 519–538.

30. WAIS built on an early version of the Z39.50 protocol, intended for use in searching library catalogs. This protocol gained widespread adoption during the 1990s. WAIS is discussed in Larry Press, "Collective Dynabases," *Communications of the ACM* 35, no. 6 (June 1992): 26–32; a lengthy introduction and tutorial can be found in Krol, *The Whole Internet User's Guide and Catalog*, 211–226. Brewster Kahle, one of the developers of WAIS, formed a company around the system and sold it to AOL in 1995 for $15 million. WAIS then vanished without trace, but Kahle used the money to start the nonprofit Internet Archive, which continues to perform an important role by preserving the long-vanished content of old Web sites.

31. The launching of AltaVista is reported in Peter H. Lewis, "Digital Equipment Offers Web Browsers Its 'Super Spider,'" *New York Times*, December 18, 1995, D4.

32. Laurie Flynn, "Making Searches Easier in the Web's Sea of Data," *New York Times*, October 2, 1995, D5.

33. Jon Auerbach, "In Search Of. Lycos Overhauls Product to Attract Users," *Boston Globe*, October 6, 1995, 57.

34. The figure on the size of the Lycos index is from Mike Holderness, "Online Searches: Where Confusion Is Still Free," *Guardian*, November 30, 1995, 6.

35. Nelson, *Literary Machines*.

36. For a summary of the optimism surrounding micropayment firms during the mid-1990s and their rather limited practical accomplishments, see Tom Seinert-Threlkeld, "The Buck Starts Here: Will Nanobucks Be the Next Big Thing, or Are We Just Talking Pocket Change?" *Wired*, August 1996, 94–97, 133–135.

37. Eric M. Jackson, *The PayPal Wars: Battles with eBay, the Media, the Mafia, and the Rest of Planet Earth* (Torrance, CA: World Ahead Publishing, 2004).

38. HotWired faded quite quickly, but its glory days are captured in Reid, *Architects of the Web*, 280–320.

39. Angel, *Inside Yahoo*, 140.

40. Doubleclick's plans to profile Web users are reported in Hiawatha Bray, "For Advertisers, Web Offers Wide Audience, Pinpoint Accuracy," *Boston Globe*, May 5, 1996, 41, 45. Its use of cookies to track use across different Web sites led to considerable controversy.

41. Anonymous, "Internet Advertisers Can 'Buy' Key Words," *Plain Dealer* (Cleveland), December 26, 1995, 3C.

42. Infoseek offered several plans for heavy or light users, with fees of ten to twenty cents per search. Trial accounts and certain limited capabilities were free. Greg R. Notess, "The InfoSeek Databases," *Database Magazine* 18, no. 4 (August–September 1995): 85–87. Infoseek's subscription model is also reported in Margot Williams, "Getting around the World Wide Web with the Help of a 'Search Engine,'" *Washington Post*, June 26, 1995, F19.

43. Ronald Rosenberg, "Godsent—and a Threat," *Boston Globe*, June 30, 1999, F4.

44. Charles Cooper, *Perspective: Paid Search? It Stinks*, News.com, 2006, available at ⟨http://news .com.com/2102-1071_3-281615.htm⟩ (accessed September 16, 2006).

45. Overture's initial success is reported in Bob Tedeschi, "Striving to Top the Search Lists," *New York Times*, December 10, 2001, C7. Overture's story is told in John Battelle, *The Search: How Google and Its Rivals Rewrote the Rules of Business and Transformed Our Culture* (New York: Portfolio, 2005), 104–121.

46. Pricing for the intranet AltaVista is reported in Eric Convey, "DEC Unveils Corporate Alta-Vista," *New York Times*, September 19, 1996, 35.

47. Flynn, "Making Searches Easier in the Web's Sea of Data."

48. A description of some of the search packages available to Web site designers is in Jon Udell, "Search Again," *Byte*, January 1997, 123–124, 126.

49. The market for corporate record indexing and search systems is actually much older than that for Web search systems. IBM offered a software package called STAIRS, developed to manage the

mountain of documents that IBM's legal team gathered for its defense against the federal antitrust suit launched in 1969. Indeed, lawyers provided an important market for this technology. Most organizations still kept their unstructured documents on paper, and even electronic documents had to be specially tagged and loaded into the index.

50. Gaston Gonnet, *Oral History Interview with Thomas Haigh, March 16–18*, 2005, oral history collection, Charles Babbage Institute, University of Minnesota.

51. Martin P. Courtois, William M. Baer, and Marcella Stark, "Cool Tools for Searching the Web," *Online* 19, no. 2 (November–December 1995): 14–27.

52. Cooper, *Perspective.*

53. AltaVista's sorry history is recounted in Jim Hu, *AltaVista: In Search of a Turning Point*, News .com, July 31, 2001, available at ⟨http://news.com.com/2102-1023_3-270869.html⟩ (accessed September 12, 2005).

54. Tim Race, "Infoseek Revises Its Internet Search Engine," *New York Times*, October 20, 1997, D15. The term portal had occasionally been used in a different sense earlier in the 1990s as an alternative to the more common "gateway" to describe an interconnection point between two networks for the exchange of e-mail or, in a few cases, mechanisms to allow users of proprietary online services like AOL to browse the Web.

55. For a contemporary description of the excitement surrounding the portal concept, see Rajiv Chandrasekaran, "One-Stop Surfing; Today's Hot Web Concept Is 'Portals.' Tomorrow, Who Knows?" *Washington Post*, October 11, 1998, H1.

56. Anonymous, "Top 25 Web Sites," *USA Today*, June 28, 1996, 4D.

57. The idea of a home page went back to Berners-Lee and the origin of the Web. Berners-Lee had imagined that browsers would include integrated editing capabilities, so that each user would have a personal home page that they could edit to include links to pages of interest as well as public messages for other visitors. (Something rather like a blog.) This explains the dual meaning of the term home page as both "the default start page for someone's browser" and "the main page holding information about a person or company." James Gillies and Robert Cailliau, *How the Web Was Born: The Story of the World Wide Web* (Oxford: Oxford University Press, 2000), 193–194.

58. David Lake, "Microsoft Dominates the World Wide Web," *Industry Standard*, August 23, 2001.

59. Suzanne Galante, *Netscape Outlines Web Strategy*, News.com, March 25, 1998, available at ⟨http://news.com.com/2100-1001-209497.html⟩ (accessed September 11, 2006).

60. Paul Festa, *Study: Netscape Share below 50%*, News.com, September 29, 1998, available at ⟨http://news.com.com/2102-1023_3-216043.html?tag=st.util.print⟩ (accessed September 15, 2006).

61. Saul Hansell, "Where Does Microsoft Want You to Go Today. The New Strategy: Keep Web Surfers Busy with a Series of MSN Sites," *New York Times*, November 16, 1998, C1.

62. David Pogue, "The Web Gets a New Dashboard," *New York Times*, October 26, 2000, G1.

63. A lively, well-informed account of Pathfinder's creation and early life is in Wolff, *Burn Rate*, 109–138. The troubled history of Time Warner's involvement with the Web is explored in Frank Rose, "Reminder to Steve Case: Confiscate the Long Knives," *Wired*, September 2000, 156–172.

64. Jim Hu, *Time Warner to Shutter Pathfinder*, CNET News.com, April 26, 1999, available at ⟨http://news.com.com/2100-1023-224939.html⟩ (accessed August 29, 2006).

65. Roger Taylor, "Disney and InfoSeek to Launch New Web Portal," *Financial Times*, December 14, 1998, 23.

66. Keith L. Alexander, "Despite Setbacks, Go's Chairman Sees Green Light Ahead," *USA Today*, December 27, 1999, 6B.

67. Danny Sullivan, *comScore Media Metrix Search Engine Ratings*, SearchEngineWatch.com, August 21, 2006, available at ⟨http://searchenginewatch.com/showPage.html?page=2156431⟩ (accessed August 27, 2006).

68. The importance of these deals to AOL is explained in Nina Munk, *Fools Rush In: Steve Case, Jerry Levin, and the Unmaking of AOL Time Warner* (New York: HarperCollins, 2004), 100–108.

69. Suzanne Galante and Paul Festa, *Lycos Up on Smaller Losses*, News.com, August 27, 1997, available at ⟨http://news.com.com/2100-1001_3-202723.html⟩ (accessed August 20, 2006).

70. Anonymous, *Lycos Reports 126% Increase in Revenues*, findarticles.com, November 22, 1999, available at ⟨http://www.findarticles.com/p/articles/mi_m0WUB/is_1999_Nov_22/ai_57758941/print⟩ (accessed September 4, 2006).

71. David Carlson, "Media Giants Create Web Gateways," *American Journalism Review*, September 1999, 88.

72. Tom Stundza, "Dot.coms Implode, Survivors Seek New Focus," *Purchasing*, August 9, 2001, 16B10.

73. U.S. General Services Administration, *FirstGov for Kids*, Spring 2003, available at ⟨http://www.kids.gov/activity.htm⟩ (accessed August 26, 2006).

74. Ben Heskett and Jeff Pelline, *Why Excite@Home Failed: A Postmortem*, News.com, September 28, 2001, available at ⟨http://news.cnet.com/news/0-1014-201-7340505-0.html⟩ (accessed October 2, 2001).

75. Verne Kopytroff, "Disney's Go.com Changes Direction with Web Site," *San Francisco Chronicle*, September 15, 2000, B1.

76. Saul Hansell, "Disney, in Retreat from Internet, to Abandon Go.com Portal Site," *New York Times*, January 30, 2001, C1.

77. David Shabelman, *Can Lycos Learn New Tricks*, News.com, September 8, 2006, available at ⟨http://news.com.com/2102-1032_3-6113790.html⟩ (accessed September 10, 2006).

78. Alec Klein, "Unconventional Transactions Boosted Sales," *Washington Post*, July 18, 2002, A1.

79. David A. Vise, "Time Warner Settles AOL Cases for $510 Million," *Washington Post*, December 16, 2004, A1.

80. Jane Black, *Is That MSN Breathing down AOL's Neck?* BusinessWeek Online, May 8, 2001, available at ⟨http://www.businessweek.com/print/technology/content/may2001/tc2001058_449 .htm?chan=tc⟩ (accessed September 12, 2006).

81. Rob Pegorano, "Any Portal in a Storm," *Washington Post*, November 3, 2000, E1.

82. Black, *Is That MSN Breathing down AOL's Neck?*

83. Bill Breen, "She's Helping Yahoo Act Normal," *Fast Company.com*, April 2003, 92.

84. Ben Hammersley, "Is Yahoo Losing the Plot?" *Guardian*, May 2, 2002, 7.

85. Amanda Spink and Bernard J. Jansen, *Web Search: Public Searching of the Web* (Dordrecht: Kluwer, 2004), 104–117.

86. For an early report on search engine spam and efforts to combat it, see Jeff Evans, "Power Searching," *Toronto Star*, December 9, 1999, 1.

87. Quoted in Paul Festa, *Web Search Results Still Have Human Touch*, News.com, December 27, 1999, available at ⟨http://news.com.com/2100-1023-234893.html⟩ (accessed September 17, 2006).

88. Sherman, "Humans Do It Better."

89. Sergey Brin and Lawrence Page, "The Anatomy of a Large-scale Hypertext Web Search Engine," in *Proceedings of the Seventh International Conference on World Wide Web* (1998), 107–117. This article includes a good description of the initial structure and operation of Google in its Stanford days. Elsevier.

90. Mitch Wagner, *Google Bets the Farm on Linux*, InternetWeek, June 1, 2000, available at ⟨http:// internetweek.cmp.com/lead/lead060100.htm⟩ (accessed September 20, 2006).

91. Google's adoption of advertising is discussed in David A. Vise and Mark Malseed, *The Google Story* (New York: Delacorte, 2005), 89–102.

92. A good summary of the current state of Internet advertising is in anonymous, "The Ultimate Marketing Machine," *Economist*, July 8, 2006, Special Repolt (2).

93. Paul Festa, *Is Google Ogling Yahoo's Crown?* News.com, 2001, available at ⟨http://news.com .com/2102-1023_3-255601.html⟩ (accessed September 25, 2006).

94. Jim Hu and Stefanie Olsen, *Yahoo Dumps Google Search Technology*, News.com, 2004, available at ⟨http://news.com.com/2102-1024_3-5160710.htm⟩ (accessed September 17, 2006).

95. comScore Networks, *Google Regains Some Ground from Previous Month's Share Decline with 0.4 Share Point Increase in August Rankings*, comScore, 2006, available at ⟨http://www.comscore.com/ press/release.asp?press=1006⟩ (accessed September 15, 2006).

96. Catherine Holahan, "Yahoo's Lost Bid Doesn't Spell Doom," *Business Week*, August 9, 2006.

97. Ben Elgin, "Google and Yahoo Rolling in It," *Business Week*, October 21, 2005.

98. PriceWaterhouseCoopers, *IAB Internet Advertising Revenue Report: 2005 Full-Year Results*, Internet Advertising Board, 2006, available at ⟨http://www.iab.net/resources/adrevenue/pdf/ IAB_PwC_2005.pdf⟩ (accessed September 8, 2006).

99. Google reported revenues of $2.69 billion for the third quarter of 2006, up 70 percent from the previous year. Sara Kehaulani Goo, "Surge in Profit Reflects Google's Widening Lead," *Washington Post*, October 20, 2006, D1. Gannett reported operating revenues of $1.9 billion over the same period.

100. Saul Hansell, "Google Wants to Dominate Madison Avenue Too," *International Herald Tribune*, October 30, 2005. It seems unlikely that Google really evaluates a million variables; Hansell may mean "a million possible combinations of variables."

101. Elinor Mills, *Yahoo: Our Ads Are Better*, News.com, May 18, 2006, available at ⟨http://news .com.com/2102-1024_3-6073504.html⟩ (accessed September 20, 2006); Saul Hansell, "Yahoo Is Unleashing a New Way to Turn Ad Clicks into Ka-Ching," *New York Times*, May 8, 2006, C1.

102. Yahoo's share price in the two years to September 2006 varied from $24.91 to $43.66. Nevertheless, because the stock split 2:1 twice between those times, the $475 peak closing value of January 3, 2000, corresponds to a price of $118.75 for each of today's Yahoo shares.

103. Hansell, "Google Wants to Dominate Madison Avenue Too."

104. Paul Sloan, "Masters of Their Domains," *Business 2.0*, December 1, 2005.

105. Charles C. Mann, "How Click Fraud Could Swallow the Internet," *Wired*, January 2006, 138–149. Click fraud is also discussed in Vise and Malseed, *The Google Story*, 240–249.

106. Elinor Mills, *Google Calls Click Fraud Estimates Overblown*, News.com, August 8, 2006, available at ⟨http://news.com.com/2100-1024_3-6103387.html⟩ (accessed September 21, 2006).

107. For an example of anti-Google backlash, see Gary Rivlin, "Relax Bill Gates; It's Google's Turn as the Villain," *New York Times*, August 24, 2005, A1.

108. Randall Stross, "AOL Said, 'If You Leave Me I'll Do Something Crazy,'" *New York Times*, July 2, 2006, Business.3.

109. Sara Kehaulani Goo, "In Strategy Shift, AOL Makes Most Services Free," *Washington Post*, August 3, 2006, A1.

110. Paul R. La Monica, *The Internet wars: A Report Card*, CNNMoney.com, May 4, 2006, available at ⟨http://money.cnn.com/2006/05/04/technology/search_reportcard/⟩ (accessed September 12, 2006); Microsoft, *Microsoft Reports Fourth Quarter Results and Announces Share Repurchase Program*, July 20, 2006, available at ⟨http://www.microsoft.com/msft/earnings/FY06/earn_rel_q4_06.mspx⟩ (accessed September 4, 2006). MSN was a division of Microsoft, so direct comparison with competitors is difficult. In mid-2006, Microsoft reorganized to merge MSN into a broader Online Services division.

111. Tom Eid, *Market Share: Enterprise Content Management Software, Worldwide, 2003–2005* (Stamford, CT: Gartner, 2006).

112. Paula J. Hane, *Autonomy and Verity Join Forces in Enterprise Search Market*, InfoToday, November 14, 2005, available at ⟨http://www.infotoday.com/newsbreaks/nb051114-3.shtml⟩ (accessed September 14, 2006).

113. Anonymous, *Google Rolls out Corporate Search*, BBC News, October 20, 2004, available at ⟨http://news.bbc.co.uk/1/hi/business/3759878.stm⟩ (accessed September 16, 2006).

114. Grigoris Antoniou and Frank van Harmelen, *A Semantic Web Primer* (Cambridge, MA: MIT Press, 2004).

115. John Markoff, "Technology: By and for the Masses," *New York Times*, June 29, 2006, C1.

6 The Rise, Fall, and Resurrection of Software as a Service: Historical Perspectives on the Computer Utility and Software for Lease on a Network

Martin Campbell-Kelly and Daniel D. Garcia-Swartz

The idea of Software as a Service (SaaS) seems to have taken the computing world by storm. Although many different meanings have been attached to the SaaS concept, the basic one involves software that resides on a server and is accessed through the Internet on a subscription basis. More specifically, for software to be delivered according to the SaaS model, at least the following has to hold: first, companies have to be able to access noncustom software through a network; and second, the management of that software has to be network-based also.[1]

The revolution that SaaS is allegedly bringing about has been compared to the one that took place at the beginning of the twentieth century, when manufacturers dismantled their waterwheels, steam engines, and electricity generators and started purchasing electric power from central suppliers. In a recent article, Nicholas Carr argued that almost a century later, information technology is undergoing a similar transformation—it is "beginning an inexorable shift from being an asset that companies own in the form of computers, software and myriad related components to being a service that they purchase from utility providers."[2]

One of the most vocal advocates of the idea that in the near future, companies will not own software as they do today but will rather have access to it on a subscription basis from an Internet site is Marc Benioff, the CEO of salesforce.com. Benioff has pointed out that his company represents "the end to software," by which he means the replacement of the current model of software distribution (i.e., up-front licensing fees in exchange for the right to install the software on a number of computers) with SaaS—software delivered at each desktop (or laptop) through the Web browser on a per user, monthly rental basis.[3]

In this chapter, we do not focus on making predictions about how long it will take for SaaS to take off (and take over). Rather than looking forward without any anchor, we look backward into the history of computing generally and software specifically. We start by noting that the SaaS concept and the computer utility idea are not one and the same but are clearly related—SaaS concentrates on the locus of software, and the computer utility on the locus of computing power. Therefore, we organize our historical excursion in light of these two issues. First, where has the locus of computing power resided at various points in time? Second, where has the locus of software resided at various points in time? And then, assuming that we find evidence of shifting loci of processing power and software, what are the technological and economic forces driving those shifts?[4]

The main thesis of this chapter is that SaaS and the computer utility are not so novel after all. In some form or another, they have been around for decades already, and to some extent they have had at least one prior commercial incarnation: the computer time-sharing industry of the 1960s and 1970s. Interestingly enough, whereas today's SaaS advocates argue that the current incarnation of SaaS and the computer utility will kill software as we know it (and perhaps the personal computer as we know it), we note that it was the advent and expansion of the personal computer in the 1980s that killed the early incarnation of SaaS. Since SaaS was tried out already in the business world in the 1960s and 1970s, succeeded for about fifteen to twenty years, and then collapsed, the question is not how long it will take for SaaS to take off (and take over) now. The real question is, What are the technological and economic factors present today (and absent in the early 1980s when the personal computer killed the time-sharing industry), if any, that will allow the current version of SaaS to eliminate software and the personal computer as we know them?

In the Beginning: Corporate Mainframes, Service Bureaus, and Software Products

In the first decade of the history of computing (1955–1965), processing power resided essentially in the mainframes that corporations owned or leased. As regards the locus of software, computer manufacturers such as IBM provided software programs to customers on request and free of charge.[5] Software, in other words, resided where computer power resided—namely, on the corporate mainframe.

Of course, from early on corporations could choose not to buy or lease a mainframe at all but rather to outsource at least some of their corporate functions to third parties that would do the data processing and computing for them. In the early days of computing these third parties were usually known as "service bureaus." (The bureaus, it should be noted, predated computers; they started processing data manually and then mechanically, and shifted to computers later on.) The bureaus, which made up what some analysts have called the "batch data processing" component of the data processing services industry, received raw data from customers via mail or messenger, processed the data according to the customers' needs and requests, and then delivered the results through the same channels.[6]

Automatic Data Processing (ADP) is one of the best-known examples of an early service bureau that became extremely successful through several adaptations over time.

ADP started in 1949 as Automatic Payrolls, Inc., a company that manually processed company payrolls; it did everything from carrying out the calculations to cutting the checks and preparing the payroll register. By 1956 the company had more than two hundred clients. In 1956–1957, it made the transition from using manual bookkeeping machines to automated punch card accounting. In 1962 ADP acquired its first computer: an IBM 1401. From the late 1960s, service bureaus offered "remote" batch data processing via a data communications link and a card reader printer. Services bureaus nonetheless, remained local in character. The batch data processing sector of the data processing services industry expanded rapidly between 1955 and 1965. The sector's revenues grew from $15 million in 1955 to $340 million in 1965.[7]

As they became computerized, the service bureaus created an alternative model of computing for American corporations. Companies that chose to hire the services of a firm like ADP could do without a mainframe; the service bureau provided both the computing (or processing) power and the software that was required for each task. For all practical purposes, then, the bureaus offered a primitive, and somewhat imperfect, incarnation of SaaS (and the computer utility) inasmuch as they allowed corporations to shift the locus of both computing power and software to an outside vendor.

The history of software supplied independently of hardware begins with the software contractors. These companies developed in the mid-1950s in response to a demand for software arising from three sources: the government-sponsored defense projects known as the L-systems, computer manufacturers, and private corporations. The software contractors' existence was predicated on bidding for and obtaining contracts executed either on a time-and-materials or fixed-cost basis. Their products were, in essence, one-of-a-kind pieces of software.

Computer Sciences Corporation (CSC) is an example of a successful software contractor. The company started with $100, which its entrepreneur founders pooled together in 1959. Between 1959 and 1963, CSC wrote systems software for every major U.S. computer manufacturer, and by 1963 it had become the world's largest independent computer services firm, with revenues of about $4 million.[8]

In terms of the issues that we are focusing on—that is, the locus of processing power and software—the advent of the software contractors was nothing close to a paradigm shift. The locus of neither computing power nor software changed with the contractors. The demand for software was strong enough to support a host of companies designing unique pieces of software for specialized applications, and the contractors took advantage of the opportunity.

Corporate software products and the companies that made them started proliferating in the mid-1960s, about a decade after the appearance of the first software contractors.

Corporate software products were not one-of-a-kind; they sold in the dozens or even hundreds. In the late 1960s, and under antitrust pressure, IBM decided to charge separately for software and other services, which clearly had an impact in terms of encouraging the entry of new companies and the development of new products.

The first successful software products firms, such as Informatics, MSA, and University Computing, typically sold a few hundred copies of their packages. The products cost some tens of thousands of dollars—and there was considerable shock and buyer resistance at what had once been "free" goods bearing such astronomical prices. Interestingly, few of the early software firms survived the merger waves of the 1970s and 1980s. Informatics, MSA, and University Computing all ended up as part of the genetic material of Computer Associates. The leading enterprise software product firms of today tended to come a little later—for example, SAP began operations in 1972, and Oracle Systems in 1977. Even then it took several years to develop sound products and effective marketing strategies. SAP did not launch its famous R/2 ERP product until 1978, while Oracle did not release an enterprise strength database until the early 1980s. But SAP and Oracle benefited as second movers by being able to charge high prices—hundreds of thousands of dollars—that would have been inconceivable a few years earlier.[9]

In terms of the issues that we are exploring, the advent and expansion of the corporate software products industry did not represent a radical departure from the predominant computing model; both processing power and software still resided on the corporate mainframe. The unbundling decision was important in the sense that it opened up the floodgates for a host of companies that started supplying software independently of the hardware, but not in the sense that it immediately transformed the predominant computing/software model—which it did not.

The Computer Time-sharing Industry

Much more radical as a challenge to the existing computing/software model was the advent of the computer time-sharing industry. The commercial time-sharing industry presented an alternative to the prevailing model because it encouraged the shift of the locus of both computing power and software from the corporate mainframe to the mainframe of an outside vendor.

A time-sharing computer was one organized so that many persons could use it simultaneously, each one having the illusion of being the sole user of the system—which in effect became one's personal machine. In the 1960s, computer time-sharing was intimately intertwined with the concept of the computer utility. Some of today's analysts seem to think that the analogy between the supply of computing power (and software), on the one hand, and the supply of electricity or water, on the other, is something new. As a matter of fact, it is not. The analogy was made often in the second half of the 1960s. And although the idea of the computer utility seems to have vanished after the computer recession of 1970–1971, the computer time-sharing industry—an early incarnation of the computer utility and SaaS—continued to thrive well into the 1980s.[10]

In terms of the computing/software model, the computer time-sharing companies had some resemblance to the service bureaus described above. In both cases, corporations shifted the locus of computing power and software from the corporate mainframe to an outside vendor. There was an important difference, however: interactivity. Service bureaus were essentially batch businesses in the sense that customers transmitted data and instructions in one block, and then waited for the complete output, which arrived considerably later. The true time-sharing services provided a "conversational" experience, in the context of which users could interact with the mainframe—they submitted instructions, received the results, and submitted new instructions.

The advent of the time-sharing companies in 1965 generated some sort of "unbundling" of computing power and software parallel to the one that took place as a result of the IBM 1970 decision. The IBM decision, inasmuch as it involved the separate pricing of hardware and software, allowed a host of companies to enter the market to supply corporate software products. In other words, it allowed corporations to purchase hardware and software from different vendors, if they so desired. The expansion of the computer time-sharing industry gave a wide variety of corporations electronic access to both raw computing power and software, in such a way that companies could choose whether they wanted both in a bundle or only one of them.

In fact, in the early years of time-sharing, corporations used the online interactive services to a good extent to access the raw computing power of the vendors' mainframes. The online interactive component of the data processing services industry (i.e., time-sharing) accounted for $310 million in 1971, right after the recession. Of this amount, $185 million (almost 60 percent) originated in customers paying for access to raw computing power, whereas the rest came from customers using the vendors' software either for calculations or file processing.[11] In the course of the 1970s, as the time-sharing industry developed, more of its revenues originated in customers accessing and taking advantage of the vendors' software libraries. For example, time-sharing generated revenues of $1,738 million in 1978. Only 44 percent ($770 million) of this amount came from customers purchasing raw computing power, and the rest originated in the customers' use of the vendors' software packages for calculations or file processing.[12]

Tymshare was one of the most successful of the time-sharing companies. It was founded in the mid-1960s, like many other interactive online companies. Its revenues skyrocketed in the 1970s, from about $10 million in 1970 to almost $200 million in 1979, and so did its profits, from about \$100,000 in 1970 to almost \$15 million in 1979.[13] One of the distinguishing features of Tymshare as a company is that it established a nationwide computer network, Tymnet, the first commercial network to deploy packet switching and extend its reach to third-party computers. Tymnet grew out of Tymshare's attempt to address the issue of the lack of dependability in the transmission of data between the central time-sharing mainframes and the terminals located at the customers' facilities. Yet it slowly became a phenomenal network connecting dozens of computers with each other. By the mid-1970s, it had already evolved into a public network linking computers and terminals of various organizations; Tymshare opened Tymnet to non-Tymshare computers in response to the demand coming from outside organizations. Toward the late 1970s Tymnet had local call access in 180 cities, encompassed 400 access nodes, had roughly 250 host computers connected to it, and supported a monthly volume of 15 billion characters.[14]

There can be little doubt that the computer time-sharing industry represented an early incarnation of the computer utility and the SaaS concept. In essence, corporations that opted for the interactive online computing experience shifted the locus of both processing power and software from the corporate mainframe to the mainframe of the outside vendor, the time-sharing company. Users of time-sharing services could sit down at their local terminals in their workplace and, through telecommunication networks, access the processing power of dozens of providers located anywhere in the country.

Based on this account, it seems that today we should be celebrating several decades of existence of the computer utility—computing power for lease—and the SaaS concept—software for rent through a network. But in fact we are not. More specifically, today's pundits hail the ideas of the computer utility and SaaS as a novelties.

What happened to the time-sharing industry? The personal computer revolution "happened" to the time-sharing industry, and from the perspective of our story, the personal computer introduced another paradigm shift: a new and radical movement in the locus of both processing power and software. The advent of the personal computer moved the locus of computing power from outside the corporation back to the inside. And more specifically, not back to the corporate mainframe but to the desktop of each corporate employee.

The personal computer not only shifted the locus of computing power from the mainframe of the outside vendor to the corporate desktop but it also created a new locus for software to reside in. The late 1970s witnessed the birth not only of the personal computer but also the personal computer software industry. Hundreds of small software firms came out of nowhere to serve the market that the personal computer had created. Some of these firms—Microsoft, Lotus, and Ashton-Tate—soon dominated the personal computer software industry and joined the ranks of the established software products vendors.

The Driving Forces

Why did this early incarnation of the computer utility and the SaaS concept—the time-sharing industry—grow in the way it did? There are several reasons, but the most fundamental one is that it provided computing power and software at a reasonable price given the cost-technology parameters of the time.[15]

In the early 1970s, businesses seeking access to computing power could, for example, resort to accounting machines, which had existed in the United States for decades before the advent of electronic computing. They offered a performance of between ten and one thousand operations per second for between $100 and $1,000 per month.[16] Starting in 1974, businesses could also use electronic pocket calculators, which offered performance of at most a hundred operations per second. On the other extreme, companies could rely on a so-called small business system, which offered performance of between one thousand and a hundred thousand operations per second. In this case, the cost of the hardware was slightly less than $1,000 per month, although the total operating cost was somewhere between $7,500 and $10,000 per month.[17]

Computer time-sharing offered performance of between a hundred and a thousand operations per second. If the machine supplying this performance was a one million operations per second mainframe computer, the total operating cost was roughly $26,000 per month. A system like that accommodated a hundred users at any given time, from one hundred full-time users to one thousand part-time users that logged on to the system for short periods of time. Therefore, the time-sharing company operating such a system could provide performance of between 0.1 and 1 percent of the system's capacity (a thousand to ten thousand operations per second) at between 0.1 and 1 percent of the system's cost ($26 to $260).

Of course, the system's overhead reduced the capacity available for customers, and the time-sharing company was in business to make a profit. The monthly price of computer capacity that users faced was therefore more in the range of $100 for a thousand operations per second to $1,000 for ten thousand operations per second. This was still much better than anything available on the market; it cost at least $7,500 to obtain that kind of performance from a small business computer.[18]

In the 1970s, in short, time-sharing systems were competitive with accounting machines, pocket calculators, and small business computers. But that cannot be the whole story. If time-sharing was indeed as efficient as it seems it was, why did it not eliminate completely all other computing alternatives? We know that all the other approaches to computing continued to exist side by side with the time-sharing companies. The computer utility and the SaaS concept thrived indeed in the 1970s, but did not displace completely the accounting machines and small business computers from the offices of the American corporation. In fact, one of the most comprehensive time-sharing reports of the late 1960s pointed out that 70 percent of the businesses that used time-sharing also had in-house computer facilities.[19] This means that a fair share of the companies that used time-sharing services had access to computing through time-sharing firms only, but it also means that most companies that resorted to time-sharing did their own in-house computing at the same time that they accessed the computer utility of the 1970s.[20]

The fact that many companies still relied on the corporate mainframe at the same time that they used terminals to log on to the time-sharing mainframe should come as no surprise. Most businesses had a wide diversity of uses for computers—companies dealt with issues related to personnel, vendors, parts, products, customers, markets, general ledgers, assets, budgets, procedures, finance, and law.[21] Furthermore, many companies performed specialized calculations of a scientific or engineering nature. It was probably not unusual to find companies that hired a service bureau like ADP to do the payroll, did most of the accounting on in-house mainframes, and hired the services of a company like Tymshare to carry out calculations requiring significant amounts of computing power and specialized software libraries.[22]

More generally, for the companies that had an in-house mainframe, the possibility of accessing additional computing power and software depending on need was useful in the sense that it gave them a lot of flexibility "on the margin." In periods of economic boom, they could expand their consumption of computing power and software by leasing them from a company like Tymshare; in periods of depression, they could just rely on the computing power they had "at home." For companies that did not own a corporate mainframe, the existence of the time-sharing companies gave them absolute flexibility in terms of how much processing power and software services they consumed. For many of them, the existence of time-sharing was what allowed them to have a computing experience in the first place.[23]

Industry reports of the late 1960s describe how companies used time-sharing services. A well-known report from 1969 tells us that the volume of time-sharing usage varied dramatically from customer to customer. The typical customer, however, was a company that had two or three terminals installed for commercial time-sharing, used each terminal approximately for two hours a day, and spent roughly $600 per terminal per month on commercial time-sharing, excluding communication costs.[24] This was indeed much lower than the $7,500 to $10,000 they would have paid for a small business computer.

The personal computer was a different story, though. Priced between $1,500 and $6,300, it offered an efficient alternative to the $10-an-hour time-sharing terminal. On any given month, companies had to decide whether to give their employees access to computing power and software through the time-sharing companies, or to acquire a personal computer with a piece of software to, say, carry out various types of calculations. Both options allowed the company to address its computing needs. The difference was that personal computers, once acquired, were assets that would be around for a number of years, whereas the time-sharing service's charge would be faced month after month.[25]

There can be little doubt that by the mid-1980s, the locus of both computing power and software had shifted from the mainframe of the time-sharing vendor to the mighty personal computer. By that time, sales of personal computers were booming, the processing speed of the machine was increasing at a fast pace, the price of computing power was declining at rates of perhaps between 26 and 32 percent per year, and a myriad of companies were producing dozens of personal computer-based products in almost every area of the software spectrum.[26] By the mid-1980s, corporate America had already participated in the funeral and burial of the early incarnation of the computer utility and SaaS, the time-sharing industry of the 1960s and 1970s.

The Resurrection of the Computer Utility and SaaS: Application Service Providers (ASPs)

In the last few years, analysts and pundits have been talking about the irruption of the computer utility and SaaS. As we have argued above, and have tried to show throughout this chapter, it is somewhat inappropriate to discuss this computing model as if it had never existed before. It is thus more adequate to talk about the resurrection of the computer utility and SaaS.

The resurrection seems to have started in the late 1990s, a period that bears a striking resemblance to the late 1960s. In both periods, there was incredible excitement surrounding the computing industry; whereas in the late 1960s the time-sharing companies were one of the main sources of the excitement, in the late 1990s the Internet start-ups played that role.

Although the advent of the personal computer destroyed the time-sharing industry in the second half of the 1980s, the other half of the online business—processing services—was relatively unaffected. In the early 1990s, the processing services industry was quite concentrated with no more than fifty international players, which included firms such as ADP, CSC, EDS, and IBM. Their offerings were commonly described as either "horizontal" or "vertical" services. Horizontal services were generic applications such as payroll processing or human resources management (in which ADP was the market leader). Vertical applications were designed for specific industries such as health care (in which CSC and EDS were market leaders) or discrete manufacturing (in which IBM was a leader).

Two major barriers to entry, network infrastructure and applications software, fueled this industrial concentration. All of the major incumbent processing services firms had developed private networks in the 1970s or 1980s. For example, CSC established its INFONET network in the early 1970s at a reported cost of $100 million. By 1975, it had become "a 100,000 mile network serving 125 cities and nearly every U.S.

government agency."[27] In 1978, it opened up satellite links extending its reach into Europe and Japan. INFONET was something of a heroic achievement at the time, battling with the aftermath of the computer recession of 1970–1971 and the contingencies of being a first mover in network building. By the 1980s, ADP was offering online claim processing for the insurance industry and dealer services for the retail auto trade. It established a data communications hub in 1988 in New York for brokerage services.[28] Furthermore, the processing services firms all offered their own mature application software, either developed in-house over a period of many years, or acquired in strategic firm acquisitions and subsequently developed in-house.

By the mid-1990s these two barriers to entry had been lowered. First, the public Internet now provided a network infrastructure as good as most private networks— indeed, many private networks were being dismantled and migrated to the common Internet, which offered comparable facilities at much lower cost, and with less managerial and technical overhead. Second, during the period 1980–1995 the enterprise software products industry had grown manifold—from $2.7 billion to $58.3 billion U.S. user expenditures.[29] For every application offered by a processing services firm, horizontal or vertical, there now existed a mature software product that other corporations ran in-house. In the same period, moreover, American industry had generally moved from costly, custom-written software to standardized application packages or enterprise resource planning software that required only modest customization.

This new computing environment created an opportunity for the ASP, a firm that offered access to a standard software product via the Internet. ASPs became one of the most hyped Internet sectors in the heady months before the dot-com crash. They were a proposition that promised irresistible benefits to corporate users. For example, ASPs would relieve users from owning and maintaining their own computing infrastructure. ASPs were thus seen as direct competitors to the existing processing services enterprises, and since the latter were highly successful industries, ASPs were expected to be likewise. Like them, ASPs offered their services on a monthly tariff compared with upfront hardware and software purchases. ASPs were supposed to be particularly attractive to midsize enterprises (e.g., those with less than one thousand users), which were reluctant to invest in a full-scale corporate information services operation and blanched at the cost of major software products (potentially in the range of $100,000 to $500,000). Lastly, as with processing services firms, the ASP model provided an ongoing relationship between vendor and customer that made for a more stable and equitable partnership than traditional onetime software sales.

There were two main entrants into the ASP market: start-up "pure-play" firms that offered access to third-party software packages via the Internet (of which USinternetworking (USi) and Corio were the best-known examples), and existing software product firms that saw the ASP model as a potential new sales channel.

USi is generally agreed to have been the first pure-play ASP. It was established in February 1998. USi was a highly capitalized venture, aimed at larger enterprises, and its start-up funds were used to establish fast and reliable hosting services. It acquired four data centers—two in the United States, and one each in Amsterdam and Tokyo. Each of these data centers was directly connected to multiple Internet backbones to guarantee fast, 24-7 access and resilience against data center outages.[30] The company established partnerships with software product vendors in several application areas, including Siebel Systems for customer relationship management (CRM) software, Broadvision for e-commerce applications, Sargent for data warehousing, and PeopleSoft for human resources software. By 2001, after two years of operation, USi claimed that it had 150 customers and annual revenues of $109.5 million.[31] USi was established at a time when the term ASP had not yet taken hold, and with its reliance on owning its own hosting services, it was more like a traditional processing services firm than many of the ASPs that followed.

Corio, also established in early 1998, was a more prototypical ASP than USi. Rather than owning its own hosting services, it used third-party hosts, Concentric Systems and Exodus, which operated their hosting businesses as classic computer utilities for major clients such as Yahoo! and Microsoft. (Incidentally, these hosting services conform closely to the vision of the computer utility as an agency that simply provides computing "power" for a third party; today's utilities host services for their clients, not themselves.)[32] Corio's business model was much more in the spirit of the virtualized world of the Internet: it owned neither hosting infrastructure nor applications; it simply brought them together and packaged them for customers. Corio began with just one application (PeopleSoft human resources software), but added new offerings similar to USi as it expanded. By 2001, it had eighty customers and annual revenues of $40 million.[33]

Of the leading software firms, Oracle Systems was the most proactive in developing an ASP-based sales channel. The main protagonist was its president and founder Larry Ellison. A few years earlier, Ellison had been a leading advocate of the networked personal computer (or "thin client") whereby software would be maintained on central servers rather than on desktop personal computers. In any event, that vision never materialized—primarily because personal computer prices fell rapidly and the networked personal computer lost any commercial advantage it might have had. Oracle On Demand opened for business in November 1998, offering only Oracle's existing products and those of PeopleSoft, which it subsequently acquired in a hostile takeover. Besides establishing its own hosting infrastructure, Oracle invested $100 million in ASP start-ups to resell Oracle products. Never one to understate a business prospect, Ellison forecast that Oracle would be making a half of its sales through on-demand services by 2010.[34]

By early 1999, the ASP concept was caught up in the Internet bubble. One reporter observed that "ASPs are emerging from every imaginable nook and cranny of the computing and network market"; another remarked that "both biggun and small-fry vendors are hurling their hats into the ring...the market for software rentals is about to explode."[35] In May 1999 the ASP Industry Consortium was founded, and it soon had nearly sixty members. Lavish projections were made for total ASP industry sales—from $150 million in 1998, Forrester Research forecast a market of $6 billion in 2001, and IDC predicted a more conservative $2 billion by 2003.[36]

Like the 1970–1971 computer recession that hit the time-sharing industry at the height of its growth spurt, so the dot-com crash brought many ASPs into crisis. It was estimated that half of all vendors ceased trading altogether. The industry leader USi— having achieved an initial public offering in April 1999 that raised $132.8 million— saw its stock price fall from a high of $72 to less than $3 by late 2000.[37] Thirteen months later in January 2001, its cash exhausted, USi filed for Chapter 11 bankruptcy. During 2002, membership in the ASP Industry Consortium fell from seven hundred to fewer than three hundred.[38]

Dramatic as the demise of the ASP industry may have appeared in print—and certainly to those caught up in it—it was not the end of the industry. Just as the time-sharing industry recovered and grew after the 1970–1971 computer recession, so the ASP industry survived the dot-com crash and prospered thereafter. For example, USi rose phoenixlike from its Chapter 11 bankruptcy, and with a cash infusion of $106 million carried on trading. At the beginning of 2005 IBM acquired Corio for $182 million, providing IBM with a sales channel for its software products and a new platform from which to position its global consulting operations. In terms of market growth, while many projections turned out to be considerably overstated, the more conservative projections turned out to be surprisingly close to the mark. For example, IDC's $2 billion projection for 2003 turned out to be a slight underestimate.[39]

Like the time-sharing industry, the ASP industry did not consist of a bunch of homogeneous firms. Rather, firms had different competencies and targeted different markets. The ASPs that survived the dot-com crash were a world apart from the speculative start-ups that were indiscriminately venture-funded in the frenzy years. In many respects, they were conservative and unexciting firms that were completely at one with the pre-existing computer services industry.

ASPs quickly realized that simply hosting a commercial software product was not enough. Traditional enterprise software products could not be used "out-of-the-box" but required at least a minimal amount of customization. ASPs that were established by individuals with a consulting background, and that additionally had technical depth, were well placed to provide this customization service. Such firms were strongly differentiated from those that merely hosted a standard software product but could not provide adequate technical support. Many ASPs adopted a vertical strategy, much as had processing services and consulting firms in earlier decades. With this strategy, ASPs obtained economies of scope by developing expertise within a single industry, to which they could supply a portfolio of standard applications. By focusing on a single industry, vendors were able to customize application software rapidly, often much faster than a customer could have deployed the same application in-house. In 2002, the leading ASP segments bore more than a passing similarity to the vertical segments served by the computer services industry—discrete manufacturing, health care, financial services, and so on.[40]

Software on Demand and Web-Native Applications

In February 2001, *Business Week* reported that "at the peak of the dot-com frenzy, the tech industry got the notion that the Next Big Thing would be replacing software as we know it with Web-based programs. Like so many other aspects of the boom, the concept of application service providers contained a solid idea within a billowing cloud of hype."[41] After the dot-com bust, the term application service provider became somewhat tarnished, and was generally replaced by terms such as software as a service and on-demand software. This was to some degree a cosmetic rebranding to freshen up a business model that had become stale, and certainly the terms were often used loosely and interchangeably. Yet software as a service and on-demand software encapsulated significant refinements of the earlier ASP concept. Web-native applications represent the most significant of these refinements.

Web-native applications grew out of the development of portal-based Web services for consumers in the early years of the commercial Internet. For example, Yahoo quickly expanded its original directory service to include Web-based e-mail and personal calendars.[42] In both of these applications, software was executed on the portal's servers and the "state" persisted between a consumer's interactive sessions. One of the most successful Web services, Hotmail, offered e-mail through an Internet browser. In 1997 Microsoft acquired Hotmail for an undisclosed sum, reported to be in the order of $400 million. These early services attracted tens of millions of consumers and were funded by advertising.

The development of Web services established tools and infrastructure management techniques that enabled a second-generation SaaS industry to develop.[43] There were significant technical and market constraints to the development of these Web-native services. First, the new wave of entrants—such as salesforce.com and NetSuite—built their software from the ground up, so that their services lacked some of the features and sophistication of the software products of their mature competitors such as Siebel Systems or SAP. Consequently, they appealed mainly to the less demanding small-to-midsize business (SMB) sector. Second, at the time that these new Web services came on to the market around the year 2000, most SMBs already owned a modest computing infrastructure and proprietary software packages. In traditional application areas, vendors were therefore competing with an installed base that had already been paid for; for users, the choice was between signing up for a new subscription or exploiting their sunk costs. Hence the new vendors prospered in two main domains: traditional application areas where on-demand software offered new benefits (such as mobile and collaborative working), and new application markets where there was no installed base.

Salesforce.com has become the poster child for on-demand software. Salesforce.com's prominence owes something to the bradaggio of its founder, Marc Benioff. Its success, however, arose primarily from the fact that it was the first vendor to identify, in 1999, a new CRM market for mobile sales professionals. CRM was a well-established software genre with major players such as Siebel and BEA Systems. Typically, in these older systems, a mobile salesperson would use a laptop loaded with CRM software, book orders on the move, and periodically synchronize with the master database in the home office. Salesforce.com reconfigured this arrangement so that the CRM software and associated databases resided on its host servers, and both mobile and office-based users accessed the service through a Web browser. This access could be through an ordinary desktop personal computer, a laptop, or even a public terminal or kiosk at an airport. The subscribing firm was thus entirely relieved from maintaining an infrastructure to support the CRM application, and the vendor also provided complementary software packages for managing and reporting on the sales operation. In 2004, salesforce.com was ranked number two in the on-demand software industry, and was the leader among some half-dozen on-demand CRM players, including RightNow, WebSideStory, and Siebel's CRM OnDemand.[44] In 2004, salesforce.com reported that it was serving over 300,000 users in 16,900 organizations.[45] Even so, with just $51 million in annual revenues, salesforce.com was a minnow in the software ocean, ranking only 242 in the 2004 *Software 500*.[46] Incidentally, salesforce.com's 16,900 customer base contrasts sharply with that of a typical ASP, which might have had only 100 customers to generate the same revenue. ASP and Web-native, on-demand software were very different markets.

Webex, ranked the largest on-demand vendor in 2004, brought mobility and simplicity of installation to the established genre of collaborative software. The Lotus Development Corporation (acquired by IBM in 1995) was primarily responsible for establishing this software genre commercially in the early 1990s. Lotus Notes enabled users to collaborate in document preparation, maintaining a single copy of a document and guaranteeing its integrity when being worked on by multiple users. The installation of collaborative software (initially on private networks) was technically demanding, particularly for multisite organizations that had to replicate local copies of documents from global databases. The on-demand incarnation of collaborative software relieved organizations from maintaining this complex infrastructure, enabling access to shared documents and calendars through a Web browser. Webex reported 2004 sales of $190 million and was ranked 130 in the *Software 500*. It has a number of direct competitors, including Microsoft Office Live, as well as several emerging low-function alternatives, including Google Office (see below).

Web-native applications also proved competitive in new application areas where there was no installed base with which to compete. One prominent example is hosted spam-e-mail eradication. From about 2000 on, when the spam problem became endemic, about a dozen players emerged in this niche. The leading firm was Postini, which secured some key patents and began operations in 1999. Its competitors included MessageLabs, FrontBridge, and Symantec.[47] To use a hosted antispam service, customers diverted their incoming e-mail stream to a dedicated server where spam was removed, and the cleaned up stream was then returned to the customer. Vendors used in-house antispam software, open-source solutions, and software licensed from third parties. Hosted antispam services proved popular because they eliminated the cost and operational overhead of installing antispam software on server and client computers. Further, antispam software required daily, sometimes hourly, maintenance to deal with the latest spam outbreaks. In 2004, Postini reported that it had over five million users. Although usually classified as a security software firm, Postini's annual sales would have put it among the top-ten SaaS vendors.[48]

From about the year 2000, software was increasingly upgraded and patched automatically over the Internet. By far the most common reason for this practice was fixing the security flaws in operating systems and office software exploited by antisocial virus writers. The on-demand software concept promised to take this process to its logical conclusion: whenever a program was invoked, a fresh copy would be drawn from the Internet. In practice, current network speeds make this scenario unlikely in the near future. For example, as early as August 1999 Sun Microsystems acquired Star Division, the maker of StarOffice, a clone of Microsoft Office. At that time Sun planned to offer the software as a free service through its "StarPortal": "If it works, computer users would be able to access their files from whatever machine they log on from—say, a terminal in a hotel room or an airport kiosk. Even better, users would't have to install new software to add new features. Instead, anytime they fired up the software from the Net, they would automatically get the latest version."[49] In any event, the StarPortal never materialized, and StarOffice was made available as traditional boxed software or via a conventional download.

At the time of this writing, on-demand software has only just begun to have an impact on productivity applications such as word processing and spreadsheet programs. The reason for this has been the poor interactivity of remotely hosted applications compared with traditional desktop software. In order to interact effectively with a program it is necessary to have an instantaneous response—in practice, an interval of less than one-tenth of a second between a user action and the system response.[50] Achieving acceptable interactivity on remote applications has been a formidable technical challenge that is not yet fully solved. Thus, both of the leading collaborative software vendors, Webex and Microsoft Office Live, use Microsoft's standard office software in the client's personal computer for document preparation and editing. This gives maximum interactive comfort and maximum functionality.

In early 2005 a technique, usually known as "ajax programming," emerged that dramatically improved the interactive experience with hosted services. Ajax (not strictly an acronym, and standing for "Asynchronous JavaScript + XML") enables code within a Web browser to communicate with the remote server without refreshing the entire Web page. The ajax technique first came to public attention with the launch of Google Maps in February 2005.[51] Google Maps provided a much more comfortable browsing experience than its incumbent competitors such as MapQuest and Yahoo! When zooming into a Google map, for instance, the image would smoothly rescale itself, compared with the earlier systems that refreshed the image discretely for each zoom level. Around the same time, early 2005, several start-up firms, including Upstartle, launched simple office applications that ran in a Web browser. These applications, primarily word processors and spreadsheets, were very simple compared with professional office software, but offered similar functionality to "Works"-type packages. The services also provided online storage and some collaboration features so that they were potentially useful for mobile workers. In fall 2005, Google acquired Upstartle and attracted considerable publicity as a potential Microsoft Office killer when it launched Google Spreadsheets in June 2006.[52]

It is an open question whether traditional desktop applications such as office suites, photo-editing programs, and computer-aided design packages will someday give way to on-demand software services. The outcome will depend on a complex trade-off between the benefits and costs, for both users and vendors. The benefits for users include the reduced cost of software installation and maintenance, the elimination of up-front costs, improved security, and the potential for mobility and collaboration. The costs include ongoing subscription costs, vulnerability to network outages, and software that is less interactive and function rich. For software product vendors, the primary benefits are the elimination of piracy and a more stable income stream from subscriptions, instead of the complex mix of up-front costs, upgrades, and maintenance. The costs include the uncertainly of moving to a radically new business model, and the technological problem of maintaining computer utilities to provide their software services.

Finally, the SaaS model has echoes of the early years of the personal computer. Just as the personal computer enabled users to engage with a new computing paradigm without the "permission" of a centralized corporate information service, so SaaS is enabling today's personal computer users to experiment with new application software genres free from significant financial and corporate constraints. It is likely that for many individuals, SaaS will be their first foray into application software selection. As with the personal computer in the 1980s, this might have far-reaching consequences over the coming decades.

The Impact of SaaS on Independent Software Vendors

SaaS has had two major impacts on existing independent software vendors (ISVs). First, SaaS is a "disruptive technology" that has undermined the long-term validity of existing software technologies and trajectories. Second, SaaS is causing reintermediation—a change in the locus and agency of software delivery to end users.

For traditional enterprise software vendors, SaaS has many of the characteristics of a disruptive technology, as described by Clayton Christensen.[53] In the typical disruptive-technology scenario, incumbent manufacturers with strong market positions in work-horse products are threatened by new entrants working in a new product paradigm. In the case of enterprise software, the existing technology consists of mature products designed for in-house deployment by computer professionals or consulting firms. For example, Oracle's database suites and SAP's enterprise resource planning software have evolved over a period of twenty-five years, provide superb reliability, and scale up to even the largest enterprise. They are, however, difficult to install and customize, and their migration to a Web-centric world has not been fully achieved. The new software paradigm is the remotely hosted, on-demand Web service. Web-native software applications—such as salesforce.com's CRM software—were written from the ground up in the late 1990s and are still maturing. By the standard of enterprise software, such products are mere youth, and they could not currently compete in terms of scalability or robustness with traditional products. It is for this reason that Web-native applications have largely succeeded in niche areas such as CRM and collaboration. For incumbent software vendors, Web services are as yet just a threat looming on the horizon.

Still, the definition of a disruptive technology is that it is one that will eventually mature and replace the established technology. In the case of Web services, it is possible that over a period of ten or fifteen years they will attain the maturity of established software products while also offering the benefits of a hosted service. At that point the market would likely switch to the SaaS paradigm. At the time of this writing, such a paradigm shift is pure speculation. Estimates of the size of the SaaS market vary, but even its most vibrant sector—CRM on-demand software—constituted only 3.4 percent of the total market for CRM software in 2004.[54] At this rate of progress it might be 2020 before SaaS becomes the "dominant mode" of software, if it ever does. On the other hand, the history of computing is peppered with examples of disruptive technologies that have reshaped whole sectors of the computer industry—for instance, the switch from core memory to semiconductor memory in the 1970s, the destruction of the time-sharing industry by personal computers in the 1980s, and the move from centralized computers to client-server systems in the 1990s.

For existing ISVs, SaaS threatens to change the agency and locus of distribution, or in some cases to introduce a distribution mechanism where none previously existed. Whereas software has traditionally been shipped in a "box," new intermediaries deliver it in the SaaS model. These intermediaries are commonly third parties (ASPs or on-demand resellers), although ISVs can also develop internal capabilities. This process of reintermediation and intermediation goes against the general trend of disintermediation observed by Nathan Ensmenger (chapter 11, this volume). In the case of traditional ISVs, remediation involves at best channel substitution (e.g., on-demand hosting instead of the retail box) or it can introduce an entirely new layer of intermediation (e.g., the replacement of direct sales with an on-demand reseller).

By the late 1990s incumbent software vendors were well aware of the SaaS concept, and responded to what some saw as a threat, others an opportunity, and a few as an inevitability. Among those that saw an SaaS model as an inevitability were computer security firms such as Symantec, for which dealing with rapidly evolving Internet-borne threats has made software development a perpetual work-in-progress, and they have had to develop novel working practices and software delivery modes in response.[55]

A detailed analysis of the impact of SaaS on mainstream ISVs is beyond the scope of this chapter, but the behaviors of the top three vendors—Microsoft, Oracle, and SAP—may perhaps stand as a proxy for the many. Their strategies were quite different: they could be described as strategic forays, embracing, and defensive, respectively.

Of the top three vendors, Microsoft is by far the most diversified and therefore the least threatened by SaaS. If there is a challenge to its Windows desktop and server operating systems, for example, this is far more likely to come from the near-term Linux threat than the long-term SaaS one. Microsoft has made two well-reported SaaS-related acquisitions: Great Plains, a maker of accounting software in January 2001, and Place-Ware, a collaborative software developer, in January 2003. These conform to Microsoft's history of securing technical competence by firm acquisition. Microsoft faces an interesting reintermediation problem in that most of its products are currently sold through resellers rather than direct to consumers. In order not to destabilize this arrangement, Microsoft has stated that it will not undertake its own hosting. For instance, its Great Plains products are sold and hosted by the ASP Surebridge. This development suggests that one future direction for software products would be for bricks-and-mortar software retailers to evolve into on-demand services.

Of the major software vendors, Oracle has been the most proactive in pursuing an SaaS strategy, having established its Oracle On Demand services in late 1998. Of the leading vendors, Oracle is the most threatened by SaaS. While Oracle is often perceived as a technical innovator, its commission-based direct-sales operation is as traditional as it is possible to be—straight from the IBM school of marketing. Oracle products, when resold through the intermediary of an ASP, are less profitable than those sold directly to a customer. The formation of Oracle On Demand can thus be interpreted as an attempt by Oracle to capture a compensatory share of the evolving on-demand software market. In order to compete with ASP vendors, Oracle has had to offer a broader range of software than the e-business suite and the database products it initially offered. The acquisition of PeopleSoft in December 2004 and Siebel Systems in September 2005 simultaneously enhanced its portfolio and Web-native offerings.

For SAP, ASPs were simply a new distribution channel. SAP has traditionally charged for its software on a per seat basis, and has cooperated with the hundreds of consulting firms that deploy and customize its enterprise resource planning software. From the SAP perspective, ASPs represented little change—they were simply another class of intermediary. Since the mid-1990s, SAP had been wrestling with rearchitecting its software for Web-based deployment—a difficult task that is being faced by all software vendors.[56] The market already demanded Web-based applications to run over a local intranet; running the same applications over the Internet did not conflict with this requirement. It was not until February 2006 that SAP introduced its first on-demand product, a CRM application to compete with salesforce.com and other vendors in the burgeoning CRM market. For SAP, this represented a major cultural and technological shift, with ramifications in technology, infrastructure, pricing, and sales. Coming so late—five years after its nearest competitors—SAPs "tepid" SaaS strategy was widely perceived as defensive.[57] It was argued that the strategy was designed to develop a competence should software products move to an SaaS world, but to do nothing to accelerate that process.

The different behaviors of the three firms bring out a rather surprising finding. All of the ISVs have had to gain exposure to and competence in Web-native applications to head off the long-term disruptive-technology threat. Their short-term responses have been markedly different, however, and entirely driven by the problems of reintermediation. Microsoft has contracted with a third-party hosting service, Oracle has established an in-house capability, while SAP has been nonreactive, leaving it to the market. The short-term impact of SaaS, so it seems, will be more about managing sales channels and supply chains than managing technological change.

Conclusions

In the fast-moving world of the Internet, the latest SaaS buzz is for integrated Web services. This concept involves something more than Web applications, which rely essentially on customers accessing software through a Web browser. Curiously enough, an old service bureau, ADP, seems to be taking the lead. Integrated Web services involve computers (and other portable devices) talking to each other, creating "a grid or cloud of electronic offerings that feed into each other."[58] One of the key issues that arises in this context is the need for a common platform on which all these services are developed—an operating system, so to speak.[59]

In recent years, companies like salesforce.com, which started as on-demand software providers focused on a specific area (CRM), have attempted to become more like a platform that other vendors use to create their own applications. Analysts have described the evolution of salesforce.com in stages. In stage one, salesforce.com built an on-demand, purely Web-based, multitenant CRM solution. In stage two, it "launched the sforce platform to integrate salesforce.com with other applications and Web services." Stage three was about facilitating customization, and stage four about expanding its partner programs and customer base. In stage five, it "introduced the concept of multiforce, an on demand 'operating system' to enable salesforce.com customers to access multiple applications from a common desktop environment, sharing one data model, one user interface and one security model."[60]

From the perspective of our story, the current incarnation of the SaaS concept and the computer utility represents a new attempt to shift the locus of software, and potentially computing power, outside the corporation. More specifically, it is an attempt to shift the power/software locus outside the corporation, and into either the servers of the hosting companies (in the case of the ASPs) the servers of the on-demand software providers (in the case of companies like salesforce.com), or "the network" (in the case of Web services that involve the interfacing of many Web service providers). At least one form of this experiment was tried before, as we have shown in this chapter, and even though it succeeded for a while, it then died in the arms of the personal computer. Does the new version of the experiment have any chance to succeed now against the personal computer? In other words, if the personal computer killed the time-sharing industry in the 1980s, why would the resurrected time-sharing firms of today kill the personal computer as we know it in the twenty-first century?

In the first place, as we have just pointed out, even though some of the SaaS companies of today are indeed the direct descendants of the computer services and software firms of the 1960s and 1970s, not all of them are. To the best of our knowledge, nothing that existed in the 1960s and 1970s is quite like the on-demand software companies (salesforce.com, for example). As far as we can tell, nobody in the 1960s designed software "for the network," and certainly nobody came up with the idea of a host of network-based services interacting with each through compatible interfaces—if somebody had the idea, the technology was not around to make it feasible. This is a truly novel component of the current incarnation of SaaS and the computer utility.

But perhaps this is pointing to a deeper issue: in the 1960s and 1970s, the network was not truly pervasive. And by this we do not just mean that the Internet as we know it was not around. We really mean that it was considerably more costly to establish a connection to the network, whatever the network was at that stage. A well-known industry report from the 1970s lists the advantages and disadvantages of time-sharing. Many of the disadvantages had in fact to do with the "cost" of connecting to the network—a cost that was in turn dictated by the technological parameters of the time. First, system crashes were not unheard of in the mid-1970s, although it seems that they were of short duration. Second, it was far from clear that the telephone companies, whose lines provided key channels of communication for the time-sharing companies and their customers, offered "the quality of service required for reliable data communications."[61] Reliability was fine for computational problem-solving purposes, but it sometimes fell short for business data processing purposes. Third, data transmission rates were often less than desirable, and limited by the technological capabilities of low-speed modems and terminals. Last but not least, communication costs could become the largest portion of the monthly time-sharing bill.[62]

Today's online experience is dramatically different from the one that the time-sharing companies offered their customers in the 1960s and 1970s. First, let us think of prices. Leaving aside telecommunication costs, the main price component of the time-sharing experience was the $10-an-hour terminal connect charge ($10 per hour in 1975 would be the equivalent of $30 per hour in 2005).[63] Today, individuals with a personal computer at home or work have a wide variety of options in terms of connecting to the network, and the price of none of them comes even close to the terminal-connect charge of the time-sharing years. People can access the Internet from home through a dial-up connection at, say, $14.95 per month, and they can also choose from a menu of options if they decide to go with a faster connection— they may be able to access the network through digital subscriber lines (DSL) at an average monthly price of between $34.95 and $59.95, a cable modem at a monthly fee of between $39.95 and $59.95, and satellite at an average monthly price of about $100.[64] All of these connection approaches are available to American businesses as well, but businesses also have additional options—they can, for example, set up a high-speed T1 connection that will provide Internet access to dozens of personal computers for between $599 and $1,799 per month.[65]

Second, let us consider the data rate (or connection speed). We can develop an understanding of the data rates that prevailed in the 1960s and 1970s from information on the modems that people used to connect to the time-sharing services. Around 1968, for example, the Bell System modems represented close to 90 percent of all installed modems in the United States. The low-end devices, the Series 400, operated at 50 bps and represented about 18 percent of all modems. The high-end devices, the Series 300, operated at a theoretical speed of 19.2 Kbps, but represented only about 1 percent of all installed modems. The two most common ones were the Series 200, which operated at 2000 bps and made up 21 percent of all devices, and above all the Series 100, which operated at 300 bps and represented 60 percent of all modems.[66] For all practical purposes, then, the typical connection speed in the late 1960s was 300 bps, and the weighted-average data rate may have reached roughly 814 bps.[67] Downloading, say, the statistical tables of the *Economic Report of the President* (7.5 Mb in 1970) at 300 bps would have consumed roughly twenty-five thousand seconds (almost seven hours). Even though faster modems were introduced in the 1970s, 300 bps seems to have remained the typical data rate through the late 1970s and the early 1980s, when the expansion of the personal computer sealed the fate of the time-sharing industry. In 1981, for example, Hayes Communications introduced the Smartmodem, considered a technological breakthrough in the industry of that time. Leaving aside its technological sophistication for the time, the Hayes Smartmodem still operated at 300 bps.[68]

Today's data rate is several orders of magnitude away from the one that prevailed through the time-sharing era. As far as home Internet access is concerned, recent reports point out that at the time of this writing, in summer 2006, about 65 percent of all adults that have Internet access from the home use a high-speed connection, and the remaining 35 percent use a dial-up connection.[69] Out of all individuals with high-speed access at home, 50 percent use DSL, 41 percent use a cable modem, and 8 percent have wireless access of some sort.[70] There is some amount of disagreement as to what the data rates for these various services are. As far as the dial-up connection is concerned, it is safe to assume that almost all adults who have one use a 56 Kbps modem.[71] According to some sources, DSL and satellite Internet-access providers advertise maximum speeds of up to 1.5 Mbps, and cable modem providers of up to 3 Mbps. The same sources report that average high speeds achieved are around 476 Kbps for DSL, 708 Kbps for a cable modem, and 400 Kbps for satellite.[72] More conservative estimates of data rates achieved with the various high-speed services record speeds of between 149 and 235 Kbps for a cable modem, and between 109 and 195 Kbps for DSL. According to these sources, the average data rate is 168 Kbps for DSL, 201 Kbps for a cable modem, and 271 Kbps for fiber-optic connections (which are still rare at the time of this writing).[73] If we take these figures as the basis for a conservative estimate of the weighted-average data rate, we conclude that currently Americans access the Internet from the home at about 137 Kbps.[74] This is essentially 457 times faster than the 300 bps rate that was typical in the time-sharing years. Downloading 7.5 Mb of data at 137 Kbps consumes about fifty-five seconds rather than seven hours.

The average data rate for Internet access in the business world is probably faster than at home. In early 2004, for example, a survey of small businesses revealed that out of all the companies with Internet access, 31.7 percent used dial-up, 30.3 percent had a DSL connection, 26.1 percent used a cable modem, 9.8 percent had a T1 connection or higher, and 2.1 percent had other connection types. In other words, at least 66 percent of small businesses with Internet access had a high-speed connection.[75] Other sources indicate that already in 2004, of those workers connected to the Internet, 77.2 percent had access to a high-speed connection. Most of them used a high-speed line, a T1 or higher, and shared bandwidth through an Ethernet network.[76]

Prices that are several orders of magnitude below the ones that prevailed in the time-sharing era, coupled with data rates that are several orders of magnitude above the 300 bps that characterized the time-sharing years, have brought about at least two fundamental changes. First, the online experience is qualitatively different in the sense that users have access to services that were inconceivable in the 1960s and 1970s. Second, the network has become pervasive; according to certain sources, about 65 percent of all American adults have Internet access at home, and about 81 percent have access from either home or work.[77]

Furthermore, the network has become pervasive in a way that is particularly relevant for the SaaS model: people can have access to it not only from home or work but also from a wide variety of locations and through a wide variety of devices. As early as 2002, about five million Americans accessed the Internet through their personal digital assistant.[78] Toward the end of 2004, about 66 percent of American adults had cell phones; a recent survey indicates that in early 2006, about 14 percent of cell phone owners used their device to access the Internet (and an additional 16 percent did not have this feature but would like to have it).[79] The number of public wireless local area network hot spot locations worldwide increased from 1,214 in 2001 to 151,768 in 2005. Over that time period the number of airport locations went up from 85 to 423, the number of hotel locations skyrocketed from 569 to 23,663, and the number of community hot spots soared from 2 to 30,659.[80] As early as 2004, about 50 percent of all hotels in the United States already offered high-speed Internet access in their guest rooms.[81]

Will all these developments spell doom for the personal computer–centric model? As we have already pointed out, the answer is perhaps, but not necessarily. And if they do, the disappearance of the personal computer as we know it is unlikely to happen overnight. It is worth recalling that the computer time-sharing industry of the 1960s and 1970s, even though it grew at a steady pace during most of its existence, only partially displaced other computing models. Part of the explanation of why time-sharing coexisted with other computing models lies in the fact that companies are heterogeneous—depending on technological and market parameters, the optimal loci of computer power and software may reside in different places for different businesses. Furthermore, the computing needs of any given business are sometimes heterogeneous enough that the company may resort simultaneously to various coexisting computing models. Another portion of the explanation is that time-sharing did not compete with a static computing model but rather with a moving target; as the time-sharing industry was developing, the mainframe was giving way first to minicomputers and then to the personal computer.

For similar reasons, the current incarnation of SaaS and the computer utility will likely coexist with more traditional models of computing at least for a number of years. Which approach will finally take the lion's share of the market is not for us to say. In fact, it may well be that computing developments that are on nobody's radar today will shift the scale one way or another tomorrow—just as the personal computer of the late 1970s and the early 1980s unexpectedly spelled doom for a time-sharing industry that in the late 1960s, pundits hailed as the future of computing.

Notes

The opinions expressed in this paper are those of the authors and do not necessarily represent those of the institutions with which they are affiliated.

1. "Software as a Service," Wikipedia, available at ⟨http://en.wikipedia.org/wiki/Software_as_a _Service⟩ (accessed June 19, 2006).

2. Nicholas Carr, "The End of Corporate Computing," *MIT Sloan Management Review* (Spring 2005): 67.

3. "Software's Jolly Iconoclast," *Economist*, June 7, 2003, 58; quoted in "Crème de la CRM," *Economist*, June 19, 2004, 63.

4. The idea that computer processing "has always been a moving target" is suggested, for example, in "Gathering Steam," *Economist*, April 14, 2001, 4–8.

5. Martin Campbell-Kelly, *From Airline Reservations to Sonic the Hedgehog: A History of the Software Industry* (Cambridge, MA: MIT Press, 2003), 6.

6. Montgomery Phister, *Data Processing Technology and Economics* (Bedford, MA: Digital Press, 1979), 28–29.

7. For the history of ADP, see Automatic Data Processing, *ADP 50th Anniversary, 1949–1999*, 1999, available at ⟨http://www.investquest.com/iq/a/aud/main/archive/anniversary.htm⟩ (accessed August 31, 2006). For the evolution of the batch data processing sector, see Phister, *Data Processing*, 277.

8. For the history of CSC, see Computer Sciences Corporation, "The CSC Story," 2003, available at ⟨http://www.csc.com/aboutus/uploads/CSCStory_WH994-2.pdf⟩ (accessed August 31, 2006).

9. Campbell-Kelly, *History of the Software Industry*, passim.

10. Martin Campbell-Kelly and Daniel Garcia-Swartz, "Economic Perspectives on the History of the Computer Timesharing Industry, 1965–1985," *IEEE Annals of the History of Computing*, forthcoming.

11. Phister, *Data Processing*, 29.

12. Ibid., 530.

13. Tymshare, Inc., *Annual Reports*, 1970–1979, Charles Babbage Institute Archives, University of Minnesota, Minneapolis.

14. Auerbach Corporation, *Tymnet, Inc. Value-Added Common Carrier* (Philadelphia: Auerbach Corporation, 1979); see also Tymshare, *A Tymshare Presentation for the New York Society of Security Analysts*, June 18, 1979, Charles Babbage Institute Archives, University of Minnesota, Minneapolis.

15. Here we draw on Phister, *Data Processing*, 164–165.

16. Ibid., 164.

17. Ibid., 155, 164–165, 543.

18. Ibid., 164–165. The cost advantage of time-sharing arose from the nonlinear relationship between the total operating cost and the performance—the larger the time-sharing system, the lower the cost per user. This relationship was usually known as "Grosch's Law," $p = kc^2$, where p = computer power, c = cost, and k is a constant. See "Grosch's Law," *Encyclopedia of Computer Science*, ed. Anthony Ralston and Edwin D. Reilly, 3rd ed. (New York: van Nostrand Reinhold, 1993), 588.

19. Auerbach Corporation, *A Jointly Sponsored Study of Commercial Time-sharing Services* (Philadelphia: Auerbach Corporation, 1968).

20. Of course, this still tells us nothing about the proportion of companies in the economy that relied on time-sharing services for at least some of their computing needs. We can develop a reasonable guess from information contained in the industry reports of the time. The 1968 and 1969 Auerbach reports note that the typical time-sharing customer was a company that had set up two or three time-sharing terminals, and spent roughly $600 per terminal per month. The 1968 Auerbach report counts ten thousand installed terminals at about $7,200 per terminal per year, which is equivalent to annual time-sharing revenues of $72 million. This seems generally consistent with the data in Phister, *Data Processing*, 277, table II.1.26, and 610, table II.1.26a. In these tables, Phister estimates the total interactive online revenue at $50 million for 1967 and $110 million for 1968. For the trade industry data, see Auerbach Corporation, *A Jointly Sponsored Study*, 2–5. If we assume that at least for a few years, the $7,200 per year per terminal and the two to three terminals per company remained roughly constant, then calculating the number of companies with access to time-sharing is just a matter of dividing the total industry revenues by $7,200 to estimate the number of terminals, and then by two or three to estimate the number of companies. So, for example, if we take the $335 million in total industry revenue for 1971 as our point of departure, we conclude that there were about 46,500 time-sharing terminals installed in that year. Thus, there were roughly 23,250 companies in the economy with access to time-sharing services at two terminals per company. According to Phister, there were 3.7 million establishments in the U.S. economy in 1971, which is roughly equivalent to 3.06 million firms, at about 1.21 establishments per firm. See Phister, *Data Processing*, 447. The bottom line, then, is that according to these estimates, about 0.8 percent of the companies in the economy had access to time-sharing services. Remember, however, that only about 1 percent of all establishments in the economy had computers installed at the time. See Phister, *Data Processing*, 447. At three terminals per company, on

average, there were 15,500 companies with access to time-sharing, which is roughly 0.5 percent of all the companies in the economy in 1971.

21. Phister, *Data Processing*, 126–127.

22. In the late 1970s, an Auerbach report commented in the section on evaluating in-house computing versus time-sharing services: "A combination of the small business computer and time-sharing may also be considered. For example, a brokerage firm may use a minisystem to process its back-office data and at the same time have time-sharing access to a stock-quoting service or a portfolio valuation service." Auerbach Corporation, *Auerbach Computer Technology Reports: Time Sharing* (Philadelphia: Auerbach Corporation, 1979), 3.

23. A simple application of Bayes' Rule suggests that depending on the assumption we make regarding the number of terminals per company, roughly between 37 and 58 percent of the companies that had an in-house mainframe installation also resorted to time-sharing services for some of their computing needs. From the Auerbach report we know that 70 percent of the companies that had access to time-sharing services also had in-house computing facilities. See Auerbach, *A Jointly Sponsored Study*, 3–7. Define M = in-house mainframe installation, and T = access to time-sharing services. Then the probability of having an in-house mainframe installation conditional on having access to time-sharing services is $P(M|T) = 0.70$. We know that the unconditional probability of having an in-house mainframe installation is $P(M) = 0.0096$. In other words, about 1 percent of all companies (in a strict sense, establishments) in the economy had an in-house computing installation. We have also estimated that under the assumption of two terminals per company, about 0.8 percent of all companies in the economy had access to time-sharing services, which means that $P(T) = 0.008$. Now, $P(M|T) = 0.70 = P(M \& T)/P(T) = P(T|M)*P(M)/P(T)$. Based on our estimates, $0.70 = P(T|M)*0.0096/0.008$, which means that $P(T|M) = 0.58$. In other words, among companies that had an in-house mainframe installation, about 58 percent also resorted to time-sharing services. Under the assumption of three terminals per company, about 0.5 percent of all companies in the economy had access to time-sharing—that is, $P(T) = 0.0051$. Thus in this case, $0.70 = P(T|M)*0.0096/0.0051$, and $P(T|M) = 0.37$.

24. Auerbach Corporation, *Auerbach Time Sharing Reports* (Philadelphia: Auerbach Corporation, 1969), 11.

25. The average price of a personal computer seems to have declined from over $10,000 in the late 1970s to around $3,700 in 1984. See Robert J. Gordon, "The Postwar Evolution of Computer Prices," in *Technology and Capital Formation*, ed. Dale W. Jorgenson and Ralph Landau (Cambridge, MA: MIT Press, 1989), 80, table 3.1. Gary Myers, a Tymshare insider, offers revealing insights on the competition that time-sharing companies faced from the rise of the personal computer. See Gary Myers, "That Pesky PC Forces a Change in Strategy," available at ⟨http://www .computerhistory.org/corphist/view.php?s=stories&id=137⟩ (accessed July 11, 2006). As Myers reports, "The early PC could do for a one-time purchase price of about $5,000 the same work that Tymshare would charge about $5,000 per month to do."

26. Gordon, "The Postwar Evolution of Computer Prices," 119.

27. Computer Sciences Corporation, "The CSC Story," 7–8.

28. Automatic Data Processing, *ADP 50th Anniversary*, 29, 32.

29. Campbell-Kelly, *History of the Software Industry*, 14–15.

30. Brian Quinton, "Kicking 'ASP,'" *Telephony*, November 1, 1999, 38ff.

31. Aberdeen Group, *ASP Market Snapshot* (Boston: Aberdeen Group, August 2001), 91.

32. Eric Nee, "Webware for Rent," *Fortune*, September 6, 1999, 215ff; Peter Burrows, "Servers as High as an Elephant's Eye," *Business Week*, 73–74. June 12, 2006.

33. Aberdeen Group, *ASP Market Snapshot*, 59.

34. Paul Keegan, "Is This the Death of Packaged Software?" *Upside* (October 1999): 138ff.

35. Mary Johnston Turner, "Migrating to Network-Based Applications," *Business Communications Review* (February 1999): 48–50; Julia King, "Users Buy in Software Rentals," *Computerworld*, February 15, 1999, 1ff.

36. Keegan, "Death of Packaged Software."

37. "Andrew Stern—CEO, USinternetworking," *Varbusiness*, November 13, 2000, 84.

38. David Friedlander, "After the Bubble Burst: ASPs Aren't New After All," *Ideabyte*, September 14, 2002, 1–2.

39. IDC, *Worldwide and U.S. Software as a Service Forecast and Analysis, 2003–2007: Beyond ASP* (Framingham, MA: IDC, March 2003).

40. IDC, *U.S. Application Server Provider Forecast, 2002–2006: A Vertical View* (IDC: Framingham, Ma., Apr. 2002).

41. Stephen H. Wilstrom, "Online Software Finally Gets Useful," Business Week Online, February 5, 2001, available at ⟨http://www.businessweek.com/2001/01_06/b3718093.htm⟩ (accessed August 30, 2006).

42. Karen Angel, *Inside Yahoo* (Wiley: New York, 2000), 126–130.

43. Jeffrey Kaplan, Software-as-a-Service Myths, Business Week Online, April 17, 2006, available at ⟨http://www.businessweek.com/technology/content/apr2006/tc20060417_996365.htm⟩ (accessed August 30, 2006).

44. Siebel's belated entry into the on-demand market was achieved by acquiring Web-native CRM vendor UpShot in 2003.

45. IDC, *Worldwide On-Demand Customer Relationship Management Applications 2004 Vendor Analysis* (Framingham, MA: IDC, September 2005), 21.

46. *Software 500* is available at ⟨http://www.softwaremagazine.com⟩.

47. James Borck, "A Field Guide to Hosted Apps," *InfoWorld*, April 18, 2005, 39–44.

48. IDC, *Worldwide Antispam Solutions 2004–2008 Forecast and 2003 Vendor Shares* (Framingham, MA: IDC, July 2004).

49. Peter Burrows, "Free Software from Anywhere?" *Business Week*, September 13, 1999, 37–38.

50. Andrew Downton, *Engineering the Human-Computer Interface* (New York: McGraw-Hill, 1994), 123.

51. "Ajax (programming)," Wikipedia, available at ⟨http://en.wikipedia.org/wiki/AJAX _%28programming%29⟩ (accessed August 30, 2006).

52. John Markoff, "Google Takes Aim at Excel," *New York Times*, June 6, 2006, C1.

53. Clayton M. Christensen, *The Innovator's Dilemma* (Boston: Harvard Business School Press, 1997).

54. IDC, *Worldwide On-Demand Customer Relationship Management Applications.*

55. See, for example: "Bunker Mentality," *Economist* (December 17, 2005): 75; Christine Y. Chen, "A Trip to the Antivirus War Room," *Fortune* (October 18, 2004): 272.

56. Stephen Baker, "The Sizzle Is out of SAP," *Business Week*, April 12, 1999, 52.

57. Steve Hamm, "SAP Gets On-Demand Religion," *Business Week*, February 2, 2006, available at ⟨http://www.businessweek.com/technology/content/feb2006/tc20060202_537653.htm⟩ (accessed August 30, 2006).

58. "Battle of the Platforms," *Economist*, April 14, 2001, 15–19.

59. Ibid.

60. AMI-Partners, "AppExchange: Salesforce.com's Next Big Bet," October 2005, available at ⟨http://www.salesforce.com/us/appexchange/resources/SalesforceBulletin.pdf⟩ (accessed August 30, 2006).

61. Datapro Research Corporation, *All about Computer Time-sharing Services*, 1972, Charles Babbage Institute Archives, University of Minnesota, Minneapolis.

62. Ibid.

63. We updated this figure using the chain-type price index for the gross domestic product from the 2006 *Economic Report of the President*, table B-7, available at ⟨http://www.gpoaccess.gov/eop/index.html⟩ (accessed August 30, 2006).

64. "High Speed Internet Access," available at ⟨http://www.high-speed-internet-access-guide .com/overview.html⟩ (accessed August 26, 2006).

65. "Covad: T1 Services," available at ⟨http://t1.covad.com/products/t1/t1.shtml⟩ (accessed August 26, 2006).

66. Phister, *Data Processing*, 22ff., 273.

67. That is, $(0.184)(50) + (0.011)(19,200) + (0.207)(2,000) + (0.597)(300)$.

68. "Modem," Wikipedia, available at ⟨http://en.wikipedia.org/wiki/Modem⟩ (accessed August 20, 2006).

69. Pew Internet and American Life Project, "Home Broadband Adoption 2006," May 28, 2006, available at ⟨http//:www.pewinternet.org/pdfs/PIP_Broadband_trends2006.pdf⟩ (accessed August 30, 2006). As of March 2006, 42 percent of all American adults had a high-speed Internet connection at home. Other sources report that around that date, about 64.5 percent of all American adults had access to the Internet from the home. See "Internet Access and Usage in the U.S.," available at ⟨http://www.infoplease.com/ipa/A0908398.html⟩ (accessed September 5, 2006). In other words, about 65 percent of all adults that had an Internet connection had a high-speed connection, and the remainig 35 percent had a dial-up connection.

70. Pew Internet and American Life Project, "Home Broadband Adoption 2006," esp. 2, 5.

71. In early 2004, when dial-up access was still more prevalent than high-speed access, out of all users with dial-up access about 81 percent had a 56 Kbps modem, about 14 percent resorted to a 28/33.3 Kbps modem, and the remaining 5 percent used a 14.4 Kbps modem. See "US Broadband Penetration Jumps to 45.2%—US Internet Access Nearly 75%—March 2004 Bandwidth Report," available at ⟨http://www.websiteoptimization.com⟩ (accessed August 24, 2006).

72. "High Speed Internet Access."

73. These speeds were recorded with *PC Magazine*'s SurfSpeed utility. The speed-testing utility is available at ⟨http://www.pcmag.com/⟩ (accessed August 15, 2006). The results are reported in John Brandon, "Find the Fastest ISP," *PC Magazine*, August 22, 2006, 83. Rather than measuring the data rate through the download of a large file through a single connection, the utility attempts to replicate more typical Internet browsing conditions.

74. As explained above, at the time of this writing 65 percent of all adults that access the Internet from home have a high-speed connection and the remaining 35 percent have a dial-up connection. Of those with a high-speed connection 50 percent use DSL, 41 percent have a cable modem, and 8 percent resort to a wireless connection of some sort. If we assume that all adults with a dial-up connection have access to the Internet at about 53 Kbps, and that all those who have a high-speed connection other than cable achieve DSL-like speeds, then the weighted-average data rate today can be obtained by adding the dial-up component—namely, (0.35)(53 Kbps)—to the DSL component—namely, (0.65)(0.59)(168 Kbps)—and finally to the cable modem component— namely, (0.65)(0.41)(201 Kbps). The weighted average is thus about 137 Kbps.

75. The information was derived from a survey of five hundred representatives of companies with less one hundred employees. See Robyn Greenspan, "Small Businesses Get up to Speed," January 27, 2004, available at ⟨http://www.smallbusinesscomputing.com/news/article,php/3303921⟩ (accessed August 23, 2006).

76. "US Broadband Penetration Jumps to 45.2%" (accessed August 30, 2006).

77. "Internet Access and Usage in the U.S.," available at ⟨http://www.infoplease.com/ipa/A0908398.html⟩ (accessed September 5, 2006).

78. "Statistics: US Internet Usage," available at ⟨http://www.shop.org/learn/stats_usnet_general .asp⟩ (accessed August 26, 2006).

79. Pew Internet Project Data Memo, April 2006, available at ⟨http://www.pewinternet.org/pdfs/ PIP_cell_phone_study.pdf⟩ (accessed August 26, 2006).

80. "Wireless Lan Statistics," available at ⟨http://www.dataquest.com/press_gartner/quickstats/ wireless_lan.html⟩ (accessed August 26, 2006).

81. "Broadband   Trends,"   available   at   ⟨http://www.fiberlink.com/release/en-US/Home/ KnowledgeBase/Resources/Stats/⟩ (accessed August 26, 2006).

III Commerce in the Internet World

7 Discovering a Role Online: Brick-and-Mortar Retailers and the Internet

Ward Hanson

The second half of the 1990s was more the day of Sam Walton than it was of Bill Gates.
William Lewis, *The Power of Productivity*

Physical retail outlets, brick-and-mortar retailers, have been slow to adopt e-commerce. This may seem surprising, given the close contact that retailers have with consumers, and the wave of publicity and interest in all things Internet during the last decade. Brick-and-mortar retailers might have embraced Internet selling enthusiastically, moving quickly to provide online information, conduct commerce, and offer both pre- and postsales services. Established retailers possess inherent advantages over retailer entrants, such as existing vendor networks, category familiarity, retailing experience, and a local presence. Instead, the most important online retailing innovations came from elsewhere.

It was venture-backed dot-coms and small firms already active in some version of electronic commerce that pioneered online retailing techniques. Among the earliest Internet retail entrepreneurs was PC Flowers, a small floral and gifts retailer making the transition from the proprietary service Prodigy to the Internet in 1994.[1] Others pioneers included celebrated dot-com start-ups launched at nearly the same time, such as the Olim brothers with CDNow, Jeff Bezos with Amazon, or Pierre Omidyar with eBay.[2] These companies merged the rapidly developing World Wide Web system of hyperlinks and Web pages with databases of product offerings, creating Web-enabled commerce platforms and notable online retailing brands.

Retailers' timidity in embracing new methods and new technology is not a recent phenomenon. Economists have long remarked on the inertia of traditional retail. Writing in 1923, Lawrence Mann observed, "Retail trade remained the field of the small business man with limited capital long after manufacturing, mining, transportation, and communication had been organized into corporations operating in large units."[3] The early twentieth century saw the rise of the department store, mail-order house, and chain store. Despite these innovations, Mann forecast the continued significance of the independent small retailer into the indefinite future:

There continue to be about 1,200,000 small retailers in the country, and their combined trade is still about three fourths of the total. Their relative importance will probably continue to decrease, but there is no probability that they will be forced out of business, as they usually have the advantage of convenience of location as compared with department stores and mail order houses, give more personal service (including credit and delivery) than chain stores, and are receiving closer cooperation from wholesale dealers who look upon the large-scale retailer as a common enemy.[4]

An analyst in 1995 could have said much the same thing, albeit somewhat more nervously, substituting Internet for mail order, and with Wal-Mart or Costco in mind instead of the A&P or Woolworth.

Generalizations about such a complex environment as the retail sector are fraught with difficulty, especially when discussing the failure of most brick-and-mortar retailers to take action and build an online commerce presence. There has been some adoption, and some brick-and-mortar retailers have established successful online businesses. Yet after a decade of growth, online retail accounts for less than 3 percent of the total retail sales. Of this 3 percent share, most is accounted for by manufacturer direct sales or through "pure-play" dot-com retailers.[5]

A variety of technical and economic factors create difficulties for retailers contemplating online commerce. There are unfamiliar technical challenges, large uncertainties regarding demand, and different skill requirements for employees. Most retailers are small and base their success on local factors—a combination ill suited to the early online world. Large retailers face problems of consistency between a national online presence and geographically based pricing and product assortment. Sizable fixed costs for Web site and supply chain creation make experimentation costly as well as risky.

Brick-and-mortar retailers evaluating an online strategy faced a rapidly evolving environment throughout the decade and escalating costs of effective entry. One consequence of the venture capital boom of the 1990s was a burst of retail start-ups. While spending much of this money on advertising and branding, these companies did ratchet up the technology and skills required for online competitive success.

As the Internet "opened for business" in the mid-1990s, many retailers throughout the United States had more immediate concerns than developing an Internet strategy. Big-box retailers, especially Wal-Mart, were in the midst of a dramatic expansion. Whereas the Internet offered retailers both a threat and an opportunity, a Wal-Mart SuperCenter opening in a market was far more bracing and one-sided.

Over the course of a decade retailers have made accommodations to online commerce. For small and midsize enterprises, much of this has taken the form of opportunistic sales through online auction houses, primarily eBay. These consignment sales occurred at the individual product level, with only a few linkages between different product sales.[6] Choices of the auction site dominated the shopping experience, rather than choices by the originating retailer.

For large national chains, technologically and financially capable of covering the fixed costs of an integrated online store, the Internet provided important quality-enhancement and useful cost-reducing opportunities. For some of these national retailers, it offered additional profitable niches and a new method of competing against leaders such as Wal-Mart. After some hesitation, some of these large retailers embraced the Internet as an effective method of establishing deeper connections with their customers. This bifurcated response, with small retailers choosing online auction houses and larger retailers developing their own online retailing presence, reflects the fundamental costs and features of online retailing.

Retail in the Economy

The Retail Sector

Retailing is a large, diverse, and complicated sector of the economy. It is the final step in a global supply chain transforming raw materials into consumer-owned products.[7] It sells basic commodities and high fashion. Some outlets are primarily single-good purchase (e.g., gasoline), others hope to sell a few items per visit (such as apparel), and some stores carry tens of thousands of items and sell shopping carts full of diverse goods for many visitors (e.g., groceries). There are bulky low-value products such as diapers, and extraordinarily high-value items such as jewelry. Small independent proprietors coexist with global chains.

Physical outlets, brick-and-mortar retailers, operate more than one million establishments in the United States, and employ more than twelve million full- and part-time employees. The North American Industry Classification (NAIC) retailing codes used by the U.S. Census Bureau illustrate the wide range of retail outlets:

NAIC 441: Motor vehicle and parts dealers
NAIC 443: Electronics and appliance stores
NAIC 445: Food and beverage stores
NAIC 447: Gasoline stations
NAIC 451: Sporting goods, hobby, book, and music stores
NAIC 452: General merchandisers
NAIC 454: Nonstore retailers (this is the category including online sales)
NAIC 442: Furniture and home furnishings
NAIC 444: Building materials and garden equipment
NAIC 446: Health and personal care stores
NAIC 448: Clothing and clothing accessories stores
NAIC 453: Miscellaneous store retailers

In the United States, the retailing sector accounts for about 42 percent of U.S. personal consumption expenditures, reaching $3.7 trillion in 2005.[8] Table 7.1 shows the

Table 7.1
Composition and growth of retailing, 1995–2005

	1995	1996	1997	1998	1999	2000	2001	2002	2003	2004	2005
Per capita spending total	8,457	8,923	9,239	9,573	10,299	10,591	10,760	10,884	11,227	11,841	12,547
NAIC code											
441 Motor vehicles and parts dealers	2,210	2,370	2,446	2,552	2,807	2,826	2,865	2,848	2,892	2,945	3,020
442 Furniture and home furnishings	242	256	272	287	310	324	321	329	335	359	375
443 Electronics and appliances stores	247	258	262	276	290	292	282	291	299	323	339
444 Building materials, garden equipment, and supplies dealers	627	667	715	750	802	813	841	864	911	1018	1103
445 Food and beverage	1,489	1,516	1,532	1,545	1,594	1,579	1,625	1,617	1,640	1,688	1,752
446 Health and personal care stores	387	413	444	480	524	551	585	626	661	676	703
447 Gasoline stations	690	734	746	710	780	886	882	871	941	1092	1310
448 Clothing and clothing accessories Stores	501	516	525	553	587	595	588	599	615	648	680
451 Sporting goods, hobby, book, and music	232	242	245	255	267	270	271	267	266	273	276
452 General merchandise	1,144	1,189	1,238	1,299	1,395	1,433	1,500	1,551	1,612	1,693	1,774
453 Miscellaneous store retailers	294	317	342	369	387	383	366	362	354	360	374
454 Nonstore retailers	395	445	472	496	557	640	634	658	701	765	840
722 Food services and drinking places	889	916	964	1,007	1,045	1,082	1,115	1,152	1,202	1,268	1,338

Source: U.S. Census Bureau, *Annual Revision of Monthly Retail and Food Services: Sales and Inventories—January 1992 through February 2006.*

total per capita retail composition and growth from 1995 to 2005. Nominal per capita spending in the United States rose by nearly 50 percent during this ten-year span, from $8,457 in 1995 to $12,547 in 2005.[9]

More than 62 percent of spending occurs in four categories: motor vehicles and parts (24.1 percent), food and beverage (14.0 percent), general merchandise (14.1 percent), and gasoline stations (10.4 percent).[10] Adding the next tier of building materials (8.8 percent), nonstore retailers (6.7 percent), health and personal care (5.6 percent), and clothing (5.4 percent) accounts for nine out of ten dollars spent on retail goods.

The vast majority of retail spending is localized and store based, with customers purchasing from establishments close to their places of residence or employment. The NAIC classifies sales on an establishment basis, so that a firm's online sales are separated out from their physical store sales. The NAIC code 453 in table 7.1 contains the main categories of nonstore-based retailing, including retail e-commerce, but also much more:

The broadcasting of infomercials, the broadcasting and publishing of direct-response advertising, the publishing of paper and electronic catalogs, door-to-door solicitation, in-home demonstration, selling from portable stalls and distribution through vending machines. Establishments in this subsector include mail-order houses, vending machine operators, home delivery sales, door-to-door sales, party plan sales, electronic shopping, and sales through portable stalls (e.g., street vendors, except food). Establishments engaged in the direct sale (i.e., nonstore) of products, such as home heating oil dealers and newspaper delivery, are included in this subsector.[11]

This relatively comprehensive list of direct sales methods totaled $395 per person in 1995, rising to $840 per person in 2005. This is only 4.6 percent of per capita retail spending in 1995, growing to 6.7 percent in 2005. Even after a decade of growth of the commercial Internet, and all the other sales that occur through mail-order sales and telephone-based direct sales, retail spending is more than 93 percent physical store–based.

The U.S. Census Bureau measures sales and employment at the establishment level. Retail is a major employer. An average county in the United States has a mean population of 120,000 (median: 42,000), with retail employing 5,000 individuals (median: 1,300). Retail outlets vary widely in size. Of the average 403 retail establishments in a county, 360 of them have fewer than 20 employees, 35 have between 20 and 99 employees, and only 8 on average have more than 100 employees per establishment.[12] Firms, especially ones with large establishments, often have multiple locations.[13]

Online Retail Sales

The U.S. Census Bureau began reporting e-commerce sales on a quarterly basis starting in the fourth quarter of 1999. At this first release, amid the height of dot-com enthusiasm and at a seasonal peak, sales were a tiny 0.2 percent of total retail. By the fourth quarter of 2005, the percent of retail sales occurring online was 2.7 percent.

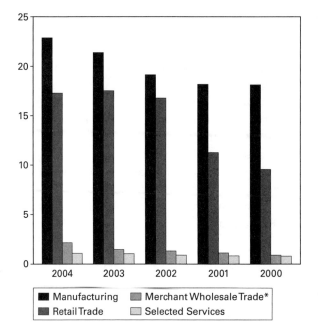

Figure 7.1
Electronic commerce within the supply chain. * Merchant wholesale trade data include MSBOs in 2002–2004 and exclude MSBOs in 2000–2001. *Source*: U.S. Census Bureau.

The same goods sold at retail begin their journey to the consumer with manufacturing and wholesaling shipments. E-commerce has made much greater inroads into these upstream shipments. E-commerce volume from manufacturers to wholesalers, or wholesalers to retailers, is more than ten times as large as the electronic sales at the final retail level. As these are the same products that eventually end up being sold at retail, although in smaller allotments, they illustrate the potential scale of electronic commerce. As in many previous examples of information technology, diffusion within businesses predates the diffusion to consumers.[14] This pattern developed early, as seen in figure 7.1.[15]

By 1997, Odyssey Research was tracking the number of online purchases made by an Internet household within the previous six months. During the second half of 1996, the number of orders was less than ten million purchases. During the second half of 1998, order volume was nearly sixty million transactions, and this included more than half of Internet households.[16]

Surveys show that individuals with Internet access experiment with online purchasing as they gain experience with the medium.[17] The diffusion of the Internet (documented in other chapters of this book) fueled growth in online shopping, as did the increasing experience of each online household cohort. Early technology adopters, almost by definition, dominate this "prehistory." The earliest group of online shoppers was disproportionately composed of employed, college-educated males with children under eighteen years old. This demographic is more sophisticated, more willing to take risks with their money, and less affected by brand names than the general population. As the Internet diffused, the potential shopping population evolved toward the general population and their shopping habits.[18]

While online sales are a small share of total retail, there are categories with substantially higher sales penetration. Computer hardware and software, driven by direct sales from manufacturers such as Dell and Gateway, have an estimated Internet penetration of 34 percent. Event ticketing, such as baseball games or movie theaters, had 20 percent of their 2003 sales occur online. Other top categories include books, toys, and consumer electronics. Some of the biggest categories in regular retail—such as food, gasoline, and auto parts—have low online sales.

The Threat of W-Day

The slow reaction of brick-and-mortar retailers to the Internet is easier to understand given the pressure to their margins and markets from the expansion of big-box retailers such as Wal-Mart.[19] From its early base in the southern United States in the 1970s, Wal-Mart's regional rollout was nearing completion by 1995. As it expanded into new states, it simultaneously deepened its market presence in areas it already served with new store formats and bigger outlets. James Brown and his colleagues succinctly capture the scale and impact of Wal-Mart "Supply Chain Management and the Evolution of the 'Big Middle'": "No retailer has ever been as dominant as Wal-Mart is today, and no retail institution seems as ascendant as today's modern discounters."[20] Wal-Mart's competitive position is based on efficiency and its ability to coordinate a global production system to provide acceptable quality at low prices. Already highly efficient in the 1980s, Wal-Mart was able to reduce its overhead selling, general administrative, and administrative expenses from 22 percent in 1980 to 17 percent by 1995, and uses its scale to negotiate low product prices. This allows Wal-Mart to aggressively reduce prices without sacrificing its margins. Wal-Mart also possesses an efficient distribution system, closely tracked through information technology, creating low inventory costs.

An expanding literature is looking at the impact of Wal-Mart or other big-box retailers entering a new market territory ("W-Day"). Several themes emerge. Wal-Mart puts pricing pressure on existing retailers while simultaneously capturing demand from them. An especially striking result is the impact of Wal-Mart opening grocery stores in an area.[21] After a period of constant or rising operating margins during the 1990s, both Safeway and Krogers saw steep falls in their operating margins following the entry into their markets by Wal-Mart. Many small retailers, especially those who are the least productive, shut down and exit the market within the first few years.[22]

Table 7.2

Percentage of consumer retail sales occurring online, 2003

Category	Percent
Computer hardware and software	34
Tickets	20
Books	15
Travel	11
Toys and video games	9
Consumer electronics	8
Music and video	7
Flowers, cards, and gifts	5
Jewelry and luxury goods	5
Apparel	4
Health and beauty	4
Home and garden	3
Sporting goods and equipment	3
Auto and auto parts	1
Food and beverage	1

Source: Shop.org

The need to maintain some form of pricing proximity to Wal-Mart places severe constraints on a local retailer. After a flurry of initial responses, many find it difficult to avoid permanently lower unit sales and prices.[23] This leads to staff reductions, avoiding extraneous expenses, and reducing other costs. Faced with an ongoing threat to their survival, local retailers find it difficult to devote major resources and time to a rapidly changing discretionary expense such as the Internet. This is especially true if the response requires multiyear investments before any anticipated profit. A far more attractive online approach is one requiring little additional capital, even at the risk of sacrificing scale and control.

Seasonally adjusted, both online sales and U.S. Wal-Mart sales show almost linear increases throughout the interval. The slopes in figure 7.2 are roughly comparable, and substantially exceed the growth in total retailing. By year-end 2005, the combined U.S. Wal-Mart store sales and all e-commerce was 8.5 percent of the total retail.

Transforming Retailing Functions to an Online Setting

Retail channels exist to satisfy shopper needs.[24] Shoppers search for information, identify alternatives, evaluate these alternatives, and eventually make selections. Shopping needs vary by the type of shopper, the product category, and the buying situation. In

Figure 7.2
The two major trends in retailing, 1995–2006. *Source*: Various Wal-Mart 10-Q reports, U.S. Census Bureau e-stats.

broad outline, these functional needs are *information, price, assortment, convenience,* and *entertainment.*[25] Shopping venues vary in their capabilities to satisfy these needs. Shoppers trade off venues based on the specific item they are shopping for, the timing of the retail activity, and the constraints on time and money they face while making choices.

An important part of the online retail learning of the past decade has been in the evolution and development of tools capable of filling these shopping needs. The failure to satisfy any of these needs, much less excel in doing so, can doom a transition to e-commerce. These challenges are central to the decision by many retailers to rely on outsourced solutions, provided and maintained by third parties.

Information

Finding products is a basic shopping task. In a physical space, decades of managerial experience and market research go into the design of stores, the layout of goods within a store, and product allocations on the shelves. National chains commonly dictate layouts and configurations from standardized designs. Shoppers learn these product locations through simple guides and repetition, intuitively understanding retail design conventions. In a grocery store, for example, we come to expect the milk in the back of the store, or candy and magazines by the checkout counters.

Newly minted online shoppers come to e-commerce sites without this background knowledge, and no online shoppers have experience comparable to their knowledge of physical retail. Even now, e-commerce sites are developing expertise and industry conventions. In the early days of online retail these were undeveloped. Effective site design and navigation remains a stumbling block for many online firms. Visitors must rely on cognitive skills and site navigation to substitute for a stroll down the aisles or a quick question posed to a salesclerk. Difficulties in satisfying these needs is clear in a popular trade press book published in 1999 for Web site design: "Imagine driving into a town where none of the road signs make sense. How would that make you feel? Would you feel unsure or wary? Now, imagine landing on a Web site that makes it hard to find your way from page to page. Would you feel as if you've driven off an exit ramp in the Twilight Zone?"[26]

As in online information-gathering activities generally, the two most important online approaches are search and browse.[27] Both present implementation challenges for retailers. A search approach relies on item-by-item keyword phrases. Each search result varies in its effectiveness, depending on the specificity or breadth of the search terms. A highly specific search, utilizing a specific brand name and product size, places a high burden on the shopper and quickly becomes tedious. A broad search leads to numerous possibilities, many of them inappropriate. Organizing such an expansive search-returns list is challenging, and to do it well requires a sophisticated understanding of both the retail category and shoppers' frames of understanding.

Browsing is an alternative to search. Effective browsing designs provide multiple pathways, depending on the shopping scenario, with the goal of routing a shopper quickly to the products they seek. Once the products of interest are located, the next step is to provide a detailed description. The Internet is inherently strong at supplying information, especially cognitively oriented detailed product attributes and specifications. It is relatively easy and inexpensive to document in great detail such items as the price or speed of a computer. Providing experiential data, such as the feel and weight of cloth apparel or the appeal of a book or song, is more challenging.[28]

Immature first-generation Web technologies made it difficult to drill down on specific items, zoom on to pictures to illustrate with greater detail, or customize the information. More modern and sophisticated sites allow retailers to make fewer trade-offs between information quality, visualization, and site performance. Testing shows that simple changes, such as the ability to zoom on to a picture image, can boost online apparel conversion rates by 7 to 10 percent.[29]

User reviews are a popular alternative to professional reviews and detailed store information, but they may only work for sites with many visitors. The challenge is to generate sufficient numbers of reviews to effectively cover the product line. For large sites, even a small fraction of contributors may provide a sufficient number of reviews over a wide range of items. For smaller sites, lacking this volume of potential contributors, some form of incentive will be required to achieve full coverage. Incentives potentially compromise objectivity, defeating the underlying value of the reviews.

Shopper physical idiosyncrasies create other difficulties. One advantage of catalog retailers moving online comes from their long experience providing suitable products and product descriptions for remote ordering. Store-based retailers, accustomed to solving this problem with salesclerks and fitting rooms, need to develop these skills for online selling.

Price

Saving money is a simple and powerful reason for shopping online. While pricing is one of the technically easiest shopping tasks to implement, traditional retailers faced pricing difficulties moving online. Rapidly changing online demographics throughout the 1990s meant a transition from early adopters to more mainstream and pragmatic shoppers. The burst of dot-com start-ups used heavy promotion to attract new customers, creating price pressure for all online outlets. There are also complications due to taxes, shipping costs, and returns. Finally, many regional and national retailers use some form of zone pricing. Zone prices are systematic variations, depending on the local competitive conditions. Zone pricing breaks down online, with the variations difficult to conceal or justify.[30]

The initial research on the Internet's impact on retail prices focused on whether prices online were lower and less variable than those in traditional outlets. Systematic large savings would indicate an emerging dominant retailing format. Less price dispersion would suggest a more efficient marketplace, which rapidly eliminates pricing variation as shoppers quickly spot and reward vendors offering deals.[31]

Research covering the first few years of online retailing found modest impacts. Empirical studies covering the late 1990s and the early 2000s found small savings for books, movies, and music, but more so for smaller online sellers than market leaders.[32] The results in other categories varied. Travel prices were somewhat lower when booked online, as were prices for automobiles.[33] Except for promotional efforts, prices for many frequently purchased goods such as groceries were higher when shipping and handling costs were included. Price dispersion studies found conflicting results, with different patterns by category. Overall, despite some belief that the Internet would force ruinous price competition, market leaders were able to develop brand loyalty and differentiating features enough to avoid this fate.

An important pricing asymmetry exists between pure-play online retailers and those combining an online presence with their brick-and-mortar operations. Early legal interpretations found that nexus was key in determining whether an online seller needed to charge sales tax for their items. This creates a strong disadvantage for traditional retailers, as their customers will draw predominantly from states where they have a physical outlet and so must charge sales tax. As early as 1998, Austan Goolsbee found tax savings an important online shopping motivator.[34] Pure-play online retailers, operating without nexus, could count on a significant price advantage in high sales tax states.

Assortment

The efficient use of shelf space is perhaps the primary skill in brick-and-mortar stores. Modern retailers use sophisticated computer programs to choose the best shelf arrangements, with algorithms determining how much product inventory to carry, and when to drop a product or category. Each product must exceed a minimum profit hurdle to remain on the shelf. Small retailers use less formal means, but must develop efficiencies of their own or fail.

These constraints are far less binding on Internet retailers. With a good user interface and effective search tools, the solutions to the information problems stressed earlier, online retailers can offer millions of items. Online retailing also has a structural inventory cost advantage. Local book and music stores must carry enough inventories to support local demand, with multiple copies for the more popular books. Internet retailers can service demand by maintaining inventory at a centralized location. Internet retailers can also use drop ship capabilities provided by distributors and suppliers. Under typical drop-ship arrangements, the retailer transmits the order. The distributor or the supplier directly ships the product to the customer. For Internet retailers these advantages become powerful in product categories that have a large selection, high configurability, rapid price changes, and perishable inventory.

Chris Anderson calls this advantage of online retailers, and the consequent expansion of product lines, the "long tail." This was the essential vision of many of the early online commerce successes, such as Amazon or CDNow. Combining centralized inventory and drop-ship capability, an online retailer can expand offerings. Profits are possible with a much wider product assortment, even though individual items sell far fewer units.

This "long-tail aggregator" model is applicable to physical goods, digital goods, advertising, information, and user-based communities.[35] Anderson observes that there are natural economic cutoff points for different retailer approaches. A physical retailer has the lowest level of economic assortment, with the needs of inventory and shelf space cited above. Next comes a hybrid retailer, operating a "pick, pack, and ship" model from a central location but selling physical items. Purely digital items offer the largest allowable assortment, potentially the full range of products available in the category. These can be "produced" on-demand and downloaded instantaneously to the buyer, with payment to copyright holders based not on their digital storage but for access to their rights.

Empirical estimates suggest that the consumer benefits from online retailing are primarily due to this dramatically increased assortment, made possible by virtual shelf space and centralized inventory. Assortment results in far more consumer surplus than the benefits of lower prices. Using books as a test case, Erik Brynjolfsson and Michael Smith estimate that approximately 90 percent of consumer surplus benefits from online retail is attributable to increased assortment.[36] If the secret of online retail is the long tail, any traditional retailer must dramatically increase their product lines when moving to online sales as a core business proposition. This injects a sizable start-up cost, and puts these traditional retailers far outside their traditional product offerings.

Convenience

Although online marketers praise the convenience of online shopping at all hours of the day, traditional brick-and-mortar retailers' strongest relative advantage is convenience. A nearby physical outlet provides immediate pickup and gratification, requires little advanced planning, and offers a presence in everyday lives.

Combining online and off-line shopping extends this convenience. A user can utilize previsit research online while retaining in-store purchase and pickup. Alternatively, a shopper can rely on the stores to provide physical exposure to the product and perhaps the first trial purchase, while using the online channel for repeat purchasing. A shopper can look at the products in the store, while ordering gifts for distant friends and family using the online channel. Returns are a costly difficulty for online firms, and an advantage for hybrid retailers.

Entertainment

Much of retailing is a discretionary activity for shoppers, engaged in not for need but pleasure and psychic satisfaction. Marketers stress that for many consumers, shopping is about "fun, fantasy, arousal, sensory stimulation and enjoyment."[37] While some retail is geared to time-saving and convenience, there is another large part of retailing that looks to shopping as a recreational time-using activity.[38]

The ability of consumers to demand entertainment and enjoyment reduces the applicability of lessons learned in the more developed supply chain and business-to-business e-commerce areas. These tools are explicitly oriented to efficiency and productivity, and the rapid solution of problems. Consumer retailing is both a task-oriented and an experiential use of time.

Modular Auctions and Minisites

An integrated e-commerce site presented many small businesses with a costly and challenging hurdle, as they lacked the relevant online and logistics skills, and faced a battle for survival in their primary market. It is also unnecessary. Early in the history of the commercial Internet, retailers discovered a less demanding approach capable of providing many of the e-commerce advantages.

Despite its founding myth of individuals selling Pez dispensers and Beanie Babies to other collectors, eBay has risen to become the eleventh-largest global retailer in the world by being the e-commerce platform of choice for small and medium retailers.[39] It has transitioned from a start-up in 1995, to a public company in 1998, to a global retailer with 181 million users worldwide and 2005 revenues of $44.3 billion.[40] Less appreciated is its central hub as a small business e-commerce solution. The story of eBay is also that of small brick-and-mortar retailers expanding into e-commerce.

Web services, the use of browser-based applications to access hosted software solutions, provides the key to e-commerce for thousands of sellers.[41] In the mid-1990s, small retailers began experimenting with online auctions to dispose of excess inventory.[42] After an early market battle between competing sites and approaches, eBay emerged as the dominant auction-hosting site.[43] For many retailers, especially small ones, their e-commerce strategy is nearly synonymous with their eBay strategy. A core element of eBay's approach is to provide easy-to-use tools to sellers, allowing them to remotely access the eBay selling platform. This modular e-commerce approach was one of the earliest successes using Web services.

EBay receives the bulk of its e-commerce revenue from the contingent fees charged to the participating merchants. In this it is like an enhanced version of the shopping mall developer of an earlier era. For each of its 1.9 billion listings in 2005, eBay charged sellers a declining block tariff auction fee (shown in figure 7.3).[44] The auction closing price determines the auction selling fees.[45] Insertion fees also follow a sliding scale, with increasing fees for higher bid minimums reflecting a form of auction insurance and the increased probability of no sale occurring. For example, if an item is listed at a beginning price of $10 and sells at $75, the fee would be $0.60 + $2.66 = $3.26. There are optional fees for upgrading the listing with a preferred listing, picture, or other upgrades.

Modularity is key to the favorable auction economics for small and midsize retailers seeking some business expansion and diversification. It allows a retailer to expand or contract offerings at will without high overhead. Many small outlets use online sales to smooth demand, as a means of lowering ordering costs through quantity discounts from vendors, and to handle imbalances between national and local tastes. Online auctions allow for e-commerce without the costs of installed software, high maintenance charges, a fixed programming staff, and many of the other volume-independent costs inherent in a Web site.

Modular auctions allow evolutionary testing of demand conditions, rather than major up-front commitments to launch a site. E-mailing a customer list or posting a few surplus items on a remote site is technologically easy, imposes few fixed costs, and does not require business adjustments. In contrast, sophisticated Web sites capable of supporting transactions have considerably higher fixed costs, skill requirements, and staffing needs.[46]

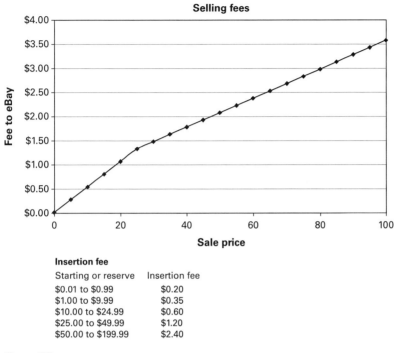

Figure 7.3
eBay 2005 setting fees by transaction price

Small firms are especially wary of major investments and creating conflict with existing customer expectations. Michele Gribbins and Ruth King find some pulled back when such conflicts appeared.[47] Similar concerns appeared in the early study of e-commerce adopters by Pat Auger and John Gallaugher. Even for these pioneers, an unusual group by virtue of their online activity already in 1995–1996, small firms faced more challenges than large firms in launching a site.[48]

Early in the history of the commercial Internet some retailers, interested in online commerce but confronting these site design and informational challenges, hit on a modular outsourcing alternative: consignment selling, especially of excess or slow-moving merchandise. The leading provider has been eBay. EBay was neither the earliest online auction house nor the biggest during the first couple of years of its existence.[49] It prospered through good management, excellent support of small retailers looking to rid themselves of excess inventory or spotting an opportunity to sell online, and effective implementation of its approaches.

One of eBay's best features was simplicity. Even a novice could follow the eBay AuctionWeb guidelines, enter a simple text description of the items, set the initial minimum bid and the length of time for the auction to run, and perhaps provide a link to external images. This simple system, which received national notice fairly quickly, drastically simplified the information problem facing retailers.[50] Gone was the need to worry about navigation systems, product display decisions, and any layout issues other than the description of their own auction items. Gone as well was the large fixed cost, in time and money, needed to launch a fully functioning commerce Web site. The auction house automatically configured and maintained the system for the sellers, with a simple listing and transaction charge for each item they uploaded. EBay augmented these site maintenance features with reporting and tracking software for its sellers.

Beyond eBay's support of retailers, an "ecosystem" of complementary suppliers, consultants, and software providers sprung up to offer independent tools and consulting. These vendors hosted images, provided e-mail systems to inform buyers, and improved the depth of information that a seller could easily supply.[51] They also provided a hedge against being locked in to a proprietary eBay approach, as a seller's listings could be transported to a different auction house if the need arose. Retailers could expand their offerings without a growing lack of control for an increasingly important sales channel.

As was the case for informational needs, early online retailers found auctions a useful method of dealing with many of their pricing problems. Auctions substitute the bidding process for a fixed and posted price. This alleviates the need to estimate market demand conditions, and risk under- or overpricing products. As long as transaction costs are low and the auction market has sufficient bidder activity, auctions will typically do quite well versus posted prices. Finally, an auction mechanism reduces channel conflict or perceived pricing inequity, as any lower price is not a policy decision but the ostensible working out of the current market conditions. It relieves the retailer from having to choose which zone-pricing level to utilize online.[52]

One of the biggest hurdles to a small retailer going online and hoping to attract customers is their lack of assortment. With a large auction house platform this concern disappears. By 1998, there were hundreds of thousands of items for sale on eBay and other auction houses. A lack of assortment was not the issue; rather it was the presence of large numbers of competitive offerings. A standardized auction house also helped satisfy many of the convenience concerns, with payment systems, tracking capabilities, and an appeals mechanism possible in the case of a shipment gone bad.

Finally, auction houses have long been one of the most successful online settings for providing entertainment. Studies of the amount of time spent on a site consistently show buyers spending much more time on an auction site than they do on a more catalog-oriented e-commerce site. Auctions have gamelike features, with an ongoing competition between bidders, a sense of a ticking clock, the potential for last-minute "snipers" to grab away an item, and the ability of shoppers to research and chat about ongoing bids. This is not always efficient or desirable, as when a shopper is merely looking for the straight rebuy of a standard item. Still, auctions are one of the few places online that have been successful in turning online shopping into a social activity rather than a privatized and individualistic task.

Integrated Multichannel Retailing

By 2001, industry studies revealed substantial hybrid purchasing behavior.[53] Rather than concentrate all purchases within a single retailing format, the same shopper would split purchases between online, in-store, or catalog retailers as best matched their immediate needs. Cross-channel information flow was even more prevalent.[54] The diffusion and growth of the Internet, with search engines capable of sorting through deals and providing the latest links to the best sources for products, led marketing theorists to proclaim the arrival of the "new hybrid consumer," utilizing the physical and virtual world to the best advantage.[55] These empowered consumers, which some authors fancifully call "centaurs" to reflect their hybrid combination of speed and intelligence, use information technology to reduce shopping costs, find in-depth comparison information, and escape the local confines of available retailers when better options appear online.[56]

Rather than cede this emerging multichannel market to the auction aggregators, and simultaneously lose control of the online shopping environment, product assortment, and customer data, large store-based retailers have increasingly invested in appropriate online content to match and augment their physical outlets. Instead of a virtual and multivendor environment of an eBay, a large site such as Best Buy or Circuit City offers its full range of products (and more) online.

Large national chains possess inherent advantages when combining physical and virtual assets. A physical presence allows for the convenience of the immediate pickup of items and in-store product testing. The ambiance of the store as well as the surrounding shops and amenities offer a social gathering place. Returns and repairs can utilize the chain's stores as drop-off and advice locations. At the same time, a virtual presence provides many opportunities for extensive information, links to manufacturer sites for product support, and the ability to carry an expanded and more specialized assortment on a store's Web site.

A national chain can tolerate much larger fixed costs to support e-commerce, as these fixed costs are spread over a large shopper base and a chain's well-known brands lower customer acquisition costs.[57] A self-contained online presence lets the retailer control the shopping experience on their site with the same care that goes into store configuration and monitoring in their physical outlets.

Despite their multichannel retailing advantages, it has taken time for e-commerce techniques to diffuse and brick-and-mortar retailers to transform into "click-and-mortar" hybrids. Retailing is risky, and the very real threats from competitors such as

Wal-Mart occupied center stage for many while nervously watching the dot-com retailing boom of the late 1990s.[58]

Venture-backed dot-com retail entrants clouded the marketplace and bid up the price of retailing inputs.[59] In 1999, dot-com firms spent more than $3 billion on consumer media, mostly geared to brand building and encouraging potential visitors to access these dot-com Web sites.[60] Consumer spending by dot-coms rose to more than $5 billion in 2000. This burst of brand-creation and customer-acquisition spending bid up marketing rates, created a glut of messages surrounding online retailing and other new online services, and served as a deterrent to similar efforts by existing retailers. Only with the market correction beginning in 2000 did consumer-oriented spending revert to more normal levels and rates.

Another source of diffusion delay for retailers is the time required to develop the necessary information technology and logistics skills. History and existing business practices shape the skills, logistics, and capabilities available to a retailer facing e-commerce. A retailer operating a national chain of physical stores lacks the back-office skills of "pick, pack, and ship" that a catalog or pure online retailer possesses. Conversely, these more remote retailers lack the brand recognition and scope of operations of the national big-box outlets.

These tensions are visible in the spending and cost patterns needed by these different retailing formats as they operate online. The trade association Shop.org surveyed 130 online retailers about their Internet operations during 2002.[61] This provides an important view of the retailing situation after the wave of pure online retailing subsided and market forces imposed more normal profit criteria. Whether the retailer was a Web-based company only, an online extension of a store-based company, or an online extension of a catalog-based direct mail company mattered greatly to current profits and the nature of the spending needed to develop online retailing capabilities.

Table 7.3 is a snapshot of 130 firms' costs and profits for their online divisions, scaled to their revenue.[62] The easiest transition was for catalog-based companies already skilled in remote ordering and shipping. For them, their online divisions were already profitable by 2002 and had an average of 22 percent earnings before income tax. While their share of incremental costs was somewhat higher than store-based outlets, their existing capacity to handle orders shows up in their low fixed costs of marketing, technology, and general and administrative expenses. Fulfillment and customer service is especially strong for catalog operations, with mail-order operations highly appropriate for the online world.

Web-based start-ups struggled due to both higher wholesale prices and the need to compete aggressively on product prices in order to entice customers to visit. The scale and buying power of store-based online operations, shown most dramatically in the cost of goods sold, provide a substantially higher contribution margin for the store- and catalog-based online operations. The survey went further, and found substantial

Table 7.3
Historical origins matter for online operations

Percentage of revenue	Web-based %	Store-based %	Catalog-based %
Revenue	100	100	100
Cost of goods sold	59	40	51
Fulfillment	12	13	8
Customer service	5	4	2
Contribution margin	*23*	*43*	*39*
Marketing	14	8	8
Site, technology, and content	7	17	4
General and administrative	8	11	4
EBIT	−16	7	22

Source: Shop.org, 2002 data

differences in marketing expenses (higher during that year for Web-based outlets), site technology and content expenses (highest for store-based companies), and general and administrative expenses (substantially higher for Web- and store-based companies than catalog-based ones). When these additional expenses were totaled, Web-based companies had negative earnings before taxes and the online operations of store-based companies were just marginally profitable.

This is not a final equilibrium position among competitors. Start-up firms needed to acquire scale and buying power, but simultaneously found their resources depleted by the marketing expenditures spent attempting to support such rapid growth. Many did not survive this process. Moving in the other direction, the profitability of store-based sites improves substantially once the core technological infrastructure for their operations are in place. Store-based outlets move toward a true hybrid, with the Internet as a coordinating mechanism, and consumers able to shop online or at their neighborhood store.

Central to this information technology spending by retailers is the creation of online shopping tools capable of handling the full range of their online products. Online shopping tools reduce the effort required and increase the accuracy of consumer decisions.[63] Shopping assistants screen a large set of alternatives based on preferences expressed by consumers and present a relevant subset. Consumers use these tools in a number of different ways. Products can be filtered through attribute cutoffs. Consumers can provide utility weights for different attributes, with the shopping assistant using its knowledge of attribute levels to rank products based on their estimated utility. In-depth questioning offers a form of personalized market research. Such a process helps the shopper comprehend their choices better, reduce the choice set to a few alternatives, and build confidence that these self-directed choices are proper.[64]

Comparison shopping tools are easier for an integrated site to provide than for an auction site, as the integrated retailer has control over the product line being offered. The retailer can request suitable information from manufacturers and wholesalers to support these information systems in a way not easily available to the much more diverse set of offerings sold by auction sites.

Assortment presents another transitional challenge, as multichannel retailers decide what products to carry in regular retail stores and online. No shelf space constraints permit an expanded assortment of specialty items. Conversely, bulky low-value items are a poor fit for shipping but can be preordered for in-store pickup. Regulation limits the sale of items such as alcohol, tobacco, and prescription pharmaceuticals.

As Brown and his colleagues argue, the core of retailing is a battle for the large and profitable "Big Middle": "The Big Middle is the marketspace in which the largest retailers compete for the preponderance of consumer expenditures."[65] Even the biggest firms are not using their online e-commerce sites as the sole venue for this Big Middle; rather, they are using their online presence to assist their traditional physical presence in providing better service for their core customers.

Conclusions

E-commerce is a much larger force in business-to-business sales than for consumer retail. Several factors explain this. Small local firms perform many of the consumer retailing functions, and they often lack the expertise and resources to operate an e-commerce site. Another factor is the historical burst of dot-com retailers, which grew rapidly during the 1990s. Despite their numerous eventual failures, these start-ups competed for retailing inputs and consumer attention. An overriding factor is the need to master consumer retailing functions online. It has taken time for retailers to develop the online skills and capacities to provide the proper information, adapt pricing strategies, offer the level of product assortment expected online, develop a delivery system that makes online shopping convenient, and provide entertainment for online shoppers.

A dual-adoption pattern developed in the first ten years of e-commerce retail. Many small and midsize retailers chose an outsourcing approach, using auction sites to offer a small subset of their products online. Companies such as eBay created an e-commerce platform and an online shopping system that allowed these small firms to go online with few fixed costs, relatively low risk, and in a manner that did not threaten their core businesses. Large national retailers chose a different strategy. They have incurred the fixed costs necessary to develop internal e-commerce capabilities, moving much more of their product line online and providing an integrated shopping environment. These efforts still account for only a small part of their total sales, although these online sites do influence a much larger share of purchases.

Emerging Internet trends may enhance the importance of online consumer retail. As Internet users gain experience, they increase their online buying. The rapid diffusion of Web-enabled cellular phones lets the Internet "escape the desktop." Consumers will be able to check prices, consult shopping guides, and receive last-minute relevant offers as they visit physical stores.

The significance of this augmented retailing, and its speed of diffusion, is hard to predict. Companies with prior experience in online retailing will be better equipped to deal with this more complicated shopping world. Another development likely to shape retailing is the continued expansion of Web services, allowing retailers to more easily add sophisticated mapping, customer tracking, customer acquisition, and other e-commerce support services through browser and phone-based software.

The core of retailing continues to be a physical activity. While retailing is certainly moving more toward a hybrid of physical and virtual, retailers can rarely force their customers to make an online transition. Rather, they must entice them with superior offerings.

Notes

1. PC Flowers sold electronically for four years prior to making the move to the Internet in 1994. It followed a strategy of utilizing a wide range of electronic platforms, including Prodigy, Minitel, interactive compact disks, and experiments in interactive television. See William Tobin, "PC Flowers: A Blooming Business," *Direct Marketing* 57, no. 8 (1994): 40–43; Mark Poirier, "Making Sense of Interactive," *Catalog Age* 10, no. 10 (1993): 77–79.

2. CDNow was an early dot-com retailer configured around a specific category (music) using the model of virtual inventory, affiliates channeling traffic, and a wide assortment. See Donna L. Hoffman and Thomas P. Novak, "How to Acquire Customers on the Web," *Harvard Business Review* (May–June 2000): 179–183. Amazon was a highly visible pioneer of many e-commerce features. See, for example, Michael Martin, "The Next Big Thing: A Bookstore?" *Fortune* 134; Bruce Knecht, "Reading the Market: How a Wall Street Whiz Found a Niche Selling Books on the Internet," *Wall Street Journal*, May 16, 1996, A1. The rise of eBay gets expanded treatment in a later section. For a history, see Adam Cohen, *The Perfect Store* (Boston: Little, Brown and Company, 2002).

3. Lawrence Mann, "The Importance of Retail Trade in the United States," *American Economic Review* 13, no. 4 (1923): 609–617.

4. Ibid., 616.

5. As we will see, the separation of pure-play and traditional selling is not so clear-cut when we consider firms such as eBay.

6. On sites with reputation systems, such as eBay, a seller does build a history across transactions. What is missing is cross-selling as well as multiproduct pricing and marketing.

7. See, for example, Joseph P. Bailey, "The Retail Sector and the Internet Economy," in *The Economic Payoff from the Internet Revolution*, ed. Robert E. Litan and Alice M. Rivlin (Washington, DC: Brookings Institution Press, 2001), 172–188.

8. This figure excludes food services such as restaurants. Personal consumption expenditures were $8.742 trillion in 2005, with $1.003 trillion spent on durable goods, $2.539 trillion spent on non-durable goods, and $5.170 trillion spent on services. See U.S. Census Bureau, *Annual Revision of Monthly Retail and Food Services: January 1992 through February 2006* (2006), Commerce, 106. Washington, D.C.

9. Ibid., 54, table 10.

10. Ibid., in 2005.

11. Ibid., appendix A.

12. Reported in Emek Basker, "Job Creation or Destruction? Labor Market Effects of Wal-Mart Expansion," *Review of Economics and Statistics* 87, no. 1 (2005): 174–183.

13. In these statistics, full- and part-time employees are counted equally. Wal-Mart, for example, had 2,400 establishments in the United States in 1998 with approximately 800,000 employees. Its average exceeds 100 per establishment, and each Wal-Mart establishment would be classified in the large category.

14. A long list of such cases would include computers, faxes, laser printers, and copiers.

15. U.S. Census Bureau, *U.S. Census Bureau E-Stats* (2006), Commerce, 26. Washington, D.C.

16. Odyssey Research, cited in Ward Hanson, *Principles of Internet Marketing* (Cincinnati, OH: South-Western Thomson Learning, 2000), 367.

17. This section and the section on retail transformations adapts some material from Ward Hanson and Kirthi Kalyanam, *Internet Marketing and e-Commerce* (Cincinnati, OH: South-Western College Publishing, 2006). The author thanks Kirthi for contributions to this material, with the usual caveat applying.

18. For a discussion of factors influencing the diffusion of purchasing, see Robert J. Lunn and Michael W. Suman, "Experience and Trust in Online Shopping," in *The Internet in Everyday Life*, ed. Barry Wellman and Caroline Haythornthwaite (Malden, MA: Blackwell Publishers, 2002), 549–577.

19. While Wal-Mart is also a form of big-box retailer, it is the "first among equals," and is driving many of the retail changes.

20. The authors also point out that even with this dominance, Wal-Mart is less than 10 percent of the total retail. James Brown, Rajiv Dant, Charles A. Ingene, and Patrick Kaufmann, "Supply Chain Management and the Evolution of the 'Big Middle,'" *Journal of Retailing* (forthcoming).

21. Wal-Mart offers groceries in their SuperCenters, making it straightforward to date their entry into a local market.

22. See the data in Jerry Hausman and Ephraim Leibtag, "Consumer Benefits from Increased Competition in Shopping Outlets: Measuring the Effect of Wal-Mart" (paper presented at the EC^2 Conference, December 2004).

23. See, for example, Mark Peterson and Jeffrey E. McGee, "Survivors of "W-Day": An Assessment of the Impact of Wal-Mart's Invasion of Small Town Retailing Communities," *International Journal of Retail and Distribution Management* 28, nos. 4–5 (2000): 170–181.

24. As a primary function, see Michael Levy and Barton A. Weitz, *Retailing Management* (New York: McGraw-Hill, 2006).

25. See Hanson and Kalyanam, *Internet Marketing*; Paco Underhill, *Why We Buy: The Science of Shopping* (New York: Simon and Schuster, 1999); Paco Underhill, *The Call of the Mall: The Geography of Shopping* (New York: Simon and Schuster, 2004).

26. Daniel Gray, *Looking Good on the Web* (Scottsdale, AZ: Coriolis, 1999), is just one of many such guides.

27. An extensive discussion of this is available in Hanson and Kalyanam, *Internet Marketing*, esp. chapters 4–7.

28. An important distinction is between search and experience goods. See Phillip Nelson, "Information and Consumer Behavior," *Journal of Political Economy* 78, no. 2 (1970): 311–330. For well-specified and branded search goods, the information requirements are simple. Price, a listing of attribute levels, and in-stock availability dominate the information needs. For experience goods, which have difficult to identify and verify product attributes, a retailer must develop a method of documenting these attributes in a credible and persuasive manner.

29. Macys.com, personal communication, November 2003.

30. Techniques such as asking for a shopper's zip code quickly lead to comparisons and complaints from residents receiving higher prices.

31. See Fabio Ancarani and Venkatesh Shankar, "Price Levels and Price Dispersion within and across Multiple Retailer Types: Further Evidence and Extension," *Journal of the Academy of Marketing Science* 32, no. 2 (2004): 176–187.

32. Erik Brynjolfsson and Michael D. Smith, "Frictionless Commerce? A Comparison of Internet and Conventional Retailers," *Management Science* 46, no. 4 (2000): 563–585.

33. One of the most interesting pricing impacts caused by online retail appears to be a partial relaxation of face-to-face pricing discrimination for some product durables. Automobiles are second only to real estate as a consumer expense, and as seen in table 7.1 are the largest retail category. While regulation blocks fully direct auto sales online, and requires some participation of a dealer, new online intermediaries such as Auto-by-Tel serve as an online comparison and ordering mechanism. These sites reduce the selling costs for dealers dramatically, allowing dealers to quote a lower price and still make an attractive return. Consumers receive a small percentage break, but may still save hundreds of dollars. These savings were statistically significantly higher

for women and minority groups compared to similar purchases negotiated in a traditional dealer setting.

34. See Austan Goolsbee, "In a World without Borders: The Impact of Taxes on Internet Commerce," *Quarterly Journal of Economics* 115 (2000): 561–576.

35. See Chris Anderson, *The Long Tail* (New York: Hyperion, 2006), esp. 83–98.

36. Some commentators dispute this magnitude, partially on technical grounds. The view is that the (standard logit) demand system used by Brynjolfsson and Smith has a bias that values assortment highly. While there is some validity to this critique, the central conclusion that assortment provides the dominant share of value is almost certainly true.

37. See Elizabeth C. Hirschmann and Morris R. Holbrook, "Hedonic Consumption: Emerging Concepts, Methods, and Propositions," *Journal of Marketing Research* 46, no. 3 (1983): 92–101; Underhill, *Why We Buy*.

38. For an interesting discussion of the power of time-using activities for consumers, see Sue Bowden and Avner Offer, "Household Appliances and the Use of Time: The United States and Britain since the 1920s," *Economic History Review* 47, no. 4 (1994): 725–748.

39. On the myth, see, for example, Sara Hazlewood, "IPO (Initial Pez Offering) Sparked eBay Launch," *San Jose Business Journal*, November 13, 1998. On eBay's global position, based on 2005 revenue, see eBay Analyst Day presentation, May 2006, 12. N.A.

40. Ibid., various pages.

41. Web services is a more recent term, coming into parlance post-2000.

42. See, for example, Pat Auger and John M. Gallaugher, "Factors Affecting the Adoption of an Internet-Based Sales Presence for Small Businesses," *Information Society* 13 (1997): 55–74.

43. The increasing returns were sufficiently strong that market concentration was inevitable and proceeded rapidly.

44. The rate falls to 1.5 percent for sales prices in excess of $1,000.

45. A reserve price or minimum bid is insurance against a failed auction, with few bidders offering a low amount for the item.

46. See Michele L. Gribbins and Ruth C. King, "Electronic Retailing Strategies: A Case Study of Small Businesses in the Gifts and Collectibles Industry," *Electronic Markets* 14, no. 2 (2004): 138–152. In their analysis, they use sales volume as the driver of costs. A combination of the breadth of product line and the volume of sales are both likely to be important.

47. Quotes cited include monitoring auctions "at times when there are no customers in the store," or "we will continue to use the Internet to complement our store's sales, as long as we don't have to hire any employees." Ibid., 134–135.

48. The sample of these early adopters is different than the smaller firms reacting to the Net three or four years later, and probably more sophisticated.

49. Onsale.com beat eBay online as well as to an initial public offering. It also received the bulk of the early media coverage, as in Edward C. Baig and Amy Dunkin, "Going Once, Going Twice: Cybersold!" *Business Week*, August 11, 1997, 98.

50. An example of the early national coverage is Fred Hapgood, "Bidder Harverst," *Inc. Tech* 19, no. 13 (1997): 58–62.

51. A recent example of these capabilities is available in Debra Schepp and Brad Schepp, *eBay PowerSeller Secrets* (Emeryville, CA: McGraw-Hill/Osborne, 2004), although many examples of listings and coverage are possible throughout the 1990s and the early 2000s.

52. Mixing a posted price (e.g., "Buy It Now") with auction selling is a relatively recent phenomenon, as documented in a patent dispute involving eBay.

53. In one study, shoppers were classified primarily as either online, a store, or a catalog based on their historical activity. For each of these groups, their propensity to buy in the other two channels was tabulated. For primarily online shoppers, 78 percent also purchased in a store and 45 percent by catalog. Primarily store buyers were less diverse, with only 6 percent of that group also buying online and 22 percent through a catalog. Of the primarily catalog buyers group, 23 percent also bought online and 36 percent in a store.

54. Using the same classification scheme as above, primarily online shoppers had their purchasers influenced by stores (25 percent) and catalogs (68 percent). Store buyers, despite their reluctance to buy online, were nonetheless influenced by these channels, with online (22 percent) and catalog (26 percent). Similarly for primarily catalog buyers, citing influence by online (39 percent) and stores (26 percent). Online information collection is the preferred solution to store-based information collection by a significant majority of both primarily online and store shoppers. For both of these groups, 73 percent cite online sources as their preferred information collection channel. Only catalog shoppers cited a slightly less than majority (49 percent) preference for online information collection.

55. Yoram Wind and Vijay Mahajan, *Convergence Marketing: Strategies for Reaching the New Hybrid Consumer* (Upper Saddle River, NJ: Prentice Hall, 2002).

56. Ibid.

57. See Hanson and Kalyanam, *Internet Marketing*.

58. For example, Sears' hesitation about cannibalizing store sales and then feeling compelled to use the Web to offset Wal-Mart and Home Depot. See Christine Zimmerman, "Partnerships Are Vital to Sears' Web Strategy; Spin-off Sites Complement Internal Efforts in Helping Retailer Freshen Its Image," *InternetWeek*, June 12, 2000, 817.

59. See, for example, Judith N. Mottl, "Brick 'N Mortar vs. Dot-Com," *InformationWeek*, June 19, 2000, 61–72.

60. From Universal McCann, N.A.

61. Reported in "The State of Retail Online, 6.0," Shop.org, 2002.

62. The distribution was fifty Web-based, sixty store-based, and twenty catalog-based firms.

63. See, for example, Gerald Haubl and Valerie Trifts, "Consumer Decision Making in Online Environments: The Effects of Decision Aids," *Marketing Science* 19, no. 1 (2000): 4–21.

64. Many retails sites including Best Buy, Circuit City, and Macy's implement a comparison matrix to allow a side-by-side comparison of products based on their attributes. Comparison matrices can compare products across price and quality tiers, or within the same tier. This shopping tool does not require consumers to store detailed attribute information in their computer's memory. Alternative formats to the comparison matrix include a brand-centric presentation where each page contains detailed information on a brand and an attribute-oriented presentation—a page for each attribute. Controlled experiments show that the nature of the format has a strong impact on how the information is processed. For example, a brand-oriented presentation leads to a strategy of processing by brands, and an attribute-oriented presentation leads to a strategy of processing by attributes.

65. Brown et al., "Supply Chain Management," 1.

8 Small Ideas, Big Ideas, Bad Ideas, Good Ideas: "Get Big Fast" and Dot-Com Venture Creation

David A. Kirsch and Brent Goldfarb

In September 1998, Jeff Pape founded WrestlingGear.com in the Chicago suburb of Franklin Park. His strategy was straightforward. In the sporting goods industry, wrestling gear represented a small, seasonal market. Every fall, young wrestlers went to local sporting goods retailers expecting to be frustrated. Limited local demand meant that stores carried only limited inventory. Few alternate retail channels existed. A former wrestler, Pape remembered this frustration only too well. With the arrival of the Internet, he saw the possibility of helping tens of thousands of wrestlers get the gear they wanted without having to settle for what local retailers happened to have in stock. For wrestling gear—and many similarly structured industries—the promise of the Internet was real. Sales and distribution in thin, fragmented markets would be transformed as the Internet allowed retailers to extend their geographic reach, aggregate demand, and centralize purchasing and fulfillment. Pape started small, initially reselling merchandise that he bought from other distributors. He gradually expanded his operation, first opening a small storefront and later borrowing $25,000 to finance inventory to shorten his fulfillment cycle. But he did not let his sales run ahead of his profits. A certified public accountant by training, Pape made sure that every sale produced a positive cash flow. Sales doubled annually, and by late 2005, Pape had hired two additional employees, with extra part-time help for the busy holiday and prewrestling season. Pape estimated that in 2006, annual revenues would exceed $1 million for the first time.

Meanwhile, in November 1997, former technology consultant Eric Greenberg and his colleagues had founded Scient Corporation in San Francisco. Part consulting firm and part Internet incubator, Scient helped firms implement Internet technology strategies. With experienced leadership and ample financial backing from prominent venture investors, Scient grew quickly. Within eighteen months, the firm employed over 260 people with offices in San Francisco, New York, and Dallas. In March 1999, Scient filed for an initial public offering, and its shares began trading on the NASDAQ on May 14, 1999. In March 2000, at its peak, Scient traded as high as $133.75 per share, yielding an implied enterprise value in excess of $7 billion. With a total head count nearing two thousand, this figure translated into almost $5 million of enterprise value for every consultant on the Scient payroll. Through the first part of 2000 revenues continued to increase, but Scient was a services firm, not a technology company. Venture-fattened margins could not be sustained, and following the stock market peak in spring 2000, the pace of revenue growth started to slow. Scient's leadership promised that the firm would be one of the few "i-builders" standing tall at the end of the shakeout and refused to trim their ambitious plans. But eventually, as new venture-funded clients evaporated and competition within the consulting business intensified, Scient stumbled. A massive layoff in December 2000 was followed by additional downsizing in 2001 and a merger with iXL, a competitor suffering from similar overcapacity. The merged firm was unable to regain its footing, and in July 2002, Scient sought protection from its creditors in federal bankruptcy court in New York. In 2006, Chief Restructuring Officer (and bankruptcy trustee) David Wood prepared to "close the books" on Scient.[1]

Both WrestlingGear.com and Scient were "typical" dot-com stories. Pape and Greenberg both identified opportunities arising from the commercialization of the Internet. Both created de novo ventures to exploit these opportunities. The two sought and acquired outside resources to purse their respective visions. Both firms might have been considered successes, by some measures at certain points in time. But their differences reveal what was unique about the process of venture creation during the dot-com era.

A third example brings these contrasts into focus. Few firms came to embody the opportunities (and excesses) of the dot-com era more concretely than Amazon. Incorporated in Seattle in July 1994 by thirty-year-old Princeton graduate Jeff Bezos, Amazon grew to become synonymous with the idea of electronic commerce: Bezos's capacious intellect and youthful self-confidence, his stumbling on the Internet while working on Wall Street, his methodical search for the best product to sell online, and finally, his relentless and unapologetic pursuit of growth defined a generation of entrepreneurs. The Amazon story quickly entered the realm of lore. More than a decade later, Amazon exemplified both the strengths and weaknesses of the strategies that characterized this cohort of firms. On the one hand, the growth of the company was, quite simply, Amazonian. At the beginning of 2006, the company employed more than twelve thousand people with offices spread across ten countries including India and China. The company Web site showcased more than thirty online "stores" selling everything from baby oil to motor oil. And annual revenues approached $10 billion, strong evidence of consistent top-line growth. At the same time, however, Amazon still bore many signs of the growing pains that accompanied this rapid expansion. Though nominally profitable on an operating basis, the firm showed relatively poor returns according to traditional accounting metrics. Competition from specialized firms in each of its submarkets was intense and growing, and the long-run sustainability of the Amazon business model remained uncertain.

Whereas WrestingGear.com followed the traditional path of small business, Scient and Amazon pursued a strategy that came to define an entire generation of Internet technology companies: "Tossing aside about every experience-honed tenet of business to build businesses in a methodical fashion, Internet businesses...adopted a grow-at-any-cost, without-any-revenue, claim-as-much-market-real-estate-before-anyone-else-moves-in approach to business. This mentality [came]...to be known as 'Get Big Fast.'"[2] As many as several thousand Internet firms received venture capital funding to pursue Get Big Fast (GBF). GBF was a single, prolonged bet on a future state of the world in which a select group of "winners" would dominate the e-commerce landscape. For Amazon, GBF seemed to have worked, but for Scient and many firms like it, GBF was not a winning strategy.

Each of the three firms discussed above represents an important thread in our understanding of the business history of dot-com era firms. GBF was not always a bad idea. A handful of Internet firms successfully pursued it, building large, modestly profitable businesses faster than ever before. These firms—Yahoo!, eBay, Amazon, and Monster—came to define the public image of the successful Internet company. At the same time, hundreds of also-rans tried GBF, but discovered that size alone was not sufficient to secure long-term profitability. The failure of firms that had dotted the covers of business magazines—companies like Webvan, Pets.com, eToys, Boo.com, the Globe, and Scient—and the painful financial losses associated with these debacles guaranteed a generally negative public perception of the dot-com era. Meanwhile, lost from view, tens of thousands of Internet start-ups followed in the footsteps of WrestlingGear.com. They started small and grew slowly. Many of these companies survived, selling products and providing valuable services online, even as public opinion continued to characterize the dot-com era as a period of unprecedented failure.

In this chapter, we establish a series of starting points for understanding the emergence of the industries associated with the commercial Internet. First, we report baseline estimates of the number of Internet technology companies created from 1994–2001. Approximately fifty thousand companies solicited venture capital to exploit the commercialization of the Internet. Of these, less than 15 percent followed the GBF model of venture-backed growth. Fewer than five hundred companies (< 1 percent) had an initial public offering. Within the larger set of initial entrants, however, the five-year survival rate was 48 percent. The survival rate is higher than most observers typically predict and similar to that associated with the introduction of other general-purpose technologies. Standing in stark contrast to the popular picture of the dot-com era consisting of a boom phase followed by an unprecedented bust, our findings suggest underlying continuity in the exploitation of entrepreneurial opportunities arising from the diffusion of a new general-purpose technology.

The persistence (and conditional success) of a broad cross section of Internet technology companies allows us to reinterpret the prevailing view of the dot-com era.

Conventional wisdom holds that Internet firms were overhyped: *bad* ideas were *over-sold* to gullible investors by entrepreneurs, venture capitalists, and investment bankers playing a multitrillion dollar game of musical chairs. When the music stopped in spring 2000, the holders of inflated securities were left standing. These ill-fated invest-ments, and the public perception of failure associated with these investments, led many to believe that nearly every Internet firm had failed. Yet the observed financial losses did not, in fact, equate with firm failure.

We therefore need a different story. In our account, the tectonic changes in the underlying entrepreneurial landscape were obscured by the financial bust. Against a highly salient backdrop of destroyed market value, we interpret the high survival rate of dot-com firms to mean that many of the business ideas that flowered during the dot-com era were basically sound. In other words, *good* ideas were oversold as *big* ideas. Most Internet opportunities were of modest scale—often worth pursuing, but not usu-ally worth taking public. Because most Internet business concepts were not capable of productively employing tens of millions of dollars of venture capital does *not* mean they were bad ideas. It does, however, imply that for most of these companies, pursu-ing GBF was not a good strategic decision.

Conventional Wisdom about GBF in the Dot-com Era

Following John Kenneth Galbraith's definition of conventional wisdom—ideas and opinions that are generally accepted by the public as true—we argue that conventional wisdom circa 1996–2000 held that GBF was the preferred strategic choice to exploit the commercialization of the Internet.[3] GBF was based on the presumption that there was a significant first mover advantage (FMA) in Internet markets. First movers, it was believed, would establish preferred strategic positions, preempt later entrants, and thereby secure above-average long-term returns. A necessary corollary of early entry was rapid expansion. Firms following a GBF strategy tried to grow aggressively and make substantial investments to both acquire customers and preempt competition.[4]

The intellectual basis for FMA and the GBF strategy it supported had been developed within academic circles over many years. This literature sought to understand the con-ditions under which a preemptive strategy was likely to succeed. Management scholars interpreted these theoretical findings for business, but the nuances of the intellectual debate did not carry over into the realm of business policy. In a study of the spread of the idea of FMA during the late 1990s, Lisa Bolton and Chip Heath found that FMA was interpreted much more positively in the business press than in the academic liter-ature from which it emerged, and dissenting views were rarely publicized.[5] Moreover, their survey research among a sample of business decision makers found a positive cor-relation between media exposure and the belief in strategic advantage from being a first mover, reinforcing the hypothesis that uncritical media coverage of FMA influenced managerial intent. In practical terms, the managerial belief in FMA was epito-mized by Toby Lenk, CEO of eToys.com, in *Business Week*: "There is all this talk about [competitors] Toys 'R' Us and Wal-Mart, blah blah blah. We have first mover advan-tage, we have defined a new area on the Web for children. We are creating a new way of doing things. I am the grizzled veteran at this thing."[6]

The irony of GBF was that it took time to grow quickly. By 1998, many e-commerce start-ups had raised venture capital to support rapid growth. In a *Newsweek* cover story titled "Xmas.com," Bezos declared, "It's going to be a web Christmas."[7] Online sales for 1998 were predicted by Jupiter Research to reach $2.3 billion, a number that was widely cited in the press. That Christmas, dot-com firms met or exceeded top-line rev-enue expectations.[8] But there was confusion about revenue and "making money" in statements such as "the $2.3 [billion] figure sent a message: Companies are making money out there in cyberspace," when of course, companies were generating revenue but losing money.[9] Importantly, profits were not used as an evaluation metric follow-ing the 1998 Christmas season. Rather, success was judged according to the number of customers and gross revenue—criteria that established whether there was general demand for online purchasing services, not whether they were profitable. Rarely did articles in *Newsweek*, *U.S. News and World Report*, *Business Week*, and similar magazines mention the costs of sales, profit margins, or any data related to the underlying eco-nomics of e-commerce during this period.

Firms were criticized, if at all, for operational failings. Web sites crashed due to excess traffic, and orders failed to arrive by Christmas Eve, suggesting poor logistics and fulfill-ment. The press reported that the takeaway lessons from 1998 were about prepared-ness, fulfillment, and meeting consumer expectations. These behaviors were entirely consistent with the pursuit of GBF. Profitability was not yet expected; entrepreneurs could credibly claim that they needed more time, more money, and greater scale to overcome operational bumps in the road and fully implement GBF. Through 1999 these claims were largely unchallenged, in part because of the fundamental uncer-tainty about the emerging industry. No one could know whether or not the GBF strat-egy would work until it was tested.[10] Following Christmas 1998, the public discussion focused on the different components of implementing a GBF strategy. This discus-sion included such issues as the "necessity" of doubling and trebling server capacity to accommodate expected increases in Web traffic, massive investments in advertising expenditures to establish a market presence, and an increasing focus on customer service capabilities to, for instance, enable real-time online support, shorten average e-mail response time, and ensure timely fulfillment.[11]

The tenor of public discussion changed as Christmas season 1999 drew near. *Market-ing News*, a trade publication, summarized the situation: "Retailers were caught off guard by last year's online Christmas crush. Many experienced site outages and prod-uct shortages, while others failed to recognize the potential of e-commerce and didn't establish an online presence in time or at all." This year, however, according to Jupiter Research analyst Ken Cassar, "They've had due warning. They have no excuses."[12] Consistent with the predictions of FMA and GBF, anticipation of a shakeout in e-commerce grew. For example, Timothy M. Halley, a venture capitalist with Institutional Venture Partners, was quoted in the November 1, 1999, issue of *Business Week* as saying, "We're interested in industry leading dominant plays. Number one is great, number two is pretty good, and number three is why bother." In the same article, Julie Wainwright, the CEO of start-up Pets.com of sock-puppet fame, predicted that "consumers are going to vote and leave a lot of businesses behind during the holidays. It's going to be a make-it-or-break-it Christmas." On December 28, 1999, Forrester Research analyst Lisa Allen was cited in the *San Francisco Chronicle* as contending that "e-commerce is past the experimental stage, but it's not completely shaken out yet." Soon, dot-com companies would no longer be able to attribute the lack of profits to difficulties in implementing the GBF strategy. These quotes appear representative of the sentiments communicated widely in the popular and trade press.[13]

Although many observers ascribed ill intent to the companies that failed at GBF, a more charitable account of the financial boom and bust that accompanied the rapid commercialization of the Internet in the 1990s attributes the rampant pursuit of GBF to the fundamental uncertainty about the advisability of pursuing GBF. The capital market was munificent because this uncertainty implied a high option value for Internet securities.[14] This, in turn, allowed companies to raise more capital by claiming that they needed to get even bigger and therefore grow even more before reaching profitability. According to industry reports, e-commerce revenues during Christmas 1999 doubled or even trebled their 1998 level, but by this time the conventional wisdom was changing. Billions of dollars had been staked in pursuit of GBF, and a lack of sufficient scale could no longer explain away the sea of red ink reported by leading dot-com companies. After several years of unprecedented capital market munificence, uncertainty about GBF was resolved. Hope for GBF gave way to a new certainty about the underlying realities of the technology: The option value of Internet securities declined, and investors demanded results.

Moreover, this uncertainty, or at least its duration, was not entirely accidental. Consider again the case of Amazon: Amazon stands out as one of the few firms that successfully pursued GBF. We speculate that one compelling explanation for the success of Amazon draws on the ways in which the firm cultivated public media to build a reputation in the emerging field of e-commerce and buy time for GBF to work. In this respect, no company took greater advantage of the uncertainty surrounding e-commerce and the prevailing capital market munificence than Amazon under Bezos. Less than a year after the company Web site opened for business on July 16, 1995, the firm had already been featured on the cover of the *Wall Street Journal*. As detailed in a comparative study by Violina Rindova and her colleagues, Amazon's actions generated press coverage that attracted new customers, and created opportunities for innovative strategic actions and additional public communications about these actions. These, in turn, allowed the firm to acquire more resources, intangible and real, which established the legitimacy of GBF and pushed back the time when investors would require GBF to yield tangible economic results. Bezos's ability to manage this complex system of strategic action and strategic communication set the firm apart.[15] The resulting virtuous cycle culminated in December 1999, when the Amazon founder was named person of the year by *Time* magazine. At age thirty-five, Bezos was literally the poster child for e-commerce.

But Amazon's ability to play the media had its limits and could not stave off the day of reckoning for GBF indefinitely. Despite the struggles, GBF was a successful survival strategy for Amazon. Other dot-com start-ups could not pull it off, however. By early March 2000, before the NASDAQ peaked, signaling the end of the boom market for technology stocks in general, prices of TheStreet.com's e-commerce index had already started to decline. By the end of the year, the specialized index would hover around sixteen, a decline of 87 percent from its peak. By comparison, when the NASDAQ bottomed out at 1,184.93 in September 2002, the larger market index was down *only* 76 percent from its March 2000 high. Regardless of the more complicated underlying reality, "massive failure" had been chiseled on to the public tombstone of Internet technology firms, especially the highly visible ones that pursued GBF.

Characterizing the Iceberg of Dot-com Venture Creation

With the preceding overview of the conventional wisdom supporting and later contradicting GBF in mind, we now characterize the entire population of firms founded to exploit the commercialization of the Internet. To understand the logic of our analysis, imagine that the entire population of new venture creation activity from the dot-com era existed as a single iceberg. The emergence of GBF as the conventional wisdom, and perhaps the media strategy of Amazon and similar companies, focused attention on the companies *above* the waterline—that is, on those that attracted the most resources, either private or public equity. These companies were visible because they managed, intentionally, to attract the attention of the business press.[16] Beneath the waterline, out of public sight, the bulk of dot-com companies remained invisible to the business press and therefore to the general public. The bursting of the financial bubble that began in 2000 and accelerated through 2001 was a phenomenon that disproportionately effected the firms that had sought and received media coverage, or the part of the iceberg that was above water. If the bubble was indeed that, a bubble, it should only imply that there was a fundamental problem with those firms in the public eye—and say little about the rest of the industry. Thus, what happened to the thousands of firms that never made it into the public eye?

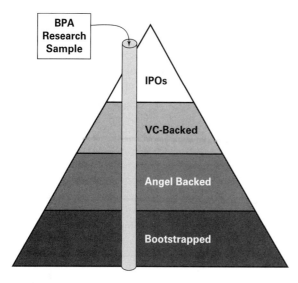

Figure 8.1
"Iceberg" of dot-com venture creation. Visual representation of new venture creation during dot-com era by type of funding showing study sample as "slice" or "core" of overall iceberg. As noted in text, media and initial scholarship focused on the top two segments, those that were most likely to be pursuing GBF.

To make statements about the entire population of dot-com firms, we sought ways to characterize this group. Definitions of industries abound, but often take product markets as given. We chose to focus on a resource-based definition by which an industry is comprised of firms competing for the same resources. We began with a collection of business planning documents submitted to a single venture capital investor in the Northeast from 1998 to 2002. This collection is housed in the Business Plan Archive (BPA, available at ⟨http://www.businessplanarchive.org⟩), a repository established in 2002 with the support of the Alfred P. Sloan Foundation to preserve business planning documents and other related digital ephemera from dot-com era technology companies.[17] The sample that we analyzed consisted of 1,165 solicitations submitted to a single venture capital fund (hereafter, the Focal VC).[18] Each solicitation in the sample represented a venture (extant or intended) that sought financial support from the Focal VC. We knew that every solicitation had been denied by the Focal VC, although some received venture funding from other investors. We wished to claim that these solicitations constituted a representative subset of the overall population of dot-com firms. In the iceberg analogy, our sample represented an ice core of the berg. If it was a representative slice, we could use it as the basis for making general claims about the entire mass.

We used various methods to evaluate the representativeness of this sample. Ideally, we would have measured characteristics of the sample and compared it to similar characteristics of the general population. This approach, however, was not possible for the very reason that we were interested in it: we did not know the characteristics of the entire population. Our study was the first that claimed to be representative of the general population of Internet firms as opposed to being representative of only VC-backed or publicly traded firms. Where others had been content to limit themselves to studying the visible layers of the iceberg, we sought to assay the entire berg.

As a second-best method, we exploited the fact that a sizable fraction of our sample received venture capital funding that was reported in a widely used industry database, Venture Economics. We compared the venture-backed companies in our sample (VC-backed BPA firms) to the comparable population of all VC-backed companies. In this way, we would be able to determine if the *funded solicitations* in our sample, as judged by the venture community, were measurably different from the general population of funded solicitations.[19]

We compared the VC-backed firms in the BPA sample to the total population of venture-backed firms in the Venture Economics database along several dimensions. First, because the European venture capital market is qualitatively different from the American market, we only included U.S.-based companies funded by U.S.-based venture capital firms.[20] Second, we selected only de novo start-ups, and excluded buyouts, roll-ups, recapitalizations, secondary purchases, initial public offerings, private investments in public equity, and debt financings. Third, we selected starting and ending dates that matched the time span of our sample. Fourth, we limited the reference sample to information technology–related businesses.[21]

The results suggested that while the VC-backed BPA companies differed in some ways from the general population of VC-backed information technology companies, they did so in ways that made our results easy to interpret. Controlling for the founding date, the VC-backed BPA sample was biased toward firms founded during the height of the bubble. Moreover, these firms raised less money overall and less in their first successful funding rounds than did firms in the reference sample. Because funding levels are indications of the relative bargaining positions of venture capitalists and entrepreneurs, the lower initial valuations and subsequent funding levels for VC-backed BPA firms suggested that the firms that approached the Focal VC were of lesser quality.[22] Finally, because the Focal VC was based on the East Coast, VC-backed BPA firms were more likely to be located in Massachusetts, New York, and Pennsylvania, and less likely to be located in California than the average firm in the reference sample. In sum, these biases indicated that bubble-focused, low-quality, East Coast firms were overrepresented in the BPA sample, implying that a general survival estimate based on this sample could be reasonably interpreted as a lower bound. Returning again to the iceberg analogy, our ice core was slightly off center, but otherwise sound.

What Happened?

Having established the representativeness of the BPA sample, we sought to use it as a point of reference to establish baseline estimates of technology entrepreneurship during the period, asking, How many dot-com ventures were created, and what became of them?

We employed a range of methods to evaluate the sensitivity of our estimates of venture creation activity. Our methods required strong assumptions; in a companion paper, we described our procedures in detail. For our current purposes, we take advantage of our finding that 13.2 percent of the solicitations recorded by the Focal VC received funding.[23] Assuming that this ratio holds across the entire population of venture-backed Internet start-ups, for every company that received funding, 7.6 companies were seeking funding but did not get it. Noting that there were 6,524 information technology companies funded by venture capitalists between 1994 and 2002, we estimate that there were 49,582 start-ups seeking capital to exploit the commercialization of the Internet during this period ($7.6 \times 6{,}524 = 49{,}582$).

What was the survival rate among these approximately 50,000 ventures? How should such an estimate be interpreted? The expected value of each business can be represented as the value of the business conditional on success, multiplied by the probability of that success. If this value is $\prod$, the probability of success is p and the expected value conditional on success is V, then this value can be represented as follows:

$$\prod = p \times V \tag{1}$$

While there is little debate that the bubble reflected an increase and subsequent decrease in $\prod$, the source of this fluctuation, in terms of p and V, has different implications for the expected survival rate. Under one explanation, the boom and bust reflected the emergence and subsequent disappearance of new business opportunities—a rise and subsequent decline of p. In this scenario, the bust represented the collective and cumulative recognition that these opportunities were at best highly uncertain, if not evanescent. Over time, in this view, investors discovered that what they believed were *good* ideas—ideas with a high probability of success—were in fact *bad* or low-probability ideas. The great explosion of new ventures formed during the run-up to the collapse was thus unsustainable, and the spate of reported failures was a consistent reaction to overoptimism and excess entry. If this explanation was correct, the failure rate of ventures formed during the Internet era should have exceeded typical rates of failure, especially as the period dragged on and the most profitable opportunities were exhausted.

A competing explanation attributes the bust phase of the Internet bubble to changes in financial markets, rather than product markets.[24] In equation (1), this scenario would be represented by a rise and decline of V. In this case, the underlying opportunity structure created by the advent of the commercial Internet was relatively un-affected by gyrations in capital markets. That is, there was a technology shock: the emergence of a commercial sector to exploit opportunities associated with a general-purpose technology, the Internet. The boom and bust reflected only "irrational exuber-ance" with respect to the valuation of new opportunities, rather than their viability. As a result of the bust, the perceived payoffs associated with new venture success declined from their previously unwarranted levels. In terms of Equation (1), investors believed they were investing in *big* ideas—ideas with a high expected value (V). The bust repre-sented the discovery that the ideas were smaller than promised. The potential growth and value of a typical Internet start-up was more limited than had been previously thought, but failure rates, under this scenario, would have been *lower* than during nor-mal periods of entrepreneurship, consistent with the Schumpeterian hypothesis of sec-ular technological change creating new entrepreneurial opportunities. Furthermore, if the bust was a reflection of a decline in business valuations as opposed to viability—driven, say, by the realization that GBF was not widely applicable—a focus on the vis-ible (i.e., financial market) part of the phenomenon would have overestimated the magnitude of the decline. The conventional wisdom about GBF and the dot-com bust described above is consistent with this view.

With this framework in mind, we researched the fate of the firms in the BPA sample. Of the 1,165 firms in the BPA sample, 214 were classified as "never entered."[25] Our sample was therefore reduced to 951 entrants. We investigated the status of the start-ups in our sample in spring 2005. First, we checked the status of the firm on the Web. We determined whether the service described in the business plan was still available. To further investigate the continuity of ownership, we compared management team profiles to those observed in the planning documents. If we suspected an acquisition, or if the service was no longer available, we consulted two additional sources: the Inter-net Archive, and LexisNexis. Using the Wayback Machine, an interface provided by the Internet Archive that offers snapshots of Web site changes over time, we determined the date of exit (if the firm exited). Where we identified Web domains that had been acquired or developed by a new team in pursuit of a different opportunity, we inferred that the original business had failed. To test for the presence of phantom firms (or "the living dead"), we used several criteria. If it was clearly not possible to procure a service, we assumed that the business had failed (there were several examples where the Web site was "under construction" for several years). Also, Web sites commonly report when they were last updated, and the Internet Archive reports when the Web site last changed. If this date was before 2003, we suspected that the business had failed. When it was more recent than 2003, but the Web site was unprofessional, we also suspected that the business had failed. We then tried to procure the services offered on the Web site (when appropriate) and/or contact the individuals who ran the Web site. Often, this latter strategy settled the issue. If it did not, and we were unable to procure a

Table 8.1

Cumulative exit rate by year-entry cohort

Year of exit	Year of entry						Cumu- lative exit rate (mean)	Total at period start	Exits	Exit rate
	≤ 1996 (54)	1997 (52)	1998 (113)	1999 (217)	2000 (231)	2001 (109)				
1998	0	0	0					219	1	0.00
1999	0.00	0.00	0.02	0.00			0.00	433	3	0.01
2000	0.04	0.02	0.07	0.12	0.03		0.06	622	42	0.06
2001	0.13	0.19	0.21	0.25	0.22	0.09	0.20	619	112	0.15
2002	0.19	0.37	0.30	0.36	0.35	0.25	0.32	536	93	0.15
2003	0.24	0.38	0.35	0.43	0.48	0.39	0.41	467	70	0.13
2004	0.35	0.42	0.42	0.50	0.65	0.56	0.52	376	91	0.19
							Total:	788	412	0.14

Note: *Cumulative exit rate* is the weighted mean of the exit rates for each cohort and represents the cumulative exit rate of firms in the sample in the various years. *Total at period start* is the total number of firms in operation during that year. *Exits* is the number of firms that ceased operating during that year.

service, we categorized the business as failed. All in all, there were forty firms with live Web sites that we categorized as failed using the above criteria.

For the sample, we report exit rates by year in table 8.1. Few firms failed prior to 2000 when the failure rate was 0.06. The failure rate increased in 2001 to 0.15. In 2002, 2003, and 2004 the failure rates were 0.15, 0.13, and 0.19, respectively. The mean failure rate across all periods was 0.14. Because of the nature of our sample, we did not observe entry after 2002, and little after 2001. In total, 48 percent of the entering firms survived through 2004.

We compared our failure rates to other studies of industry survival. In a study of nearly 300,000 U.S. manufacturing firms over the 1963–1982 period, Timothy Dunne and his colleagues found exit rates similar to ours. The 1963–1967 cohort of firms had a 42 percent cumulative exit rate after four years. Similarly, the five-year cohorts from 1967 to 1982 had exit rates of 58, 64, and 63 percent, respectively. Taking a finer-grained look at the plant data also shows comparable failure rates among firms that entered through the construction of new plants, a category arguably most comparable to our sample of new dot-com firms: from 1967 to 1982, the three five-year cohorts had cumulative exit rates of 64, 57, and 64 percent, respectively, all of which are somewhat higher than the five-year exit rate in our sample.[26] In a follow-up study of over 200,000 U.S. manufacturing plant entrants in two five-year cohorts, 1967–1972 and 1972–1977, Dunne and his colleagues found that on average, 40 percent of new plants (aged one to five years) had exited after five years. In other words, 60 percent of the new entrants in the manufacturing industry survived for at least five years—a number larger than our 48 percent.[27] More recently, Rajshree Agarwal and David Audretsch examined over 3,400 firms in 33 U.S. manufacturing industries. The one-year average failure rate was 6 percent, identical to our exit rate for dot-com firms. Over five years, manufacturing entrants exited at an average rate of 32 percent, compared to 52 percent for dot-coms in our sample.[28] Finally, cumulative four-year exit rates of 3,169 Portuguese manufacturing firms that were founded in 1983 were 22, 32, 41, and 48 percent.[29] These data compare favorably to the cumulative four-year failure rates of our sample reported in table 8.1 (10, 25, 37, and 44 percent).

The comparisons with results from classic studies of firm entry and exit suggest that the exit rate among the dot-com era firms in our sample was not extraordinary. Cross-industry comparisons are necessarily suspect, though, and may be confounded because the events we are comparing occurred in different time periods and dot-com ventures in an emerging industry are not necessarily comparable to a broad cross section of manufacturing plants. With these concerns in mind, we examined the survival rates of new firms in four emerging industries: automobiles, tires, television, and penicillin. Exit rates for autos from 1900 to 1909 were 15 percent, then 21 percent during the 1910–1911 shakeout, and 18 percent from 1910 to 1919. The exit rate from the tire industry from 1905 to 1920 was 10 percent, then 30 percent during the shakeout in 1921, and 19 percent from 1922 to 1931. The exit rate from the television (production) industry was 20 percent from 1948 to 1950, and 18 percent from 1951 to 1957. Finally, the exit rate from the penicillin industry was 5.6 percent from 1943 to 1954, and 6.1 percent from 1955 to 1978.[30]

From these comparisons we can conclude two things. First, with the exception of televisions, the first shakeout for dot-com firms occurred earlier (after five years) compared to other emerging industries. Second, with the qualification that we only observe survival through 2004 and with the exception of the penicillin industry, the average 14 percent exit rate among information technology entrants is lower than other industries. Finally, we note that our failure estimates, especially those of 2003 and 2004, are biased upward, as we do not observe entry after 2002. Histories of other emerging industries suggest that entry is a constant phenomenon.[31] If we were to take into account this unobserved entry, the observed failure rates would be even lower.

Recalling the relationship set forth in equation (1), the relatively high survival rate, p, is consistent with the arrival of a technological shock. The rise and fall of $\prod$ only reflected changes in the perceived value of dot-com ventures (V). The spread of the commercial Internet heralded a secular shift in the underlying opportunity structure, while public market gyrations represented an irrational increase and subsequent decrease in the perceived value of these opportunities, perhaps relating to evolving beliefs about the viability of the GBF business strategy. The bust reflected a decrease in valuations to more realistic levels.

Conclusions

The closing years of the twentieth century produced a critical moment for entrepreneurial capitalism. Beginning in the mid-1990s and lasting through the stock market peak in 2000, this period saw unprecedented levels of technology entrepreneurship, venture capital investment, initial public offerings, and finally, wild price gyrations in the public markets on which the shares of these new companies were traded.

Returning to WrestlingGear.com, Scient, and Amazon, we suggest that another principal distinction between these three firms lay in the fact that a typical reader of the business press during the heyday of the Internet boom might have heard of Scient and certainly had heard of Amazon. But unless that person was also a wrestler, or the parent or coach of a wrestler, they would have never known that WrestlingGear.com existed. This contrast applies more broadly: the Icarian arcs of a handful of high-flying Internet companies occupied the bulk of public attention on both the way up and the way down. In the public eye, these stories came to represent the totality of Internet entrepreneurship in the 1990s, even as thousands of successful, if less spectacular, Internet companies followed a more traditional growth trajectory, survived, and even thrived.

This study has allowed us to see the ways in which WrestlingGear.com, Scient, and Amazon were typical dot-com start-ups. Scient typified the venture-backed gazelles that captured the public imagination and ultimately cost investors many billions of dollars. Today, its principal narrative of rise and fall is the prevailing story—the conventional wisdom—that most observers associate with the dot-com era. By contrast, WrestlingGear.com typified the counternarrative: a traditional, behind-the-scenes story of entrepreneurial opportunity identification and exploitation that is remarkable for its normalcy. Amazon stands out as one of the few firms that successfully pursued GBF, but as we have seen, its success has obscured the many viable and *small* Internet businesses enabled by the Internet.

Exploiting a unique database of dot-com era business planning documents, we have estimated the scale of entrepreneurial activity during the period. Approximately 50,000 start-ups were founded in the United States between 1998 and 2002 to exploit the commercialization of the Internet. The survival rate of dot-com ventures founded during the height of the bubble in late 1998, 1999, and 2000 was a surprisingly high 48 percent, in line with, if not higher than, that observed in prior instances of industry emergence. To be clear, we do not suggest that one out of every two dot-com companies was successful, defined as meeting investor expectations, achieving sales and growth targets, or delivering on promises made in their original business plans. But they did not fail. Over time, census data and other studies may further refine this estimate, but for the moment, many dot-com entrepreneurs can share the sentiment expressed in Mark Twain's famous quip, "The report of my death was an exaggeration."

Taken together, these findings—the concentration of resources in too few large ventures pursuing GBF, the normal to higher-than-normal survival rate, and the full extent of companies created—suggest that previous accounts of venture creation in the dot-com era have understated the extent of the phenomenon. Technology entrepreneurship in the dot-com era was more successful than people imagine today, and there was more of it than originally reported. To return to the formal relationship presented in equation (1), the probability of success (p) for a given dot-com era venture was normal or slightly higher than normal, but the valuation associated with that outcome (V) was inflated by external gyrations in the financial markets. If the dot-com era had been the result of an irrational cascade of bad business ideas, the observed failure rate would have been higher, not lower, than the average in other emerging industries. Regardless of the wild swings in the perceived value of new Internet ventures, their high survival rate underscores the idea that the ventures were created in response to real changes in the underlying opportunity landscape. Thus, in the mistaken pursuit of GBF, many *good* opportunities were oversold to investors and the public as *big* opportunities. As the bubble burst, valuations were brought in line with the realistic scale of the typical online venture, but the underlying, exogenous change in Schumpeterian opportunities persisted, enabling many *small* technology companies to survive and grow.

Notes

1. In one of his last acts as trustee, Wood signed a waiver allowing the digital archive called the Birth of the Dot-Com Era, directed by one of the authors, to collect digital materials from the wider Scient community (personal communication with the author, April 3, 2006).

2. Robert H. Reid, *Architects of the Web* (New York: Wiley, 1997), 37.

3. John Kenneth Galbraith, *The Affluent Society* (New York: Houghton Mifflin, 1958).

4. Allan Afuah and Christopher Tucci, *Internet Business Models and Strategies Text and Cases* (New York: McGraw-Hill, 2002).

5. Lisa E. Bolton and Chip Heath, *Believing in First Mover Advantage*, working paper, Wharton School, 2004.

6. Quoted in Heather Green et al., "The Great Yuletide Shakeout," *Business Week*, November 1, 1999, 28.

7. Quoted in Steven Levy, "Xmas.com," *Newsweek*, December 7, 1998, 50.

8. See, for example, Marilyn Geewax, "For Online Stores, It's All Over But the Shipping and Counting," *St. Louis Post-Dispatch*, December 19, 1999, five-star lift edition: E7.

9. Daniel Roth, "My, What Big Internet Numbers You Have!" *Fortune*, March 15, 1999, 114–115.

10. There are, of course, instances of naysayers speaking out before the crash in spring 2000; see, for instance, Anthony B. Perkins and Michael C. Perkins, *The Internet Bubble* (New York: Harper-Collins, 1999).

11. Green et al., "The Great Yuletide Shakeout"; Susan Kuchinskas, "Shop Talk," *Brandweek*, December 6, 1999, 64; Richard Karpinski, "IT Haunted by Ghost of Christmas Past," *Internet Week*, August 16, 1999, 1.

12. Quoted in Dana James, "Merr-E Christmas!" *Marketing News*, November 8, 1999, 1.

13. See also Stephen Lacy, "E-Tailers Initial Public Offering Plans Hinge on 1999 Christmas Sales," *Venture Capital Journal* 40 (January 2000): 5–6.

14. Lubos Pastor and Petro Veronesi, "Was There a NASDAQ Bubble in the Late 1990s?" *Journal of Financial Economics* 81, no. 1 (2006): 61–100.

15. Violina P. Rindova, Antoaneta P. Petkova, and Suresh Kotha, "Standing Out: How New Firms in Emerging Markets Build Reputation," *Strategic Organization* 5, no. 1 (2007): 31–70.

16. Violina Rindova, Timothy Pollock, and Mathew Hayward, "Celebrity Firms: The Social Construction of Market Popularity," *Academy of Management Review* 31, no. 1 (2006): 50–71.

17. The archive contains metadata on more than thirty-five hundred companies assembled from various overlapping samples of dot-com era firms.

18. We are careful to use the language "solicitation" as opposed to "firm" or "entrant" as many of the groups that solicited funding never moved beyond the planning stage of their ventures nor engaged in commercial activity, and hence should not be considered entrants. While the solicitations that we consider did not receive support from the Focal VC, a significant fraction of them did receive venture financing from its competitors. According to the terms under which the sample was given to the BPA, we are not permitted to reveal the identity of the Focal VC. Researchers are encouraged to their direct inquiries to the BPA, available at ⟨http://www.businessplanarchive.org⟩.

19. Note that in this way, we are actually taking advantage of the decision making of the entire community of venture investors, not the potentially idiosyncratic decisions of the Focal VC. To assess the representativeness of the Focal VC, we undertook additional benchmarking reported in Brent D. Goldfarb, David A. Kirsch, and Michael D. Pfarrer, "Searching for Ghosts: Business Survival, Unmeasured Entrepreneurial Activity, and Private Equity Investment in the Dot-com Era," Robert H. Smith School working paper no. RHS 06–027, October 12, 2005, available at ⟨http://ssrn.com/abstract=825687⟩.

20. Steven N. Kaplan, Frederic Martel, and Per Strömberg, "How Do Legal Differences and Learning Affect Financial Contracts?" *NBER Working Paper* 10097 (November 2003).

21. More than 95 percent of the VC-backed firms in the BPA were categorized by Venture Economics as information technology–related, suggesting that our sample was accurately drawn

from our study population. Information technology includes funds categorized by Venture Economics in one of the following categories: communications and the media, computer hardware, computer software and services, Internet specific, and semiconductors/other electronics.

22. Roman Inderst and Holger M. Mueller, "The Effect of Capital Market Characteristics on the Value of Start-Up Firms," *Journal of Financial Economics* 72, no. 2 (2004): 319–356.

23. As noted in the companion paper (see note 19, above), we acquired two related data sets from the Focal VC. One consists of a larger, low-information sample of which 15.3 percent received funding. The funding level in the second (high-information) sample, which serves as the principal basis of the analyses reported in the paper, is 11.1 percent. We use the arithmetic average of these two numbers (13.2 percent) in the exercise described in the text.

24. Eli Ofek and Matthew Richardson, "The Valuation and Market Rationality of Internet Stock Prices," *Oxford Review of Economic Policy* 18 (2002): 265–287.

25. In the iceberg analogy, the nonentrants might be seen as loose ice around the bottom of the berg.

26. Timothy Dunne, Mark J. Roberts, and Larry Samuelson, "Patterns of Firm Entry and Exit in U.S. Manufacturing Industries," *RAND Journal of Economics* 29, no. 4 (1988): 495–515.

27. Timothy Dunne, Mark J. Roberts, and Larry Samuelson, "The Growth and Failure of U.S. Manufacturing Plants," *Quarterly Journal of Economics* 104, no. 4 (November 1989): 671–698.

28. Rajshree Agarwal and David B. Audretsch, "Does Entry Size Matter? The Impact of the Life Cycle and Technology on Firm Survival," *Journal of Industrial Economics* 49, no. 1 (March 2001): 21–43.

29. Jose Mata and Pedro Portugal, "Life Duration of New Firms," *Journal of Industrial Economics* 42, no. 3 (September 1994): 227–245.

30. Kenneth Simons, "Shakeouts: Firm Survival and Technological Change in New Manufacturing Industries" (PhD diss., Carnegie Mellon University, September 1995).

31. Ibid.

IV Industry Transformation and Selective Adoption

9   Internet Challenges for Media Businesses

Christine Ogan and Randal A. Beam

There are a number of reasons for our inertia in the face of this advance. First, newspapers as a medium for centuries enjoyed a virtual information monopoly—roughly from the birth of the printing press to the rise of radio. We never had a reason to second-guess what we were doing. Second, even after the advent of television, a slow but steady decline in readership was masked by population growth that kept circulations reasonably intact. Third, even after absolute circulations started to decline in the 1990s, profitability did not. But those days are gone. The trends are against us. Fast search engines and targeted advertising as well as editorial, all increase the electronic attractions by a factor of 3 or 4. And at least four billion dollars a year is going into R&D to further improve this process.
CEO Rupert Murdoch, speaking to the American Society of newspaper editors, April 13, 2005.

Murdoch, the entrepreneurial media baron who controls News Corporation, has a track record of placing savvy bets on ventures he believes in and making them pay off handsomely. In the mid-1980s, he surprised television industry analysts by stitching together ninety-six local stations, creating the FOX television network.[1] Few of those analysts thought the FOX network would be around at the end of the century. But it's still here and has, in its *American Idol* franchise, the two most popular shows on television.[2] Murdoch had doubters again when he launched the FOX News Channel on cable in 1996, facing off against Time Warner's entrenched CNN and newcomer MSNBC, a joint venture of NBC and Microsoft. Today, the FOX News Channel has more viewers than the other two cable news channels combined.[3]

Murdoch, who has made billions of dollars in newspapers, magazines, radio, television, books, and films, has been a new media skeptic.[4] He made a couple of online investments in the late 1990s, and they did not pan out.[5] But as his remarks to newspaper editors suggest, he's a skeptic no more. In 2005, he proved that by spending an estimated $1.4 billion to buy Internet companies.[6] Murdoch's News Corporation, along with virtually all other traditional mass media organizations, has witnessed both opportunity and disappointment in the face of the expansion of the Internet.

This chapter will detail the experiences of media businesses as they have struggled to adjust to the growth of the Internet and particularly the introduction of Web browsers with a graphic-interface capability. It discusses the early efforts by longtime media companies to establish a presence on the Web; the business strategies that these organizations have followed to try to make their Web ventures successful; the impact that the "digital revolution" has had on the structures and cultures of these media companies; the influence that the Web has had on the product that they make—the news, information, and entertainment that we see, hear, and read; and some of the challenges that confront these companies as they seek to remain profitable enterprises in a business environment that's being transformed by the Web. Though the Internet has led to major changes in the structures, cultures, business models, and content of these organizations, we suggest that not all of these changes may allow for the continued diffusion of high-quality news, information, and entertainment to consumers. We conclude this chapter by calling for a renewed examination of what it will take for media organizations, which have been so vital to our social, political, and commercial lives, to continue to be viable choices for future audiences.

Early Mass Media Online

Print Media

It is hard to document exactly when the first mass medium ventured online. CNN and the *Chicago Tribune* delivered an online product through America Online's (AOL) dial-up service beginning in 1992. Other media, such as *Florida Today* and *U.S. News and World Report*, chose CompuServe for electronic delivery in 1993. Still other media used Prodigy or Freenets. And some newspapers used bulletin board systems to disseminate news even before that date. When the Mosaic browser appeared in late 1993, it gave media producers a tool that made it easy to distribute information and display images. Yet the media industries were slow to see the value in creating Web sites to deliver information, and when they did go online, executives viewed their presence in cyberspace as supplementary or experimental rather than as part of their core business. It was often the larger chain-owned newspapers that went directly to online production and bypassed the dial-up AOL-type interfaces. In 1995 those papers included the *San Francisco Examiner*, the *San Francisco Chronicle*, the *St. Petersburg Times*, and *USA Today*. International papers also were added that year.[7] But even media that made the early bold move to online publications were concerned that they would cannibalize their print readership.[8] At the Interactive Newspapers conference in 1998, Chris Neimeth, then director of sales and marketing for the New York Times Electronic Media Company, noted that only 20 percent of the readers of the online product were subscribers to the print version.[9] The Internet option for publishing appeared to be a rerun of the threat that newspapers saw when Videotex came along in the 1970s. Worried that competitors would erode the classified advertising base, newspapers had made defensive moves into electronic publishing only to later drop out—sometimes after sizable investments—when Videotex proved not to be viable. Publishers were caught in a bind. If they put the same content online as was in their traditional product, they feared undercutting themselves. If they failed to show up online, they feared that the classified advertising would be ceded to some other online business that offered a better opportunity. So print publications—both newspapers and magazines—ventured cautiously into the online environment.

Some early efforts illustrated a great amount of innovation; the Nando Times, site of the *Raleigh News and Observer*, comes to mind. It included games, chat rooms, the Newspaper in Education program, and interactive holiday Web sites. A few news organizations showed an interest in exploiting the interactive capability of the Internet by including reader forums, e-mail addresses of reporters, and interactive histories of players on local sports teams. But most publications merely used repurposed or un-repurposed stories that appeared in the daily off-line version. Soon, the not-so-nice term "shovelware" was being used to describe the content of online media, particularly print media. In 1994, Jon Katz, then editor of *Wired* magazine, criticized the print media in a widely cited column, "Online or Not, Newspapers Suck." Referring to the great number of large newspapers that had appeared on AOL that year, Katz said, "Watching sober, proper newspapers online stirs only one image: that of Lawrence Welk trying to dance at a rap concert. Online newspapers are unnatural, even silly. There's too much baggage to carry, too much history to get past. They never look comfortable, except on some of the odd community message boards, when the paper ends up offering just another BBS, instead of a reinvention of itself." Katz gave the papers no credit for their interactive efforts either, observing that "online papers pretend to be seeking and absorbing feedback, but actually offer the illusion of interactivity without the reality, the pretense of democratic discussion without yielding a drop of power. The papers seem careful about reading and responding to their e-mail, but in the same pro forma way they thank readers for writing letters. They dangle the notion that they are now really listening, but that's mostly just a tease—the media equivalent of the politically correct pose."[10]

In fact, the editors probably didn't know how to adapt their product for this new environment. And given newspapers' solid bottom line at the time, they weren't much interested in investigating how they might do that. Nevertheless, that didn't dissuade them from going online, and by the end of 1995, at least sixty newspapers had appeared on dial-up services or had their own Internet sites.[11]

Pablo Boczkowski, a historian who studies online newspapers and their predecessors in Videotex, wrote that online papers largely followed the pattern they had used in the 1980s when adopting teletext (which transferred pages of text to a television with a decoder) and Videotex (a somewhat interactive service that transferred data to television receivers with a graphic interface). "Despite scholarly and anecdotal evidence suggesting that such content [shovelware] was not well received by users of those services, a decade later online newspapers on the Web—at least in their first years—have been following the same path."[12] Just as Boczkowski and others have written that the motivation for newspaper executives to adopt Videotex technology was the fear that it might replace traditional newspapers, the same scenario was being played out in newspapers online.[13] Having been burned by the cost of the previously failed technology, newspapers did not embrace the online format with aggressive investment, or by seriously taking advantage of its interactive and other unique features.

Broadcast Media

Broadcast initially had a harder time online, as bandwidth didn't make it practical to watch streamed video or even hear more than a short audio clip. So the local television stations and the national networks were content to provide news text and program schedules, and use their Web sites mainly to promote their broadcast content. A study by Sylvia Chan-Olmsted and Louisa Ha found that local television stations used their Web ventures in a support role—primarily to help cement relationships with viewers.[14] At the outset, using the Web to raise revenue was not the key concern of station managers, though they anticipated that as time passed their Internet ventures might be able to contribute to the bottom line. Exactly how that was to happen, however, seemed uncertain.

Radio stations began using the Internet to distribute their programming almost from the outset. Radio Webcasts were on the Internet as early as 1994, allowing listeners from around the world to hear "local" radio stations. But in *Promises to Keep*, William Fisher III describes how disputes between radio broadcasters and record companies over royalties hobbled the growth of these Webcasts.[15] Record companies were unhappy that radio stations did not have to pay them royalties on the songs that they broadcast, so they convinced Congress to pass copyright legislation that imposed fees for Webcasting music. (Radio broadcasters had been paying royalties to music composers, though not to the record companies.) The copyright legislation was complicated, but Fisher said its impact was straightforward. Commercial Webcasts that streamed music noninteractively to listeners—akin to an online radio "broadcast"—were subject to royalties. Ultimately, those royalties were set quite high for all but small and noncommercial Webcasters. As a consequence, only the largest radio stations could afford to continue streaming music on to the Web, so many stations stopped their Webcasts.

The challenges confronting television broadcasters were initially more technical than legal. Because most Internet users had only dial-up modem connections at that time, streamed video was out of the question for regular use. But both technology and media companies were already looking ahead. In 1996, Microsoft teamed up with NBC to form MSNBC and MSNBC.com, bringing together the video expertise of a broadcaster to the Internet.

Niche for Mass Media Online

The "theory of the niche" provides a useful prism through which to see how the Internet has been affecting traditional media businesses. John Dimmick and Eric Rothenbuhler have used this theory, borrowed from the field of ecology, to explain how media industries either displace one another or adapt to compete with one another.[16] A fundamental tenet of the theory is the premise that if a new species enters the territory of an established one, and if the two species share a taste for the same items in their diet, they have to either adjust parts of their diets to survive or one becomes extinct. Dimmick and Rothenbuhler describe "niche breadth" as the range of items in that diet, and "niche overlap" as the degree to which the two species feed from identical foods in the environment.[17] The authors argue that it is useful to apply this theory to media competition because it explains what happens when a new medium is introduced to the public and that new medium serves some of the same functions for consumers as an old medium. That happened in the case of radio and newspapers, and then among radio, newspapers, and television. At first, newspapers tried to fight off radio through legal means by trying to limit radio's news broadcast abilities. Later, when those legal strategies grew impotent, it became clear that both could prosper if they chose different niches. Radio carried up-to-the-minute developments, and newspapers focused on writing more reflectively about the news. Because radio broadcasts couldn't deal easily with in-depth stories, listeners would turn to newspapers for longer and more detailed reports on the same topic. (Newspapers also discovered that it was even profitable to carry radio program schedules; listeners would turn to the newspaper to learn what programs were planned for a particular day.) Television, in turn, had the same tug-of-war with radio, magazines, and newspapers. Eventually commercial radio became a predominantly local medium devoted to various music genres, talk programs, sports, and headline news. Broadcast television, which serves both local and national audiences, focused on news during peak news-interest times and lighter feature news programs in prime time, along with its entertainment programs. Magazines, once a national mass medium, found that television was more effective in reaching large, heterogeneous audiences. So magazines began identifying smaller niche audiences and supplying more specialized content to serve them. Newspapers have changed their approach repeatedly, though for the most part they have maintained their focus on serving local or regional audiences. First their emphasis was on length and depth—things that their broadcast competitors could not sufficiently accomplish. Then they added more features, consumer-interest news, color, and graphics as their younger readers drifted away, many to the Internet. The current period has been characterized by shorter stories, smaller news holes, and staff cuts (to prop up profit margins). Each medium had settled more or less comfortably into its niche until the Internet came along.

Dimmick, along with other colleagues, has published several studies related to the niche that apply to the issue at hand. In particular, he has measured conditions for media exclusion or displacement where the new medium takes on some of the roles of the old one, and gratification opportunities for consumers of media.[18] By gratification opportunities, he means "consumers' beliefs that a medium allows them to obtain greater opportunities for satisfaction, more specifically, the perceived attributes of a medium relating to time use and expanded choice of content."[19] Indeed, these are the very features of online news, information, and entertainment that drive people away from the rigid schedules of print and broadcast media. And they are likely to be the Internet advantages that will drive what remain of the "mass" media to "niche" status in the future. In Dimmick's and his colleagues' study of 211 people in the Columbus, Ohio area, the respondents reported less use of traditional media—especially television and newspapers—following their adoption of the Internet.[20] When the authors examined the degree of niche overlap involving the Internet, cable television, broadcast television, radio, and newspapers, the highest overlap with the Internet occurred with cable television, followed by broadcast television and newspapers. They also found that the Internet's degree of competitive superiority (in providing gratifications to the audience) over broadcast television, newspapers, and cable was greater than any superiority those media might have over the Internet. Though the results of this study cannot be generalized because of the small sample, they should be sufficiently alarming to traditional media to pursue this research in more detail and with larger samples in order to assess the options for more successful competition with the Internet.

Online Media Users Grow, Off-line Media Recede

It is no secret that there is a correlation between the growth of the Internet and the decline of traditional media. But what nobody yet knows is how strong that relationship will become. The best statistical source available for Internet use over time comes from the Pew Internet and American Life Project. Collecting data through national surveys since the mid-1990s, the Pew studies have found that the percentage of U.S. adults online has grown from about 15 percent in 1995 to about 73 percent in 2006.[21] Other studies support that finding. Harris Interactive reported that 77 percent of adults were online in April 2006, an increase of 3 percentage points from spring 2005. Home access to the Internet was reported by 70 percent of U.S. adults, also up 4 points over the previous year.[22] And the survey found that the Internet population was increasingly resembling the U.S. population in its demographic profile, though it is still a little younger and a little more affluent than the nation at large. Though the Harris poll puts the percentage of users at only 9 percent in 1995, whatever early figure is used, it is clear that in just over a decade the Internet has grown from being a niche technology used largely by young and more affluent males to a mass consumer technology diffused to every part of the country.

Meanwhile, all the mass media are witnessing the shrinkage or splintering of their audiences.

Daily Newspapers

The daily newspaper has suffered what appears to be the largest declines over time. It is generally acknowledged that 1987 was the peak year for newspaper readership, when 64.8 percent of Americans read the newspaper on weekdays. The percentages have been sliding every year since then. At the end of 2006, 49.9 percent of Americans read a weekday paper.[23] The rate of decline has been accelerating. In the six months from October 2005 to March 2006, the aggregate daily newspaper circulation declined 2.5 percent.[24] To be sure, newspaper circulation was dropping long before the Internet came along. The reasons are numerous, including a lack of interest from younger readers, competition from television news, cuts in staff resulting in reductions in news content, and the increasing costs of marketing and distribution. But the Internet has posed a particularly serious challenge to newspapers because online readers have come to expect that content will be offered at no cost, and few have been willing to pay for it. A Pew Internet survey found that although more than half of online news readers are willing to register on a news site, only about 6 percent have ever paid for news content.[25]

Television

Broadcast news on network television has also faced pre-Internet competition in the form of cable news. In the twenty-seven years since CNN appeared, the number of people watching the networks' nightly news has fallen by more than 48 percent, or about twenty-five million viewers. The peak of nighttime network news viewing came in 1969 when the news programs had a combined rating of 50 (percent of households viewing the news) and an 85 share (percent of television receivers switched on at the time of the news). And as with daily newspapers, the pace of the decline appears to be increasing.[26] Declining viewership is also evident in many local television news markets.[27] It simply has become harder to hold the audience's attention for the evening news since the advent of DVDs, broadband Internet, and digital video recorders such as TiVo.[28]

News is not the only kind of network programming that has experienced audience erosion. The broadcast television networks once drew 90 percent of the prime-time audience. Today, the six broadcast networks together get less than a 50 percent share.[29] Of course, a lot of that audience loss has been to cable, but the Internet is expected to draw an increasing number of eyeballs as more people migrate to video content on their computers, iPods, and even cell phones. How should the networks respond? One suggestion has been to deliver more programs to desktops so that people can watch prime-time or daytime television at the office.[30] In the desirable eighteen- to thirty-four-year-old age category for women, daytime viewing was down 5 percent between 2004 and 2005.[31] If women could watch their favorite daytime program over their lunch hour in the office, it might be a way to regain some of that audience loss.[32]

Other Media

Consumer magazines and radio have also been affected by the Internet. Magazines have witnessed their most serious decline in readership in the newsweekly category, where they appear to be searching for an identity that has so far eluded them. *Time* reached its peak circulation in 1988, while *Newsweek* and *U.S. News and World Report* peaked in 1998 and 1993, respectively. But the trends over time for all three have been just holding steady or in decline. Double-digit drops in advertising also occurred for all three magazines in 2005, and *Time* laid off more than a hundred employees.[33]

Women's service magazines (*Ladies Home Journal*, *Good Housekeeping*, and *Better Homes and Gardens*) are also struggling. *Media Week* reported that the advertising page growth was less than 1 percent between April 2005 and 2006, while both newsstand sales and subscriptions are either flat or losing circulation.[34] And teen magazines, once the hot category, are losing circulation, including the long-term staple *Seventeen*.[35] The appearance of celebrity Internet magazines is one of the causes for the decline. Overall, annual circulation in the magazine industry is projected to decline to 320 million units in 2008 from about 372 million units in 1999.[36] Rebecca McPheters, in a column written for magazine and media marketers, framed the future of magazines this way:

According to Veronis Suhler Stevenson, consumer magazines' share of the communications industry has declined from 3.6 percent to 2.9 percent from 1999 through 2003. During this same period, the Internet grew from a 2.6 percent share to a 3.9 percent share, while garnering a dramatically larger share of consumer time, attention, and spending. Among consumer media, our industry's growth rate of 3 percent exceeds only that for newspapers. Clearly, unless we want to go the way of gaslights or buggies, we must find ways to increase the relevance of our brands to our two principal constituencies.[37]

Over-the-air radio has had to confront multiple delivery platforms in recent years. The Web offered only the first of these, whereby off-line stations would stream their content to a dedicated Web site. But more recently, the advent of iPods and other MP3 players with accompanying "podcast" opportunities, personalized-syndication feeds, and the newest competitor satellite radio have allowed listeners to choose selected broadcasts, or decide how they will receive streaming content. One of the problems in determining how the Internet has affected radio listening is the method used to estimate audience size. A good discussion of this was presented in the 2006 report "Radio" by the Project of Excellence in the News Media.[38] As the report points out, radio listening information has traditionally been collected through listener diaries and surveys. It is therefore difficult to draw accurate conclusions on self-reports of media use. According to the report, Arbitron research indicates that the radio audience is holding strong. Where in 1998, 95.3 percent of the population age twelve and over said they listened to the radio weekly, in 2004 that number was 94.2 percent.[39] But it is hard to know what to make of these statistics because radio listening is often passive and the broadcast serves mostly as background noise. Portable People Meters are being introduced to measure actual listening behavior that might provide more accurate information. A 2005 Arbitron report suggests that Internet radio may be having some impact on traditional radio, as twenty million Americans reported weekly use and 8 percent said that they had listened to it in the last week. Indeed, one survey found that Internet radio listening was up 10 percent over reported listening in 2000.[40] That figure grew even more by the end of 2005, with 12 percent (or thirty million) of Americans aged twelve and over reporting weekly use.[41] Satellite radio has come along recently and may have a larger effect on both over-the-air and Internet listening because the content can be tailored to adult themes. Podcasts are being embraced by both broadcast and print media, but it's too early to tell how much of radio's content advertisers will be interested in sponsoring.

Organizational Transformation

It's hard to imagine that any "old media" organization has remained unaffected by the growth of "new media" such as the Internet. As digital communication technologies have evolved, most of the organizations that have prospered by publishing magazines and newspapers, or by carrying radio and television programming, have struggled to redefine their missions and goals. That struggle has forced changes to the structures and processes of these organizations as they've tried to chase audiences—and revenues—that have been migrating online. As the pace of this migration increases, the rate of organizational change seems likely to rise along with it.

Traditional media businesses have varied widely in the extent to which organizational change has accompanied their online endeavors. They have acted like amateur poker players. Some companies put down large wagers on the Web, and they have drastically reshaped their organizations to reflect the high priority that they've put on their new media initiatives. Others have placed much smaller bets on the Web, and as a result have undertaken only minor organizational remodeling. The organizational changes precipitated by online ventures have tended to fall into two broad categories: structural and cultural changes. Structural changes are revisions in the way an entity organizes its tasks or activities, and defines relationships among the parts of the organization. Cultural changes are modifications in the customs, norms, and values that drive the work of the organization. At many media organizations, the extent of structural change brought about because of the Internet has been, to a great extent, a function of the perceived centrality of online ventures to the future of the organization. That, in turn, has influenced how management defined the relationships between their new and old media products, and how extensive the cultural changes have been.

Steve Outing, a journalist who has written extensively about online ventures in the newspaper industry, has been among those to point out that many traditional media organizations were initially slow to appreciate the potential of the Web and Internet to both wreak havoc on and enhance their missions to provide consumers with news, information, and entertainment. "A handful of publishers recognized early on that the Web was to dominate the online scene and set about to learn how to publish there," he wrote in a short monograph, published in 2000, that discussed the evolution of online editions of newspapers. But he goes on to say that "far more sat back idly and watched with curiosity."[42] In retrospect, the senior managers of many traditional media companies just didn't get the Web. Their online strategies were designed largely to bolster (or protect) existing lines of business rather than to establish entirely new ways to serve consumers and advertisers. In some cases, managers were concerned that their Web sites would never make money; in other cases, they were fearful that a successful Web site would undermine their "legacy" media products, though those concerns began to fade as they gained experience in the online business.[43] Their postures have been reflected in the structural arrangements that media organizations made to accommodate their Web operations. At organizations where Web ventures were clearly seen as defensive or ancillary to the old media products, the structural changes made might be better characterized as tinkering. In those situations, the Web staffs tended to be "add-ons." Web editors or producers were grafted on to the existing editorial or advertising staffs, with only a minimal effort made to integrate them into the existing operations. At some newspapers, for example, perhaps two or three employees were added to the company payroll, and then given the job of shoveling "print" stories on to the Web once the publication went to press or trying to sell advertising for the online edition.[44] At others, the Web site was deemed so "noncentral" that it became the responsibility of the firm's marketing department or was farmed out to a third-party vendor.[45]

Other media organizations were more excited about the potential of the Web, and their level of investment reflected that enthusiasm. *USA Today*, the Washington Post Company, the New York Times Company, CNN, and Time Inc. were among the high-profile media companies that bet heavily on the Web.[46] For these organizations, the structural changes were more extensive. Each created separate divisions or operations to accommodate their online activities, and poured tens of millions of dollars into these ventures.[47] Time Inc. invested an estimated $75 million in its Pathfinder portal before closing it in 1999 after only five years in operation.[48] The Washington Post

Company hired more than a hundred employees as it created its digital division.[49] The *Raleigh News and Observer*, a midsize regional publication, was among the first to have a relatively large online operation for a business of its size.[50] Though the online investments of firms like these may have been substantial and were intended to position these organizations to develop successful online businesses, all these companies left the fundamental structures of their organization intact. For them, organizational change consisted of adding on an Internet operation rather than integrating online staffs more fully into the existing information- and revenue-producing departments of the company. The creation of a separate online division avoided the messy process of combining the staffs of the Web, print, and broadcast operations, which often operated with different values and priorities. And it had another potential benefit: it could have made it easier to spin off the online division into a separate company. That possibility looked promising to many media companies until the dot-com bubble burst in 2001.

The decision to use a divisional structure was based on the assumption that online ventures were fundamentally different animals from the legacy media businesses. Those assigned to the online ventures operated more or less independently. Certainly, senior managers hoped for coordinated efforts among those working in their old and new media shops, but the online employees were not fully integrated into the staffs of the newspaper, magazine, or broadcast news operation. The strategy was, it appears, to repurpose content from the legacy products and add value to the online publication through the creation of some additional unique content. Divisional structures remain common today. One circumstance in which a divisional structure seems to be preferred is when an acquisition is made. During the last two years, companies most frequently associated with traditional media products have become more aggressive about buying Web sites. Dow Jones and Company, the publisher of the *Wall Street Journal*, bought MarketWatch, a business news and finance site; the Washington Post Company purchased the online magazine *Slate* from Microsoft; News Corporation acquired the social networking site MySpace; the New York Times Company scooped up About.com, a network of Web sites that provides information on a variety of topics; and the Tribune Company acquired the real estate site ForSaleByOwner.com.[51] All these Web sites became separate operations within the parent company. Another circumstance in which a divisional structure is chosen is when a partnership or new venture is started involving entities outside an organization. That situation occurred in 2006 when NBC Universal and its 213 affiliated stations formed the National Broadband Company. This "newer NBC" will distribute news, lifestyle, sports, and weather videos over the Internet.[52]

If add-on restructuring reflects a defensive approach to moving online, and if divisional restructuring reflects a view of the Internet as a separate creature that needs special care and handling, a third kind of organizational restructuring fully embraces the

Web as yet another in a portfolio of "platforms" that a media company can use to distribute news, information, and advertising to consumers. This third kind of organizational restructuring, full-blown staff integration, suggests management has come to view the Internet as among its core products. Since roughly 2000, a growing number of organizations have started to integrate their new and old media staffs into one news and entertainment operation. One of the pioneers of integration was Media General, a company that owns newspapers, television stations, and online enterprises. In Tampa, Florida, Media General built a $40 million "News Center" to house the combined operations of its daily newspaper (*Tampa Tribune*), local television station (WFLA), and regional Web site (TBO.com).[53] Though Media General did not fully integrate the staffs of the three media outlets, it took a big step in that direction by creating a physical and cultural environment in which a high degree of cooperation among those staffs was encouraged. Reporters were asked to "cross over" to create content for all three of the distribution platforms (newspaper, television, and online) when it made sense to do so, and managers from all three outlets jointly planned news coverage. Media General's CEO at the time, J. Stewart Bryan III, was a strong advocate for restructuring to facilitate this kind of convergence. "I think it will show that by combining the strengths of three different mediums, you will be able to cover the news for the consumer better," Bryan told *Editor and Publisher* magazine.[54] He promised that Tampa would become the model for integration at other Media General properties. By many accounts, the convergence experiment in Tampa has not met all the expectations of Media General's management, but it seems to have set the stage philosophically for the kind of organizational integration that is increasingly becoming the model that many media businesses are following. The Tribune Company, a major newspaper publisher and broadcaster, made one of the largest investment with an eye toward the greater integration of old and new media. It spent more than $8 billion in 2000 to acquire the Times Mirror Company, another large newspaper publisher, so that it would own print, broadcast, and online media in the three largest markets in the United States.[55] The objective was to sell advertising and share content across all three platforms, though those goals were never fully realized.

In early 2001, CNN was perhaps the first national news organization to completely integrate its news-gathering operations.[56] Its separate television, radio, and interactive divisions were combined, and the organization waded into the contentious process of cultural change. Television reporters were told that they'd need to file stories for radio and the Web; Web journalists were told that they'd no longer be able to focus only on online content. More recently, the *New York Times*, the *Boston Globe*, the *Wall Street Journal*, the *Financial Times* (London), and *USA Today* all have moved toward the full integration of their online and off-line news-production operations. At *USA Today*, editor Ken Paulson said the goal was to create a single news organization that would provide information to readers on multiple platforms. "That means going beyond arm's length collaboration," he explained in announcing the decision.[57] The *New York Times* offered a similar explanation. The restructuring reinforces the goal of Arthur Sulzberger, chair of the New York Times Company, to create a "platform-agnostic" media organization that is comfortable providing consumers with information and advertising using print, the Internet, and video.[58] In a similar vein, Time Inc. is moving back toward greater integration of its Web and print operations after allowing its stable of magazines to more or less go their own ways online since Pathfinder was abandoned. *Time* is requiring that writers produce more copy for its Web sites.[59]

As media organizations embrace Web publishing more vigorously, they are confronted not only with the need to alter their structures but also with cultural change. In a 2006 *American Journalism Review* article that is provocatively titled "Adapt or Die," Rachel Smolkin describes the tumult at newspaper companies where journalists are trying to figure out how to adapt to a workplace environment in which the long-standing core product—the newspaper—is losing favor to the new media upstart—the online publication.[60] The bottom line, literally, is that revenue growth in the newspaper industry has slowed to a crawl as circulation has slipped and advertising in key sectors has declined. The exception is online advertising revenue, of course. Though it's expected to represent only about 6.5 percent of a newspaper's ad revenue in 2006, its growth rate is in double digits.[61] "I think the handwriting is kind of on the wall that there is a large migration to the Web," Colby Atwood, a newspaper analyst with Borrell Associates, told the *New York Times*. "Increasing amounts of revenue and focus should be on the online properties."[62] That shift in focus is precipitating a change in workplace norms, values, and customs. Some employees are finding the change invigorating, but others are finding it disconcerting.

One cultural difference is that the online staff members are gaining status and influence within these organizations. Jeff Gaspin, who handles digital ventures for NBC Universal, tells a story about attending meetings a few years back for the Bravo cable channel at which he had trouble remembering the name of the person who directed online ventures. "For an entire year, I just couldn't remember [the man's name].... Now," Gaspin says, "the online guy is the most important guy in the room."[63] Indeed, at all the major broadcast networks, online ventures have gained significance in both the news and entertainment divisions. One consequence has been greater demands on employees, both in the kinds and volume of work that employers expect. And this has led to tension. At NBC, for example, writers for its situation comedy *The Office* were asked to produce additional "Webisodes" of that series for summer 2006—without more compensation. Those expectations were leading to complaints from writers unions.[64] At the *Washington Post*, the paper's Newspaper Guild chapter filed a complaint about the extra demands being put on journalists to provide information and expertise to the paper's online edition, arguing that journalists should get extra pay for extra responsibilities.[65]

But the cultural changes go well beyond the expectations for taking on more tasks. In print and broadcast newsrooms, journalists who have been trained in one craft now are being asked to create content for distribution platforms with which they have little familiarity, leading to more workplace stress. In *Managing Media Convergence*, Kenneth Killebrew examined this as well as several other challenges facing journalists and their managers in media organizations where print, broadcast, and online distribution are converging.[66] He identified the following cultural changes:

• Journalists who previously faced only one print or broadcast deadline a day are being forced to adjust to the more stressful 24–7 publishing cycle of the Web.

• Internal coordination and communication are often more challenging in organizations that distribute information across multiple platforms, again increasing the potential for workplace stress.

• Employees who identify more strongly with one distribution platform may find themselves competing with coworkers who identify more strongly with another distribution platform.

• Journalists are having to adjust to getting direct, immediate feedback from the online audience—some of it coming in the form of the number of "hits" that signal the popularity of their stories with the audience.[67]

• New ethical challenges are arising. For example, should book reviews include links to online bookstores, or is that a breach of the traditional separation between editorial and advertising departments? Should online editions publish the names of people—particularly young people—who have committed minor legal infractions given that employers could easily discover these minor misdeeds with a simple Google search?

• News organizations are trying to determine how to balance the online world's demand for getting information online quickly with traditional journalistic norms that call for carefully vetting information before letting it loose in public.

Business Models Affected

The growing importance of online ventures at media companies also is leading to far-reaching changes in the ways these organizations are conducting their business. Analysts who follow the media believe that a fundamental change is occurring because digital technologies, led by the Internet, are undermining the business models for traditional media organizations.[68] The media have been cyclic businesses, so their managers have come to understand that their fortunes tend to rise and fall with economic cycles. But now structural, not cyclic, change appears to be occurring at companies that produce newspapers, magazines, and broadcast content. "This is different from before," Mike Burbach, the managing editor of the *Akron Beacon Journal* newspaper, recently said in reflecting on the major staff reductions that have occurred at many daily newspapers. "The business has changed dramatically."[69] And while the new media ventures of newspaper, magazine, and broadcast companies have been growing at a healthy rate, they are not expanding fast enough to offset the declines that are occurring with the legacy products.[70] The damage that technologies such as the Internet and digital video recorders are inflicting on traditional media is neither hard to detect nor understand. Fundamentally, these new technologies are disrupting the relationship between advertisers and the media that those advertisers have relied on for decades to reach potential customers. The media organizations' role as an intermediary linking those who peddle products and services with those who buy products and services is threatened because the Web can more precisely link content to audiences.[71]

For newspapers and magazines, the Internet is drawing away audiences and advertisers even as production costs rise, and as Wall Street investors insist that publicly held media companies protect their traditional 20 percent profit margins.[72] While it is the case that the decline in the print audience, particularly for newspapers, predates the Internet, the pace of that decline has increased as more people have turned to the Web for their news and information. Advertising rates are based, in part, on the size of the audiences that these media organizations can attract. So with audiences shrinking quickly, it has become more challenging for newspaper and magazine publishers to maintain ad volume and revenue growth. Advertisers have been balking at paying more to reach fewer readers. In fact, neither daily newspapers nor consumer magazines have succeeded in maintaining their ad volume in the face of the circulation trends reported above. Figures from the Newspaper Association of America show that the total ad volume for U.S. dailies peaked in 2000, and by late 2005 had declined about 4 percent. The Magazine Publishers of America report that the number of ad pages in consumer publications dropped almost 10 percent from 2000 to 2005.[73]

The trouble for print media doesn't stop there, however. Within the newspaper industry, perhaps the most ominous technology-based change has been the emergence of Web sites that provide free or low-cost searchable classified advertising. Sites like Craig's List, Monster.com, cars.com, and Oodle represent a frontal assault on the most profitable segment of the newspaper business. A McKinsey and Company report released in April 2005 estimated that between 1996 and 2004, the Internet had deprived newspapers of almost $2 billion in revenues in one classified category alone: help wanted.[74] And the report warned that other critical categories such as real estate, automotive, and general merchandise may be threatened as well. Newspaper publishers have responded by buying or starting Web sites that specialize in classified advertising, and trying to build up the online classified component of their publications' Web sites.[75] Some have even started to offer free online classified ads to try to thwart the competition.[76] But the erosion of this crucial revenue stream for newspapers seems irreversible and perhaps irreplaceable.

Within the consumer magazine industry, the nature of the Internet threat may be even more fundamental. Consumer magazines have tended to follow segmentation-oriented business strategies—*Cosmopolitan* for young women, *Field and Stream* for outdoorsmen, or *Rolling Stone* for music enthusiasts.[77] The segmentation strategy began emerging a half-century ago as it became clear that television could more efficiently and effectively reach mass national audiences. And the segmentation strategy worked well for magazines because it corresponded to the needs of many product makers and service providers as they switched from mass-market to more targeted brands.[78] But magazine industry analysts now say that the Internet and cable television are threatening to undo the segmentation strategy. In a sense, those newer technologies can "out-segment" the magazines by efficiently linking niche audiences to advertisers. In a 2005 article in *Publishing Research Quarterly*, industry analyst Thomas R. Troland gave this assessment: "In competitive terms, the magazine industry now finds itself in-between two powerful media that command increasing expenditures, attention, involvement and usefulness for consumers: Cable TV and Internet access. Each is eroding the hard-won role of magazines in both the consumer and advertiser media marketplaces. The bottom line is that we are a large—$25 billion—business which isn't likely to grow a great deal bigger in the future."[79]

Though the handwriting about the Internet has been on the wall for some time for senior managers at newspapers and magazines, it's increasingly clear they're now reading and reacting to it. Some of the response has been to try to reframe perceptions about the businesses that they run and the business environment in which they operate. In late 2005, newspaper journalists and their employers were stunned when Knight Ridder, the nation's second-largest newspaper publisher in circulation, put itself up for sale. Investment firms that owned shares in Knight Ridder forced the sale after they became anxious about another important downward trend at newspaper companies—the price of their stock. Less than two years later, the publicly held Tribune Company decided to take itself private after its share prices disappointed both Wall Street and a significant block of owners. (Uncertainty about the future of print in an Internet-oriented media world robbed publicly held newspaper companies of about 20 percent of their market capitalization in 2005.)[80] Another publicly held newspaper company, McClatchy, bought Knight Ridder for about $6.5 billion—a price that would have been a bargain only a few years earlier.[81] Within days, McClatchy's chief executive, Gary Pruitt, argued in a commentary published in the *Wall Street Journal* that the newspaper industry was actually doing reasonably well, particularly compared with other traditional news media, all of which have been shedding audience. And in an acknowledgment of the growing importance of the company's Web ventures, he claimed that if newspapers' online and off-line readers were combined, "newspaper" audiences are actually growing, not shrinking.[82]

Pruitt's touting of newspaper Web sites reflects another change that has occurred in both the print and broadcast industries: they have embraced the enemy. The print media have been motivated, in part, by a desire to reach the younger consumers whom many advertisers covet. Those younger consumers seem more inclined to seek their information using digital technologies, which is one of the reasons that the News Corporation bought MySpace.[83] In his 2005 speech to the American Society of Newspaper Editors, News Corporation CEO Murdoch said, "I'm a digital immigrant. I wasn't weaned on the Web, nor coddled on a computer. Instead, I grew up in a highly centralized world where news and information were tightly controlled by a few editors, who deemed to tell us what we could and should know. My two young daughters, on the other hand, will be digital natives. They'll never know a world without ubiquitous broadband Internet access."[84] Murdoch isn't alone in his view. The "MySpace generation," largely adolescents from ages twelve to seventeen, are said to live comfortably and simultaneously in the real and virtual worlds using the social networks as "virtual community centers," while adults spend their social lives rooted in telephone calls or face-to-face interactions.[85] Advertising is also blurred with content on these sites, displaying itself in such a subtle manner that the users can't tell it from the content. Almost defying a definition beyond that of social networking, the functions of MySpace include "a multi-level entertainment opportunity involving blogs, instant messaging, classifieds, peer voting, special interest groups, user forums and user-created content," and a means of launching new and emerging recording artists.[86] The challenge that Murdoch and others face with social networking sites is to figure out how to get reasonable returns on their investment. As a recent *Wired* magazine profile of Murdoch pointed out, these sites are fundamentally more about social and cultural activities than commercial activity.[87]

Both broadcast and print media are experimenting with other ways to attract audience members who are resistant to traditional media, or to blend old with new media. These include podcasts and vodcasts, which are audio and video reports that can be downloaded to personal digital devices like iPods; souped-up search engines for their Web sites; personalized syndication that allows readers to get customized news reports through Really Simple Syndication (RSS) feeds; blogs created both by journalists and citizens; Web sites directed at niche audiences, such as those for sports fans or young mothers; and news updates and other information sent to cell phones.[88] Publishers also are exploring vehicles for downloading electronic versions of their print publications to new "E-newspaper devices," such as a thin plastic reader that can be rolled up when not in use.[89]

Though their online and other digital ventures are benefiting from the boom in Internet advertising, organizations that own traditional media have yet to find reliably profitable business models for services like these. Many of their Web operations are no longer operating in the red, technically speaking, but they still do not generate enough revenues or profits to keep the doors open by themselves because they rely on their old media siblings for so much content.[90] Media companies continue to struggle over basic business decisions, such as whether it's desirable—or even feasible—to charge consumers for access to some or all of their online information or services.[91] Newspapers that have imposed charges for their content have seen traffic to their Web sites plummet, which undermines efforts to sell advertising on the sites.[92] Newspapers that have imposed charges for their content have seen traffic to their Web sites plummet, which undermines efforts to sell advertising on the sites. The *Wall Street Journal* is a major exception to that trend. In the first quarter of 2007, the paid subscriptions to the online version increased 20 percent to 931,000.[93] Developing sound Internet business strategies has been an even thornier matter than deciding whether to charge for content. When newspaper and magazine publishers first ventured on to the Internet, they embraced a subscription model for their content by teaming up with fee-based Internet service providers such as AOL, Prodigy, or CompuServe.[94] Those liaisons were short-lived, however, as was that business model. Publishers were eager to establish their own Web presence under their own brands, although Susan Mings and Peter White point out that many did so without a clear sense of how they'd make money on the Web.[95] Since then, publishers have tried permutations of various business models in no particular order and with no clear success. In a 2000 article in the *International Journal on Media Management*, media economist Robert Picard identified four business models that at the time, had either failed or been abandoned: Videotex, paid Internet, free Web, and Internet supported with "pushed" advertising.[96] He identified two others, both based on the portal concept, that he argued held promise. The strategies that have been tried seem to be based on one or more of the following basic models:[97]

• *Subscriptions* Among the print news media, the *Wall Street Journal* has been the most visible of the handful of news organizations that have developed a successful subscription model. Otherwise, Web readers continue to be notoriously resistant to paying for information, though some publishers have imposed fees for access to most of their content.[98] Many sites do charge for some kinds of content, such as access to their archives or columnists.

• *Advertising* The revenue stream has taken many forms, including banner ads, sponsorships, searchable classified ads, interactive ads, and most recently, ads associated with searches.

• *Transactions* Under this model, the media organization seeks to create a virtual marketplace, bringing together sellers and buyers. The organization aims to profit from the transactions that occur in this market.

• *Bundling* The news organization enters into partnerships, supplying content and agreeing to participate in revenue-generating opportunities that arise out of the partnerships. In 2006, for example, several major newspaper companies formed a partnership with Yahoo! to share content, advertising, and technology.[99]

Of late, some media companies have been revisiting the idea of advertising-based business models because Web advertising appears to be gaining greater acceptance. And advertising that is linked to search functions is also drawing attention. The success of Google and Yahoo! appears to be the motivation for trying to capitalize on search-related advertising. But in general, the thinking today tends toward the view that no single business model—advertising, subscriptions, transactions, or whatever—will dominate for publishers. Rather, the prediction is that if newspapers and magazines are to find significant profits on the Web, this will happen by cobbling together multiple revenue streams based on providing both content and services.[100]

If the Web has been an unforgiving environment for publishers, it has been an equally challenging place for broadcasters.[101] Among the operators of local television stations, the pattern early in the evolution of the Web was to use it more to build audience relationships than to try to generate revenue. But that may be changing. Local television stations are facing many of the same kinds of problems as local newspapers—declining audience size, revenues, and profits. And as with newspapers, the Internet's appeal to viewers and advertisers is blamed.[102] Increasingly, those stations may be turning to their online ventures to provide additional revenue to protect their profit margins, which can easily reach 40 percent.[103] No widely accepted Web business model exists for local broadcasters either, so a variety of things are being tried—video Webcasts with ads that cannot be skipped, Web sponsorships for local businesses, and regional portals.[104] If any consensus exists among local newspapers and broadcasters about how they should go about making money online, it's that it is too risky to wait for clearly viable business models to emerge.[105] The sense is that if significant, sustained profitability is to come, it probably will be from experimenting with many different things.[106]

Experimentation is also under way within the television industry at the national level after years of adopting something of a wait-and-see approach to the Web. Certainly the major broadcast networks long ago established a presence on the Web, using their sites to recycle news and information as well as to promote programming.[107] Two of the major networks—Disney/ABC and NBC—funded ambitious efforts to develop Web portals in the late 1990s. But when those sites failed to live up to expectations, their owners became more cautious with their Web investments.[108] The News Corporation set out in 1997 to acquire PointCast, a Web technology that "pushed" content to Web users. But when its bid failed—PointCast itself eventually failed too—Murdoch's company turned its attention elsewhere.[109] Within the last couple of years, however, national television companies have lost much of their reticence about investing in the Web. They announced one new initiative after another beginning in 2005. The

initiatives with the greatest potential to affect the business models in the television industry have come from the major broadcast networks. ABC, CBS, Fox, and NBC each is
testing the distribution of some entertainment programs over the Web. For example,
ABC and NBC have struck deals with iTunes to allow programs to be downloaded
for viewing for $1.99. In addition, ABC has plans to make available a few programs
free on its own site, embedding advertising that cannot be skipped. CBS and Fox shows
can be downloaded from their own Web sites or those of their partners. CBS, for example, is selling its programs through Google Video. Fox plans to use MySpace, its social-
networking site, to distribute at least one hit show. The local broadcast stations
affiliated with the networks have found such ventures unnerving.[110] Few think that
these arrangements will have much impact immediately on the economic health of
the hundreds of local network affiliates. But those affiliates are in the business of being
middlepeople—of supplying a link between the network's programs and the audience.
Now, it appears, the networks are testing an alternate route—a more direct one—to
consumers. Like the broadcast networks, the local affiliates generate virtually all their
revenue by selling advertising within network and local programs, such as news. Their
fear is that over time, downloads of network programs could undermine their ability to
draw a local audience that is large and desirable enough to appeal to advertisers. That
could make it tougher to sell local advertising, and that is one of the reasons that the
affiliates have agitated for the networks to share revenue earned from Web downloads.
Fox was the first of the four largest networks to agree to such a plan.[111]

The networks did not embark on this new business path lightly. Digital technology
has forced their hands. The main culprits are digital video recorders such at TiVo,
which are undermining the foundation of advertising-supported television—the
thirty-second spot advertisement. One research company predicts that by 2010, about
17 percent of spot television advertising will be skipped.[112] Web downloads may afford
an additional revenue stream for the networks and other organizations that provide
programming to consumers. Web technology also offers a couple of other advantages:
it can make it impossible to skip advertising in programs that are downloaded or
streamed from Web sites, and it could present new opportunities for e-advertising and
e-commerce. For instance, one Web technology that networks are evaluating lets
viewers "bookmark" scenes in television shows that are being watched on the Web.
Those scenes—and the advertising accompanying them—can be viewed again. The
technology also lets viewers click on an item, such as a piece of clothing being worn
by an actor in a program, and buy an item like it instantly.[113]

The cable television and film industries are also trying to figure out where the Internet fits into their business strategies. Like the broadcast networks, cable channels such
as MTV and Comedy Central are using the Web to distribute programming and other
content. MTV, which is frantically trying to hang on to its teen and preteen audience,

is selling music, showing movie trailers, and telecasting concerts through its MTV Overdrive Web sites. It also announced plans to start an audience-controlled Web "channel," providing tools to help users create their own multimedia content.[114] Television programs and film are available online through services such as Akimbo, Vongo, Movielink, Guba, and the Google video store. The E. W. Scripps Company is taking advantage of its cable television brand and video expertise to launch a business-to-business venture via the Web. Scripps, the owner of both the Home and Garden and the Food Network cable channels, has created HGTVPro, a site that targets online video content to professional builders. *Media Business* has singled out this initiative as an example of how new competitors can enter the business-to-business market and at the same time help create different video forms.[115]

The critical question that remains to be answered is whether all this investment will have a payoff. As a 2005 *Business Week* article points out, it's not yet clear whether this is a dot-com era replay, with established companies and investors sinking huge sums into fast-growth start-ups with no viable business models, or an inflection point at which traditional media businesses are making prudent investments for the future. *Business Week* notes that Facebook is only a couple of years old, is run by a twenty-one-year-old student on leave from Harvard, and is so far profitless.[116]

Observable Content and Format Changes

In the first years of mass media's use of the Internet, little new content was being created. Today, competition from independent online media, user-generated content, and new competitors like Google and Yahoo! have forced traditional media to think about both new ways to deliver old media and new types of content to include. Robert Picard reminds us that there will likely be few possibilities for content that is fundamentally new. "New [Internet] technologies cannot revolutionize content because they provide no real new communication capabilities. They are not affecting communications in such fundamental ways as did the arrival of the printing press, telegraph and telephone, photography and motion pictures, and broadcasting, which provided the abilities to move text, sound, and images with or without terrestrial lines."[117] The major impact of the Internet on communications is in increasing the speed and flexibility of transmission, and the shift in control of the content, Picard argues. He looks to users of media content to create the demand for additional services that provide more of the same, not for new content to be generated.

The story of change in media is not so much that the content is fundamentally new; instead, it is a question of who is allowed to produce this content. Ordinary people have had opportunities never before imagined to produce products that look like news, editorials, documentaries, investigative reporting, and various types of entertainment media products. Many of these products have been able to pay for themselves or even be profitable by attracting advertising to their Web sites. These products range from entrepreneurial journalism, where the individual does reporting on an issue, and may raise funds from users and advertisers to sponsor the project, to formal independent media sites where citizen journalists volunteer their time to report on issues central to their political and social interests. Political blogs fall more in the editorial content category, where the blogger writes opinion pieces that might otherwise appear on an op-ed page in a newspaper or a listener-response program on the radio. Newer to the scene are the video presentations created by cell phones and other video-recorder devices, and then placed on a Google or Yahoo! video space. These videos are also sometimes sold to stations for over-the-air clips often related to storms or accidents that a citizen might have witnessed.

In the aftermath of Hurricane Katrina, a great variety of citizen journalist activities appeared on the Web. Flickr, the online personal photography site, was a location for organizing photographs of the disaster. Wikipedia and Craig's List were used to help victims and their families locate each other. The traditional media in the region used citizen content as part of neighborhood forums.[118] And still others created maps of the affected areas with information on the damage done to specific houses. None of these examples represent firsts, however. But the Internet made it possible for people to provide and disseminate information more quickly with more detail, and include graphics and images to assist in the recovery effort.

Some media have taken advantage of that citizen capability. *OhmyNews* is a South Korean online newspaper that employs only sixty reporters and editors, but takes citizen-generated stories from forty-one thousand people around the world. At least seven hundred thousand repeat visitors go to the site every day.[119] Founded in 2001 by Korean journalist Oh Yeon-ho, the news site has been profitable since 2003. Though *OhmyNews* has been influential in electing the country's president, one of the reasons for its success has been the public's rejection of the traditional media and its ultraconservative approach to news.[120] The international director of the site, Jean Min, cites the combination of professional journalists and their control over the final product as key to the organization's success.

In his book on the role of citizen journalists, *We the Media: Grassroots Journalism by the People, for the People*, Dan Gillmore writes that audiences will take an increasing role in producing journalism. His view of the Web is that it has created the opportunity for media to move from a lecture to a conversation. "That's a shift I consider absolutely essential for all of journalism's constituencies," he said in a speech at Columbia University.[121] Or as Shayne Bowman and Chris Willis put it in a report on the future of media, "The audience is now an active, important participant in the creation and dissemination of news and information, with or without the help of mainstream news media."[122]

Journalists have long rejected the idea of allowing the audience to be a partner in the production of the daily news. It is believed that amateur reporters who don't have the benefit of a credible news organization behind them, and don't have editors to find and correct errors in the news that these people produce, will not be able to replace or even compete with professional news media. As much as news media would like to put their product on a pedestal, the consumers of the news media don't have the same confidence in that product. A Gallup Poll taken in 2005 found that only 28 percent of Americans say they have "a great deal" or "quite a lot of" confidence in either television news or newspapers.[123] In describing the value of blogging to journalism, J. D. Lasica states that bloggers can build up a "publishing track record" to achieve the credibility of journalists. "Reputation filters—where bloggers gain the respect and confidence of readers based on their reputation for accuracy and relevance—and circles of trust in the blogosphere help weed out the charlatans and the credibility-impaired. If the blogs are trustworthy and have something valuable to contribute, people will return."[124]

All of the above examples relate to shifts in who is producing the content. It also appears that new commercial players are producing traditional media content online—or may have aspirations in that direction. Yahoo! has created a couple of focused coverage sites—"Kevin Sites in the Hot Zone," a series of reports on the Iraq war, and "Richard Bangs Adventures," which are travel stories and information. Following his move from ABC Entertainment to Yahoo!, Lloyd Braun originally thought he might produce television programs on the Web, but later abandoned his ambitious plans when he realized what that would require. For the moment at least, Yahoo! plans to offer more user-generated content alongside existing traditional media content purchased from the networks and cable stations, and to limit the amount of original content it produces. Braun now sees original content as the "salt and pepper on the meal," rather than the "engine driving it."[125] Film studios and television networks had expressed worry that Yahoo! would turn into a competitor and not just a distributor. That might have led the traditional entertainment media to offer content to Google rather than Yahoo!, forcing Yahoo! to retreat from its original plans.[126]

Current TV, an Al Gore initiative, is the latest idea for participatory media. On its Web site (⟨http://www.current.tv⟩), it is described as a "national cable and satellite channel dedicated to bringing your voice to television." The "viewer created content" on the channel is presented in "pods," or short segments largely produced by the audience.

Mainstream media have had a variety of reactions to the new competitors and distributors. In some cases, they have acknowledged the shift and allowed the audience to play some part in the production of their product. In other situations, legacy media have continued to shut out users, thereby forcing them to their own spaces. And in still others, they have provided a limited separate space on their sites, creating a narrow role for audience participation.

Whether traditional media allow audience participation or not, some rules for comments and contributions have had to be created. The *Los Angeles Times* tried an approach called "wikitorial" (patterned after Wikipedia) in 2005 where it put an editorial about the Iraq war online, and asked readers to comment and insert information in the online copy. Using open-source software to allow input from readers, the editors were not able to stop the addition of hard-core pornography that some users put on the site, and finally ended what they called a "public beta" experiment in citizen participation in their online newspaper.[127]

To prevent such content from being posted on its site, the BBC audience participation area, "Have Your Say," asks readers to monitor the comments and file a complaint when a reader believes the author has violated the house rules. The filing of the form alerts a moderator who reviews the complaint and takes action. Because all those who write comments on the site are required to register, breaking the rules can result in the suspension of the participant's membership.[128] Another British publication, the *Guardian Unlimited*, has a news blog written by members of its own staff, but also allows readers to comment on the blog entries if they are registered with the site. The *News-Record* (Greensboro, NC) online edition is an example of a news organization that is combining the traditional newspaper with reader input. In a special section called Town Square, readers are invited to submit their own stories to the site. They are told that their stories will be edited for grammar, spelling, and the elimination of profanity and libel, but that the stories or comments will be the readers' own. Contributions are limited to two per week per person. A link on the home page takes readers to this section of the site (⟨http://www.news-record.com⟩). The paper is also trying to deliver its product through most multimedia available—podcasts, RSS feeds, streaming video, and photo galleries with audio narratives. On parts of this newspaper's site, the bylines are attributed to people called "content producers" instead of reporters. The biggest content change over time online is from a single way of delivering content—from text, audio, or video in traditional media—to multimedia platforms. All media have been experimenting with various ways of packaging their content to remain viable to users as well as advertisers.

Fisher examines the media content that is being created through the alteration of the original format of that content in films, music, and television programs produced by mainstream media businesses. He argues that this kind of creativity contributes to "semiotic democracy," allowing ordinary citizens to free themselves from the limited options offered them by the small number of cultural industries in the world.[129] When John Fiske coined the term in his book, *Television Culture*, he was referring to the process that audience members use when interpreting media texts.[130] He said that audiences that watch films or television programs are not limited to understanding those cultural products from the meanings intended by the producers or directors but can inscribe their own meanings to those texts. Fisher is taking that idea one step further when he describes the ways that audiences can electronically edit the original films or recordings to produce new meanings.

In addition to the shift from only media-produced content delivered to a mass audience (from the few to the many) to audience combined with professional media worker-produced content delivered to a mass or niche audience, one other important change has taken place in recent years. That shift has occurred among mostly young people communicating in social spaces on the Web. As Bowman and Willis describe this social environment, "There are participants who serve different roles in the creation, consumption, sharing and transformation [of media]. This is giving rise to information ecosystems, such as the blogosphere."[131]

Social-networking sites have grown 47 percent year over year, reaching 68.8 million in April 2006.[132] The current leader in social-networking sites is MySpace. It grew 367 percent from April 2005 to April 2006, reaching nearly 38.5 million people. According to Nielsen/NetRatings, the interactive characteristic of these sites is what draws people to return regularly. MySpace realized a 67 percent return rate in April over March 2006, while MSN Groups and Facebook, two other popular social-networking sites, had 58 and 52 percent return rates, respectively.[133] But does MySpace or any of the social-networking sites really have any new content to offer? Jon Gibs, senior director of media at Nielsen/NetRatings, sees social networking as the reality television of the Internet. And like reality television, the sites will proliferate. Yet Gibs says that also like reality television, they will eventually have to "provide consumers with distinct content they can identify with" to sustain themselves.[134] They are just an extension of the larger trend enabled by the Internet, digital expression, according to Robert Young.[135] And social networks, rather than providing new content, are just new delivery devices for that expression, Young believes. He attributes great power to digital expression, however, by labeling it a new medium to compete with traditional mass media.

This medium is dedicated mostly to a certain age group, and the implications for the future of traditional and online media are uncertain. Danah Boyd, a researcher who has studied MySpace since its inception, is also uncertain about the long-term impact of social networks on youth but offers this observation:

By going virtual, digital technologies allow youth to (re)create private and public youth space while physically in controlled spaces. IM serves as a private space while MySpace provides a public component. Online, youth can build the environments that support youth socialization.

Of course, digital publics are fundamentally different than physical ones. First, they introduce a much broader group of peers. While radio and mass media did this decades ago, MySpace allows youth to interact with this broader peer group rather than simply being fed information about them from the media. This is highly beneficial for marginalized youth, but its effect on mainstream youth is unknown....

What we're seeing right now is a cultural shift due to the introduction of a new medium and the emergence of greater restrictions on youth mobility and access. The long-term implications of this are unclear. Regardless of what will come, youth are doing what they've always done—repurposing new mediums in order to learn about social culture.[136]

It would be easy to dismiss social networks that engage adolescents, especially since the short history of these spaces indicates that youth move quickly from one to another as soon as they tire of them, or if another site offers something newer and edgier. MySpace has been replacing Friendster, Facebook, and MSN Groups, for example. And Tagged.com and Bebo.com are the up-and-coming sites that are challenging MySpace. This movement may be important for traditional media to watch more closely rather than dismiss the social-networking phenomenon as a fad.

What's Next

In the introduction to this chapter, we quoted Rupert Murdoch on the reasons that news media (and perhaps entertainment media as well) at first rejected the changes in communication that have taken place on the Internet. But will traditional media executives continue to stand by and watch while their audiences and authority dwindle? News editors have always thought that their role was to make sure that the information the public received was accurate and trustworthy. The introduction of citizen journalists, bloggers, and other social-network participants may be viewed as further proof that editors are needed. The Wikipedia case has demonstrated both that citizen editors can correct information and that erroneous information can appear on collaborative Web sites. Josef Kolbitsch and Hermann Maurer, in an article on the Web's transformation, also note that Wikipedia-type information is often incomplete, thus presenting another trust issue for audiences.[137] They also caution that on many sites where users contribute information, it is difficult to know whom to hold responsible when the information is inaccurate because complete identification is frequently not required. They suggest that this implies a new obligation for users, a role that used to be assigned only to the editors of mass media: to check other sources to determine the accuracy of any information they find on the Internet. On the other hand, perhaps users will value traditional media more for that function.

Whatever happens, Murdoch is right to say that media managers "may never become true digital natives," but they "can and must begin to assimilate to their culture and way of thinking."[138] Vin Crosbie has written that the mass media and the interpersonal media have combined on the Web to overcome the limitations of each of their weaknesses: "1. Uniquely individualized information can simultaneously be delivered or displayed to a potentially infinite number of people. 2. Each of the people involved—whether publisher, broadcasters, or consumer—shares equal and reciprocal control over that content."[139] With a lot of luck and a changing mind-set, the individuals at the top of the media world today will appreciate these changes and act accordingly.

The Mutual Shaping of the Internet and Social Changes

It is easy to see that some changes in U.S. society over time have created conditions for the Internet to flourish. The United States has witnessed increased levels of real per capita income and educational attainment. One analysis of income change from 1929 to 2003 reported that real income grew by a factor of five while income disparity declined.[140] The U.S. Census Bureau reported that in 1990, 75.2 percent of Americans held high school diplomas and 45.2 percent had attended some college. In 2000, those figures had risen to 80.4 and 51.8 percent, respectively.[141] Higher levels of education are required to make full use of the Internet, while increased income is needed to be able to afford the purchase of a computer and pay the Internet access fees.

We have also seen Americans working longer hours and taking fewer days of vacation. Americans average about ten vacation days plus public holidays each year while they work about two thousand hours every year, more than workers in most other countries.[142] All of this work means that the Internet is useful in quick access to goods and services, fast communication to reach others for business or personal reasons, and locating information from a vast database wherever you happen to be.

So it could be argued that the social change created the demand for the Internet—that the Internet was created to serve the needs of a more affluent, educated, and workaholic population. Changes in social conditions provide a climate for substituting some of the traditional mass media use with Internet use when viewed from that perspective. Or we could claim the technologically deterministic position—that because of the Internet, Americans work more, make more money, and stay in school longer. Neither of these statements is true, of course. A mutual shaping contention is likely more accurate, one in which changes in society and the new media are interdependent. As Boczkowski has observed, transformations in society and the corresponding ones in the development of a technological artifact over time have to be accounted for in a broader view of study of the ways in which change occurs.[143]

Following the creation of the Internet and graphic interfaces in Web browsers, creative and enterprising designers began to develop content and programs that served the needs of busy people who had less time to read traditional newspapers and magazines, and wanted to watch video-based content wherever they happened to be. The narrowcasting of content in media was occurring long before the Internet came along. News media managers were noting the loss of young readers, attempting to create products that would be of more interest to them. And cable television had developed narrowly focused content to cater to particular interests. The Internet just advanced that process to the extreme because of its ability to deliver highly personalized content to smaller and smaller audiences. It even comes down to audiences dictating content choices and platforms to deliver that content. As Russell Davies, global consumer planning director for the Nike campaign put it, "We make content available to consumers via mobiles—if they find our brand and communications compelling enough, they'll seek them out and, ideally, pass them on. We can't target consumers any more, they have to target us."[144]

Notes

1. "Fox Broadcasting Co. History," available at ⟨http://www.answers.com/topic/fox-broadcasting⟩.

2. Brian Steinberg and Suzanne Vranica, "Fox Leads in a Tepid 'Upfront' Push," *Wall Street Journal*, May 14, 2006, B3.

3. Project for Excellence in Journalism, "The State of the Media 2006: Cable TV," available at ⟨http://www.stateofthemedia.org/2006/narrative_cabletv_intro.asp?cat=1&media=6⟩.

4. Ibid.

5. "Can Rupert Murdoch Adapt News Corporation to the Digital Age?" *Economist*, January 21, 2006, available at ⟨http://web.lexis-nexis.com⟩.

6. Ibid.

7. "New Media Timeline (1996–2004)," Poynteronline, January 10, 2005, available at ⟨http://poynter.org/content/content_view.asp?id=75953&sid=26⟩ (accessed May 27, 2006).

8. "Online Newspapers Forge Ahead," Seybold Report on Internet Publishing, March 1998, available at ⟨http://seyboldreports.com/SRIP/subs/0207/IP020701.HTM⟩.

9. Ibid.

10. Jon Katz. "Online or Not, Newspapers Suck," September 1994, available at ⟨http://www.wired.com/wired/archive/2.09/news.suck_pr.html⟩ (accessed June 1, 2006).

11. "New Media Timeline (1996–2004)."

12. Pablo Boczkowski, "The Development and Use of Online Newspapers: What Research Tells Us and What We Might Want to Know," in *Handbook of New Media*, ed. Leah Lievrouw and Sonia Livingstone (London: Sage, 1989), 274–275.

13. John Carey and John Pavlik, "Videotex: The Sword in the Stone," in *Demystifying Media Technology: Readings from the Freedom Forum Center*, ed. John Carey and Everette Dennis (Mountain View, CA: Mayfield, 1993), 163–168.

14. Sylvia M. Chan-Olmsted and Louisa Ha, "Internet Business Models for Broadcasters: How Television Stations Perceive and Integrate the Internet," *Journal of Broadcasting and Electronic Media* 47, no. 4 (2003): 597–617.

15. William W. Fisher III, *Promises to Keep* (Stanford, CA: Stanford University Press, 2004), 102–110.

16. John Dimmick and Eric Rothenbuhler, "The Theory of the Niche: Quantifying Competition among Media Industries," *Journal of Communication* 34, no. 1 (1984): 103–119.

17. Ibid.

18. John Dimmick, Yan Chen, and Zhan Li, "Competition between the Internet and Traditional News Media: The Gratification-Opportunities Niche Dimension," *Journal of Media Economics* 17, no. 1 (2004): 19–33; John Dimmick, Susan Kline, and Laura Stafford, "The Gratification Niches of Personal E-mail and the Telephone," *Communication Research* 27, no. 2 (2000): 227–248.

19. Dimmick, Chen, and Li, "Competition"; John Dimmick and Alan Albarran, "The Role of Gratification Opportunities in Determining Media Preference," *Mass Communication Review* 21, nos. 3–4 (1994): 223–235.

20. Dimmick, Chen, and Li, "Competition," 27.

21. Mary Madden, "Internet Penetration and Impact 2006," Pew Internet and American Life Project, available at ⟨http://www.pewinternet.org/PPF/r/182/report_display.asp⟩ (accessed June 1, 2006).

22. "Poll Shows More U.S. Adults Are Going Online at Home," *Wall Street Journal Online*, May 24, 2006, available at ⟨http://online.wsj.com/public/article/SB114840389678260791-IREjYVgN_rGLeE3_6Djin1jeJZc_20070523.html?mod=rss_free⟩.

23. National Newspaper Association, "Daily Newspaper Readership Trend—Total Adults (1998–2006)," available at ⟨http://www.naa.org/marketscope/pdfs/Daily_National_Top50_1998-2006.pdf⟩ (accessed May 22, 2007).

24. Julie Bosman, "Online Newspaper Ads Gaining Ground on Print," *New York Times*, June 6, 2006, C3.

25. John Horrigan, "Online News: For Many Home Broadband Users, the Internet Is a Primary News Source," Pew Internet and American Life Project, available at ⟨http://www.pewinternet.org/PPF/r/178/report_display.asp⟩ (accessed May 25, 2006).

26. Project for Excellence in Journalism, "State of the Media 2006: Network TV," available at ⟨http://www.stateofthemedia.org/2006/narrative_networktv_intro.asp?media=5⟩.

27. Brooks Barnes, "Local Stations Struggle to Adapt as Web Grabs Viewers, Revenues," *Wall Street Journal*, June 12, 2006, A1.

28. Ibid.

29. "How Old Media Can Survive in a New World," *Wall Street Journal*, May 23, 2005, R1.

30. Ibid.

31. Cynthia Littleton, "TV Reporter: Downtrend in Daytime Reflects Biz Changes," *VNU Entertainment Newswire*, January 23, 2006.

32. "How Old Media Can Survive."

33. "Magazines," Project for Excellence in Journalism, 2006, available at ⟨http://www .stateofthenewsmedia.org/2006/narrative_magazines_intro.asp?media=8⟩ (accessed May 25, 2006).

34. "Out of Service? As Competition from Lifestyle Players Intensifies, Women's Service Titles Struggle," *Media Week*, April 3, 2006, 36.

35. "Girl Trouble: The Once Red-Hot Teen Category Cools as *Elle Girl* Closes and *Teen People* Struggles," *Media Week*, April 17, 2006, 36.

36. Thomas R. Troland, "Seeing Ahead: Underpinnings for What Is Next for Magazine Publishing," *Publishing Research Quarterly* 20, no. 4 (2005): 3–13.

37. Rebecca McPheters, "This Magic Moment. Major Challenge #3: Embrace Change," McPheters and Company, available at ⟨http://www.mcpheters.com/news/challenge3.htm⟩ (accessed August 3, 2006).

38. "Radio," Project for Excellence in the News Media, 2006, available at ⟨http://www .stateofthenewsmedia.org/2006/narrative_daymedia_radio.asp?cat=9&media=2⟩ (accessed August 3, 2006).

39. Ibid., available at ⟨http://www.stateofthenewsmedia.org/2006/narrative_radio_audience.asp⟩ (accessed August 3, 2006).

40. "Arbitron Survey Sees Growth of Internet Radio," *Silicon Valley/San Jose Business Journal*, March 24, 2005, available at ⟨http://www.bizjournals.com/sanjose/stories/2005/03/21/daily39 .html⟩ (accessed June 5, 2006).

41. "Weekly Internet Radio Audience Increases by 50%," Direct Marketing Association, April 18, 2006, available at ⟨http://http://wwwl.the-dma.org/cgi/dispnewsstand?article=4798⟩ (accessed June 15, 2006).

42. Steve Outing, *Newspapers and New Media: The Digital Awakening of the Newspaper Industry* (Pittsburgh: GAFTPress, 2000), 7.

43. Iris Hsiang Chyi and George Sylvie, "Online Newspapers in the U.S.: Perceptions of Markets, Products, Revenue, and Competition," *International Journal on Media Management* 2, no. 2 (2000): 69–77; Steve Outing, "Making the Most of Digital Dollars," *Editor and Publisher*, September 18, 2000, 18–19; John V. Pavlik, *Journalism and New Media* (New York: Columbia University Press, 2001).

44. Everette E. Dennis and James Ash, "Toward a Taxonomy of New Media: Management Views of an Evolving Industry," *International Journal on Media Management* 3, no. 4 (2001): 26–32; Mark Fitzgerald, "Newspapers Go It Alone in Cyberspace," *Editor and Publisher*, February 22, 1997, 7–8.

45. Chan-Olmsted and Ha, "Internet Business Models," 597–617.

46. Outing, *Newspapers and New Media*; Steve Outing, "An Inevitable Mix: Old and New Media," *Editor and Publisher*, January 29, 2001, 115–116; Matthew Karnitschnig, "Time Inc. Makes New Bid to Be Web Player," *Wall Street Journal*, March 29, 2006, B1.

47. Outing, *Newspapers and New Media*; Outing, "Making the Most of Digital Dollars"; Erica Iacono, "NY Times to Combine Its Print, Online Newsrooms," *PR Week*, August 8, 2005, 2.

48. Karnitschnig, "Time Inc."

49. Outing, *Newspapers and New Media*.

50. "About Us," *Raleigh News and Observer*, available at ⟨http://www.newsobserver.com/443/story/200547.html⟩.

51. James Bandler, "New York Times Buys About.com from Primedia for $410 Million," *Wall Street Journal*, February 18, 2005, B3; Julia Angwin, "MySpace to Offer Downloads of TV Show '24,'" *Wall Street Journal*, May 15, 2006, B4; Reuters, "Tribune Company Buys Real Estate Web Site," *New York Times*, available at ⟨http://www.nytimes.com⟩.

52. "NBC, Affiliates Form Venture to Sell Video via the Internet," *Wall Street Journal Online*, available at ⟨http://online.wsj.com⟩ (accessed April 20, 2006).

53. Joe Strupp, "Three-Point Play," *Editor and Publisher*, August 21, 2000, 18–24.

54. Quoted in ibid.

55. Joseph Menn, "Chandlers Demand Breakup of Tribune," *Los Angeles Times*, available at ⟨http://www.latimes.com/business/la-fi-tribune15jun15,1,6071888.story?coll=la-headlines-business⟩ (accessed June 18, 2006).

56. Outing, "An Inevitable Mix."

57. Quoted in "USA Today Combines Online and Print Newsrooms," available at ⟨http://www.cyberjournalist.net/news/003099.php⟩ (accessed December 12, 2005).

58. Anthony Bianco, "The Future of the New York Times," *Business Week*, January 17, 2005, 64.

59. Karnitschnig, "Time Inc."

60. Rachel Smolkin, "Adapt or Die," *American Journalism Review*, available at ⟨http://www.ajr.org/Article.asp?id=4111⟩.

61. Julie Bosman, "Online Newspaper Ads Gaining Ground on Print," *New York Times*, available at ⟨http://www.nytimes.com/2006/06/06/business/media/06adco.html?ex=1307246400&en=b2e717aff6afca05&ei=5088⟩.

62. Quoted in ibid.

63. Quoted in Brooks Barnes, "The New Reality," *Wall Street Journal*, May 15, 2006, R3.

64. Ibid.

65. Smolkin, "Adapt or Die."

66. Kenneth C. Killebrew, *Managing Media Convergence: Pathways to Journalistic Cooperation* (Ames, IA: Blackwell Publishing, 2005).

67. Jube Shiver Jr., "By the Numbers," *American Journalism Review* (June 2006), available at ⟨http://www.ajr.org/article_printable.asp?id=4121⟩ (accessed May 20, 2007).

68. Paul Maidment, "Stopping the Presses," *Forbes*, February 22, 2005, available at ⟨http://www.forbes.com/2005/02/22/cx_pm_0222news print.html⟩; Jon Fine, "Net to Newspapers: Drop Dead," *Business Week*, July 4, 2005, available at ⟨http://www.businessweek.com/magazine/content/05_27/b3941024.htm⟩ (accessed May 20, 2007).

69. Quoted in Paul Farhi, "Under Siege," *American Journalism Review* (February 2006), available at ⟨http://www.ajr.org/Article.asp?id=4043⟩.

70. Farhi, "Under Siege"; Lori Robertson, "Adding a Price Tag," *American Journalism Review* (December 2005), available at ⟨http://www.ajr.org/Article.asp?id=4004⟩ (accessed May 25, 2006).

71. Dennis and Ash, "Toward a Taxonomy of New Media."

72. "All the News That's Fit to...Aggregate, Download, Blog: Are Newspapers Yesterday's News?" *Strategic Management Wharton*, available at ⟨http://knowledge.wharton.upenn.edu/article.cfm?articleid=1425⟩ (accessed March 23, 2006).

73. Gary Pruitt, "Brave News World," *Wall Street Journal*, March 16, 2006, A12, "The Magazine Handbook: A Comprehensive Guide 2006/2007," Magazine Publishers of America, p. 18, available at ⟨http://www.magazine.org/content/Files/MPAHandbook06.pdf⟩ (accessed on May 22, 2007).

74. Fine, "Net to Newspapers."

75. Reuters, "Tribune Company Buys Real Estate Web Site"; Dale Kasler, "Online Job Site Coveted Asset," *Sacramento Bee*, May 31, 2006, available at ⟨http://www.mediainfocenter.org/story.asp?story_id=93901688⟩.

76. Julia Angwin and Joe Hagan, "As Market Shifts, Newspapers Try to Lure New, Young Readers," *Wall Street Journal*, March 22, 2006, A1.

77. Troland, "Seeing Ahead."

78. Anthony Bianco, "The Vanishing Mass Market," *Business Week*, July 12, 2004, 61–68.

79. Troland, "Seeing Ahead," 4.

80. Lieberman, "Papers Take a Leap Forward."

81. Pete Carey, "Knight Ridder Sold to McClatchey," *Mercury News*, March 13, 2006, available at ⟨http://www.mercurynews.com/mld/mercurynews/14084153.htm⟩.

82. Pruitt, "Brave News World."

83. Angwin and Hagan, "As Market Shifts."

84. Rupert Murdoch, speech to the American Society of Newspaper Editors, April 13, 2005, available at ⟨http://www.newscorp.com/news/news_247.html⟩ (accessed June 9, 2006).

85. Jessi Hempel and Paula Lehman, "The MySpace Generation," *Business Week Online*, December 12, 2005, available at ⟨http://www.businessweek.com/magazine/content/05_50/b3963001.htm⟩ (accessed August 3, 2006).

86. Scott G, "The Virtual world of Music Marketing: MySpace.com puts it All Together," *PostNuke*, June 1, 2004, available at ⟨http://www.lamn.com/print.php?sid=22⟩ (accessed May 22, 2007).

87. Spencer Reiss, "His Space," *Wired*, July 11, 2006, available at ⟨http://wired.com/wired/archive/14.07/murdoch.html⟩.

88. Smolkin, "Adapt or Die"; Angwin and Hagan, "As Market Shifts."

89. A. S. Berman, "E-Editions: The Next Step," *Presstime* (July 2006): 42–45.

90. Farhi, "Under Siege."

91. Robertson, "Adding a Price Tag."

92. Ibid.

93. "Defying Industry Trends; Wall Street Journal Circulation Increases 4.5 Percent," *Web Wire*, April 30, 2007, available at ⟨http://www.webwire.com/ViewPressRel_print.asp?ald=34388⟩ (accessed May 22, 2007).

94. John Motavalli, *Bamboozled at the Revolution: How Big Media Lost Billions in the Battle for the Internet* (New York: Viking, 2002), 7.

95. Barrie Gunter, "Business Implications of Internet News," *News and the Net* (Mahwah, NJ: Lawrence Erlbaum, 2003), 35–54; Susan Mings and Peter B. White, "Profiting from Online News: The Search for Viable Business Models," in *Internet Publishing and Beyond: The Economics of Digital Information and Intellectual Property*, ed. Brian Kahin and Hal R. Varian (Cambridge, MA: MIT Press, 2000).

96. Robert G. Picard, "Changing Business Models of Online Content Services: Their Implications for Multimedia and Other Content Producers," *International Journal on Media Management* 2, no. 2 (2000): 60–68.

97. Mings and White, "Profiting from Online News."

98. Hsiang Iris Chyi, "Willingness to Pay for Online News: An Empirical Study on the Viability of the Subscription Model," *Journal of Media Economics* 18, no. 2 (2005): 131–142.

99. Miguel Helft and Steve Lohr, "176 Newspapers to Form a Partnership With Yahoo," *New York Times*, November 20, 2006, C1.

100. Robertson, "Adding a Price Tag."

101. John Caldwell, "The Business of New Media," In *The New Media Book*, ed. Dan Harries (London: British Film Institute, 2002): 55–68.

102. Barnes, "Local Stations Struggle to Adapt."

103. Ibid.

104. Ibid.

105. Lieberman, "Papers Take a Leap Forward."

106. Smolkin, "Adapt or Die."

107. Caldwell, "The Business of New Media," 63.

108. Julia Angwin, "Media Firms Dig into War Chests for Latest Assault on the Internet," *Wall Street Journal*, September 28, 2005, A1.

109. Reiss, "His Space."

110. *Eliminate* 2005; Barnes, "Local Stations Struggle to Adapt."

111. Brooks Barnes, "Fox to Share Revenue from Reruns on Web with Affiliated Stations," *Wall Street Journal*, April 14, 2006, A13.

112. Joe Mandese, "See Spot Run—or Not," *Broadcasting and Cable* 1, November 1, 2004, available at ⟨http://www.broadcastingcable.com/article/CA483256.html⟩.

113. Brooks Barnes and Kevin J. Delaney, "'Clickable' Web Video Ads Catch ON," *Wall Street Journal*, May 18, 2006, B4.

114. Tom Lowry, "Can MTV Stay Cool?" *Business Week*, February 20, 2006, available at ⟨http://www.businessweek.com/magazine/content/06_08/b3972001.htm⟩ (accessed June 7, 2006). Gareth Jones, "TV Firms Decide Interactivity Is the Future," *New Media Age*, May 18, 2006, 14.

115. Sean Callahan, "Media in Motion; Business Media Companies see Video Content as a way to Attract Viewers and Advertisers," *Media Business*, March 1, 2006, p. 16.

116. Jesse Hempel with Paula Lehman, "The MySpace Generation; They Live online. They Buy Online. They Play Online. Their Power is Growing," *Business Week*, December 12, 2005 p. 86.

117. Picard, "Changing Business Models of Online Content Services," 60.

118. "Reporting Katrina," 2005, available at ⟨http://www.journalism.org/resources/research/reports/Katrina/default.asp⟩ (accessed June 8, 2006).

119. Stuart Biggs, "Citizen Journalists Reshape the Media," *South China Morning Post*, March 7, 2006, available at ⟨http://www.asiamedia.ucla.edu/article.asp?parentid=40358⟩ (accessed May 27, 2006).

120. Ibid.

121. Dan Gillmore, "My NYC Talk on Journalism Principles and the Future," Dan Gillmore's blog, available at ⟨http://sf.backfence.com/bayarea/showPost.cfm?myComm=BA&bid=2271⟩ (accessed June 9, 2006).

122. Shayne Bowman and Chris Willis, "The Future Is Here, but Do News Media Companies See It?" *Nieman Reports* (Winter 2005): 6–10.

123. Lydia Saad, "Military Again Tops 'Confidence in Institutions' List," June 1, 2005, available at ⟨http://poll.gallup.com/content/default.aspx?ci=16555&pg=1⟩ (accessed August 4, 2006).

124. J. D. Lasica, "Blogs and Journalism Need Each Other," *Nieman Reports* (Fall 2003): 73.

125. Saul Hansel, "Yahoo Says It Is Backing away from TV-Style Web Shows," *New York Times*, March 2, 2006, 1.

126. Ibid.

127. Alicia C. Shepard, "Postings of Obscene Photos End Free-Form Editorial Experiment," *New York Times*, June 21, 2005, p. 8.

128. "BBC News/Have Your Say," available at ⟨http://www.bbc.co.uk/blogs/worldhaveyoursay/⟩ (accessed August 4, 2006).

129. Fisher, *Promises to Keep*, 28.

130. John Fiske, *Television Culture* (London: Routledge, 1987).

131. Bowman and Willis, "The Future Is Here," 10.

132. Nielsen/NetRatings, "Social Networking Sites grow 47% Year over Year, Reaching 45% of Web Users," May 2006, available at ⟨http://nielsen-netratings.com/pr/pr_060511.pdf⟩ (accessed August 4, 2006).

133. Ibid.

134. Quoted in ibid.

135. Robert Young, "Social Networks Are the New Media," May 29, 2006, available at ⟨http://gigaom.com/2006/05/29/social-networks-are-the-new-media/⟩ (accessed June 12, 2006).

136. Danah Boyd, "Identity Production in a Networked Culture: Why Youth Heart MySpace," speech to the American Association for the Advancement of Science, February 19, 2006, available at ⟨http://www.danah.org/papers/AAAS2006.html⟩ (accessed June 12, 2006).

137. Josef Kolbitsch and Hermann Maurer, "The Transformation of the Web: How Emerging Communities Shape the Information We Consume," *Journal of Universal Computer Science* 12, no. 2 (2006): 207.

138. Murdoch, speech.

139. Vin Crosbie, "What is New Media?" *Rebuilding Media*, April 27, 2006, available at ⟨http://rebuldingmedia.corante.com/archives/2006/04/27/what_is_new_media.php⟩ (accessed June 1, 2006).

140. Paul Gomme and Peter Rupert, "Per Capita Income Growth and Disparity in the United States, 1929–2003," Federal Reserve Bank of Cleveland, April 15, 2004, available at ⟨http://www.clevelandfed.org/Research?Com2004/0815.pdf⟩ (accessed June 7, 2006).

141. "Educational Attainment: 2000," U.S. Census Bureau, August 2003, available at ⟨http://72.14.203.104/search?q=cache:OY_ctMKZpYkJ:www.census.gov/prod/2003pubs/c2kbr-24 .pdf+census+bureau+educational+levels&hl=en&ct=clnk&cd=2⟩ (accessed June 7, 2006).

142. Mortimer B. Zuckerman, "All Work and No Play," *U.S. News and World Report*, September 8, 2003, available at ⟨http://www.usnews.com/usnews/opinion/articles/030908/8edit.htm⟩.

143. Pablo Boczkowski, "The Mutual Shaping of Technology and Society in Videotex Newspapers: Beyond the Diffusion and Social Shaping Perspectives," *Information Society* 20, no. 4 (2004): 257.

144. Quoted in "The Future of Media," *Campaign*, March 17, 2006, 22.

10 Internet Challenges for Nonmedia Industries, Firms, and Workers: Travel Agencies, Realtors, Mortgage Brokers, Personal Computer Manufacturers, and Information Technology Services Professionals

Jeffrey R. Yost

The Internet and the World Wide Web have become ubiquitous throughout the industrialized and increasingly the developing world, inspiring substantial and ever-growing interest in the impact of these technologies on economics, management, politics, and culture. Though the Internet and American business is a topic that has attracted widespread attention, often scholars, journalists, and others have written about it unevenly. Certain topics—the rise of e-commerce, "the dot-com collapse," intellectual property, and corporate "e-strategies"—have been examined frequently. In business journalism, the transformation of media firms and industries has been a common subject, while the more specialized trade press has focused on functional applications of the Internet (and earlier Electronic Data Interchange systems) to supply chains, customer relationship management (CRM), marketing, and other topics.

Over the past dozen years, no electronic commerce firm has received more attention than Amazon.com.[1] Understandably, some analysis of Amazon has extended beyond the insular concentration of the firm to explore its meteoric rise (in sales volume) contextually within the competitive environment of its original and still primary industry: bookselling. While e-commerce firms, such as Amazon, or other culturally transformative electronic media businesses, such as music downloading enterprises (Napster and iTunes), are intriguing and important to examine, the degree to which nearly all businesses have been impacted to a greater or lesser extent by the Internet and the World Wide Web has often been overlooked. In telling and retelling the stories of the Internet and American business, the highfliers (Amazon, Yahoo!, Google, and eBay), and to a lesser extent the major failures (Webvan, Pets.com, eToys, and Go.com), have been in the spotlight, reducing to the shadows the less exciting firms and industry segments that were harmed by the Internet but often continue to exist, albeit in a different form.

There are many commonalities to how industries have been impacted by the Internet, but fundamental differences are often present as well. Most trades have sufficient nuance to their history to warrant considerable, independent attention. In an era in which the Internet and the World Wide Web are lauded every day as wondrous new tools for work and play, perhaps no topic is more important to balance perspectives than industry segments and firms adversely impacted or altered by the Web. This is a subject far broader than but hitherto almost exclusively explored with regard to the media sector—small local bookstores destroyed by Amazon, the recording industry's intellectual property protection challenges in a file-sharing world, cable television stations' plight to continue to attract viewers as more and more eyes are glued to computer rather than television screens, and competitive pressures forcing many newspapers to modify revenue-generation strategies.

This chapter will examine several nonmedia industries and industry segments, highlighting the challenges and adverse impacts for some participants. It will stress how the Internet and the Web changed the playing field in ways that simultaneously created major challenges for some firms and industries, but concomitantly presented opportunities for new or appropriately positioned existing enterprises to tailor their capabilities to gain or extend competitive advantage. For some trades, the Internet had little overall impact on industry dynamics—it was used in similar, straightforward ways by many firms.[2] For other industries, the new environment was transformational. It led to significant market share shifts, where one or a small group of companies benefited, and others were hurt. A prime example of a beneficiary of the Internet is Dell Computer. This company utilized its strong capabilities in direct marketing and sales, and used information technology and the Web to provide quicker, better, and more customized service than its personal computer competitors or retail computer distributors. While possessing a number of unique attributes, Dell is part of a broader story of how new computer-networked tools have facilitated opportunities to reduce or eliminate the level of value creation by traditional "middlemen," or intermediaries. This has been true in many industries in the United States and throughout the world. Simultaneously, the Internet and the Web have reduced transaction costs, facilitating greater modularity in complex value chains.[3] This has allowed firms to focus on core competencies and outsource (increasingly offshore) many or all noncore functions. At times, alongside this growing disintermediation, there has also been substantial "reintermediation," where different types of intermediaries take part in value chains in new ways.[4]

The following analysis will explore these and other issues by taking a detailed look at fundamental changes in the travel reservations industry. Early in the new millennium this became the leading e-commerce industry in the United States. It has continued to hold this distinction, and over the past few years the ratio of U.S. travel revenue booked online has grown from roughly a quarter to a third. Total online travel revenue in the United States in 2005 was estimated at $68 billion, a number expected to grow to $104 billion by 2010.[5] The discussion of travel agencies is followed by shorter examinations of other nonmedia industries and topics—realtors and mortgage brokers, personal computer firms and divisions, and offshoring information technology—to provide perspective on similarities and differences in how the Internet and the Web have altered or adversely impacted nonmedia trades, segments, and companies.

Travel Agencies

Among nonmedia firms, no industry has been more adversely affected by the Internet and the Web than traditional travel agencies. These typically small-scale enterprises long depended on location, walk-in customers, print advertising, referrals, and institutional clients (account control) to create and extend their base. The personal touch these companies had with their clients helped insulate them from competitors, especially nonlocal ones. National and international travel agencies existed, including giants such as American Express, but many customers preferred the local, personalized service received from smaller travel agencies nearby. These agencies navigated the relatively complex travel world, building relationships with and using particular information technologies—including computer reservation systems (CRSs)—to interact with different direct-service providers (airlines, hotels/motels, car rental agencies, etc.) and serve as value-adding intermediaries.

The relationship between travel and computer networking is long and profound. The coordination of air travel reservations, along with bank and retail data processing, was the impetus behind and initial beneficiary of private sector real-time computer networking applications. Computer networking technology had previously been the exclusive domain of universities and corporate contractors for the Department of Defense. IBM and System Development Corporation (SDC), respectively, were the lead hardware and software contractors for the Air Force's Semi-Automatic Ground Environment (SAGE) radar and computer-based air defense system during the second half of the 1950s. SDC, MIT, and MIT's spin-off, the MITRE Corporation, provided system integration for SAGE. In 1953, however, a couple of years prior to IBM's receipt of the primary computer contract for SAGE, Big Blue had already begun to work with American Airlines on computer networking solutions to the passenger reservation problem— the necessity of maintaining seat inventories in real time to avoid costly inefficiencies (unnecessary empty seats and difficulties coordinating discriminatory pricing).

The partnership between these firms began after a chance meeting between American Airline's president Cyrus R. Smith and a senior IBM sales representative, R. Blair Smith, on a flight from Los Angeles to New York. The resulting networked system, the Semi-Automatic Business Research Environment (SABRE), was developed by IBM over the following decade and became fully operational in 1965—several years after the completion of SAGE and the same year that the ARPANET project was being initiated by the Department of Defense's Advanced Research Projects Agency's (ARPA)

Information Processing Techniques Office. SABRE, which replaced the former hand-written passenger reservation systems, was by far the largest, most advanced private real-time networked computer system of its time.

American Airline's SABRE was a quick success. By the end of 1965 it could handle eighty-five thousand phone calls, forty thousand passenger reservations, and twenty thousand ticket sales per day.[6] In 1972 the SABRE system was moved to a consolidated center in Tulsa, Oklahoma, that was built to house American Airlines' data processing operations. SABRE, and later other such systems, increasingly became tools of the trade following the first installation of SABRE with a travel agency in 1976. By year's end 130 agencies were connected to SABRE, and the numbers continued to grow rapidly in the succeeding years. In this market, United Airlines' CRS, Apollo, was not far behind. Though Apollo was not fully operational until 1971, like SABRE, it was first marketed to and used by travel agencies in 1976. More generally, CRS enterprises of the second half of the 1970s—SABRE, Apollo, Trans World Airline's PARS, and others—gave birth to a thriving CRS industry. These systems expanded into additional travel reservation areas, and specialty systems and firms emerged in the succeeding years and decades such as Dallas-based Pegasus Solutions, Inc., an emerging global giant in the hotel reservations area.

A wave of consolidation occurred in the travel reservations industry in the 1990s, with no higher-profile acquisition than the purchase by American Express of Lifeco, a corporation with airline ticket sales revenue exceeding $1 billion annually. Many smaller and midsize firms sought suitors. Nevertheless, consolidation did not change the overall nature of the industry, which continued to be composed of many small enterprises. In a 1991 study, the leading trade association in the industry, the American Society of Travel Agents (ASTA), estimated that there were thirty-eight thousand travel agencies in the United States and two hundred thousand agents. Owing perhaps to the work flexibility that some agents often enjoyed as well as institutional barriers to other occupations, many women became travel agency owners, managers, and agents. The ASTA study estimated that 60 percent of the full-time agencies were owned by women, while 84 percent of the managers and 81 percent of the agents were women.[7] Another characteristic of the industry was the formation of consortia of smaller agencies. This allowed firms to remain independent, but enjoy some of the economic benefits of consolidation in terms of combined purchasing power, efficiency, and knowledge transfer. During the 1990s, more than half of all U.S. travel agencies were affiliated with a consortium.[8]

While travel agents made portions of their revenue from booking hotels, cruises, car rentals, rail tickets, and other reservations for clients, most derived the majority of their revenue and earnings selling plane tickets and collecting commissions from the airlines. Commissions were typically on a percentage basis, and varied over time, but were at or near the 10 percent level for many years after U.S. airline deregulation in 1978.

In 1985, American Airlines' easySABRE allowed individual users with personal computers and modem connections to tap into SABRE for the first time. The system, however, was not used by many travel customers. The use of such systems continued to reside primarily with travel professionals in the 1980s and the early 1990s. With the advent of the World Wide Web in the early 1990s, and its rapid growth following the widespread dissemination of Netscape Navigator and other browsers beginning in the mid-1990s, consumers' usage of easySABRE and similar systems began to escalate, and arrangements between travel agents and airlines began to change.[9]

In the second half of the 1990s, the major airlines in the United States imposed a $50 cap on the total commission paid to travel agents for selling round-trip domestic tickets to their clients. At this time, airlines sought to invest increasingly in direct sales capabilities, both on the Internet and through other means. They created Web sites for ticket information and sales, boosted the size of call centers, and initiated special "hotline" numbers to provide expedited service for loyal customers. The growing interest in and expansion of frequent-flier plans to allow miles to accumulate more rapidly (through partnerships with travel and nontravel businesses), along with other marketing and promotions, further fueled traffic to airline Web sites.[10] At this time, some customers made purchases online using credit cards, while many others used airline Web sites only for information. After finding a desired itinerary, they telephoned the airline to book their flights.

With the airlines increasing their commitment to a direct sales model, in 1998 major U.S. carriers reduced the commission rate paid to travel agents to 8 percent and imposed a cap of $100 for round-trip international tickets.[11] The former harmed all travel agents, and the latter especially hurt agents focused on international travel involving more expensive itineraries, such as flights between the United States and Asia. Many travel agents began to charge $10 to $15 to clients to offset part of their loss in commission revenue.[12] Yet this tended to erode their base, as an increasing number of people were moving away from traditional travel agents, and toward direct purchasing online or by telephone.

The scale and nature of the terrorists' attacks on the United States on September 11, 2001—using hijacked planes as missiles to strike symbols of the nation's military and economic strength—had an immediate, detrimental impact on the entire travel and hospitality industry, but especially on air travel. This coupled with a recession, followed by an anemic recovery, led to major declines in higher-margin business fares. In light of both the trend toward direct sales and the struggling overall travel industry, in March 2002 Delta Airlines eliminated commissions to travel agents.[13] In an industry that tends to change fares, services, and other policies based on signaling, a number of the other major U.S. airlines quickly took the opportunity to partially or fully follow Delta's lead.[14] Overall, in 2002 only an estimated 26 percent of travel agencies received commissions on bookings, down from an estimated 37 percent the previous year.[15] This downward trend continued in the succeeding years.

In addition to geopolitical factors and the poor economic environment, the evolving functionality, convenience, and efficiencies gained from the Internet and various information technology applications also accelerated the trend toward the customer use of airline Web sites instead of traditional travel agents. Electronic tickets, or e-tickets, were introduced broadly by airlines in the mid to late 1990s and saved the airlines roughly $7 per ticket.[16] Though some passengers were resistant to this change, and many problems were reported at first, within a half decade most air travelers switched to and many came to prefer e-tickets. In 2003, more than 90 percent of commercial air carrier travelers in the United States utilized e-tickets.[17] Such a shift not only depended on the push by the airlines for this cost-saving option but also the convenience offered to customers—a high percentage of people who frequently travel on planes are regular users of the Internet and could check itineraries remotely. Soon customers could also use airline Web sites to change seat assignments, and even check in and print out boarding passes the day before the flight. Overall, the basic services offered, coupled with the different conveniences provided on Web sites, took away central elements of the value proposition supplied by traditional travel agents. These included ready access to scheduling and fares, past knowledge of customer preferences (now remembered by Internet cookies), and writing up and delivering tickets.

In addition to the airlines push forward with an Internet sales channel in the mid to late 1990s, many other types of travel suppliers did as well. In 1998, Forrester Research reported that only 37 percent of travel companies could facilitate Internet sales in 1997; the following year this was above 75 percent.[18] Furthermore, the number of individuals who actually purchased travel after visiting a site, the "look-to-book" ratio, was increasing steadily. Airlines also began to offer Internet-only fares and mileage bonuses in the late 1990s to encourage visitors and promote online sales, resulting in condemnation from travel agents and their trade associations.[19]

In short, during the first years of the twenty-first century, the primary revenue base for travel agencies—commissions—took a major hit. Further fueling the problem, this was at a time when overall travel, and especially air travel, plummeted. Many travel agencies boosted their commissions, usually to $20 to $35 per round-trip ticket, to address the situation. This tended to result in additional losses to their client base, however, as many customers were unwilling to pay that amount when they felt they could just as easily book their own tickets online or by phone. These factors led to a precipitous decline in the number of travel agencies in the United States. In 1997 there were 29,322 travel reservations businesses in the United States, generating a total of $9.98 billion in revenue according to the U.S. Census Bureau.[20] A half decade later, in 2002,

the number of enterprises had dropped to 21,679, while the total receipts, despite inflation and the growth of the travel sector, dropped to $9.35 billion.[21] The loss of jobs
was equally striking. There were 183,178 employees in the travel agent industry in
1997 and only 147,069 in 2002.[22] The concentration of volume among fewer firms
was also a trend, as the four largest firms held 20.7 percent in 2002 of a once extremely
decentralized industry.[23] A panel of experts on the travel industry has estimated that
there will only be 13,282 travel agency businesses in the United States by 2007—
down more than two-thirds from the early 1990s.[24]

This substantial decline in travel agencies in the United States, industry revenue, and
jobs is in many respects a typical example of disintermediation, or the dissolution of
intermediaries with the growth of the Web as an efficient, direct channel for marketing
goods and services. The ubiquity of the Web has also fundamentally altered the way
that the remaining travel agencies operate as well as created opportunities for new
intermediaries in the industry.

In 2002, the travel industry in the United States surpassed the computer industry as
the leader in online revenue generation. In 2003 sixty-four million Americans, or 30
percent of the adult population, used the Internet for travel information and two-
thirds, or forty-two million Americans, purchased travel online—an 8 percent gain
over 2002.[25] Increasingly, the Web has not only been a source of value creation but
also value capture. This not only extended from people purchasing airline tickets from
the Web sites of America's major air carriers but also as a result of reintermediation,
new firms or divisions of existing corporations entering as Web-based intermediaries,
or "cyberintermediaries."

At the end of the 1990s, one cyberintermediary, Priceline.com, was in the spotlight
as a result of the excitement over its reverse-auction model for airline tickets, hotel
rooms, and rental cars; its skyrocketing stock; and its effective marketing. Priceline
declined rapidly in the dot.com collapse of 2000–2002, but has since recovered and
added traditional reservation sales to complement its "name your price" or reverse-
auction model. Other firms and corporate divisions, however, have led the way as
new intermediaries, including online travel booking businesses such as Expedia, Travelocity, and Orbitz.

At the heart of travel agencies and cyberintermediaries effectively being able to compete with airline Web sites or call centers is their ability to create value that exceeds
their fees. With cyberintermediaries, the fees are quite modest. Expedia, which was
launched by Microsoft in 1995, acquired by USA Network/InterActiveCorp in 2001,
and spun off as an independent in 2005, instituted a $5 fee in 2002.[26] This fee rate
has remained at this level to the present. In its early years, Expedia benefited substantially from Microsoft's MSN pointing to it at every opportunity. The firm sought
to add value for clients by saving them time in comparative shopping between the
rates of all the major airlines (the discount airlines are not included in the reservation

systems or search mechanisms of the leading cyberintermediaries).[27] Furthermore, Expedia invested in software that allowed it to "remember" customers, and then suggest travel plans when consumers returned to the site or by e-mail marketing messages. Expedia has also instituted frequent-customer hotlines to allow clients to be able to speak to a "real person" and avoid long hold times. Like many Internet-based businesses, it benefits from the input of customer reviews and ratings to aid other consumers in making their travel decisions (with hotels/motels)—a useful and extremely low-cost tool. Expedia and the other major cyberintermediaries have suffered reduced commissions from airlines in the first years of the new century, but unlike the case with traditional travel agencies, special deals are sometimes struck, so commissions are not entirely eliminated.[28] Equally important, the surcharges can be small due to the high volume of the major cyberintermediaries—something that is not possible with the small local travel agency.

The other two major cyberintermediaries in the United States also possess deep pockets and operate in large volume. Travelocity, owned by Sabre Holdings, merged in 2000 with the third-largest online travel company, Preview Travel. Preview Travel was launched in 1985, and became the primary travel service of America Online (AOL) and the co-branded travel site of browser corporation Excite, Inc. in the early Web era.[29] In 2005 Travelocity bought Lastminute.com, a leading British online travel reservation site, for £577 million.

Orbitz, a Chicago firm, was formed in early 2000 by a group of airlines as a way to sell to customers directly. It was sold for $1.25 billion in 2004 to Cendant, a travel and property company based in New York. Cendant also owns Galileo, a competing CRM and a descendant of the original United Airlines Apollo as well as rental car businesses (Avis and Budget) and hotels (Ramada Inn and Days Inn). This mix of businesses allows for the in-network bundling of travel packages. Along with the cyberintermediary giants, there are some large specialty firms catering to certain sizable market segments, such as Biztravel.com, geared specifically to business travelers. This company provides compensation for flight delays, sends last-minute flight updates to pagers, and offers various tools for businesses to remain within their travel budgets.[30] Increasingly, some of the largest cyberintermediaries are providing similar services.

While cyberintermediaries have grown steadily in recent years, many travel customers remain loyal to one or a few airlines where they are enrolled in frequent-flier programs. Though miles can be collected using cyberintermediaries, some find it easier or preferable to do all their travel booking through airline sites. Others go to cyberintermediaries to check schedules and for comparative shopping, but avoid the fee by going to the airline site to purchase the tickets.[31] Both cyberintermediaries and airline sites have promoted travel packages, where customers can book their hotel, rental car, and plane tickets together, sometimes bundled at a meaningful discount. By offering packages, these services have further cut into the perceived and real value afforded by traditional travel agents.[32]

These changes hurt the smaller travel agencies that catered especially to individual travelers and smaller businesses, and the larger travel agencies that served corporations. Though Expedia, Travelocity, Orbitz, and others do not have large sales forces, their marketing has been effective, and corporations are increasingly seeing opportunities to save by using these firms. Corporate travel department managers are increasingly listening to their colleagues throughout their companies who are finding cheaper travel on their own using cyberintermediary sites. This resulted in businesses such as Texas Instruments and Charles Schwab facilitating more of their reservations through cyberintermediaries by the mid to late 1990s.[33] It also has led to formal deals between corporations and cyberintermediaries, such as McDonald's and Knight Ridder signing on with Orbitz in 2003. At the same time, some downsized U.S. corporations of the twenty-first century, with fewer secretaries and administrative assistants, do not want managers spending their time booking travel and have kept travel centralized.[34]

The pie for Internet travel bookings is continuing to grow rapidly, but so is the competition in the field. The barriers to entry are low, yet there is a high failure rate. Not only are customers often changing who or what they use to book travel but also their purchasing behavior. Both business and other travelers are booking closer to travel dates than ever before—partially disrupting the long-established discriminatory pricing model of airlines based on the notion of business travelers' propensity for later booking and lower price sensitivity compared to other travelers. While major cyberintermediaries have extended their customer base in recent years, many of the smallest Internet travel booking businesses no longer exist. Some have gone bankrupt, and others have been acquired. Meanwhile, leading cyberintermediaries have not only faced ever greater competition from the sites of direct providers—airlines, hotels and motels, and car rental firms—but also from metasearch travel firms, including Farechase, Kayak, Mobissimo, QIXO, Sidestep, and others.[35]

The generic metasearch business strategy is to point customers to direct-service providers or cyberintermediaries, and collect a fee on the completed transactions. A recent study by Nielsen/NetRatings reported that 54 percent of travelers start with a cyberintermediary, 37 percent begin with direct suppliers (airlines and hotels), and 9 percent initially visit the sites of metasearch firms.[36] While cyberintermediaries are drawing the most initial attention, direct sellers are the best at turning Web site visits into bookings, whether the ultimate purchase is made on the Internet or by phone. With the latter, privacy and a lack of trust in the online channel are major factors, but research has also indicated that for some customers it is their lack of trust in their own Internet skills (the fear of booking incorrectly, double booking, etc.).[37] Also, customers have different expectations, and evaluate value and quality along different lines with regard to cyberintermediaries and direct suppliers.[38] For the cyberintermediaries, information content is most critical, while the ease of use tends to be the main factor with direct supplier sites.[39]

The impact of the metasearch firms, while considerably smaller than standard cyber-intermediaries or direct sellers to date, should not be ignored. Recent trends indicate escalating competitive pressure from metasearch companies. Currently, online bookings are about evenly split between the cyberintermediary distributors and the travel suppliers, with a growing amount of fees ending up in the hands of the metasearchers.

A minority of substantial commercial air carriers choose not to list their flight inventory and prices with major CRS enterprises, which increasingly are now being referred to as global delivery systems (GDSs). GDSs sell reservation data services to large cyberintermediaries and subscribing agencies. This involves business-to-business-to-consumer applications to facilitate smooth interoperability, and often, to combine content-based filtering technologies, interactive query management, case-based reasoning, and rank suggestions from structured catalogs or databases. Airlines that choose not to list inventories with GDS providers tend to be the smaller to midsize enterprises (JetBlue) and discount airlines (Southwest) looking to capitalize on direct sales.[40] In 2006 Air Canada, one of the larger traditional airlines, with revenue of $9.83 billion (Canadian dollars) in 2005, bucked the trend and discontinued providing inventory to GDSs. This prompted ASTA to condemn the action, and issue a statement on how Air Canada's decision hurt consumers, travel agencies, and ultimately the airline itself.[41]

Also worrisome to the leading cyberintermediaries, travel providers, and traditional travel agents is the fact that a number of metasearch businesses have been acquired recently by some leading Internet and information technology firms interested in quickly entering into or expanding their online travel booking business. For instance, in July 2004 Yahoo! acquired Farechase, and AOL invested substantially in Kayak. These established, well-branded, and high-traffic Internet sites pose a threat to the big three cyberintermediaries and other industry participants. Yahoo!, in particular, seems committed to the travel area as never before and is trying to taking advantage of its massive traffic to promote travel sales. Clearly, consolidation will continue and travel will likely trend in the same direction as Internet search—toward a few major players, and perhaps in time, one major dominant brand (like Google in online search). In the meantime, competition will be increasingly fierce, margins will likely decline for cyberintermediaries, and this will create an even more challenging playing field for the traditional local-based travel agencies.

In the end, it comes down to whether traditional travel agents can offer value that justifies their higher fees. This, of course, will increasingly involve using the Internet to interface with customers, but also will involve substantial off-line communication depending on the community and the clientele. While a significant number of traditional travel agencies have gone under, and overall the industry has been hurt by the massive trend toward online bookings, some traditional travel agencies have persevered. This has occurred because their business model has been less vulnerable to online competitors all along, or they have transformed their business to more appropriately meet today's different and highly competitive travel reservations environment. The latter has typically involved greater specialization, and concentrating on more complicated travel arrangements as well as types of travel destinations and experiences. This may include a multidestination international trip, or traveling to parts of the world where the agent has some expertise in the locality, be it Albania or Zambia, but many customers do not. This is also true of theme trips based on ecotourism and other concepts, where recreation and education are combined to create unique travel products that appeal increasingly to a discriminating clientele that expects expertise from the agent. This type of booking business is lower volume and limited to only certain market segments, but it tends to be for more expensive trips and yields significantly higher fees than traditional travel agents collected in the past.

One such travel agency that successfully transformed after the reduction and often elimination of airline commissions is Sunflower, a firm run by Barbara Hansen in Wichita, Kansas. Hansen, who has traveled extensively throughout the world, established a business relationship with a wholesaler that operates in Australia. She has used the Web to advantage, adding a special packaged tour highlighted on her South Asia Web site that complements her main Sunflower Web site. Hansen has utilized her connections to create tours in Australia based on the themes of vineyards (tours and wine tasting) and classical music performances to entice and add value to clients' excursions to the "Land Down Under."[42] The Sunflower case also underscores the fact that smaller agencies have to decide how best to create and define products, and market them on the Web.[43]

While such specialization has provided a haven for some small travel agencies, as customers have overall shifted to the online giants, it has not gone unnoticed by the large firms. Cyberintermediaries are beginning to test the waters of offering more personalized attention and service for certain types of travel, including employees who are available to assist on group travel excursions. In February 2006, for the first time, some travelers booking cruises through Travelocity were assisted throughout their trip by a Travelocity guide. Travelocity representatives sailed on over forty cruises on a number of different cruise lines in 2006.[44] Such service is rarely possible for small travel agencies—unless they want to completely change their business model to become travel guides—and further challenges brick-and-mortar travel agencies.

One thing is certain in the travel reservations industry: both individual and corporate consumers are using the Internet for travel as never before. PhoCusWright estimates that the online portion of the leisure and employee booked (reservations that are not made by a central department) travel will be $78 billion in 2006, while corporate travel office bookings will exceed $36 billion.[45] The traditional travel agent will come under increasing pressure to be more resourceful and innovative in creating and capturing value. They will need to be highly knowledgeable travel consultants as opposed to basic facilitators of reservations. This is something that many are already doing. Yet they might find it increasingly difficult to differentiate the value they add as cyberintermediaries move into higher-margin niche areas and use sophisticated market segmentation strategies to complement their broad core businesses. The cyberintermediaries also are facing a challenge as year-to-year traffic was down 14 percent in February 2006 as direct suppliers, especially airlines, succeeded in increasing visits to their Web sites and online sales. Industry-leading Expedia, however, defied the trend with traffic to its combined sites—Expedia.com, TripAdvisor, and Hotels.com—up 11 percent during 2006.[46] Needless to say, the trend toward further consolidation among the cyberintermediaries, metasearch firms, and traditional travel agencies will continue. Furthermore, some in the latter group will face particularly daunting challenges and need to be ever more strategic in their approach. Despite these challenges, all three groups will continue to use the Web to varying degrees and in varying ways, and will coexist for years to come in the travel reservations business.

Realtors and Mortgage Brokers

The challenges posed to traditional travel agents by the Internet were accentuated and increasingly transparent as the airline and other travel industries faced difficult times in the aftermath of the terrorist attacks on September 11, 2001. In other industries, including the real estate and real estate mortgage businesses, prosperity in the overall market has masked some of the existing and potential deleterious structural impacts of the Internet to the relative performance, opportunities, and long-term outlook for certain industry segments and firms.

Real estate and mortgage broker businesses are different enterprises, but there is a deep symbiosis between them.[47] Mortgage brokers provide the means for consumers to shop for real estate, while realtors market the real estate products that underlie the demand for mortgage products. The two businesses are highly dependent on larger economic and social factors, especially interest rates, but also national and local unemployment, income, and crime levels. Both businesses, albeit in different ways, have faced changes and challenges with the Internet in the Web era.

The proliferation of the Web has resulted in the broad dissemination of real estate information to potential consumers. Previously, such information was controlled solely by real estate agents and carefully protected. In 1995, *Kiplinger's Personal Finance* reported that some realtors "put a sample of their Multiple Listing Service [MLS] ads on the Net, but it was just to tease: brokers don't want to give away their lifeblood."[48] An

MLS is an organization that requires licensing and trade association membership. They have existed for decades as a tool to share information among realtors.

By the decade's end this changed. Most MLS listings are now freely available on the Web. The main MLS in the United States is controlled by the National Association of Realtors. Prior to the advent of the Web, it was easy for the association to maintain a monopoly on listings, which helped to sustain high commission rates for its members. Debates continue among real estate brokers and scholars of real estate management on the optimal amount of information disclosure, and the relative benefits and costs of providing or withholding MLS listings.[49] Though MLS listings are now broadly on the Web, certain information, particularly the time on the market and the listing agent, is generally withheld. Research has indicated that homes listed on the Internet are on the market slightly longer and sell for slightly more.[50]

Some real estate agents have suffered with this transformation, but others have adjusted well. Agents who are part of larger real estate firms often have access to more technological support for creating virtual tours of homes with photographs and video. Meanwhile, independents, and those with smaller companies, have seen both success and failure in implementing Internet strategies. The propensity and extent of usage of the Internet by real estate brokers not only varies with the firm size but also with many demographic factors. More education, the number of firms the agent has worked with, marriage, franchise affiliation, ownership interest, and hours worked are all positively correlated with the level of Internet marketing. Conversely, Internet marketing usage rates are lower among women, nonwhites, and older agents.[51] Overall, in 2000, 23 percent of all real estate agents in the United States were using the Internet as a tool to market individual properties.[52]

If done effectively, the Internet can be a means of significant cost savings and an avenue to reach more customers, not merely a force toward disintermediation.[53] Research has consistently demonstrated a small positive correlation between realtor income and the degree to which the realtor uses the Internet to market their listings.[54] Consumers typically like having more information available. Despite this, research has shown that Web listings do not reduce buyers' search time. Instead, they result in more intensive searches, where more properties are evaluated within a given period of time.[55] Usage of the Internet in searching for real estate has escalated rapidly with the advent and dissemination of the Web in the second half of the 1990s and beyond. Only 2 percent of buyers used the Internet to look for properties in the United States in 1995, escalating to 23 percent in 1999, and 55 percent in 2001; that number is typically estimated at over 70 percent today.[56]

In addition to the question of listing MLS data on the Internet and other Web-based real estate marketing, the technology can also encourage and facilitate other mechanisms for the transfer of properties. Electronic real estate auctions, while still a small percentage of the overall transactions, are becoming more common. To address the growth of electronic auctions, Denver-based Rbuy, Inc. was launched by realtors to allow other realtors to run successful electronic auctions for their clients.[57] Overall, however, the nature of the typically differentiated high-end product of real estate tends to limit the ability to take advantage of reduced transaction costs and heightened market speed through Web auctions or other Internet-based real estate transactions.[58] Furthermore, in this area, to an even greater extent than many other areas of e-commerce, the use of the Internet and serving a broader geographic clientele raises the legal issue of personal jurisdiction, or in what jurisdiction harmed parties can seek restitution.[59] Many challenging legal questions continue to be adjudicated in the courts with regard to the Internet and real estate transactions, alongside the efforts of some to better manage risks and take advantage of certain efficiencies in electronic real estate transactions.[60]

There are definitely many challenges as well as some significant opportunities faced by realtors. In the near-to-intermediate term, the overall economic and housing environment will likely be a far more crucial factor in the performance of real estate agents and firms than what appear to be ongoing, but gradual pressures from the Internet as a result of broadening access to real estate information and new tools for conducting business. During the last half-dozen years, many factors other than the Internet have aligned in favor of real estate sales and realtor revenue. The favorable environment that realtors have enjoyed may change dramatically in the future when the market either stagnates or declines, and only then will the impacts of the Web be more readily apparent.

Needless to say, the Internet has become a critical tool for selling real estate over the past half decade, as Web access and high-speed Internet connections have proliferated. Such access to information has contributed to the boom in real estate prices and transactions. Far more important to this boom, however, has been the Federal Reserve's reductions to the federal funds rate between May 2000 and June 2003 that brought the rate from 6.5 to 1 percent, and the associated fall in mortgage rates to historic lows—thirty-year no-point mortgages near 5 percent. While mortgage rates have risen modestly over the past couple of years, they have not grown in line with the federal funds rate increases over this period. This has continued to make real estate purchases and refinancing attractive. With the stock market decline associated with the dot-com collapse at the start of the new millennium, an increasing number of people have seen real estate as a safe, highly attractive investment—something more tangible and secure than the value of stocks.

Over the past half-dozen years many have bought first homes, and others have purchased vacation properties or higher-priced primary homes. At the same time there has been an increase in participation in the rapid buying and selling, or flipping, of properties as a short-term investment vehicle. Regions have varied substantially in all of these market areas and practices, especially the flipping of properties, but nationwide the trend has been decidedly upward. Overall, real estate prices increased 57.68 percent in the United States between the end of 2000 and 2005.[61] All of these factors contributed to a real estate boom, and concern and debate about a real estate bubble. Between 1997 and 2002, real estate agent establishments increased from 60,620 to 76,166, and employees from 219,633 to 284,827.[62]

The real estate market and low mortgage rates have not only been a major positive factor for realtors but also for the mortgage broker business. The booming real estate market of the first half decade of the new millennium and the prevalence of the Internet has also brought many changes as well as challenges to some of these enterprises. Mortgage brokers have had to rapidly adjust to higher customer expectations with regard to both the access and speed of delivery of services. Specifically, customers have come to expect asynchronous communication (through e-mail) in addition to phone contact and interactive Web sites. In this environment, some mortgage brokers have adjusted fairly well while others have suffered.[63] Though complex title search, property appraisal, and other factors favor local brokers, large primary lenders have rapidly entered nonlocal markets using the Web as a tool to offer service in most or all states. In the mid-1990s the Web was primarily a tool for brokers to advertise, although a small number were providing complete loan applications online at this time. Several years later, the Web became an increasingly common channel for mortgage loan applications.

Forrester Research reported that $482 million in loans had originated online during 1997, the majority of which were for mortgages, and predicted the number would go up approximately eighty times by 2001.[64] By the end of the 1990s, a growing number of mortgage firms had begun to offer the ability to be preapproved for a loan online.[65] Like many e-commerce areas, the ability to see the terms of many lenders led to increased competition and downward pressures on rates and fees. Some established well-branded mortgage enterprises (including a number of large banks) avoided the Internet through the mid to late 1990s to try to maintain their higher prices and larger margins. Yet escalating online competition forced them to change their ways, and nearly all the giants now have loan products on the Web.[66] With this, the opportunities for the large national firms to take business away from local brokers have been considerable for basic generic transactions. On the other hand, specialized clients and packages, such as for customers with poor credit, often continue to be handled by smaller firms.

Like the travel agency business, mortgage brokers have also become vulnerable to cyberintermediaries that connect clients to certain lenders, leaving others out in the cold. One of the leading cyberintermediary firms in the mortgage broker area is LendingTree, which was conceptualized in 1996 by Doug Lebda, and formally launched in July 1998. LendingTree connects customers to lenders by providing quotes from a number of lending organizations that have become a part of the LendingTree network. Lenders in the network bid against each other to be referred by LendingTree to a customer. Before the end of the year Lebda's firm had struck deals with over twenty-one national lenders, including Citibank, Advanta, and BankOne, and signed an important advertising agreement with Yahoo! to enable its banners to reach millions of people daily.[67] LendingTree continued to grow its volume during the dot-com collapse and exceeded $90 million in annual revenue by 2002.

An even larger player in the online lending industry is E-Loan, which was formed in 1997 by Janina Pawlowski and Christopher Larson. While mortgages and home loan refinancing are its primary areas, it also makes other types of loans, such as vehicle loans. In 1998, E-Loan led the online industry with a mortgage origination total of $1.6 billion.[68] Its closest online competitor, QuickenMortgage, a firm that partnered with a network of about two dozen lenders, originated $1.2 billion in loans that year.[69] This business and the overall online origination field represented but a small portion of the total mortgage origination market. This market was led by Norwest with $110 billion in loans. The online origin segment, however, has grown much faster than the traditional off-online one. E-Loan and other sizable Web-based firms can offer lower origination fees than the big banks and small brick-and-mortar companies that have to support offices, other physical infrastructure, and more employees relative to their loan volume.[70] Also, optimization software to process online loan applications sped up work flow, decreased labor requirements, and reduced overall costs.[71]

Internet-based loans have tended to hurt smaller mortgage brokers far more than the well-branded giants such as Norwest, Countrywide, Chase Manhattan, Citigroup, and Bank of America.[72] It is difficult to say if the Internet is a significant factor in the longer-term trend toward industry consolidation of lenders and brokers.[73] As major traditional lenders have developed an online presence, most of them also became part of the networks of leading cyberintermediaries such as LendingTree.[74] In the case of Bank of America, it extended its Internet know-how and capabilities in mortgage lending by trading an 80 percent ownership stake in its auto lender CarFinance.com for a 5 percent stake in E-Loan.[75] Whether large mortgage broker firms and businesses obtained online sales capabilities through acquisitions, equity stakes, and alliances, or developed them internally, most major lenders had a solid online operation in place by early in the new millennium. They overcame the internal politics that often made this addition difficult.[76] These companies also enjoyed the benefit of recognition and trust that extended from their long-established brand. Conversely, word of mouth and referrals tended to be more important for small businesses, but this was often a challenge given that most people are unwilling to recommend their lender to a friend.[77] Getting product information out to the public is a major hurdle for small enterprises. These organizations simply do not have the funds for advertisements on the high-traffic Web sites or in major newspapers.[78]

Moreover, privacy and trust have always been concerns with entering financial information online, and this is heightened for loan providers given the type of financial information that is required by brokers, lenders, and others in the financial industry. In light of this situation, many Internet-based mortgage firms have gone to great lengths and expense to ensure the security of their systems. This is costly and more easily accomplished by large firms—major banks and well-branded Web-based firms like E-Loan—than small ones.[79]

In March 1999, E-Loan filed its initial public offering and was valued at over a half billion dollars.[80] The company has continued to expand. In 2006 E-Loan, which had originated more than $27 billion in consumer loans since its inception, entered an alliance with Mortgage Initiatives Inc., a division of RE/MAX Regional Services launched as an agent initiative program to provide agents a quicker means to obtain loans for their clients.[81] With such alliances, the distinctions between traditional mortgage brokers and Internet ones are becoming blurred.

One thing is certain in the rapidly changing environment for mortgage and real estate brokers: consumers are increasingly the beneficiaries of the higher competition facilitated by the Web. Realtors and smaller traditional mortgage brokers that operate out of offices might find a tougher road ahead as the real estate market cools off. Realtor sales commissions, which have typically been 5 to 7 percent, will likely come under increasing pressure, as mortgage brokers and others become more involved in managing different parts of the real estate transaction process, and seek to make their money on volume rather than high fees. Peter Sealey, a professor of technology and marketing at the University of California at Berkeley, recently emphasized that "slowly but surely technology is coming into play in the real estate market; as it does, [real estate] brokers will lose their stranglehold on the process, commissions will ultimately be cut in half."[82] Whether the results are this severe is open to question. Still, it is probable that increased competition and the use of the Web will have a continuing downward impact on commissions and fees. By the beginning of 2005, the average commission on a home sale had dropped to 5.1 percent from 5.5 percent a few years earlier.[83] With regard to mortgages, LendingTree, E-Loan, Quicken, and others are pressuring down fees, but the savings over brick-and-mortar infrastructure only goes so far. Much of the same paperwork is involved, as is the time of professionals. Both realtors and mortgage brokers will face increasing Internet and other Web-based challenges in the future—challenges that will be better understood when the real estate market cools or slumps. Most likely it will hit smaller firms disproportionately hard, especially mortgage broker businesses. Overall, however, the impact on the two industries will potentially be modest and gradual, rather than immediately transformational as was commonly predicted a few years ago.

Personal Computers

The personal computer business was significantly altered by the Internet and the Web. In the mid to late 1990s, the growth of the Web provided a substantial incentive for Americans to purchase computers for their homes and businesses. It also inspired longer-term computer users to upgrade to new, more powerful machines. The network effect of e-mail, as it moved beyond just higher education, some businesses, and government to the public at large, also extended the demand for personal computers. The Web has unquestionably lent substantial momentum to the growth of the personal computer business, yet it is a tool that has been used most effectively by a few personal computer businesses to gain and extend competitive advantage at the expense of others in the industry.

Intel Corporation made the personal computer possible by its development of microprocessors in the early 1970s. In the mid-1970s, entrepreneurs took the leap to use Intel microprocessors to build and sell the first personal computers. These computers were often sold as kits to hobbyists and had only a small market. It was in 1977, when Apple Computer's second machine, the Apple II, was released that the personal computer field began to broaden. Software Arts' "killer app" spreadsheet VisiCalc substantially boosted demand for the Apple II. It demonstrated that personal computers could be far more than just a platform for video games.

The IBM brand lent further legitimacy to the personal computer as a business tool in 1981 when it came out with its IBM PC. Like the Apple II, the IBM PC benefited from a software firm, Lotus Development, that sold a spreadsheet program for its platform, Lotus 1–2–3. Popular word processing programs, such as MicroPro's Wordstar, also gave businesses and the broader public a reason to purchase personal computers at the end of the 1970s and the early 1980s. Furthermore, networking technology, particularly local area networks, led to increased demand for personal computers within organizations by facilitating file sharing and the common use of expensive peripherals such as laser printers.

It was IBM's reputation in computing that gave the firm early momentum to become the leading personal computer manufacturer, and it was software firms' incentive to produce products first and foremost for the leading platform that helped Big Blue extend its early leadership. This would be short-lived, though. IBM believed that publishing and obtaining a copyright for its Read-Only Memory–Basic Input-Output System code (ROM-BIOS) would protect its platform from imitators. Houston-based Compaq Corporation, a firm formed by three former Texas Instruments managers, proved otherwise. In 1982 Compaq reverse-engineered IBM's ROM-BIOS, using different code but replicating the functionality, to produce an effective IBM PC clone that did not violate IBM's copyright. Compaq's revenue from its IBM PC clone exceeded $100 million in 1982.[84] Other IBM PC clones soon followed, but Compaq had an advantage of being the first to sell a comparable, lower-priced, IBM-platform computer. Compaq was also more effective at working with the top microprocessor supplier, Intel, to beat Big Blue to market with an Intel 386-based computer. With these advantages, Compaq passed IBM in the second half of the 1980s and became the leading personal computer enterprise in the world.

While IBM's long computer industry leadership and brand equity was critical to giving its personal computer the early lead, its existing sales infrastructure of highly skilled sales engineers (ideal for higher-end mainframes and mid-range systems) was ill suited for the personal computer business. Once consumers became satisfied that the quality of established competitors' machines, such as those from Compaq, were comparable, the Houston-based firm and other clone manufacturers had a substantial advantage over IBM. They benefited from selling personal computers at a lower cost than Big Blue in the emerging computer retail chains, such as CompUSA.

Compaq surpassed IBM on product development, the time to market, price, and its more appropriate infrastructure for the early dominant distribution channel of retail stores. Another personal computer firm, however, would surpass Compaq on the same criteria, and its business model, capabilities, and structure were ideally suited for the Web environment of the second half of the 1990s.

In 1983, Michael Dell, a first-year University of Texas student, successfully launched Dell Computer. He recognized the opportunity extending from the fact that the industry-leading IBM PC retailed for about $3,000, but was composed of off-the-shelf components and software totaling roughly $700.[85] Dell Computer began purchasing components and assembling computers, and then selling directly to avoid the substantial cut taken by retailers. The firm became the first significant personal computer company to institute such a model. By 1985, it was operating in a thirty-thousand-square-foot facility and generating $70 million in revenue.[86]

In the mid-1980s Dell designed its computers to have not only cost but also performance advantages over IBM and others. At the 1986 Comdex, it offered a twelve-megahertz personal computer for $1,995 while IBM was selling a six-megahertz one for $3,995.[87] Unlike most other personal computer-compatible producers, Dell was issuing catalogs and using the phone to sell directly to customers. This not only circumvented the retail market but also led to major cost advantages from being able to maintain far less inventory than firms that supplied retail chains. Dell also utilized a specialized sales staff that targeted corporate, government, educational, and other markets. The direct contact with customers proved beneficial in allowing Dell to better understand their needs.

Dell Computer's success continued in the second half of the 1980s, but it was not free from missteps. In 1989 a decrease in demand, holding excessive inventory, a misguided, overly ambitious project to produce a product that spanned the desktop, workstation, and server areas (code-named Olympic), and the addition of using a retail model led to a substantial stumble, and the firm posted an annual loss. The following year Dell abandoned Olympic. Michael Dell learned that building to customer needs, rather than trying to define them, was critical to achieving and sustaining competitive advantage in the personal computer field. In 1993 he reevaluated the retail channel move, and concluded that it was only providing a modest increase in volume and generated no additional profits. That year he abandoned using retailers to return to an exclusive direct sales model. The miscalculations at the end of the 1980s had taught him the importance of careful inventory management, not to be ambitious with products, and the advantage of an unwavering focus on direct sales. All three elements would be fundamental to the firm in using the Web to surpass Compaq.

By the advent of Netscape Navigator and other Web browsers in the mid-1990s, the personal computer business had become mature. Margins had dropped considerably, and the near monopoly of primary suppliers (Microsoft in operating systems and common applications software, and Intel in microprocessors) meant that they captured much of the profit from the sale of personal computers, not the computer "manufacturers." Personal computers had become a commodity. Service became a lone area for differentiation (with the exception of Apple, which had a different operating system, a small but loyal customer base, and sold at a premium). Cost containment and cost leadership became even more critical.

Dell's direct sales channel gave it a fundamental advantage, but the firm did not rest on this. In 1995 and 1996, it aggressively developed and expanded its Web capabilities as a direct sales vehicle, added or extended a combination of Internet, phone, and in-person services that dwarfed those of its competitors, and focused on a mass-customization strategy. As part of this strategy, all computers were built to order, allowing customers to make choices within some twenty different specification areas.

In addition to this well-conceived strategy, Dell continued to thrive as a result of its continuously strong execution. In an age of increased outsourcing, something Dell has done with great skill, it recognized opportunities where internal capabilities would prove essential and built its value-chain software from scratch. Dell's just-in-time practices, achieved in part with its proprietary software, facilitated hourly updates of orders to suppliers—something unheard of in the industry. At the end of the 1990s, Dell on average only maintained personal computer product inventory for roughly six days, versus twenty-seven days for Compaq and thirty-seven days for IBM.[88] Meanwhile, Dell's strong execution allowed it to gain substantial competitive advantage over its primary long-time rival in the direct sales of personal computers: Gateway, Inc.

Gateway, Inc., a firm that was launched on a small scale by Ted Wiatt in an Iowa farmhouse in the late 1980s as Gateway 2000, was among many enterprises seeking to sell computers bypassing intermediaries. Its initial success and later struggles highlight the broad capabilities and execution needed to succeed long-term in direct, Internet-based personal computer sales. Gateway moved to Sioux City, South Dakota, in the early 1990s and went public in 1993, a year in which it achieved sales of more than $1 billion. The firm was successful in the mid to late 1990s. In 1994, it was the first personal computer firm to make compact disc read-only memory standard on all its computers.[89] The following year, it was the first company to sell computers on the Web (Dell followed soon thereafter).[90] Nevertheless, Gateway quickly was and remained in Dell's shadow, in large part because it lacked the services capabilities and execution of the Texas-based firm. Gateway did not have the efficiency necessary to compensate as margins in the personal computer business began to decline precipitously in the early years of the new millennium. Gateway posted losses in four consecutive fiscal years, 2001 to 2004 inclusive, before rebounding to profitability in fiscal year 2005.[91]

Between 1995 and 2000 annual revenue at Dell grew from $3.48 billion to $25.27 billion, and earnings from $0.21 billion to $1.86 billion.[92] In 1999, Dell surpassed Compaq to become the leading personal computer business in the world. Compaq, acquired by Hewlett-Packard in 2002, like most other personal computer firms besides Dell was severely hurt by the decline in demand with the aftermath of Y2K (which had depleted many corporate information technology budgets) and the dot-com collapse.[93] Meanwhile, IBM had struggled in the personal computer field, and was starting a transition in which information technology services would be its future core business and revenue generator. In 2004, IBM sold the vast majority of its personal computer business to Chinese-based Lenovo Group.

Though many factors were at play, the Internet, and particularly the advent of the World Wide Web and popular browsers such as Netscape and Explorer, provided a major boost to personal computer demand. This led to an environment that favored certain organizational capabilities and strategies over others. Gateway and Dell were able to take advantage of cost reductions from direct sales as never before by focusing on online sales. Dell was especially strong at using the Web to its advantage by developing a mass-customization strategy, superior online and off-line services to its competitors, and unparalleled logistics. Its Web pages were far easier for customers to navigate and graphically more appealing than those of its rivals. As a result, the previous giants in the personal computer business—IBM, Compaq, and others—suffered a loss of market share. Their organizational capabilities were geared to a different environment. In the words of Michael Dell, "Every part of their organizational mechanism was trained in a radically different discipline from the one we practice. That is a difficult change to make. It's like going from hockey to basketball. They're both sports ... but they are very different games."[94]

IBM's strong sales force for mainframes was of little help in the personal computer world, and Compaq relied on retail outlets to sell machines. A Compaq senior vice president estimated that the retail markup, cooperative marketing costs, and price protection for retailers added more than 10 percent to the overall costs of Compaq's personal computers; these are all costs that have long been avoided by Dell.[95] Dell's success also hurt specialty retail stores such as CompUSA, as more and more people purchased online.

In U.S. markets, despite recent earnings setbacks and a major recall, Dell has the potential for a much brighter future than other personal computer firms as a result of its direct model and logistics capabilities. Though the U.S. market for personal computers is the largest in the world, it is now growing far slower than in many other countries. Dell has had some success in international markets, including a number of European countries, and is now becoming increasingly aggressive in the fast-growing markets of China and India. It remains to be seen how Dell's model will ultimately play out in the developing world, though. Given that India shares the common language of English (for conducting commerce), coupled with the interaction as well as business and political alliances that exist between the two nations, the Indian personal computer business might hold a number of similarities to its earlier growth in the U.S. market. The Indian market currently represents just 2 percent of the world personal computer market, but it is expected to grow by 40 percent to 6.8 million personal computers in 2006, and 15.3 million by 2009.[96]

Dell has had a presence in India since the start of the new millennium, but has primarily been drawing on Indian workers for its call center operations. The firm has call centers in Bangalore, Hyderabad, and Maholi as well as global software development and product-testing centers in Bangalore.[97] Thus far, Dell has had only modest success in India with personal computers—it accounts for less than 4 percent of the Indian personal computer market. Hewlett-Packard has faired much better and has roughly 18 percent of the personal computer unit sales in India.[98]

Though continually expanding call center operations, with plans to reach fifteen thousand employees in India by 2008, Dell long avoided establishing a manufacturing base in India due to India's inverted duty structure.[99] As a result of expectations for future rapid growth, Dell revisited this decision in 2005 and announced plans for a manufacturing facility in India at the end January 2006. Its Chennai facility will be operational in 2007. The decision is part of Dell's goal to become the number one personal computer firm in the Indian market by reducing costs to customers and shortening delivery times.[100] Dell should benefit from its past experience and extensive presence in India in services. This has exposed the firm to cultural differences, and addressing the ongoing challenges of employee retention and developing business relationships in the country. Yet manufacturing in India has a number unique challenges and risks, including the poor transportation infrastructure. Perhaps the most significant hurdle for Dell in its goal to be India's leading personal computer supplier is that the country currently has a regulation to prevent the sale of computers through online channels. Dell CEO Kevin Rollins, however, is hopeful that this policy will be reversed. Whether and when Dell's Web-based direct model will become possible in India remains to be seen.

In China, there is an even greater opportunity for Dell. In 2004 IDC ranked China as the world's second-largest personal computer market behind the United States.[101] At the same time, there is substantial uncertainty in the Chinese market regarding what will be the dominant distribution channels for the future. In 1998 Dell initiated the first model for the direct sales of personal computers in China, opening a manufacturing and sales operation in Ximen, and allowing consumers to purchase using either the Internet or phone. Customers in China were not used to buying high-end products sight unseen. Dell addressed this by having hands-on, promotional events at malls and an aggressive print marketing campaign. Another challenge was the fact that Chinese consumers were not used to paying with credit cards. Dell responded by routing payments through Chinese banks and thus gained a substantial foothold in this attractive market by 2004.

IBM took a far different route to addressing its challenges in the personal computer field and, to a degree, participating in the fast-growing Chinese market. In 2004, as mentioned earlier, the firm sold its personal computer business to the Chinese firm Lenovo, receiving $1.75 billion (including an assumed debt of $0.50 billion) and retaining an 18.9 percent stake in the enterprise.[102] Lenovo's parent firm, the New Technology Development Company, also referred to as Beijing Legend, was formed in 1984 and has led the Chinese personal computer market since the mid-1990s. It achieved these results through building strong relationships within communities and meeting local needs. At the decade's end, the Internet took off in China, growing more than 100 percent between 1999 and 2000. Legend came out with its first-generation "Internet PC" in November 1999 and held more than a quarter of the country's personal computer market.[103] In 2003 Legend adopted the Lenovo brand name.

While IBM backed away in the personal computer area in selling its division, through its minority stake and ongoing alliance with Lenovo, it is seeking to take part in the expected rapid growth in the Chinese personal computer market. The deal allows Lenovo to benefit from IBM's brand during a transition period—a critical factor given Lenovo's goal of growing its market presence and reputation outside of China.

In 2003, Gartner Group reported that Dell had achieved the number two market share position in China in personal computers at 6.8 percent, well below Lenovo, which had approximately 27 percent.[104] It remains to be seen whether Dell's model, extremely successful at achieving higher margins as well as creating industry leading growth in the United States and other developed nations, is positioning itself effectively to replay this history as the Chinese personal computer market matures. A further indication of what is at stake and the great opportunity for future growth in China is the fact that as of 2004, laptops—a higher-margin product than desktops—represented only 13 percent of personal computer revenue in China versus more than

50 percent in the United States.[105] While the opportunity for Dell is great, cultural differences, language barriers, politics, and highly localized sales channels may continue to make the environment more conducive to Lenovo for many years to come.

Offshoring Call Centers, Business Process Outsourcing, and Research and Development

In recent years, debates have raged over the outsourcing overseas or "offshoring" of jobs. For decades, blue-collar manufacturing jobs have been sent abroad to Asia, Central America, and elsewhere in large numbers by some of the leading firms in the country. Offshoring has received extensive attention of late because a growing number of white-collar information technology jobs have been sent overseas. To some, this is seen as the benefits of globalization, where labor arbitrage and talent abroad allows firms to produce goods and services more efficiently; America is thereby freed from the drudgery of tedious jobs, and allowed to focus on more innovative activities and lucrative segments of the value chain. Labor becomes just one more form of trade. Alternatively, others emphasize that a rapidly growing number of higher-paying jobs of an ever broader range are being sent overseas, that American workers are being displaced, and that available alternatives are either not feasible or attractive to the newly unemployed. Some also stress the deleterious social, cultural, and political impacts of offshoring on developing nations. While jobs are going overseas in a broad range of information technology fields, perhaps none has received more attention than call centers. This has occurred in part as a result of widespread media attention as well as American consumers' direct experience receiving service from call centers abroad. Offshoring also became a significant issue in the 2004 presidential campaign, as John Kerry and John Edwards highlighted the problem of information technology and other jobs being outsourced overseas, and offered protectionist legislation as a possible remedy. The issue of offshoring has remained a popular one in part because of continuing media coverage, including a best-selling book on the topic: *New York Times* columnist Thomas L. Friedman's *The World Is Flat: A Brief History of the Twenty-First Century*.[106]

In large part, the trend to offshore call center services was a consequence of the massive decline in long-distance telecommunication charges brought about by the overinvestment in infrastructure during the bubble of the late 1990s. This overinvestment was the product of unrealistically high projections regarding the demand for the transmission of voice and data (via the Internet), and from unexpected efficiencies gained through advances in technology. Large volumes of low-wage, educated English speakers allowed people in India to move into call center work. This has adversely impacted future growth and opportunities for some call center services firms in the United States as well as resulted in the offshoring of some internal corporate call center operations.

The overall growth of call centers in the past decade has been pronounced. Call center business generated approximately $2.5 billion in India in 2003, but continues to grow rapidly, and is expected to reach $15 billion by 2008.[107] U.S.-based firms and Indian companies serving U.S. corporations make up the greatest portion of this business. Regardless of the challenges in understanding the full impacts of these trends for different types of corporations, a substantial number of American call center workers have suffered job losses.

While call centers have received perhaps the largest share of attention, U.S. corporations have been outsourcing a broadening range of back-office functions to India and elsewhere, often referred to as business process outsourcing. This has included accounting services, legal services, human resource management, and other functions. Furthermore, the vast talent pool of highly educated people in India is increasingly being used not just for cost savings and more mundane business process outsourcing labor but to tap stellar talent for research and development as well as the development of intellectual property. By the start of the new millennium, U.S. firms were outsourcing scientific and engineering design work in a range of fields, but especially information technology. Software development has become a major area where U.S.- and European-based multinationals such as IBM, Hewlett-Packard, EDS, Oracle, Accenture, SAP, Capgemini, and others have taken advantage of the tremendous talent of educated Indian workers that have been trained at world-class universities such as the Indian Institutes of Technology. While only a small number of the hundreds of universities and thousands of colleges in India are top caliber, more and more Indians receiving bachelors and graduate technical degrees at leading U.S. and European schools are now returning to India as a result of the increased opportunities. In addition to Indian operations of American information technology companies, Indian corporations in the information technology services area, including Tata Consulting, Infosys, Wipro Technologies, HCL Technologies, and Satyam, have grown to become large corporations employing tens of thousands in India and serving corporate clients around the world.[108]

Sparked by polemics from pundits spanning the political spectrum (including conservative journalist Lou Dobbs) as well as high projections from some analysts on future information technology job losses, offshoring is increasingly presented and perceived as a crisis. Executive and legislative branches have introduced a range of protectionist bills impacting work visas, tax codes, and penalties to corporations that offshore. Though it is difficult to accurately count information technology jobs sent overseas, and even more difficult to predict future job losses, the best estimates place the annual rate of offshoring at 2 to 3 percent of the information technology workforce—a rate lower than the level of information technology job creation in the United States. There are many risks to offshoring. These include political, infrastructural, data security, privacy, cultural, and legal risks. These have and will continue to quell unrestrained offshoring. More sober analyses tend to present the benefits of offshoring—maintaining global competitiveness, cheaper goods and services for U.S. consumers, and opportunities to innovate and create new jobs higher up the value chain—alongside the negative impact of lost U.S. jobs. Educational investments and policies that encourage and facilitate scientific and technical study, as well as the attraction and retention of top talent from around the world, appear the best means to meet the ever-critical goal of extending America's base for technical innovation in the future.

Film and Film Development, Florists, and Bike Messengers

The Internet and the Web have often been just one factor impacting industries alongside many others. This makes establishing cause and effect difficult, and sometimes impossible. In certain cases, even where the impact of the Internet and the Web is unquestionable, it is one of lending momentum rather than setting change in motion. One such instance is the film and film development segments of the photography industry. The worldwide growth of the digital camera business has transformed the photographic equipment, film, and film development businesses.

Digital cameras grew out of the same basic technology used for television and videotape recorders in the 1950s. Most subsequent advances in digital-imaging technology were achieved by the U.S. government or government contractors for space probes, spy satellites, and other scientific and defense purposes. In 1972, Texas Instruments pioneered the first digital camera. In 1981, Sony released the first commercial digital camera. Kodak, a longtime leader in film and developing services, was also a leader in digital camera technology and bringing this technology to the public. In 1986 the firm produced the first megapixel sensor, and in 1991 released the first digital camera system targeted for the photojournalist market. It also pioneered, with Kinko's and Microsoft, digital image-making software workstations and kiosks that allowed customers to produce photo compact discs.

Digital camera technology was rapidly advancing in the late 1980s and the early 1990s. In the mid-1990s, when digital cameras were first engineered to work with personal computers via serial cables, the Web was first coming into widespread use. Even if it had not been possible to post images on Web sites or send them as e-mail attachments to others, there would have been significant demand for digital cameras. Yet it would have been substantially less than the level readily apparent today. It is the ability to take digital pictures and then send them to friends and family or post them on a site—the network effect—that has driven the rapid demand for digital cameras, and investment and innovation in the industry. Kodak was not blindsided by digital camera technology, but the trend has clearly hurt the firm's previously lucrative film and film development businesses. In both areas, it was the industry leader, whereas it is just one of many players in the various businesses associated with digital cameras.

Kodak lost $1.4 billion in 2005 and is only midway through its restructuring to transition resources out of its traditional businesses into the digital field.[109] While Kodak held an early lead in digital cameras, its rapidly declining film and developing cash cow businesses overwhelmed the firm. Recently, Canon and Sony have surpassed Kodak in the digital camera market. Digital photo kiosks looked to be a bright spot for the Rochester, New York–based firm, but they face increasing competition from kiosks from Fuji, Hewlett-Packard, and others.[110] One thing that would have greatly aided Kodak is a more gradual market transition from film to digital. This was not possible given the excitement and demand for digital cameras and photos in the Web era (but may have been were it not for the Web). In September 2003 when Kodak executives announced the company's transformation to focus on digital products, they never envisioned the firm could lose roughly a quarter of its consumer film business each of the following few years. Even China, a country that Kodak had hoped would be an important market for traditional film products and services for years to come, as more and more Chinese consumers possessed discretionary income, is bypassing this medium for digital cameras.[111]

Whether it is the result of labor arbitrage, new complementary technologies, or high capital or infrastructure costs, a substantial number of U.S. firms and industry segments are vulnerable to the changes and further proliferation of the Internet in American business. The latter category includes a broad range of brick-and-mortar businesses—including flower shops and neighborhood video stores—that have overhead that makes it difficult to compete with their Internet-based competitors (such as Flowers.com and Netflix). But abandoning the storefronts for an Internet model can be a risky strategy for retailers and suppliers alike, and brand continues to matter, whether it is earned and extended off-line or online. This is a lesson one flower supplier, Sunburst Farms, learned the hard way. This company sought to cut out the intermediary channel to sell flowers over the Web, as FlowerNet, and use FedEx for next-day delivery. It was able to significantly undersell most competitors and thought this would be enough with a commodity product like flowers. Many off-line competitors as well as online giants that had long-established brands, such as FTD and 1800Flowers, succeeded as FlowerNet struggled.[112] For most products and services, brand continues to be highly relevant. It can lower search costs, convey quality, and inspire trust.

In some industries and situations, brick-and-mortar businesses building and extending relationships with leading cyberintermediaries may be an effective strategy, as a number of mortgage brokerage firms and other enterprises have done. In other cases, focusing on greater differentiation and creatively developing distinctive products and services might be the most effective path to long-term success.

For some, often smaller services industries, there may be far less room to maneuver to try to create greater differentiation. One such industry is bicycle messenger and delivery services. This trade, which exists primarily in the downtown areas of large urban centers of the United States, and has long been part of the urban landscape of the streets of Manhattan and San Francisco, has been particularly hurt in recent years by the Web. This is not the first technology to cut into this business—faxes have long existed as a substitute—but the effects have been measured as a result of the poor quality and limitations of faxes. Since the turn of the millennium, with increasingly ubiquitous high-speed Internet, digital photography, and especially PDF files, the need for bicycle messengers has diminished. Estimates place losses in industry revenue at 5 to 10 percent a year since 2000. Wages and employment have fallen in step. In San Francisco in 2000, bicycle messengers made $20 per hour on average. Despite inflation, the average has now dipped to $11 per hour.[113] In New York there were an estimated twenty-five hundred bicycle messengers in the late 1990s, and in 2006 the number was but eleven hundred. Even this industry, though, will not be eliminated completely. Certain documents and articles, such as legal filings, architectural drawings, and original artwork, not to mention deli sandwiches, will continue to be delivered by cyclists. Furthermore, the extent and speed with which the bicycle messenger trade has been hurt, in terms of percentage drops in revenue, profits, employment, and wages, and the relative lack of strong alternatives, tends to be the exception rather than the rule in American business.

Conclusions

Despite the significant and real threats posed to some businesses and industry segments by the Internet, it is important to remember that many predictions of Internet-based businesses destroying their brick-and-mortar counterparts have been misguided. Notions of traditional grocers, pet supply stores, and pharmacies being severely hurt or put out of business by Internet-based providers such as Webvan, Pets.com, and PlanetRX.com were common at the height of the dot-com bubble.[114] These firms no longer exist, and the challenges faced by companies in these industries are primarily from their traditional competitors, not the Internet or Internet-based businesses. In general, the adverse impacts and forced changes tend to be much more subtle in non-media industries than their higher-profile, media trade counterparts.

Nevertheless, in some major industries such as travel reservations, the Internet has been transformational. It has opened opportunities for some, while forcing many to alter the nature of their business. This is sometimes by adoption, integration, or innovation of the latest networking and software systems, and at other times it is redefining the business along different parameters, where much of the value is added off-line and personal relationships still matter. Even with travel, it is crucial to remember the pre-Internet infrastructure, both technological (SABRE) and the existing structure of the trade, shaped key elements of industry dynamics in the later Internet and Web era. Change has been significant, but it has been on top of this existing infrastructure. A far different situation exists in a country such as China that is building a transportation planning and travel reservations infrastructure anew, and will have unique opportunities and challenges with the Internet.[115]

Overall, the Internet and the Web have altered nearly every business and industry, but to far differing degrees, and in extremely different ways. The most important single trend of the Web in American business is the reduction of traditional intermediaries. This has hurt those in certain enterprises, while it has been beneficial to many firms, industries, and individuals in reducing costs. At times, it has also brought into the fold new types of intermediaries—cyberintermediaries—that operate on lower cost structures and charge far less per transaction than traditional intermediaries, making up for it in volume. In certain respects, the Internet is analogous to earlier communication (and even transportation) technologies where the business and economic landscape undergoes critical change. In these new environments, businesses develop different approaches, possess and cultivate different capabilities, and succeed or fail long-term based on both their strategies and execution, not merely the underlying technological transformation that is occurring.

Notes

1. Amazon.com Inc. was launched in 1994, several years earlier than most other e-commerce firms that have had long-standing success. There are more than 1,600 article references on Academic Search Premier and over 167 million hits for "Amazon.com Inc." on Google.

2. Overall, there has been substantial homogeneity to the use of the Internet within industries, and heterogeneity in the use of the Internet between industries. Chris Forman, Avi Goldfarb, and Shane Greenstein, "Which Industries Use the Internet?" in *Organizing the New Industrial Economy*, ed. Michael Baye (Amsterdam: Elsevier, 2003), 47–72.

3. The advent of the Web and new possibilities to reduce the cost of transactions has spawned a heightened interest in the long-influential work of economist Ronald Coase, who argued that substantial transaction costs are the primary reason why all transactions are not market transactions. It has also led scholars to revisit the more recent theoretical work of economist Oliver Williamson and others on transaction cost economics. Ronald Coase, "The Nature of the Firm," *Economica* 4, no. 16 (November 1937): 386–405; Oliver E. Williamson, "Transaction Cost Economics: The Governance of Contractual Relations," *Journal of Law and Economics* 22, no. 10 (1979): 233–261.

4. Don Tapscott, David Ticoll, and Alex Lowy, *Digital Capital: Harnessing the Power of Business Webs* (Boston: Harvard Business School Press, 2000).

5. Juniper Research, "Juniper Research Forecasts the U.S. Online Travel Industry Will Reach $104 Billion by 2010," press release, November 3, 2005.

6. Martin Campbell-Kelly, *From Airline Reservations to Sonic the Hedgehog: A History of the Software Industry* (Cambridge: MIT Press, 2003), 45.

7. "Travel Agencies Gradually Move under the Aegis of Women—Special Report: Women-Owned Business-Industry Overview," *Los Angeles Business Journal*, June 22, 1992. Available at ⟨http://findarticles.com/p/articles/mi_m5072/is_n25_v14/ai_12429433⟩.

8. "SIC 4272 Travel Agencies," in *Service and Non-Manufacturing Industries*, vol. 2 of *Encyclopedia of American Industries*, ed. Lynn Pearce, 4th ed. (Detroit: Gale, 2005), 446–453. Available at ⟨http://findarticles.com/p/articles/mi_m5072/is_n25_v14/ai_12429433⟩.

9. Jane L. Levere, "Hitting the Road without Stopping at the Travel Agency," *New York Times*, August 6, 1995, sec. 3, 8; Walter Baranger, "Booking a Trip on the Internet," *New York Times*, September 10, 1995, sec. 5, 4.

10. Paul Burnham Finney, "Business Travel," *New York Times*, November 1, 1995, D5.

11. Donald J. McCubbrey, "Disintermediation and Reintermediation in the U.S. Air Travel Distribution Industry: A Delphi Study," *Communications of the Association for Information Systems* 18, no. 1 (June 1999): 470.

12. Ibid.

13. Paulette Thomas, "Travel Agency Meets Technology's Threat," *Wall Street Journal*, May 21, 2002, B4.

14. Ibid.

15. Michele McDonald, "My, My, How Things Are Changing," *Travel Weekly*, November 17, 2003, 101–107.

16. Robert Fox, "Flying into Glitches," *Communications of the ACM* 41, no. 8 (August 1998): 10.

17. "Travelers Saw Tech Take a Bigger Role in 2003," *USA Today*, December 30, 2003, B5.

18. Forrester Research, "On-line Leisure Travel Booking Is Booming," 1998.

19. Edwin McDowell, "Travel Agents Express Anger on Internet-Only Plane Fares," *New York Times*, August 5, 1999, C8.

20. U.S. Census Bureau, *Economic Census. Travel Arrangement and Reservation Services: 2002* (Washington, DC: U.S. Census Bureau, 2002).

21. Ibid.

22. Ibid.

23. Ibid.

24. McCubbrey, "Disintermediation and Reintermediation in the U.S. Air Travel Distribution Industry," 466.

25. Hannes Werthner and Francesco Ricci, "E-Commerce and Tourism," *Communications of the ACM* 47, no. 12 (December 2004): 101.

26. Expedia press release, July 16, 2001, available at ⟨http://press.expedia.com/index.php?s=press_releases&item=142⟩.

27. Justin Hibbard, "Airlines, Online Agencies Battle for Customers," *InformationWeek* 708 (November 9, 1998), 30.

28. C. E. Unterberg, Towbin, *Effects of Airline Commission Cuts on Online Travel Agencies*, April 3, 2001.

29. Preview Travel Inc. company analysis, 2006, available at ⟨http://mergentonline.com⟩.

30. Reid Goldsborough, *Back Issues in Higher Eductation* 17 (November 9, 2000): 33.

31. Sharon Machlis, "Flying by Different Rules," *Computerworld* 33, no. 11 (March 15, 1999): 12.

32. "Wishing They Weren't There," *Economist* 372, no. 8387 (August 7, 2004): 51–52.

33. Mitch Wagner, "Booking Businesses Travel on Net," *Computerworld* 31, no. 2 (January 13, 1997): 3.

34. "More Executives Book Travel on the Internet, but Issues Linger," *USA Today*, April 27, 2004, B7.

35. Scott H. Kessler, "Online Travel: All over the Map," *Business Week*, January 14, 2005, available at ⟨http://search.ebscohost.com/login.aspx?direct=true&db=aph&AN=15854995&site=ehost-live⟩.

36. "Flying from the Computer," *Economist* 377, no. 8446 (October 1, 2005): 65–67.

37. Jonna Jarvelainen and Jussi Puhakainen, "Distrust of One's Own Web Skills: A Reason for Offline Booking after an Online Information Search," *Electronic Markets* 14, no. 4 (December 2004): 333–343.

38. Shohreh A. Kaynama, Christine I. Black, and Garland Keesling, "Impact of the Internet on Internal Service Quality Factors: The Travel Industry Case," *Journal of Applied Business Research* 19, no. 1 (Winter 2003): 135–146.

39. Woo Gon Kim and Hae Young Lee, "Comparison of Web Services Quality between Online Travel Agencies and Online Travel Suppliers," *Journal of Travel and Tourism Marketing* 17, nos. 2–3 (2004): 105–116.

40. Sean O'Neill, "Be Sure to Look before You Book," *Kiplinger's Personal Finance* 58, no. 8 (August 2002): 110.

41. "ASTA Releases Remarks concerning Air Canada's Withdrawal of Inventory from GDS Systems," Press Releases, 2006 archive, available at ⟨http://www.astanet.com/news/releasearchive06/050506.asp⟩.

42. Thomas, "Travel Agency Meets Technology's Threat."

43. Jenny Ji-Yeon Lee et al., "Developing, Operating, and Maintaining a Travel Agency Website: Attending to E-Consumers and Internet Marketing Issues," *Journal of Travel and Tourism Marketing* 17, nos. 2–3 (2004): 205–223.

44. Avery Johnson, "Travel Watch," *Wall Street Journal*, February 28, 2006, D7.

45. PhoCusWright, Inc., "Online Travel Overview: Market Size and Forecasts, Fourth Edition, 2004–2006," available at ⟨http://store.phocuswright.com/phontrovupma.html⟩.

46. Maya Roney, "Online Travel Sites Feeling Pressure," *Forbes*, March 7, 2006, available at ⟨http://www.forbes.com⟩.

47. Though some realty firms have mortgage brokerage divisions, the vast majority of mortgage loans are made by separate mortgage enterprises (banks, savings and loans, etc.).

48. Suzanne Sullivan and Elizabeth Razzi, "House Hunting on the Internet," *Kiplinger's Personal Finance Magazine* 49, no. 7 (June 1995): 96–100.

49. Carl R. Gwin, "International Comparisons of Real Estate E-Nformation on the Internet," *Journal of Real Estate Research* 26, no. 1 (January March 2004): 1–23.

50. James Schott Ford, Ronald C. Rutherford, and Abdullah Yavas, "The Effects of the Internet on Marketing Residential Real Estate," *Journal of Housing Economics* 14 (2005): 92–108.

51. John D. Benjamin et al., "Technology and Realtor Income," *Journal of Real Estate Finance and Economics* 25, no. 1 (2002): 51–65.

52. Daniel W. Manchala, "E-Commerce Trust Metrics and Models," *IEEE Internet Computing* 4, no. 2 (2000): 36–44.

53. Waleed A. Muhanna and James R. Wolf, "The Impact of E-Commerce on the Real Estate Industry: Baen and Guttery Revisited," *Journal of Real Estate Portfolio Management* 8, no. 2 (2002): 141–152.

54. Benjamin et al., "Technology and Realtor Income."

55. Leonard V. Zumpano, Ken H. Johnson, and Randy I. Anderson, "Internet Use and Real Estate Brokerage Market Intermediation," *Journal of Housing Economics* 12, no. 2 (June 2003): 134–150.

56. Waleed A. Muhanna, "E-Commerce in the Real Estate Brokerage Industry," *Journal of Real Estate Practice and Education* 3, no. 1 (2000): 1–16; B. Brice, "E-Signatures in the Real Estate World," *Real Estate Issues* (Summer 2001): 43–46.

57. Julia King, "E-Auctions Change Real Estate Model," *Computerworld* 33, no. 27 (July 5, 1999): 40.

58. J. Christopher Westland, "Transaction Risk in Electronic Commerce," *Decision Support System* 33, no. 1 (2002): 87–103.

59. Robert J. Aalberts and Anthony M. Townsend, "Real Estate Transactions, the Internet, and Personal Jurisdiction," *Journal of Real Estate Literature* 10, no. 1 (2002): 27–44.

60. Lawrence Chin and Jiafeng Liu, "Risk Management in Real Estate Electronic Transactions," *Journal of Real Estate Literature* 12, no. 1 (2004): 53–66.

61. U.S. Senate, Joint Economic Committee, "Fourth Quarter Housing Report," 2005, available at ⟨http://jec.senate.gov/⟩.

62. U.S. Census Bureau, *Economic Census. Activities Related to Credit Intermediation: 2002* (Washington, DC: U.S. Census Bureau, 2002).

63. Anthony M. Townsend and Anthony R. Hendrickson, "The Internet and Mortgage Brokers: New Challenges, New Opportunities," *Real Estate Finance Journal* 14, no. 4 (Spring 1999): 11–15.

64. Katie Hafner, "New Internet Loan Services Are Not for the Faint of Heart," *New York Times*, July 2, 1998, G3.

65. "Internet Mortgage Shopping, Too," *Consumer Research Magazine* 79, no. 1 (January 1996): 32–33.

66. David Drucker, "Lenders Take Processes Online," *InternetWeek*, December 11, 2000, 10.

67. Yahoo!, "LendingTree Signs Agreement with Yahoo!," press release, December 8, 1998.

68. Scott Woolley, "Should We Keep the Baby?" *Forbes*, April 19, 1999, 222–227.

69. Ibid.; Cathy Charles, "Monthly Clicks," *Forbes*, May 22, 2000, 120.

70. Woolley, "Should We Keep the Baby?"

71. David Lewis, "Mortgage Lending Optimized," *InternetWeek*, April 23, 2001, 54.

72. Lawrence Richter Quinn, "Brokers: Don't Count Us Out in the Internet Age," *American Banker, Supplement* 1655, no. 155 (August 14, 2000): 6A.

73. Muhanna and Wolf, "The Impact of E-Commerce on the Real Estate Industry."

74. Larry Armstrong, "Click Your Way to a Mortgage," *Business Week*, March 6, 2000, 174–176.

75. Gregory Dalton, "Bank Swaps Stakes in Web Lenders," *InformationWeek*, August 30, 1999, 28.

76. "Inside the Machine," *Economist*, November 9, 2000, Special Section, 5–10.

77. Monte Burke, "Net Game," *Forbes*, October 10, 2005. A significant factor in the tendency not to recommend mortgage brokers to friends and close associates may be the nature of financial information disclosure and privacy concerns.

78. Quinn, "Brokers: Don't Count Us Out in the Internet Age."

79. Bomil Suh and Ingoo Han, "The Impact of Customer Trust and Perception of Security Control on the Acceptance of Electronic Commerce," *International Journal of Electronic Commerce* 7, no. 3 (Spring 2003): 135–161.

80. Ibid.

81. E-Loan, "E-LOAN Expands RE/MAX Regional Services Strategic Alliance," press release, May 8, 2006.

82. Quoted in Daniel Kadlec, "The Commission Squeeze," *Time*, January 31, 2005, 50–51.

83. Ibid. The degree to which this drop is the result of Web listings, or higher real estate prices and realtors' willingness to accept lower commissions to get business in a competitive environment, is uncertain.

84. Jeffrey R. Yost, *The Computer Industry* (Westport, CT: Greenwood Press, 2005), 183.

85. Michael Dell with Catherine Fredman, *Direct from Dell: Strategies That Revolutionized an Industry* (New York: HarperBusiness, 1999), 8–9.

86. Ibid., 18–19.

87. Ibid., 8–9.

88. Tapscott, Ticoll, and Lowy, *Digital Capital*.

89. Ira Sager and Peter Elstrom, "A Bare-Bones Box for Business," *Business Week*, May 26, 1997, 136.

90. Steven V. Brull, "Gateway's Big Gamble," *Business Week*, June 5, 2000, available at ⟨http://www.businessweek.com/2000/00_23/b3684027.htm⟩.

91. Robert Levine and Jia Lynn, "The Cow in Winter," *Fortune*, April 17, 2006, 55–56.

92. Dell, Inc., *Annual Reports*, 2000 and 2005. Available at ⟨http://www.dell.com/content/topics/global.aspx/corp/investor/en/annual?c=us&l=en&s=corp⟩.

93. Dell suffered a decline as well, but was one of the major computer firms not to have its revenue, profits, and stock price decimated between 2000 and 2002.

94. Quoted in Tapscott, Ticoll, and Lowy, *Digital Capital*, 109.

95. Yost, *The Computer Industry*, 196.

96. "Dell Plans a Factory for India; Plant to Join Its Call Centers, including One to Open in April," *Austin American Statesman*, January 31, 2006, D1.

97. "Dell to Set Up Plant in India," *Hindustan Times*, May 26, 2006, 14.

98. IDC, "2005 India PC Shipments Cross 4 Million Unit Landmark to Close the Year at a Record 4.3 Million Shipments," *IDC India Quarterly PC Market Programme, 4Q 2005* (February 2006). Available at ⟨http://www.idcindia.com/Press/17feb2006.htm⟩.

99. "Dell to Make Big Investment in India," *Hindustan Times*, January 29, 2006. Available at ⟨http://www.hindustantimes.com/StoryPage/StoryPage.aspx?id=c01d1739-4ba9-48d4-bcd1-18d94c93a3ba⟩.

100. "Dell Unit in India by Year-end," *Statesman*, May 26, 2006, 1.

101. "China PC 2004–2008 Forecast and Analysis," July 2004, available at ⟨http://www.idc.com⟩.

102. Mike Musgrove, "IBM Sells PC Business to Chinese Firm in $1.75 Billion Deal," *Washington Post*, December 8, 2004, A1.

103. Yigang Pan, "Lenovo: Countering the Dell Challenge," Asia Case Research Center, University of Hong Kong, HKU356, 2005.

104. Ibid.

105. Zhu Boru, "China's Laptop PC Market Remains Attractive," *China Business Weekly*, August 3, 2004. Available at ⟨http://www.chinadaily.com.cn/english/doc/2004-03/08/content_312858.htm⟩.

106. Thomas L. Friedman, *The World Is Flat: A Brief History of the Twenty-First Century* (New York: Farrar, Straus and Giroux, 2005).

107. "The Place to Be," *Economist*, November 11, 2004, 10–12.

108. A rapidly expanding literature has been published over the past several years on information technology offshoring. The best overview and analysis of offshoring in software and information technology services is William Aspray, Frank Mayadas, and Moshe Y. Vardi, eds., *Globalization and Offshoring of Software: A Report of the ACM Job Migration Task Force*, 2006, available at ⟨http://www.acm.org/globalizationreport/⟩.

109. "Down with the Shutters," *Economist*, March 25, 2006, 68–69.

110. Ibid.

111. David Henry, "A Tense Kodak Moment," *Business Week*, October 17, 2005, 84–85.

112. John M. Gallaugher, "E-Commerce and the Undulating Distribution Channel," *Communications of the ACM* 47, no. 7 (July 2002): 89–95.

113. "Soft-Pedaled," *Economist*, July 1, 2006, 30.

114. Marty Jerome, "Is Your Company Next? The Internet Is Crushing Whole Industries," *PC Computing* 13, no. 2 (February 2000), 90.

115. "From Scratch," *Economist*, September 2, 2000, 63.

11 Resistance Is Futile? Reluctant and Selective Users of the Internet

Nathan Ensmenger

"Happy families are all alike; every unhappy family is unhappy in its own way." This famous opening line of *Anna Karenina*, suitably modified, might apply also to the study of the Internet and its influence on American commerce. It is relatively easy to describe the shared characteristics of those markets and industries that have readily embraced Internet technologies. We can do so using the seemingly imperative logic of economic rationality: reduced transaction costs, efficient distribution channels, disintermediation, and economies of scale and scope. Understanding why some users and industries might resist the Internet, or at least adopt it reluctantly or selectively, is more difficult. It requires us to consider a much larger, more complex, and often idiosyncratic set of motivations, rationales, and structures. Which brings us back to Leo Tolstoy: although we can fruitfully generalize about the reasons that the Internet has succeeded, its failures require us to tell more particular stories about specific industries, professions, and users.

Of course, talking about resistance to the Internet in terms of failure is misleading. There is a constant temptation when studying the adoption of new technologies to categorize potential users as either sages or Luddites—those who have the foresight and courage to embrace new technologies, and those who do not.[1] Such simplistic dichotomies are rarely intellectually productive. The dismissal of reluctant users of technology as being ignorant, recalcitrant, or backward is a rhetorical strategy, not an analytic device.[2] Recent scholarship in the history of technology has shown that most users respond selectively to new technologies, embracing those aspects that they find appealing or useful, and rejecting those that they do not.[3] In fact, the study of resistance, rejection, and other so-called failures is often a most valuable tool for understanding the larger process of technological innovation: the negative response of users to new technologies often reveals the underlying assumptions, values, and power relationships that are embedded in those technologies.[4]

All this being said, however, the rapid and widespread adoption of the Internet in the past decade, its seemingly ubiquitous presence in American business, and the apparently inexorable march of Moore's Law toward smaller, less expensive, and more powerful computing makes talk of reluctance and resistance seem quaint as well as irrelevant. Perhaps there are a few groups that are not yet regularly online—the poor, the elderly, or the technophobic—but the Internet is clearly becoming the dominant infrastructure for communications, commerce, and recreation. As James Cortada has suggested, for any business not to have a Web presence or e-mail address in today's economy would be like not having a Yellow Pages listing a decade ago.[5] There might be a few holdouts, but the vast majority of businesses are either online or have plans to be.

And yet even within a commercial landscape that has undeniably been transformed by Internet technology, we can identify not just pockets but vast territories in which reluctant users have successfully resisted technological innovations. In this chapter, I will explore three major industries or industry groups in which the Internet has had limited or unexpected influence. These include the health care industry, higher education, and what I am calling indispensable intermediaries. These are not insignificant industries; health care, for example, is a $1.7 trillion industry that absorbs almost 15 percent of the American gross domestic product. Among my indispensable intermediaries are included such sales and service industries as automobile dealerships, residential real estate, and fashion retailing. My point is not that the Internet has had negligible influence on these industries but rather that its influence has been highly mediated by the actions of reluctant users. These users have not rejected the Internet altogether but instead have adopted it selectively. University professors, for instance, have embraced e-mail, which serves their purposes well, and fits neatly into established patterns of work and authority. On the other hand, they have proven extremely reluctant users of Web-based instructional technologies, which threaten their traditional control of the classroom environment. Physicians, by contrast, regularly make use of the Web for research and educational purposes, but have rejected e-mail in the context of their professional practices.

So what makes physicians like real estate brokers like automobile manufacturers like university professors? It is not entirely clear. Like Tolstoy's unhappy families, it is not their similarities but their differences that make them interesting and deserving of further study. By reflecting on the ways in which idiosyncratic professional, economic, and legal concerns shape the responses of these various groups and industries to emergent Internet technologies, I hope to introduce additional nuance and historical specificity into a conversation that has long been dominated by technological or economic determinism.

The E-Health Revolution

Telemedicine. Telehealth. Health informatics. Interactive health communications. Electronic medical records. E-health. From the late 1950s to the present, these various efforts to effectively integrate electronic computing and communications technologies have captured the imagination of visionaries, entrepreneurs, health care benefits managers, insurance companies, hospital administrators, public health officials, and government agencies—and to a lesser extent patients and physicians. The appeal of these systems appeared self-evident to their promoters. Telemedicine would extend the reach of physicians and specialists into rural or otherwise-underserved areas.[6] Expert systems promised to standardize medical practice and encourage better-informed decision making on the part of physicians.[7] Interactive health communications tools could be used to educate patients, promote healthy behaviors, and manage the demand for health services.[8] Health informatics, electronic medical records, and other forms of computerized medical data processing would increase efficiency and lower costs through the enhanced oversight of practices, spending, and costs. And electronic communications networks would improve the quality of medical care for all by making possible vastly improved data sharing between patients, physicians, benefits providers, and medical researchers.[9] Although each of these individual initiatives attracted some attention and garnered some successes, it is safe to say that prior to the 1990s these broader goals of integration, efficiency, cost reduction, and improved access and care had not been achieved through the introduction of new computing and communications technologies. In recent years, however, the emergence of the Internet as a low-cost, high-speed, and widespread electronic communications infrastructure has prompted a resurgence of interest in medical computing. In fact, in the heady days of the late 1990s, no industry seemed as amenable to Internet-based transformation as the U.S. health care industry. Not only was health care the single-largest industry in the United States—$1.5 trillion in 1996 alone, as Wall Street analysts were fond of reminding potential investors—but it was also "the ultimate knowledge business."[10] Many of the most significant problems facing the industry were perceived to be informational in nature. As much as one-third of the spending in health care was believed to be wasted shuffling paper between patients, providers, and third-party payers—waste that could be neatly eliminated by making such transactions electronic.[11] In addition, the combination of increasing costs, an aging population, and an apparently worsening shortage of nurses and certain medical specialists seemed to demand a more efficient allocation of scarce resources.

Under the broad umbrella of e-health, many of the earlier visions of telemedicine and health informatics have been resurrected as e-mail or Web-based services. E-health systems would allow physicians and nurses to perform remote consultations, manage patient records, and process benefits claims via electronic clearinghouses. Inexpensive Webcams and digital cameras would be used to make high-quality specialist care available to the homebound, isolated, and poor. Patients would be able to access health-related information and records, communicate with physicians via e-mail, participate in online support groups, and use the Web to make appointments, refill prescriptions, and purchase health care products. Within a "few years," the economies of scale of the Internet would ensure that "every physician will choose to connect his or her office to a community health information network based on the World Wide Web."[12]

By the turn of the twenty-first century, it appeared that an Internet-based transformation of American medicine was desirable, imminent, and inevitable. The rapid expansion of the Internet into other areas of life and commerce were cited as precedents for a similarly rapid shift toward e-health services; as one representative editorial in the *New England Journal of Medicine* predicted, "On-line, computer-assisted communication between patients and medical databases and between patients and physicians promises to replace a substantial amount of the care now delivered in person."[13] Physicians would use e-mail to treat common diseases and would provide highly customized Web-based services to patients. Some of these services would be offered by their in-house staffs, and some by partnering with external dot-com providers.[14] Following this compelling dream of improved, efficient, and consumer-oriented health care, venture capital funding in health care in the late 1990s shifted rapidly toward Internet-based services, rising from $3 million in the first quarter of 1998 to $335 million by the fourth quarter of 1999.[15] In that year more than twenty-one e-health start-ups went public—including Netscape founder Jim Clark's Healtheon, whose initial valuation topped $1 billion. Clark predicted that within a few years Healtheon would control $250 billion of the $1.5 trillion health care industry.[16]

And yet despite massive investment in e-health initiatives by private firms, government agencies, and even medical professional societies, the e-health revolution has been slow in coming. The predicted convergence on Web-based standards for the coordination and exchange of medical records, laboratory results, billing information, and patient outcomes has not happened, nor has the widespread use of digital cameras or videoconferencing for patient monitoring. This is not to say that the Internet has had no effect on health care practices. Eight out of ten Internet users have accessed health information on the Web. The health information portal WebMD.com received eleven million unique hits in January 2006 alone.[17] Of those who have used the Internet to gather medical data, almost 12 percent (seventeen million) report that the Internet played a crucial or important role as they helped another person cope with a major illness.[18] More than 97 percent of physicians use the Internet, many on a daily basis, for clinical research and communication.[19] In 2004, more than 423,000 physicians went online to pursue continuing medical education credit.[20]

Nevertheless, the overall influence of the Internet on medical practice has been remarkably—and quite unexpectedly—limited. With the exception of information gathering, prescription refilling, and the occasional purchase of health-related equipment, most patients do not, and cannot, access traditional medical services online. Many of the early entrants into the e-health arena died in infancy or went bankrupt, with the few survivors being forced to dramatically adjust their business plans to accommodate more traditional patterns of patient-physician interaction.

Why the slow and fitful adoption of Internet technologies in one of the nation's largest and most information-centric industries? The answer to this question is almost as complex as the health care industry itself, and illustrates the many ways in individual technological innovations, even one as seemingly ubiquitous and powerful as the Internet, cannot be fully understood outside the context of their larger sociotechnical environment. The short answer, however, is that physicians, seemingly one of the principle beneficiaries of e-health initiatives, have proven reluctant to adopt them as a tool for interacting with, diagnosing, or monitoring patients.[21]

The evidence of this reluctance is undeniable. The majority of physicians do not provide even basic clinical services or even the means of scheduling appointments over the Internet; fewer than 6 percent of all patients have reported ever having communicated with their doctor via e-mail (a figure that has remained remarkably unchanged over the past decade).[22] Of the 34 percent of physicians who do have a Web site, the vast majority of these sites are little more than "online business cards."[23] Only a small number of institutions support "telemedical" technologies for monitoring or follow-up care. The up-and-coming health care Internet turned out to be "vaporware," in large measure because skeptical physicians resisted its implementation.[24]

Explaining physicians' resistance to Internet technologies is a little more difficult. After all, today's physicians are hardly opposed to technology on principle; physicians were early adopters of the personal computer as well as cell phones. Most physicians are actually highly Internet savvy: 97 percent have Internet access, with 71 percent spending time online daily.[25] Modern medicine is for the most part exceedingly (perhaps excessively) high-tech, with new diagnostic and therapeutic technologies being introduced and adopted on a regular basis. Physicians' continued reluctance to embrace e-health initiatives is clearly not a result of latent neo-Luddism, an inability to learn new technologies, or insufficient access or training.

One obvious explanation is a lack of economic incentives: in the current third-party payer system, physicians are almost never reimbursed for Internet-based activities. This is certainly a powerful disincentive. And yet reimbursement is rarely cited by physicians as their principal reason for avoiding the Internet. Rather, concerns about privacy, liability, and patient safety and well-being are described as being primary.[26] Even allowing for a certain degree of calculated disingenuousness on the part of physicians, it seems clear that more than just economic factors have influenced their collective wariness of Internet-based medicine. A more complete and satisfying explanation of their behavior requires that we situate the history of physician resistance to the Internet in a larger economic, legal, professional, and ethical context. Doing so allows us to move beyond the simplistic economic and technological determinism that often dominates discussions about the history and future of Internet commerce.

Telemedicine

The influence of technological innovation on medical practice in the past century cannot be overstated. The introduction of new clinical tools for diagnosis and therapy as well as new instruments for scientific and biomedical research, the development of mass production techniques for pharmaceutical production, widespread improvements in sanitation, transportation, and public health infrastructure, and even the development of new survey and advertising technologies have all significantly shaped the burgeoning twentieth- and twenty-first century health care industry. One of the unintentional side effects of the increased importance of technology in medicine, however, has been the centralization of medical practice around sites of technological innovation and capital investment: hospitals, laboratories, and specialized diagnostic treatment centers.[27] This process of centralization and specialization has, in turn, led to problems of access and resource distribution, particularly among rural populations, the poor, and the elderly.

In order to counter the centralizing effects of high-tech, capital-intensive medicine, hospitals, medical schools, and government agencies began experimenting, in the late 1950s with the use of information and communications technologies aimed at expanding the reach of medical practitioners. These systems of telemedicine—quite literally "medicine at a distance"—allowed physicians to use telephone, videoconferencing, and remote-control technology to consult with colleagues and patients in remote areas. In 1959, for example, a group of psychiatrists at the University of Nebraska Medical Center made use of a campuswide interactive television network to link groups of off-site patients with on-site psychiatrists. Finding little difference in therapeutic efficacy or patient satisfaction between "real" and "virtual" consultations, in 1965 they introduced a production telepsychiatry system that linked via microwave the psychiatrists in Omaha with patients at the Norfolk State Mental Hospital, 112 miles distant.[28] Funded by a grant from the National Institutes of Mental Health, the program lasted for six years and logged three hundred hours of clinical telepsychiatry sessions.

Over the next several decades telemedicine programs, typically funded through grants from government agencies, were tested in medical schools, state psychiatric hospitals, municipal airports, jails, and nursing homes as well as on Native American reservations.[29] For the most part, these systems were used to provide high-quality or specialist medical services to rural or otherwise-remote areas. Although a broad definition of telemedicine did not imply the use of any particular communications medium—telephones, fax machines, radio, or even the conventional postal system could all serve as mechanisms for the provision of services—in the United States the focus has historically been on interactive video, which often required participating sites to install fixed, studio-quality video equipment.[30] The high cost of such equipment—as much as $50,000 per installation, even as recently as 1995—limited the applicability of telemedicine, and necessitated a "hub-and-spoke" topology that linked rural or otherwise-remote areas with an urban tertiary care center. Patients were still required to travel to suitably equipped medical centers, and the real-time demands of video-based telemedicine meant that the valuable time of consulting physicians had to be carefully coordinated in advance.

Perhaps because of this bias toward videoconferencing, or because much of the funding for experimental telemedicine came from NASA and the Department of Defense—both agencies having a particular interest in providing medical care to otherwise-inhospitable or inaccessible areas—the focus of telemedicine research has been on the provision of access where it was not available, rather than on cost-effectiveness.[31] In 1997, a Department of Commerce study showed that despite there being more than 150 telemedicine sites in 40 states, only 5,000 patients were being treated remotely using telemedicine technologies.[32] The majority of telemedicine occurred within a limited set of medical problem domains: radiology, cardiology, orthopedics, dermatology, and psychology, in that order.[33] These specialties were either image or interaction oriented, and had traditionally used technology to operate at a distance. Perhaps most important, their remote contributions had been approved for reimbursement by most major third-party benefits providers. In any case, the broader promise of telemedicine for providing more mundane services on a cost-effective basis remained unrealized.

The emergence of the Internet as a more economical architecture for electronic communications promised an opportunity to transform telemedicine from the treatment-option-of-last-resort into the mainstream of contemporary medical practice. Not only was the Internet a lower-cost and more widely available network infrastructure for delivering telemedical services but its "store-and-forward" architecture helped solve the second most pressing problem for telemedicine: namely, the difficulties inherent in coordinating the activities of multiple, busy medical specialists. Instead of requiring these specialists (and their patients) to always gather together for "live" video consultation, physicians could gather lab results, radiological images, patient histories, and other medical records, and forward them to a multimedia consultation "folder" that a specialist could examine at their leisure. The specialist would add their interpretation to the growing folder, and a notification would be sent to the primary physician. Not only was this electronic mediated system of store-and-forward faster and less expensive than shipping physical documents but it did not require either physician to be present on a live television screen.[34]

The potential of the Internet reinvigorated the telemedicine community. As early as 1995 NASA, along with private companies such as Inova Health Systems, began experimenting with pilot programs that used personal computers, inexpensive video cameras (Webcams in today's parlance), and Multicast Backbone, an experimental videoconferencing-oriented subset of the Internet.[35] In 1996, the National Library of Medicine announced the award of nineteen multiyear telemedicine projects intended to serve as models for the following:

- Evaluating the impact of telemedicine on cost, quality, and access to health care
- assessing various approaches to ensuring the confidentiality of health data transmitted via electronic networks
- testing emerging health data standards

These projects moved beyond the traditional tools and problem domains of telemedicine to include information dissemination, chronic disease management and home care services, systems for the management of patient records, and the use of "home-based personal computers connected to the National Information Infrastructure."[36]

The use of a public network to transmit medical information raised questions about security and privacy, however, as well as a potential digital divide in access to Internet-based health care. While in 1997 more than one-third of all American household had home computers, less than 15 percent were connected to the Internet. In addition, access to computers varied greatly by race, gender, and socioeconomic status; fewer than 10 percent of people with an annual income of less than $10,000 had home computers, only 1 to 2 percent of which were networked, while two-thirds of Americans with incomes over $75,000 had home computers, 60 percent of which were networked.[37] Unfortunately, the former were the underserved population most in need of the benefits provided by telemedicine. And even the fortunate few with Internet access suffered from the "last-mile" problem that limited the speeds at which they could connect to network services.

The principal problem confronting telemedicine—in the early years of the Internet as well as today—was not technological or even economic.[38] The problem was not even with patients, or patient access to the Internet. The real problem, again, was the physicians. Outside of a small group of specialists, physicians have proven extremely reluctant to embrace Internet-based telemedicine. In order to fully understand this reluctance and the many reasons for its persistence, it is necessary to first describe the fate of a second great hope of Internet-based medicine: e-mail.

E-mail

The practice of medicine has always been limited by geography—that is to say, by the ability of physicians to have physical access to patients. Traditionally this required the movement of physicians, since travel in the preautomobile era was too stressful or dangerous for patients. Physicians were therefore always generally willing to adopt new technologies of transportation and communication. This became particularly true during the nineteenth century as medicine became increasingly specialized, dependent on complex (and immobile) equipment for diagnosis and therapy, and centralized around the hospital. The growth of cities, the emergence of railroad networks, and the introduction of the telegraph enabled individual physicians to practice medicine over large territories while still maintaining their ties to hospitals and other physician specialists.

As Alissa Spielberg has suggested in her insightful analysis of the use of e-mail in patient-physician communication, the invention of the telephone in 1876 along with its rapid integration into community and regional networks "marked a radical change in patient access to individual physicians."[39] Physicians were early adopters of the new technology. The first telephone exchange connected several Connecticut physicians to a central drugstore. Individual patients used the telephone to contact physicians in emergencies. Increasingly they expected immediate telephone access to their physicians, even in nonemergency situations. An 1878 advertisement from one physician noted that "he may be summoned or consulted through the telephone either by night or day."[40] While this ready access was perhaps a boon to some physicians and their patients, it could also become a burden. Some physicians felt that they were becoming slaves to their anxious patients. They also expressed concern about privacy (a real problem in the age of party lines and operator-assisted calls), reimbursement, a decline in professional standing, and the possibility that the telephone would lead patients to forego necessary physical examinations and even cause themselves harm by "misinterpreting muffled prescriptions." In response to these issues, physicians began using the telephone more strategically, relying on intermediates to screen calls and assess their priority, and declining to provide a diagnosis based solely on phone-based information. Nevertheless, the ability of patients to interact with physicians over the phone from their own homes dramatically altered the nature of the physician-patient relationship, bringing with it increased expectations of access, immediacy, and privacy.

It is in light of this longer historical tradition of patient-physician communication that we can best understand the physician response to the growing popularity of e-mail. Physicians' readiness to embrace the telephone as a tool for communication with patients has not been mirrored in their response to e-mail technology.[41] Given the low cost, simplicity, and ubiquity (particularly among physicians) of e-mail, resistance to it is perhaps the most unexpected and seemingly inexplicable aspect of a larger pattern of resistance to Internet technologies.

At first glance, the use of e-mail for patient-doctor interaction seems to simply represent a subset of the larger topic of telemedicine. And using the broadest definition of telemedicine—again, the use of information and telecommunications to support medicine at a distance—this would indeed be true.[42] But as we have seen, in the United States at least, telemedicine acquired in practice a specific and constrained set of sociotechnical meanings: video rather than text based, dependent on expensive equipment and trained personnel, and as such limited in use to highly paid specialists rather than general practitioners. Electronic mail, on the other hand, was the most widely available, easy to use, and familiar of the new Internet-based technologies. While not every patient had access to the Internet, the vast majority of those who did had access to e-mail, even if they did not have a permanent or broadband connection.

The use of e-mail in medicine was widely lauded in the popular and professional press for having "revolutionary" potential for restructuring traditional relationships in health care.[43] The low cost and ready availability of e-mail promised to open up new channels of communication between all participants in the system: physicians, patients, benefits providers, hospitals, and pharmacies. E-mail would make physicians more accessible, and the intimate nature of the medium would strengthen relationships between them and their patients.[44] At the same time, the asynchronous nature of e-mail would allow physicians to balance their workload and respond more thoughtfully to patient queries. Evidence suggested that patients might be more willing to discuss via e-mail sensitive topics that they might otherwise avoid in person.[45] And by reducing the prevalence of unnecessary office visits, over- and underbooking appointments, and playing phone tag, the use of e-mail offered to reduce direct and overhead costs, personal frustration, and possibly even medical errors. Patients could potentially use e-mail to book appointments, obtain test results, ask minor follow-up questions, request repeat prescriptions, and submit charts for monitoring chronic conditions.

In fact, e-mail offered as much in terms of comfortable continuity as radical change; for patients and physicians already accustomed to communicating via telephone, e-mail seemed to provide incremental improvements to traditional medical care. Patients still had to work within the context of the third-party payer system, and despite having access in theory to a wide range of service providers and consultants, in reality most e-mail-based consultations would still have to be routed through one's primary-care physician. And since these physicians had long been accustomed to interacting with their patients via telephone, it seemed quite natural that they would transition readily to e-mail. Anecdotal evidence suggested that using e-mail did not significantly increase a physician's workload or reduce the number of in-office patient visits.[46] And yet despite all this, physicians have consistently refused to communicate with patients via e-mail.[47] At no point during the past decade has the rate of e-mail interaction between physicians and patients increased beyond 6 percent.[48] This is despite the fact that national surveys show that as many as 90 percent of respondents would "welcome the opportunity to communicate with their doctors by e-mail," with 37 percent indicating that they would be willing to pay for such access.[49]

So why have physicians not yet taken to e-mail? The most frequently cited reasons are concerns about privacy, liability, maintaining standards of care, and being overwhelmed by a deluge of new work.[50] The more cynical answer is that they have not yet figured out how to get paid for it. Reimbursement has been a traditional problem for telemedicine. Prior to the late 1990s, private benefits providers rarely had specific policies about paying for telemedical services. The Medicare program did cover some services that did not require face-to-face contact, such as radiology (which explains in large part radiology's prominent historical role in telemedicine initiatives). Although the 1997 Balanced Budget Act changed the reimbursement situation somewhat, it is still not clear where electronically mediated consultations fit into traditional reimbursement schemes.

Since the economic argument against using e-mail has such a powerful reductionist appeal, it is worth examining in some detail. There is no question that in a health care system dominated by third-party benefits providers, the reimbursement policies of these providers, private or public, have an enormous influence on the practice of medicine.[51] Physicians make decisions about which patients to accept, which tests to order, and which therapies to prescribe based on what insurance providers are willing pay for. And it is not clear that these providers have much incentive to cover telemedical services of any sort, particularly e-mail-based ones that would be widely accessible, broadly applicable, highly likely to be utilized, and difficult to monitor.[52] It is true that in 1999, the provisions of the 1997 Balanced Budget Act that increased coverage for telemedicine under Medicaid went into effect, but these provisions applied only to patients in federally designated rural Health Professional Shortage Areas, and deliberately excluded store-and-forward systems of participation.[53] Only consultations in which a patient was "present" (via videoconferencing) would be eligible. In addition, although under the new system fees were split 75–25 between the consulting and referring physician, the accounting system used by the Health Care Financing Administration (as of 2001, the Center for Medicare and Medicaid) was incapable of handling split payments. Participating physicians would only receive a portion of the reimbursement, but would be liable for tax and auditing purposes for the total fee.[54]

The situation improved somewhat with the passage in late 2000 of the Medicare, Medicaid, and State Childrens' Health Insurance Program Benefits Improvement Act, which became effective October 1, 2001. This act greatly expanded coverage to include all nonmetropolitan statistical areas, and included in its definition of telemedicine not just professional consultations but also office and outpatient visits, medication management, and individual psychotherapy. It also eliminated fee splitting. It did not, however, explicitly include store-and-forward systems such as e-mail, with the exception of two federally funded demonstration programs in Alaska and Hawaii.[55] The act also did not address some of the liability and licensure issues posed by telemedicine.

Although the rules for reimbursement as they apply to e-mail and other telemedical systems are complicated and constantly changing, the lack of clear guidelines does appear to have an inhibiting effect on their use in clinical practice. This is particularly true of e-mail, which often serves as a supplement to more traditional office visits or treatment regimes, as part of what are generally categorized as "case management"

activities. These activities include time spent on pre- or postservice patient management, coordination of care, and follow-up. Unless these case management services
involve (well-documented) high-level medical decision making, they can be difficult
to bill through to third-party payers.[56]

And herein lies the rub: although it appears from the above evidence that it would be
obvious that physicians would avoid using e-mail out of purely economic reasons, the
same basic economic argument could also be used against the use of the telephone, a
technology that physicians do use extensively. This is particular true of pediatrics,
where as much as 20 percent of all clinical care and 80 percent of all after-hours care
occurs over the telephone.[57] And yet pediatricians, as well as most physicians generally, have reconciled themselves to the fact that time spent on the telephone, although
often not directly billable, is an important component of providing high-quality medical care, maintaining patient relationships, and balancing workloads. And as was
mentioned earlier, the available evidence suggests that e-mail interactions do not take
more time or result in fewer office visits than do telephonic consultations. For those
physicians participating in HMOs or other programs whose patients are insured under
capitated contracts, avoiding office visits actually has positive economic benefits.[58]
Other physicians are implementing mandatory "administrative" or "access" fees to
cover unreimbursable services such as e-mail or telephone consultations.[59] The point
again being that relying overmuch on economically determinist explanations can be
misleading. It seems clear that physician aversion to e-mail cannot be explained purely
in terms of reimbursement. Nevertheless, when the lack of direct economic incentives
is combined with other factors, such as legal and moral ambiguity, or concerns about
status and authority, then this aversion becomes much more explicable. When considered within the larger context of practice, patient-physician relationships, and legal
and sociotechnical systems, e-mail represents much more of an extension of older
technologies of communication.

One of the potential advantages of e-mail over other forms of communication is that
as a text-based medium, it is inherently self-documenting; that is, by its very nature
e-mail becomes part of the medical record.[60] This seemingly innocuous feature of
e-mail differentiates it in fundamental ways from purely spoken forms of communication such as a telephone conversation, and has enormous implications for its use
by physicians. E-mail not only enables but in fact demands a more detailed, thoughtful, and guarded response than a telephone call usually permits.[61] This runs counter
to the generally casual conventions of e-mail communication. Whereas for patients
e-mail might appear impermanent and erasable, from the point of view of physicians
they are permanent (often even when deleted) and, more significantly, legally discov-
erable documents.[62] For some physicians the unique legal status of e-mail is a positive
benefit, providing additional documentation that could be used to protect against malpractice suits.[63]

Because e-mail correspondence automatically becomes part of a patient's medical record, it also becomes subject to increasingly stringent requirements for privacy protection. Even prior to the passage of the Health Insurance Portability and Accountability Act (HIPAA) in 1996, which greatly extended the privacy rights of health care consumers, the burden to ensure patient confidentiality has always been borne by the record holder.[64] Under HIPAA, e-mail messages that contain protected health information—both incoming and outgoing—are required to be secured. What exactly constitutes protected health information, or what technologies and procedures are necessary to protect this information, is unclear.[65] The HIPAA provisions for e-mail went into effect in 2003.

Given that the Internet in general and e-mail in particular are notoriously open and insecure, the HIPAA requirements pose challenges for physicians. There are, of course, powerful encryption systems available that could be used to ensure privacy and security. But encryption technologies have not yet been widely integrated into the e-mail practices of the average Internet user. Requiring the use of cumbersome encryption schemes by patients seems to defeat the whole purpose of e-mail. Yet under existing regulations, physicians who use e-mail must take "reasonable precautions" to limit unauthorized access to electronic communications.[66] Needless to say, the phrase "reasonable precautions" is both legally and technically ambiguous, especially as it applies to Internet-based commerce. The burden of deciding which precautions are appropriate as well as the financial burden of implementing and administering them appear to fall on individual practitioners.

Closely related to the problem of privacy is that of authentication. How can a physician be reasonably certain that the person who they are communicating with via e-mail is really who they say they are? How can a patient be sure that the person who responds to their e-mail is really their physician, and not a nurse, a physician's assistant, an office manager, or even a complete stranger? Once again, it is possible to use technologies such as digital signatures to authenticate identity on the Internet. But the infrastructure for managing digital identities is not well developed, and is unfamiliar to most users.[67] And even if online identities could be perfectly managed and authenticated, what would this imply for the work of medical practitioners? Physicians have traditionally managed their workloads using a variety of intermediaries. In the office, the work associated with a patient visit is divided among the front-office staff who triage patients and gather information, the nurses who perform routine evaluations and procedures, and the physician, whose actual interaction with the patient is frequently quite limited. Even telephone contact can be managed using a combination of answering machines or services, front-office staff, and nurses or physician's assistants. The unstructured nature of e-mail (as opposed to, say, a paper-based form) makes automatic routing or processing difficult, and in any case, the expectation is that an e-mail address to a physician will be responded to by that physician and not his or her support staff. The wonderful convenience and directness of e-mail communication does not lend itself well to the traditional division of labor within medical practice.

For all of these reasons and more, the use of e-mail by physicians has not been widely adopted. What at first glance seems to be a straightforward progression from one set of communications technologies to another—this progression has occurred so naturally in other industries that it might reasonably expected to have happened in the health care industry as well—turns out in practice to be much more complicated than most observers anticipated. In many ways, e-mail is a very different technology for physicians than it is for their patients or other professionals. The characteristic features of e-mail—its intimate and casual nature, asynchronous mode, text orientation, and general lack of security and authentication mechanisms—acquire new and professionally significant meanings in the context of medical practice. In terms of physician-patient communication, e-mail is not a generic replacement for face-to-face encounters or even telephonic conversations; for these specific users in this specific context its specific characteristics are tremendously important. Obviously many of these features are incidental, historically contingent, and even socially constructed. One could easily imagine e-mail systems designed with different technological characteristics, operating in different legal and social contexts, and embedded in different sociotechnical and economic systems. But in the current system of clinical practice, third-party reimbursement schemes, privacy and medical malpractice legislation, and health care labor organization, e-mail as it is presently configured is a technology of questionable utility. At the very least, physicians on the whole do not at present find it useful and productive, and unlike many other users of Internet technologies, physicians are a powerful and well-organized group of users.

Medicine and the Web

The focus of this chapter is on the reluctant users of the Internet, and so my discussion of the health care industry has focused on individual physicians and their generally negative, or at least ambivalent, response to Internet-based telemedicine and e-mail consultation. But there are other players in the health care industry, some of whom seem to have adapted readily to the Internet. The WebMD.com health portal, for example, was mentioned earlier as one of the success stories of the Internet-based e-health revolution. In fact, the term e-health was coined in the late 1990s as an umbrella term to describe the broad array of consumer and health care provider activities—including but not limited to telemedicine and e-mail communication—that make use of the Internet, particularly the World Wide Web.[68] In addition to capitalizing on the marketing buzz of e-commerce, e-health represents a shift in emphasis from the patient-physician relationship toward broader, industry-oriented systems and technologies, particularly those that linked business-to-business and business-to-consumer.

In many ways the story of e-health begins and ends with Jim Clark and his Health-eon start-up. Clark, the founder of Silicon Graphics and the cofounder of Netscape, was one of the media darlings of the dot-com boom of the late 1990s. In 1996, after retiring from Netscape and while being treated for a blood disorder at a Silicon Valley hospital, Clark reflected on the inefficiencies inherent in the fragmented, rigidly bureaucratized, and paper-based health care industry. Such a highly inefficient industry—particularly such a highly inefficient, $1.5 trillion industry—seemed the perfect candidate for Internet-based consolidation. As much as one-third of the waste in health care, he believed, could be almost immediately eliminated through the use of electronic clearinghouses.

Clark quickly drew a sketch of the various players in the health care market—patients, physicians, payers, and providers—and added in their midst a "magic diamond," the key intermediary that would link all of these entities together in a seamless web of Internet integration. That same year he founded Healtheon to play the magic diamond role, and predicted that within a few years Healtheon would control $256 billion of the industry. Healtheon went public in 1998—and immediately collapsed as a result of the bursting of the dot-com bubble. The next year it tried again, and this time raised almost $1 billion in capital. In the first quarter of 2000, Healtheon lost $471 million.

Michael Lewis, in his book *The New New Thing: A Silicon Valley Story*, ably tells the story of the rise and fall of Healtheon.[69] Lewis describes it as a tale of technology-driven hubris: a group of entrepreneurs and investors, none of whom knows the slightest thing about the health care industry, take on the largest and most complicated bureaucratic system in the world, and fail miserably in the trying. His story is quite correct, as far as it goes. But Healtheon is unique in that it survived the dot-com explosion. In 1999 it merged with WebMD (founded in 1998) to form Healtheon/WebMD, acquired several of its major competitors, and in 2005 was renamed Emdeon. Those of its competitors that it did not acquire either went bankrupt (for example, DrKoop.com in 2002) or were left by the wayside (in 2006, the WebMD portal attracted nearly three times as many hits as its nearest competitor, Microsoft's MSN Health). Although at this point its business plan no longer resembled that of the original Health-eon, Emdeon had become a $1 billion business, the largest clearinghouse of medical claims, whose customer base included twelve hundred payers, five thousand hospitals, and three hundred thousand physicians.[70]

The success of WebMD.com and other health information portals seems to indicate that at least some elements of the e-health program have succeeded. And indeed, recent surveys show that as many as 80 percent of all Internet users, particularly women, have used the Internet to research health-related topics. Users searched for information on specific diseases (66 percent), diet, nutrition, vitamins, or nutritional supplements (51 percent), health insurance (31 percent), alternative treatments or medicine

(30 percent), environmental health hazards (18 percent), experimental treatments and medicines (23 percent), and Medicare or Medicaid (11 percent), among other topics.[71] Even more surprisingly, 58 percent reported using the Internet preferentially, meaning that they would use it before any other source, and only 35 percent said that they would look to a medical professional first.[72] In addition to doing research, users are participating in health-related support forums, purchasing health equipment online, and ordering pharmaceuticals.[73] Although there are debates within the medical literature about the accuracy and safety of Internet-based information sources, it is clear that the majority of Internet users access health-related information online.[74]

What is not so obvious, however, is whether or not the use of the Internet for health-related research has fundamentally altered the structures or practices of the medical community. For instance, the WebMD health division, which runs the WebMD.com portal, although successful in relative terms, represents only a small fraction ($50.1 million) of the parent company Emdeon's first-quarter revenues ($339.1 million) for 2006. Some of its revenue came from advertising and subscription fees (following the purchase of Medscape in 2001, WebMD Health is now the leading provider of online continuing medical education for physicians). Yet the majority of Emdeon's revenue, however, derives from its electronic claim clearinghouse and practice divisions, both of which are largely based on technologies acquired through purchase and that predate the World Wide Web. Contrary to popular belief (at least among e-health enthusiasts), a large percentage of medical claims—45 percent of all commercial claims, 80 percent of Blue Cross claims, and 97 percent of all hospital claims to Medicare—were already being processed electronically well before the e-health revolution.[75] They are just being processed using proprietary electronic data interchange systems rather than the Internet.

There is, in fact, little incentive for any of the major players in the current system to open up access to outside parties via the Internet. As J. D. Kleinke has suggested, the real reasons that it takes so long for medical claims to be processed has nothing to do with whether or not they are processed electronically, but rather with the network of state and federal regulations, insurance provider regulations, and fraud and abuse protections such as antikickback and Stark self-referral laws that make human intervention into claims processing inevitable. "The obstacles to achieving long-sought integration," observes Kleinke, "have nothing to do with IT and everything to do with the modern health care system."[76] This is perhaps an overly cynical position, but it does highlight the legal and economic dimensions of health care reimbursement rarely taken into account by purely technologically oriented "solutions."

In any case, the increased availability of health information on the Internet has not succeeded in opening up the marketplace for health-related services. Most Americans receive health insurance through their employers, and have limited opportunity to choose between benefits providers. Within a given provider's network of physicians, consumers do have some semblance of choice, although this is constrained by the usual limits of availability, geographic distance, and so on. In this sense the lack of widespread access to telemedicine and e-mail consultations, and the physician's role in limiting such access, contributes directly to the larger stagnation of e-health initiatives. If the value of e-health is dependent on the existence of a robust network of services and information, the failure of individual elements of that network contributes to the failure of the entire network.

Concerns about privacy affect the potential users of e-health networks, albeit for slightly different reasons than those that preoccupy physicians. A recent study of Internet users found that three-quarters are concerned about the privacy of their health-related data; 40 percent will not allow their own doctor online access to their medical records; 25 percent will not purchase or refill prescriptions online; and 17 percent will not even go online to seek health information because of concerns about privacy.[77] A number of highly public instances of health providers—including Global Healthrax, Kaiser Permanente, and the University of Michigan Medical Center—inadvertently revealing sensitive patient data, along with even more numerous security breaches among e-commerce firms more generally, have only heightened fears about potentially lax privacy standards.[78] It is also not yet clear how, or even whether, the rigorous HIPAA standards that apply to physicians and other, more traditional medical providers apply to the intermediaries of the e-health network.

Finally, it is difficult in constructing any sober prognosis for the future of e-health to avoid running up against the brick wall of the third-party payer system. The private third-party benefits providers that pay for most medical care in this country have little incentive to rationalize or speed up claims adjudication. Like most insurance companies, they make money on the "float"—the pool of prepaid premiums that they invest prior to paying back out in claims.[79] In addition, the developers of proprietary information technology systems have no interest in moving toward open Internet standards that might threaten the "lock-in" value of their particular offerings. We have already seen that individual physicians have little financial incentive to participate in e-health networks—and strong legal and ethical arguments against doing so. The only groups with a compelling interest in e-health services are entrepreneurial information technology firms and pharmaceutical companies. In 2005 pharmaceutical industry spending on Internet advertising, directly targeted at the many users searching for information about specific diseases and conditions, rose 30 percent to $53.9 million, while spending on television advertising remained the same.[80] In a health care system whose "fundamental problems" already stem from "irrational consumer behavior, uneven patterns of utilization, and runaway costs," it is not clear what, if anything, this limited constituency for e-health development implies for the future of the Internet and medicine.[81]

The Professor and the Internet

Of all the industries that have been fundamentally changed by the invention of the Internet, nowhere were these changes so early or so readily apparent as in higher education. Universities were early adopters of the Internet, and indeed, many core Internet technologies were developed by, or at least for, academic researchers. Three of the first four original nodes of the ARPANET, one of the precursors to the modern Internet, were located at universities.[82] Many of the key figures driving the development of the ARPANET were university faculty.[83] These faculty, and their graduate students, were instrumental not only in defining how the ARPANET, NSFNET, and Internet would be constructed but also in shaping how it would be used. E-mail, file sharing, and the World Wide Web were all developed and popularized at academic institutions.[84] Until home broadband access became widely available, universities stood at the center of the Internet universe, and trained generations of software developers, entrepreneurs, and users.

Universities continue to serve as important centers of Internet activity. The vast majority of university students own their own computer (85 percent) and regularly go online (74 percent). Almost three-quarters use the Internet more than the library for studying and research. Students use the Internet to meet in virtual study groups (75 percent), socialize (95 percent), download music (60 percent), and entertain themselves (78 percent). Compared to the rest of the population, college students are more likely to use instant message, online chat, and file-sharing software. It is safe to say that students are perhaps the most active and enthusiastic of all users of Internet technologies.[85]

What is true of students is also true of their professors—to a more limited degree. Most college professors are also regular users of computer technology, with a surprising number (90 percent) having been early adopters (since at least 1994).[86] Nearly two-thirds (60 percent) of faculty are online from four to 19 hours per week, and 40 percent twenty or more hours per week.[87] Internet use among faculty varies by age, gender, and discipline, but is generally high and increasing.[88] Faculty use the Internet to communicate with colleagues and students, do research, and to a lesser extent, disseminate knowledge and publish electronically.[89]

Given the widespread adoption of the Internet by both university students and their professors, why would we include professors in our discussion of reluctant users? The answer is that professors, like physicians, have embraced some uses of certain Internet technologies—e-mail, for example—but have rejected others, such as Web-based distance learning, electronic publishing, and course management software. That they have continued to do so in the face of considerable pressure from students, administrators, funding agencies, and legislators suggests that not only are professors selective users of technology but also that they have some power to resist the technological and economic imperatives imposed on them by others. And as in the case of physicians, professors are an intriguing group of reluctant users because, for the most part, they make frequent use of the Internet in their personal and professional lives. The seeming pervasiveness of the Internet in the modern academy, however, conceals those aspects of scholarly production and distribution that have remained fundamentally unchanged by technological innovation.

It is important to note that there is perhaps no occupational group more difficult to generalize about than the university and college professorate. By definition, the members of this group are affiliated with a fairly limited range of institutional forms—either a research university or teaching college, or some combination of both—and presumably most share responsibility for some degree of teaching and research. Yet within the loose confines of academic society, individual disciplines often cultivate very different disciplinary cultures, values and reward systems, tools and methodologies, and increasingly even career paths. It is not always clear, for example, what, if anything, a tenured materials science professor at a major research university shares with a Spanish language instructor at a local community college. To make broad generalizations across institutions and disciplines even more difficult, one of the few academic values that does seem fairly universal is a tendency toward idiosyncrasy and iconoclasm.

Nevertheless, in this section I will seek to describe general patterns in the response of the professorate to the Internet. The focus will be on the faculty of traditional research universities and teaching colleges. Although in recent decades these institutions and their faculties have been challenged by a series of structural and demographic changes in higher education, including the rise of online alternatives, for the time being they remain the standard by which all other forms of higher education and academic teaching are evaluated.

E-mail

Without question, the most widespread use of the Internet by faculty is for e-mail communication. According to a recent study by Steve Jones and Camille Johnson-Yale, nine-tenths of all faculty access e-mail regularly at work, and an almost equal number also access e-mail from home. Many check their e-mail from multiple locations, and as large a percentage of faculty use wireless-enabled laptop computers to access the Internet as does the tech-savvy population in general. Only 14 percent of faculty reported that they check their e-mail only once per day—while almost a third do so almost continuously.[90]

One obvious faculty use of e-mail is to communicate with colleagues. As such, e-mail simply extends the traditional "community of letters" that has defined the academy for centuries. The significance of such social networks (or "invisible colleges," as the historian Derek de Solla Price famously called them) has been one of the grand themes of the sociology of knowledge for decades.[91] In addition, the use of e-mail listservs makes e-mail the ideal tool for disseminating information among widely dispersed professional communities.[92]

E-mail also facilitates communication with students. This is in fact one of the largest uses of e-mail among faculty. Faculty communicate with students to make class announcements (95 percent), arrange appointments (97 percent), handle attendance matters (62 percent), discuss assignments (71 percent), and field complaints about classes and assignments (52 percent).[93] Nearly 90 percent of college students have communicated with their professors via e-mail, and almost half (49 percent) initiate contact with their professors at least every two weeks.[94] Two-thirds of faculty feel that e-mail has improved their communication with students, and nearly four-fifths of all students agree.[95]

To the extent that e-mail does encourage interaction between faculty and students, though, it often does so by reinforcing existing social hierarchies. E-mail communication between faculty and students generally occurs within the context of the extended classroom (in which students are being graded), and faculty frequently have greater expectations of formality and respect than is conventional in e-mail communication.[96] E-mail allows faculty to control the interaction, serving alternatively as a tool for establishing intimacy and a means of maintaining social distance.[97] Students feel that they have access to faculty in new and unprecedented ways; faculty are relieved of the need to meet with students in office hours. In this respect, the particular technological features of e-mail suits the needs of professors quite effectively. Not only is e-mail easy to use and widely available but it is also text based and asynchronous. The former quality means that e-mail fits neatly into the existing work patterns and value systems of academia; the latter means that unlike the telephone or instant messaging, e-mail communication can easily be deferred, ignored, or delegated to others.[98] Faculty have generally not adopted instant messaging or other chat-oriented technologies, which although superficially similar, do not offer the same benefits.

Cybereducation

If e-mail is the success story of the academic Internet, then the wired classroom is its greatest failure. Like the failure of e-health initiatives, that of universities to fully embrace Web-based educational technology represents something of a paradox. Once again, as was true with physicians and online medicine, university professors have played a central role in limiting the adoption of online instructional technology.

Since the advent of the networked computer and the microcomputer, analysts have predicted a computer-based revolution in the classroom. From Christopher Evans's 1979 *The Mighty Micro: The Impact of the Computer Revolution* to Parker Rossman's 1992 *The Emerging Worldwide Electronic University*, computer networks have always been seen as the vanguard of educational reform. The rapid emergence in the mid-1990s of the World Wide Web promised to accelerate and extend the revolutionary reach of computerized learning. The Web also promised to make access to higher education universal, promote improved learning, and control rising costs.[99] In the late 1990s, these costs had risen so dramatically that a National Commission on the Cost of Higher Education was drafted to help "lift the veil of obscurity" that lingered over college education. And Internet technology seemed the ideal answer to the problem. As Frederick Bennett declared in his 1999 *Computers as Tutors: Solving the Crisis in Education*, the use of such technology was imperative: "Schools can use technology more effectively, and for the welfare of students, teachers and the nation, they must do so."[100]

The seemingly sudden emergence of successful and lucrative online-oriented educational institutions such as the University of Phoenix appeared to confirm the early potential of instructional technology. By 1998 the University of Phoenix had become the nation's largest private university, enrolling more than forty-two thousand students at sixty-five locations in twelve states and Puerto Rico.[101] Perhaps even more important, it had become an educational e-commerce phenomenon: within three years of its going public, the stock price of the Apollo Group, which owns the University of Phoenix, split twice and tripled in price.[102] Despite the fact that most learning at the University of Phoenix happens in a traditional classroom setting rather than online, the success of this and other educational technology-related initial public offerings encouraged a rush of online education initiatives, even among Ivy League universities.[103] The most famous of these is MIT's OpenCourseWare initiative, launched in 2001. The goal of OpenCourseWare, according to MIT, is to make its entire curriculum—lecture notes, assignments, discussions, and quizzes—available online.[104]

The political, pedagogical, technological, and economic discussions that roil around the subject of Internet-based learning are too complex to summarize adequately here. As John Seely Brown and Paul Duguid have suggested, visions of the "electronic university" are part of a larger historical conversation about distance learning, the democratizing effects of education, the changing role of the university in industrial and postindustrial society, and the entry of for-profit enterprises into a traditionally nonprofit educational environment.[105] What is crucial for my purposes here is that despite the fairly substantial investment that was made in developing online course materials, the influence of such materials on the pedagogical practices of university professors has been extremely limited. While an increasing number of professors—particularly those in business, engineering, and medical schools—make use of digital images and presentation software in the classroom, there has not been a widespread shift toward using more revolutionary forms of online teaching resources, such as interactive discussion, computer-aided instruction, or even course Web sites.[106] In fact, a growing number of faculty are concerned that their students spend too much time on the Internet and are looking for ways to limit access, at least in the context of the university classroom. These include bans on laptops, and the installation of "kill switches" that allow instructors to close off access to e-mail and the World Wide Web.[107] This curious retreat from the Internet revolution is in part due to concerns about plagiarism and other forms of cheating, but is largely a response to students using the Internet during class to surf the Web, e-mail their friends, and even watch videos.

There are a number of reasons why professors are reluctant to incorporate computers into the classroom. Some are intellectual or pedagogical in nature: professors are skeptical about the reliability of information available on the Web or are concerned about their students becoming overreliant on only digital sources.[108] Still more are wary of being dragged into the business of technical support, or have concerns about spotty or unreliable classroom access to computers, digital projectors, and Internet connections. But the real reason seems to be the lack of professional or financial incentives. For many professors, particularly those at research universities, investments made in teaching can yield negative returns. What is valued is research and publication, not pedagogical innovation. Creating useful online teaching resources is time-consuming and expensive, and the constantly changing nature of the Internet means that such resources must be continually updated.[109] And electronic publication, whether informally on a course Web site or more formally in an online journal, was (and is) in most disciplines not considered "real" publication when it came to tenure or promotion.[110] To put it more succinctly, for most professors the costs of online teaching are high and the rewards are low.[111]

Although in the late 1990s university administrators and venture capitalists still saw great promise in online education, the response among professors remained largely ambivalent. And then in fall 1998, the historian David Noble began circulating the first of a series of articles (later collected into a book, provocatively titled *Digital Diploma Mills: The Automation of Higher Education*).[112] The impetus was an effort at Noble's own institution, York University, that required untenured faculty to put their courses on video, CD-ROM, or the Internet, or lose their jobs. Then, according to Noble, these same faculty were fired and rehired, this time "to teach their own now automated course at a fraction of their former compensation." In the meantime, the York University administration had established, in collaboration with a consortium of private-sector firms, a subsidiary aimed at the commercial development of online education. These actions precipitated a two-month strike by York faculty, who eventually won "direct and unambiguous control over all decisions relating to the automation of instruction." A small and temporary victory, declared Noble, in a struggle whose "lines had already been drawn" between university administrators and "their myriad commercial partners" and those who constituted the "the core relation of education"— namely, students and their professors. York was not the only university mandating course Web sites and commercializing online education; the University of California at Los Angeles had recently launched its own Web-based Instructional Enhancement Initiative, which also required professors to post online course materials.

The push for online education was just another step in the long march toward the commercialization of the university, suggested Noble. The first step had been the development of correspondence schools in the 1920s—an effort also driven by the cynical demands of industry and university administrators. The second was the cultivation, in the late 1970s, of strong ties with commercial corporations—ties aimed at developing an infrastructure for conductive, lucrative, commercially viable research. The final step would be the commodification of instruction into mass-distribution, corporate-friendly electronic courseware. "As in other industries," contended Noble (himself a well-known historian of industrialization), "the technology is being deployed by management primarily to discipline, de-skill, and displace labor." By representing faculty "as incompetent, hide-bound, recalcitrant, inefficient, ineffective, and expensive," administrators promoted instructional technology as a panacea, one allegedly demanded by students, parents, and the public.[113]

Although the harsh tone of Noble's Marxist polemic was off-putting to some readers, his essay clearly touched a nerve within the academic community. In an academic job market that had been constricting for decades, in which tenure-track positions were being increasingly eliminated and replaced by temporary adjunct appointments, the specter of technologically driven unemployment loomed large indeed.[114] Even the true believers in the Internet revolution worried that many cybereducation initiatives were "top-down" efforts driven more by the desire to cut costs than by the real pedagogical potential of the Web.[115] It was difficult to deny that many of the commercially driven initiatives that Noble had identified—including the York and University of California programs, the emergence of educational management organizations, and the formation of virtual universities—were very real phenomenon, and carried with them enormous implications for the work of university professors. These last initiatives, the virtual universities, were consortia of state governments, educational publishers, local employers, and high-tech firms. The largest of these, the Western Governors' Virtual University Project, was quite explicit about its goal of circumventing the traditional university: "The use of interactive technology is causing a fundamental shift away from the physical classroom toward anytime, anywhere learning—the model for post secondary education in the twenty-first century."[116]

This transformation, made possible by "advances in digital technology, coupled with the protection of copyright in cyberspace," would create a glorious future in which "an institution of higher education will become a little like a local television station," as one of the consortium's directors, then Utah governor Mike Leavitt, proudly declared. It was unclear for whom he thought this vision would be appealing.[117]

Noble's essay raised uncomfortable questions about the goals and purposes of Internet-based innovation as it applied in the classroom. Faculty began to wonder, perhaps for the first time, about who owned the rights to their classroom materials. For decades universities had been assuming more and more control over the products of a professor's research, but never before had control over course materials, syllabi, and lecture notes come into question. The legal issues involved are quite complex, and I will not discuss them here.[118] The point is that for the first time, professors were faced with the real possibility that their courses could be taken from them. And in the strange economy of the academic world, courses are one of the few intellectual products that translate directly into income. For the most part academics do not get paid directly from the primary product of their labor, which is scholarly productions (books, articles, and conference presentations). Instead, in a process that Yochai Benkler calls "indirect appropriation," these products are transformed first into a reputation, and ultimately (hopefully) into a tenured university teaching position.[119] The teaching itself is not highly valued, but in a sense, this is what academics actually get paid for. It is certainly their only activity that translates directly into revenue.

In addition to this financial stake in traditional classroom learning, there are also powerful sociological and psychological factors why professors might be loath to cede control of the classroom. As David Jaffee has suggested,

The classroom institution has historically centralized power and influence in the hands of the instructor. When faculty walk into the classroom the learning begins; faculty are the source of knowledge; faculty communicate information and influence the students; faculty determine what will be taught, who will speak and when; faculty determine the correct or incorrect answer; and faculty determine when it is time for students to "stop learning" and leave the classroom.[120]

And not only do faculty often insist on maintaining a dominant, authoritative role, but students frequently agree. One of the common objections to interactive or student-oriented assignments is that students want to learn from the expert, not from each other.[121]

Finally, it is not at all clear that there is much of a pedagogical payoff to using technology in the classroom, or even whether such use results in tangible cost savings.[122] Online-only courses are less expensive to administer, but are a sufficient number of students interested in taking such courses? A recent study showed that only 6 percent of students have taken online courses for college credit, and of those only half (52 percent) thought the online course was worth their time.[123] The University of Phoenix has thrived not because it saves money by offering courses online but because it caters to the largely untapped market of noncollege-age, nontraditional, fully employed workers in search of professional advancement.[124] For the vast majority of more traditional students, college is as much a social as an educational experience, and online universities offer little by way of coming-of-age adventure.[125] As Brown and Duguid have suggested, universities serve valuable social functions that involve more than just the transfer of knowledge.[126] These functions are difficult to re-create in an online environment.

For all of these reasons and more, the promise of the electronic classroom has thus far not been fully realized. Professors continue to successfully resist the use of Internet technologies, particularly the World Wide Web, that do not "count" in the academic credit system or that interfere (such as instant message) with more highly valued activities such as research.[127]

Indispensable Intermediaries

This last section describes a range of industries in which reluctant users have forced businesses to forego the use of the Internet for direct sales to individual consumers. In doing so, these businesses were unable to take advantage of one of the most compelling features of Internet-based e-commerce: disintermediation, or the elimination of intermediaries, distribution channels, and other barriers to "frictionless" commerce. Disintermediation was supposed to doom a host of distributors, retailers, wholesalers, and other intermediaries that stood between manufacturers, service providers, and customers. In some industries this process worked just as expected; witness the decimation of travel agents and independent booksellers described in the previous chapters. But in other key industries that seemed equally suited for direct-to-consumer Internet commerce, the real story is the "disintermediation that wasn't."[128]

Because of the diversity of firms and industries in which indispensable intermediaries have successfully resisted Internet commerce, this section will be broad rather than deep. Unlike the previous two case studies, my focus will be on general themes rather than detailed historical analysis.

Channel Conflict and the Internet

In 1995, the clothing manufacturer Levi Strauss & Company introduced a flashy new e-commerce site that included, among other things, the first use of animated graphics on the Web. In 1998, it began selling more than three thousand products directly to consumers. Two years and $8 million later, the site was quietly closed down. It is now no longer possible to purchase jeans online directly from Levi's.

Just as you cannot purchase your jeans via the Internet directly from Levi's, you also cannot go online to buy insurance from Allstate. Or motorcycle parts from Kawasaki. Or a Toyota Prius directly from Toyota (or for that matter, any automobile from any automobile manufacturer). Depending on where you live, DrugEmporium.com may be forbidden from selling you pharmaceuticals—even in states in which online pharmaceutical sales are perfectly legal. You can purchase tools online from Ryobi, but only at prices that are higher than those at the local Home Depot.[129]

The reason that you cannot purchase any of these products has nothing to do with a lack of technology or capital, high shipping costs, or state or federal regulations. The reason is that each of the products and companies listed above has voluntarily (with the exception of DrugEmporium.com, which was forced by an arbitrator) agreed not to compete over the Internet with its real-world agents, franchisors, and distribution partners.[130]

Why have some businesses turned their backs on the most revolutionary promise of Internet-based commerce: the ability to eliminate intermediaries and interact directly with consumers? In most cases, it is because selling directly to consumers via the Internet causes conflicts with other valuable marketing and distribution channels. This is particularly true of businesses that operate on a franchise model; for the most part local franchisees are contractually guaranteed exclusive access to particular territories. In this case, Internet sales violate these exclusivity agreements, threatening the existence of an existing distribution channel. This is what happened with Drug Emporium, when local franchisees responded by suing the parent company. A similar suit has been filed against the tax services provider H&R Block. In the case of Drug Emporium, an arbitrator ruled in favor of the local franchises, and DrugEmporium.com was barred from selling directly via the Internet in certain markets.

Even when there is no formal contractual relationship barring companies from competing with existing distribution channels there are compelling reasons to avoid channel conflict. Automobile manufacturers, for example, have long cultivated strong relationships with their network of local dealers. These dealers serve several important functions for the manufacturers: they maintain the local inventories that allow consumers to view, test drive, and purchase vehicles; they allow immediate access to financing; and they provide long-term service and support. In short, dealers play an essential role in the marketing and distribution of products, and in fact assume a number of the costs and risks associated with automobile sales. If the manufacturers were to compete too directly with the dealers and put them out of business, they would have to re-create these local networks of sales and support in some other forms. Although consumers might have an interest in purchasing their vehicles directly on the Internet, neither the manufacturers nor the dealers have much incentive to do so. Some dealers are also franchises (and are therefore legally protected from competition), but for the most part such protections are simply not necessary; the business model itself is enough to deter Internet-based encroachment.[131] Auto dealers have resisted any incursion of the Internet into the auto business—even manufacturer-provided information about options and pricing is seen as being detrimental—and thus far have greatly limited its disintermediating potential.

Even for companies with less direct ties to their distribution channels, the reluctance of distribution partners to participate in Internet-based sales and marketing programs can prohibit their implementation. In the case of Levi Strauss, it was conflict with retail chains such as JC Penney and Montgomery Ward that forced it to withdraw from e-commerce. When faced with direct competition from supplier-based Internet sites, retailers respond by withholding information about sales and inventory, refusing to process exchanges, or threatening to remove products from shelves. Home Depot sent the Ryobi Group, which makes the Craftsman line of tools, a letter warning Ryobi not to undercut Home Depot prices on its direct-to-consumer Web site. Tower Records sent a similar message to the vice president of sales at Warner Brothers Records. In both cases, the retail chains were able to use their size and influence to control the ways in which the Internet would affect their businesses. Other, smaller retailers have not always been so successful.

Obviously, there are ways in which businesses can successfully use the Internet and still avoid channel conflict. The point of this section is to suggest that even in the realm of e-commerce, groups of reluctant users—in this case, marketing and distribution partners—have been able to shape the ways in which Internet technologies have been implemented and adopted. Once again, it is the details that matter: certain industries have adapted readily to direct-to-consumer Internet sales, often at the expense of intermediaries. In other cases, these intermediaries have shown themselves to play a much more significant and perhaps indispensable role in the distribution chain.

Real Estate

Residential real estate is another example of an industry that was expected to be entirely transformed by Internet technology.[132] Real estate has traditionally been an industry dominated by intermediaries. In the previous chapter, Jeffrey Yost addressed the impact of the Internet on the real estate industry as a whole; this section will describe the ways in which a particular group of users—real estate brokers—have mediated and influenced this impact.

The average home purchase has historically involved at least sixteen participants: real estate brokers (for both the buyer and seller), mortgage brokers, bank agents, appraisers, inspectors, and title company researchers, among others. The transaction costs associated with such a purchase were significant—more than 6 percent of the total purchase price—most of which went to the real estate agents. If ever there was an industry ripe for disintermediation, it was residential real estate. Through its control of the Multiple Listing Service (MLS) database, however, the National Association of Realtors (NAR) was able to limit competition and maintain high rates of commission for its members. Like their analogues in the travel industry, real estate agents relied on their proprietary access to information to assure their central role in the transaction chain.

By the early 1990s, new technologies and markets were emerging that threatened to eliminate the NAR's monopoly control of the industry. In particular, the increasing availability of Internet-based listings seemed to make agents irrelevant: "If buyers and sellers can sit at their personal computers and gather enough information about each other's offerings—and even make offers—why should they pay an agent?"[133] Industry observers predicted that the Internet would have "profound" implications for the industry, and bring with it reduced commissions, lower incomes, and downsizing.[134] In his 1996 *The Road Ahead*, Bill Gates himself declared that the real estate industry would be "revolutionized" by technology.[135] Internet-induced disintermediation seemed imminent.

By the end of the decade, the Internet had indeed eliminated the real estate agent's monopoly access to information about the housing stock. Sites such as Yahoo! Real Estate, MSN's HomeAdvisor.com, Homeseekers.com, Homestore.com, and even the NAR's own Realtor.com made MLS data widely available, and in addition provided visitors with data about neighborhoods, schools, taxes, and the cost of living as well as tools for financing and insuring a home.[136]

And yet all of this new information made available by the Internet has had remarkably little impact on employment in the real estate industry. Although as Yost has suggested in the previous chapter, the average commission earned by agents has decreased slightly in recent years (from 5.5 to 5.1 percent), both the total number of real estate agents and their median income have increased steadily. Agents still remain central to the purchasing process, with Internet-based "for sale by owner" sales actually decreasing in the years between 1999 and 2001.[137] Despite the widespread availability of technologies that promise what still seem to be gross inefficiencies in the traditional real estate market, real estate truly represents the disintermediation that wasn't.[138]

So how were real estate agents able to avoid the potentially negative effects of the Internet? Unlike university professors and physicians, individual real estate agents have little power in the marketplace. The barriers to entry in real estate are low, and the competition in most local markets is heavy. It would seem that although agents would be reluctant to embrace the Internet, they would have little control over whether or not, or even how, it might eventually be adopted in their industry.

To begin with, real estate is a complex product that does not lend itself well to Internet purchasing.[139] Buyers might use the Internet to gather basic information about the location, lot size, price, and number of rooms, but other forms of information require hands-on, qualitative evaluation that can only be gleaned from an on-site visit. Homes are not like plane tickets, as one insightful observer has noted.[140] Not only are they much more expensive, making the risk associated with an ill-informed purchase much more significant, but each home is also a unique entity. Even in hot markets, most buyers are still unwilling to purchase real estate directly over the Internet. Local agents are still able to provide value by gathering and presenting information that cannot be readily captured on a Web site listing.

Real estate agents have also been able to successfully transform themselves from purely information brokers into providers of "process support."[141] Real estate purchases are intricate legal and financial transactions, and real estate agents have become increasingly active participants in the transaction process.

Some business-to-business aspects are moving toward standards like XML to smooth work flows between, say, mortgage lenders and title insurers, but conceiving of the process as analogous to even car buying ignores the coordination and other roles played by a trusted party in a complicated, emotional, and large purchase.[142]

By guiding buyers and sellers through a difficult process, agents add value beyond their ability to broker information about the housing stock. In this new role agents actually embrace information technology, because in this context it enables new forms of work rather than threatening monopoly control.[143] Although cell phones and digital cameras have thus far been more useful to agents than the Internet, increasingly they are turning to e-mail and the Web for communications and marketing purposes (including the use of personalized information portals and blogs).[144]

Finally, although individual real estate agents rarely have much economic or political power, NAR is well funded and influential. In many states, NAR has effectively limited attempts to create alternative business models in real estate—models that involve more than no-frills "for sale by owner" listings but less than full-service, agent-mediated transactions.[145] As we have seen in Yost's chapter, travel agents were not so effectively organized.

Conclusions

Although the Internet is increasingly well integrated into the modern commercial and communications infrastructure, its effect on American business is not always immediately apparent, at least in certain industries. Rather than dismissing these industries as being exceptional or their participants as backward neo-Luddites, this chapter has attempted to focus on their reluctance as a means of provoking a more nuanced discussion of the role of technological innovation in shaping American business practice. In fact, as we have seen, these reluctant users are perhaps not so much reluctant as selective: like most users, they are simply attempting to limit or influence the way in which technological innovation undesirably affects their work practices, professional authority, or individual autonomy. And so professors embrace e-mail but not instant messaging, and physicians use the World Wide Web but not e-mail. In both cases these are users with influence, and the ability to explicitly and successfully resist change. But as Nelly Oudshoorn and Trevor Pinch have recently suggested, all users matter: collectively considered, users "consume, modify, domesticate, design, reconfigure, and resist" technological innovations.[146] This is particularly true of such an amorphous and protean technology as the Internet. And just as we must be aware that the selective users of the Internet have interests and agendas, we should recognize the same of enthusiasts and advocates. In this way we can better situate the commercial Internet in terms of a larger context of economic transformation, social change, organizational politics, and professional development.

Notes

1. Larry L. Morton and Christopher J. Clovis, "Luddites or Sages? Why Do Some Resist Technology/Technique in Classrooms?" *Simile* 2 (2002).

2. Gregory C. Kunkle, "Technology in the Seamless Web: "Success" and "Failure" in the History of the Electron Microscope," *Technology and Culture* 36, no. 1 (1995): 80–103.

3. Ruth Schwartz Cowan, "The Consumption Junction: A Proposal for Research Strategies in the Sociology of Technology," in *The Social Construction of Technological Systems*, ed. Wiebe E. Bijker, Thomas Parke Hughes, and T. J. Pinch (Cambridge, MA: MIT Press, 1987), 261–280.

4. Nelly Oudshoorn and T. J. Pinch, eds., *How Users Matter: The Co-Construction of Users and Technologies* (Cambridge, MA: MIT Press, 2003).

5. See James Cortada's chapter in this book.

6. Marshall Ruffin, "Telemedicine: Where Is Technology Taking Us?" *Physician Executive* 21, no. 12 (1995): 43.

7. Bonnie Kaplan, "The Computer Prescription: Medical Computing, Public Policy, and Views of History," *Science, Technology, and Human Values* 20, no. 1 (1995): 5–38.

8. Thomas R. Eng and David H. Gustafson, *Wired for Health and Well-Being: The Emergence of Interactive Health Communication* (Washington, DC: U.S. Government Printing Office, 1999).

9. Ibid.

10. Jeff Goldsmith, "How Will the Internet Change Our Health System?" *Health Affairs* 19, no. 1 (2000): 148–156.

11. Michael Lewis, *The New New Thing: A Silicon Valley Story* (New York: W. W. Norton, 2000).

12. Marshall Ruffin, "Why Will the Internet Be Important to Clinicians?" *Physician Executive* 22, no. 10 (1996): 53.

13. Jerome Kassirer, editorial, *New England Journal of Medicine* 332, no. 25 (1995): 52–54.

14. Jerome Kassirer, "Patients, Physicians, and the Internet," *Health Affairs* 19, no. 6 (2000): 115.

15. James Robinson, "Financing the Health Care Internet," *Health Affairs* 19, no. 6 (2000): 72. See also Lewis, *The New New Thing*.

16. Lewis, *The New New Thing*.

17. Arlene Weintraub, "Will WebMD's Healthy Glow Last?" *Business Week Online*, February 23, 2006, 13–13. http://www.businessweek.com/technology/content/feb2006/tc20060223_957756 .htm.

18. Susannah Fox, "Health Information Online," Pew Internet and American Life Project, May 17, 2005, available at http://www.pewinternet.org/PPF/r/156/report_display.asp.

19. Vinod K. Podichetty et al., "Assessment of Internet Use and Effects among Healthcare Professionals: A Cross Sectional Survey," *Postgraduate Medical Journal* 82 (2006): 274–279.

20. ⟨http://www.allbusiness.com/periodicals/articlc/532690-1.html⟩.

21. Tyler Chin, "E-Health Fails to Fulfill Promise," *American Medical News*, August 21, 2000; J. D. Kleinke, "Vaporware.com: The Failed Promise of the Health Care Internet," *Health Affairs* 19, no. 6 (2000): 57–71.

22. Alissa Spielberg, "On Call and Online: Sociohistorical, Legal, and Ethical Implications of E-mail for the Patient-Physician Relationship," *JAMA* 280 (1998): 1353–1359; Podichetty et al., "Assessment of Internet Use and Effects among Healthcare Professionals."

23. ⟨http://www.physiciansweekly.com/pc.asp? issueid=54&questionid=60⟩.

24. Kleinke, "Vaporware.com."

25. Podichetty et al., "Assessment of Internet Use and Effects among Healthcare Professionals."

26. Tom Ferguson, "Digital Doctoring: Opportunities and Challenges in Electronic Patient-Physician Communication," *JAMA* 280 (1998): 1361–1362.

27. Stanley Joel Reiser, "Medical Specialism and the Centralization of Medical Care," in *Medicine and the Reign of Technology* (Cambridge: Cambridge University Press, 1978).

28. Cecil L. Wittson et al., "Two-Way Television in Group Therapy," *Mental Hospitals* 12 (1961): 22–23; L. Baer et al., "Telepsychiatry at Forty: What Have We Learned?" *Harvard Review of Psychiatry* 5 (1997): 7–17.

29. Jim Grigsby and Jay Sanders, "Telemedicine: Where It Is and Where It's Going," *Annals of Internal Medicine* 129, no. 2 (1998): 123–127.

30. Mary Gardiner Jones, "Telemedicine and the National Information Infrastructure: Are the Realities of Health Care Being Ignored?" *Journal of the American Medical Informatics Association* 4, no. 6 (1997): 399–412.

31. NASA and the Department of Defense provided hundreds of millions of dollars in funding for telemedicine.

32. Mickey Kantor and Larry Irving, *Telemedicine Report to the Congress*, U.S. Department of Commerce, January 31, 1997.

33. Ibid.

34. Marshall Ruffin, "Telemedicine: Where Is Technology Taking Us?"

35. Ibid.

36. National Library of Medicine, "Secretary Shalala Announces National Telemedicine Initiative," press release, October 8, 1996, available at ⟨http://www.nlm.nih.gov/archive/20040831/news/press_releases/telemed.html⟩.

37. Jones, "Telemedicine and the National Information Infrastructure."

38. Although the initial cost of video-based telemedicine was quite high, there is evidence that overall cost savings could be achieved.

39. Spielberg, "On Call and Online."

40. Quoted in ibid.

41. Dean Sittig et al., "A Survey of Patient-Provider E-mail Communication: What Do Patients Think?" *International Journal of Medical Informatics* 61, no. 1 (2001): 71–80.

42. Marilyn Field, ed., *Telemedicine: A Guide to Assessing Telecommunications in Health Care* (Washington, DC: National Academy Press, 1996).

43. Jerome Kassirer, "The Next Transformation in the Delivery of Health Care," *New England Journal of Medicine* 332 (1995): 52–54.

44. Spielberg, "On Call and Online."

45. S. G. Millstein and C. E. Irwin Jr., "Acceptability of Computer-Acquired Sexual Histories in Adolescent Girls," *Journal of Pediatrics* 103, no. 5 (1983): 815–819; Stephen M. Borowitz and Jeremy C. Wyatt, "The Origin, Content, and Workload of E-mail Consultations," *JAMA* 280 (1998): 1321–1324.

46. Lee Green, "A Better Way to Keep in Touch with Patients: Electronic Mail," *Medical Economics* 73, no. 20 (1993): 153.

47. Paul Starr, "Smart Technology, Stunted Policy: Developing Health Information Networks," *Health Affairs* 16, no. 3 (1997): 91–105.

48. Sittig et al., "A Survey of Patient-Provider E-mail Communication"; Tom Delbanco and Daniel Z. Sands, "Electrons in Flight: E-mail between Doctors and Patients," *New England Journal of Medicine* 350, no. 17 (2004): 1705–1707.

49. Josip Car and Aziz Sheikh, "Email Consultations in Health Care: 2—Acceptability and Safe Application," *British Medical Journal* (2004): 439–442.

50. Ibid.

51. Dana Puskin, "Telemedicine: Follow the Money," *Online Journal of Issues in Nursing* (September 20, 2001), available at http://www.nursingworld.org/ojin/topic16/tpc16_1.htm.

52. Kleinke, "Vaporware.com."

53. Rural Health Professional Shortage Areas generally suffer from a shortage of primary-care providers.

54. Puskin, "Telemedicine: Follow the Money."

55. Ibid.

56. Sanford Melzer and Steven Poole, "Reimbursement for Telephone Care," *Pediatrics* 109 (2002): 290–293.

57. Ibid.

58. Green, "A Better Way to Keep in Touch with Patients."

59. Anonymous, "Access Fees: Worth the Risk? What to Charge?" *Medical Economics* 81, no. 14 (2004): 50.

60. Delbanco and Sands, "Electrons in Flight."

61. Spielberg, "On Call and Online."

62. Ibid.

63. Ibid. For others, this additional degree of accountability and potential exposure makes e-mail a risky proposition. See also Doreen Mangan, "Save Time and Patients with E-mail," *Medical Economics* 76, no. 13 (1999): 155–159.

64. Lawrence Gostin, "Health Information Privacy," *Cornell Law Journal* 80 (1995): 451–527.

65. Janlori Goldman and Zoe Hudson, "Virtually Exposed: Privacy and E-Health," *Health Affairs* 19, no. 6 (2000): 140.

66. Spielberg, "On Call and Online."

67. Lawrence Lessig, *Code and Other Laws of Cyberspace* (New York: Basic Books, 1999).

68. Vincenzo Della Mea, "What Is E-Health (2): The Death of Telemedicine," *Journal of Medical Research* 3(2) (2001): c22.

69. Lewis, *The New New Thing*.

70. Karen Southwick, "WebMD May be Due for a Checkup." CNet News.com. (2004), available at http://news.com.com/WebMD+may+be+due+for+a+checkup/2100-1011_35198935.html.

71. Fox, "Health Information Online."

72. Joe Flower, "American Health Care, Internet Style," *Physician Executive* 30, no. 3 (2004): 69–71.

73. Ibid.

74. Alejandro Jadad and Anna Gagliardi, "Rating Health Information on the Internet: Navigating to Knowledge or to Babel?" *JAMA* 279, no. 8 (1998): 611–614; Anthony Crocco et al., "Analysis of Cases of Harm Associated with Use of Health Information on the Internet," *JAMA* 287, no. 21 (2002): 2869–2871.

75. Kleinke, "Vaporware.com."

76. Ibid.

77. Goldman and Hudson, "Virtually Exposed."

78. Ibid.

79. Kleinke, "Vaporware.com."

80. Weintraub, "Will WebMD's Healthy Glow Last?"

81. Kleinke, "Vaporware.com."

82. The University of California at Los Angeles, the University of California at Santa Barbara, and the University of Utah. The fourth was the Stanford Research Institute, which although not itself a university, is an academic research institution in its own right.

83. J. C. R. Licklider and Lawrence Roberts were on the faculty at MIT, Robert Taylor was at the University of Utah, and Leonard Kleinrock taught at both MIT and Stanford.

84. Janet Abbate, *Inventing the Internet* (Cambridge, MA: MIT Press, 1999).

85. Steve Jones, *The Internet Goes to College: How Students Are Living in the Future with Today's Technology*. Pew Internet and American Life Report, September 15, 2002, available at ⟨http://www.pewinternet.org/⟩.

86. Even more surprising, a third report having used the Internet since the 1980s. See Steve Jones and Camille Johnson-Yale, "Professors Online: The Internet's Impact on College Faculty," *First Monday* 10, no. 9 (2005), available at ⟨http://www.firstmonday.org/issues/issue10_9/jones/index.html⟩.

87. Ibid.

88. Srinivasan Ragothaman and Diane Hoadley, "Integrating the Internet and the World Wide Web into the Business Classroom: A Synthesis" *Journal of Education for Business* 72, no. 4 (1997): 213.

89. Paul David Henry, "Scholarly Use of the Internet by Faculty Members: Factors and Outcomes of Change," *Journal of Research on Technology in Education* 35, no. 1 (2002): 49.

90. Jones and Johnson-Yale, "Professors Online." Only six of the twenty-three hundred respondents in their survey reported checking e-mail only a few times per week.

91. Derek J. de Solla Price, *Little Science, Big Science . . . and Beyond* (New York: Columbia University Press, 1986).

92. Jeanne Pickering and John King, "Hardwiring Weak Ties: Interorganizational Computer-Mediated Communication, Occupational Communities, and Organizational Change," *Organization Science* 6, no. 4 (1995): 479–486.

93. Jones and Johnson-Yale, "Professors Online."

94. Jones, *The Internet Goes to College*.

95. Ibid.

96. Jonathan Glater, "To: Professor@University.Edu Subject: Why It's All about Me," *New York Times*, February 21, 2006, A14.

97. Many students report that they feel more comfortable asking questions or discussing course material via e-mail than in person; Jones, *The Internet Goes to College*. On the other hand, an increasing number of professors are finding that this approachability comes at a cost: students are also more likely to use e-mail to provide excuses for absences, request extensions, complain about grades, and even harass or threaten their professors; ibid.

98. Kathryn Wymer, "The Professor as Instant Messenger," *Chronicle of Higher Education* 52, no. 23 (2006): C2.

99. Ronald Owston, "The World Wide Web: A Technology to Enhance Teaching and Learning?" *Educational Researcher* 26, no. 2 (1997): 27–33.

100. Frederick Bennett, *Computers as Tutors: Solving the Crisis in Education* (Sarasota, FL: Faben, 1999).

101. ⟨http://www.wweek.com/html/mcniche1032598.html⟩.

102. ⟨http://www.fastcompany.com/magazine/68/sperling.html⟩. As of 2005, the University of Phoenix was the largest private university in the United States, with two hundred thousand adult students enrolled, seventeen thousand faculty on staff, and $2.25 billion in revenues.

103. ⟨http://news.com.com/2100-1017-269067.html⟩.

104. ⟨http://chronicle.com/free/v49/i15/15a03101.htm⟩.

105. John Seely Brown and Paul Duguid, *The Social Life of Information* (Cambridge, MA: Harvard Business School Press, 2000).

106. Tina Kelley, "Virtual-Classes Trend Alarms Professors," *New York Times*, June 18, 1998.

107. See John Schwartz, "Professors Vie with Web for Class's Attention," *New York Times*, Junuary 2, 2003, A1; Maia Ridberg, "Professors Want Their Classes 'Unwired,'" *Christian Science Monitor*, May 4, 2006, 16.

108. Jeffrey R. Young, "Professors Give Mixed Reviews of Internet's Educational Impact," *Chronicle of Higher Education* 51 (2005): A32.

109. Andrew Trotter, "Too Often, Educators' Online Links Lead to Nowhere," *Education Week* 22, no. 14 (2002): 1.

110. Lisa Guernsey, "Scholars Who Work with Technology Fear They Suffer in Tenure Reviews," *Chronicle of Higher Education* 43, no. 39 (1997): A21.

111. Jon Marcus, "Online Teaching's Costs Are 'High, Rewards Low,'" *Times Higher Education Supplement*, January 21, 2000, 14.

112. David Noble, "Digital Diploma Mills: The Automation of Higher Education," First Monday 3 (1) January 5, 1998, available at ⟨http://www.firstmonday.org/issues/issue3_1/noble/⟩.

113. Ibid.

114. Benjamin Johnson, Patrick Kavanagh, and Kevin Mattson, eds., *Steal This University: The Rise of the Corporate University and the Academic Labor Movement* (New York: Routledge, 2003).

115. Frank White, "Digital Diploma Mills: A Dissenting Voice," *First Monday* 4, no. 7 (1998), available at ⟨http://www.firstmonday.org/issues/issue4_7/white/index.html⟩.

116. Jonathan Newcomb, CEO of Simon & Schuster, cited in Noble, *Diploma Mills*.

117. ⟨http://www.firstmonday.dk/issues/issue3_1/noble/⟩.

118. Matthew D. Bunker, "Intellectuals' Property: Universities, Professors, and the Problem of Copyright in the Internet Age," *Journalism and Mass Communication Quarterly* 78 (2001): 675–687.

119. Yochai Benkler, "Coase's Penguin, or, Linux and the Nature of the Firm," *Yale Law Journal* 112 (2002): 369–446.

120. David Jaffee, "Institutionalized Resistance to Asynchronous Learning Networks," *Journal of Asynchronous Learning* 2, no. 2 (1998).

121. Clyde Freeman Herreid, "Why Isn't Cooperative Learning Used to Teach Science?" *Bio-Science*, Vol. 48, No. 7. (Jul., 1998), pp. 553–559.

122. Ronald Owston, "The World Wide Web: A Technology to Enhance Teaching and Learning?"

123. Mary Madden and Steve Jones, "The Internet Goes to College: How Students are Living in the Future with Today's Technology," Pew Internet and American Life, 15 September 2002, available at ⟨http://www.pewinternet.org/PPF/r/71/report_display.asp⟩.

124. Brown and Duguid, *The Social Life of Information*.

125. Gary Wyatt, "Satisfaction, Academic Rigor, and Interaction: Perceptions of Online Instruction," *Education* 125, no. 3 (2005): 460–468.

126. Brown and Duguid, *The Social Life of Information*.

127. Dan Carnevale, "Never, Ever Out of Touch," *Chronicle of Higher Education* 50, no. 41 (2004): A29–A30.

128. Guidewire Group, "The Disintermediation That Wasn't," available at ⟨http://www.guidewiregroup.com/archives/2006/01/the_disintermed.html⟩.

129. Luisa Kroll, "Denim Disaster," *Forbes* 164, no. 13 (1999): 181.

130. Stuart Gittleman, "Franchisees Win Landmark Internet Arbitration Ruling" American Lawyer Media, September 12, 2000, available at ⟨http://www.franatty.cnc.net/art32.htm⟩.

131. Diane Katz and Henry Payne, "Traffic Jam: Auto Dealers Use Government to Build Internet Roadblocks," *Reasononline* (2000), available at ⟨http://www.reason.com/news/show/27766.html⟩.

132. John S. Baen and Randall S. Guttery, "The Coming Downsizing of Real Estate: Implications of Technology," *Journal of Real Estate Portfolio Management* 3, no. 1 (1997): 1–18.

133. M. Rosen, "Virtual Reality—Real Estate Agents Face Extinction in an Information Rich Century," *Dallas Observer*, January 18, 1996, 6.

134. Baen and Guttery, "The Coming Downsizing of Real Estate."

135. Bill Gates, *The Road Ahead* (New York, Penguin, 1996), 196.

136. Waleed A. Muhanna and James R. Wolf, "The Impact of E-Commerce on the Real Estate Industry: Baen and Guttery Revisited," *Journal of Real Estate Portfolio Management* 8, no. 2 (2002): 141.

137. Ibid.

138. Guidewire Group, "The Disintermediation That Wasn't."

139. Ibid.

140. Ibid.

141. Kevin Crowston et al., "How Do Information and Communication Technologies Reshape Work? Evidence from the Residential Real Estate Industry" (paper presented at the International Conference on Information Systems, 2000).

142. Guidewire Group, "The Disintermediation That Wasn't."

143. Ibid.

144. Kimberly Blanton, "Realtors Get Their Hands on Technology," *Boston Globe*, November 28, 2005.

145. Guidewire Group, "The Disintermediation That Wasn't."

146. Oudshoorn and Pinch, *How Users Matter*.

V New Technology: Old and New Business Uses

12   New Wine in Old and New Bottles: Patterns and Effects of the Internet on Companies

James W. Cortada

Discussions about the role of the Internet in businesses are often loud, passionate, and indeed strident. Not since the deployment of computers across the U.S. economy in the 1950s and 1960s, or even personal computers in the 1980s, has there been so much attention and gushing hubris devoted to an information technology as Americans have experienced with the Internet. A search on Google for businesses and the Internet, or any other combination of similar terms, can generate anywhere from a hundred thousand "hits" to well over a million. The word Internet alone spins off nearly 2.5 billion citations. Every major industry in the United States that had a trade magazine, association, or annual conference discussed the role of the Internet in business, invariably in laudatory language, promoting the benefits of using the Internet to improve productivity, reach new customers, and redefine one's "business model."

Historical Significance of the Internet in the Private Sector

Like those who came before them in the 1950s, 1960s, and 1980s, management often felt initially an enormous and later growing pressure to use this technology so as to be seen as modern and progressive. As the mantra had it, one ignored adoption of the Internet at their peril, risking ruination as the economy evolved into an information-based postindustrial market that was both global and fast changing. One had to use the Internet as a strategic component of their business plan.[1] This gospel almost drowned out discussions about more widely used forms of information technology, such as the nearly Victorian-sounding "legacy systems," which still made up the majority of information technology in use around the world. So, one might ask, what effect did all this attention to the Internet have on the way companies worked? More specifically, did the nature of business change as a result of using the Internet?

The reality of Internet use proved quite different from the hype. Thousands of companies of all sizes in every American industry responded in essentially one of two ways to the availability of the Internet, although increasingly in both ways. First and most important, they slowly integrated the Internet into their existing ways of doing business, in many cases replacing proprietary networks, or using the Net in parallel with these networks, such as intranets and the earlier Electronic Data Interchange (EDI), the latter used to communicate among manufacturing plants, suppliers, and large retailers. Second, as management came to understand the nature of the Internet and what new tasks could be done with it, they modified work practices, business strategies, and their offerings (products and services). An example of the first process was making it possible for customers to acquire information about products, and later for employees to submit their business expenses for reimbursements. An instance of the second path transforming internal business operations involved IBM. In the 1980s and 1990s, the firm moved a majority of its manufacturing either to multiple new sites around the world or outsourced them to other subcontractors, such as for disk drives and personal computers. In addition to having other firms do much of its core work, IBM had to develop a highly efficient set of processes for integrating demand, back orders, components, and manufacturing plans, and all that required changing fundamentally every prior version of these work streams, linking them into a combination of private networks and intranet-based systems. IBM saved billions of dollars in operating costs while also reducing the percentage of employees involved in manufacturing and distributing products. By the end of the century, for example, one could order small mainframes (now called servers) over the Internet, much as people had been ordering personal computers by the mid-1990s. In each case, much of the order processing had been shifted to software accessed over the Internet, with an ever-declining involvement of direct sales or call center personnel. Package delivery services made parcel tracking an Internet-based task in which customers accessed the Internet to find out the status of their package as it moved through the delivery process, without having to talk to an employee. They used the same Internet-based system to update the status of shipments either personally by scanning boxes or through automated scanning devices. Both IBM's approach to redesigning work flows and that of package delivery services spread across all large manufacturing, wholesale, and retail industries. Now this approach is appearing in service industries doing work for clients, in which tasks or whole processes are outsourced over the Internet and tracked using software through its various stages. Tax preparation firms, insurance companies, and in some European countries, government social services are now handled this way.

A less dramatic, yet nonetheless profound example involved brokerage firms allowing clients to place orders directly for stocks over the Internet without engaging the services of a human broker. The range of possible uses of the Internet expanded all through the 1990s and the early 2000s as the technology evolved and became capable of performing new functions, and as firms learned how to leverage them.

The private sector traveled down both paths almost simultaneously, with the experience of one informing the other. The journey was iterative, incremental, and often slow, standing in sharp contrast to the declarations of techno-enthusiasts. Myriad surveys tracked the extent of deployment of the Internet across many industries, all announcing the speed of change as the nation's economy was supposedly transforming rapidly into a multitrillion dollar e-economy. While the data on deployment varied, by the early years of the new century, deployment was extensive.[2] The Internet had become a key component of all medium and large businesses—that is to say, companies with over a hundred employees. At a minimum, almost every smaller enterprise had a Web site by the early 2000s, but that observation is only slightly more exciting than saying every business was listed in the telephone Yellow Pages. No firm operating in manufacturing and no retailer of any size did its work without using the Internet in the United States. Furthermore, nearly 70 percent of all residents of the United States now used the Internet, with over a third for the purpose of conducting business of one sort or another.[3]

Thus, a central question is not so much how many enterprises used the Internet—the historical record clearly shows that the proverbial "everyone" had a presence on the Net by the early 2000s—but rather what effects did this technology have on the internal operations of companies?[4] It is a new question that we could not even have asked at the dawn of the twenty-first century. Indeed, one might even challenge our ability to begin addressing it in 2007–2008. Yet sufficient historical perspective exists to describe the role of the Internet, the central task of this book. This chapter focuses largely on the period of the Internet's early adoption by the private sector (the mid-1990s forward).

Stages of Adoption of the Internet

There are discernible stages (or phases) in the adoption of the Internet by the private sector useful in cataloging both uses of this new tool and the effects such changed ways of working had on firms, industries, and the economy at large. To start that process of identifying the stages of adoption, we first need to acknowledge that many industries had become extensive users of telecommunications with which to conduct their business long before the Net became an attractive technology for companies to use. That prior experience informed management on how to respond to the Internet. For instance, since the 1970s General Motor's proprietary EDI network linked thousands of suppliers to its design, manufacturing, and distribution processes. This use of combined telecommunications and databases provided General Motors with new levels of efficiency and coordination over prior practices, so much so that all other automotive manufacturers in Western Europe and Japan either followed suit or had concurrently developed similar uses of information technology. Every bank in the United States had used telephone and proprietary networks to clear checks before World War II, driven largely by federal regulators. To be sure, there were companies and even whole industries that were relatively ignorant of networking because their businesses did not require such use. Small service enterprises are the obvious example; barbers and bakers did not need EDI or other proprietary networks. Nor did moviemakers or film distributors.

Where networks had been used for many years, they had already transformed how firms functioned. Large petroleum companies had shrunk their workforces used to monitor pipelines; all major manufacturers had created ecosystems of suppliers and distributors. Large retailers had shifted the balance of power of who decided what goods should be sold from manufacturers to retailers, thanks to a combination of point-of-sale data and just-in-time replenishment of inventories long before deployment of the Internet; Wal-Mart became the operative model of this trend at work as the largest retailer in the world.[5]

So the arrival of the Internet did not immediately lead to a rapid exploitation of this relatively new form of telecommunications.[6] Although the Internet had been around for a quarter century before businesses began to use it, it was not until the Internet became both easier to access (thanks to browsers arriving in the mid-1990s) and increasing numbers of consumers began using it (thanks also to declining communications costs and cheap personal computers) that it began to make sense to look at this new technology in a serious way. The availability of broadband communications made the whole process of using the Internet more convenient (by increasing the response time), and made it possible to transmit more varied content and functions (such as interactive transactions or video). Once these conditions emerged, management at various levels started learning about the technology; before then they understood little, if anything, about the Internet. To be more precise, they only began to learn about its evolving capabilities and determine how best to leverage the new technology in the 1990s.[7]

Existing enterprises went through stages of deployment of the Internet, most of which had minimal impact on their work. Simply put, between 1994 and 1998, managers went through a period of discovery, creating their first Web sites, populating them with information about their firms, products, and services, and providing contact information. The initial information about products, for example, mirrored the "look and feel" of printed brochures of the 1980s and the early 1990s. As companies learned how to present materials on screens and to use hypertext capabilities in an effective, user-friendly way, the presentation of material evolved away from the wordy, highly compact look of paper-based publications and more toward the format evident today of brief statements, the use of bullets, and links to additional levels of details.

All major industries exhibited this pattern of adoption. Increasingly, customers would go to these sites for information about products as well as to others devoted solely to presenting comparative data about classes of products. *Consumer Reports* ranked products; Progressive Insurance provided comparisons for automotive insurance, while others did the same for loans. Customers learned quickly to use such sites, along with those of product and service providers. This happened particularly early in the automotive industry, for instance, where customers armed themselves with information about new cars. By the end of the century, consumers were so used to collecting this information before visiting a dealership that they had fundamentally altered how negotiations took place over the price of a new car by eliminating the asymmetrical condition that had always better armed the dealer with more data than the customer. Bargaining quickly devolved to price; functions, features, and fashion—the historical points of debate in such negotiations and marketing—declined in importance in the negotiations since they had been resolved to the satisfaction of consumers before entering the dealership.[8] By the end of the 1990s, similar experiences began affecting firms in other manufacturing industries that sold primarily to consumers and more slowly to transactions between companies.[9] This was also the case in retail industries, hotels, and all manner of transportation; increasingly in banking as well as life and automotive insurance; but less so with more complex offerings, such as health insurance or medical services. The more individualized a service or product became (such as the construction of an addition to one's home) the less such online searches made sense, since such activities required a high degree of customization along with interpersonal dialogue and negotiations.

After companies completed passage through their initial phase of establishing Web sites, they next worked their way through a second stage, beginning in approximately 1998–1999 (the start date varied by industry and company) as the quality of software for securing transactions over the Internet improved. That turn of events made it possible for firms to begin offering customers the capability of conducting business over the Internet, such as viewing a catalog, then placing an order for a product, and paying for it with a credit card—in effect, now doing business anytime, day or night. Catalog vendors were some of the earliest to add this capability alongside their preexisting toll-free telephone call centers, such as Land's End and L. L. Bean, as discussed more fully by Jeffrey Yost in an earlier chapter. Brick-and-mortar retailers followed suit by the early 2000s. Companies added the Internet channel for distribution to their existing ones for selling consumer goods, such as stores and mail orders. In short, they did not use the Internet as a replacement for other channels. Tables 12.1 and 12.2 catalog various sample uses of the Internet by firms in key industries during the first and second waves of adoption. Observe that deployment spread widely with the result that as use increased, companies could not ignore that trend, and needed to take action of some sort either to slow or exploit it. In other words, as time passed inaction with respect to the Internet was not a strategic option.

Online sales grew slowly late in the century and during the early years of the new decade, with the most optimistic data suggesting that as much as 3 percent of all sales went through this channel by 2005.[10] But the volumes of business, the percentage of participation, and so forth varied by industry. It is still not clear whether sales through the Internet reflected incremental (new) business or transactions that would otherwise

Table 12.1

Sample uses of the Internet, 1994–1998

Purchaser inquiries	Automotive and manufacturing industries
Product knowledge	Catalog sales, such as Land's End
Online auctions	eBay
e-mail	Manufacturing industries
Global product design	Semiconductor manufacturers
Brochures and contact information	Insurance, banks, retailers, automotive
Schedules	Passenger airlines, railroads, bus companies
Book sales	Amazon.com, Barnesandnoble.com

Table 12.2

Sample uses of the Internet, 1999–2005

Electronic trading markets	MetalSite, e-Steel
Virtual product design	Automotive industry
E-commerce	Retailers of all kinds
Relations with partners	Chemical, pharmaceutical firms
Product configuration	Automotive firms
Software sales	IBM, Microsoft, video game producers
Account management	Brokerage
Internet banking	Retail banks
Fleet scheduling	Trucking companies
Rentals	Automotive and equipment rental firms
Dating	eHarmony.com
Supply chain management	Wal-Mart, large retailers, wholesalers
Music retail sales	Amazon.com, iTunes, recorded music providers
Internet subscriptions	Newspapers, magazines
Licenses	State governments
Tax filing	Federal and state governments

have gone through preexisting catalog, retail, and 800-number channels. Extant evidence indicates that Internet sales were done at the expense of others that would have gone through preexisting channels, with the result, however, that the cost of sales per transaction declined. This occurred because Internet-based transactions often proved less expensive to conduct since they rarely involved employee involvement.[11] Table 12.3 provides evidence of how much business was being done over the Internet.

These patterns are highly generalized. Companies and whole industries arrived at, went through, and entered a new phase at different speeds, as they learned from competitors, responded to consumer requests for Internet-based services, and came to understand the potential effects of the Internet on their business. Companies that had

Table 12.3
Value of shipments of goods and services, total and e-commerce, 1999–2003 ($ billions)

Year	Total manufacturing	E-commerce	% e-commerce
1999	$4.03	$0.730	18
2000	4.21	0.756	18
2001	3.97	0.724	18
2002	3.98	0.752	19
2003	3.98	0.843	21
	Total wholesale trade	E-commerce	% e-commerce
1998	2.44	0.188	8
1999	2.61	0.223	9
2000	2.80	0.269	10
2001	2.78	0.310	11
2002	2.82	0.343	12
2003	2.95	0.387	13
	Total retail sales	E-commerce	% e-commerce
1998	2.57	0.005	Miniscule
1999	2.80	0.014	Miniscule
2000	2.98	0.027	1
2001	3.07	0.034	1
2002	3.14	0.048	1
2003	3.30	0.056	2

Note: All figures rounded up to clarify and simplify the presentation of trends.
Source: U.S. Bureau of the Census, historical tables, ⟨http://www.census.gov/eos/www/historical/20003ht.pdf⟩.

preexisting networks tended to continue using them, especially suppliers and others providing them with services. Such firms normally used the Internet to reach out to potentially new customers to augment work already done by existing channels of distribution of goods and services.

Then there were the exceptions. For example, secondhand book dealers began posting their inventories on the Internet in the late 1990s, thereby reaching so many new customers that hundreds of retail dealers simply closed their physical shops, stopped publishing paper catalogs, and now sold their wares online. They did this either through their own Web sites, or by way of other consortia and vendors on the Net, such as Alibris, Powell's, and Amazon.[12] But as a general statement, from the early 1990s through the mid-2000s, such uses of the Internet had not yet profoundly affected the work of most firms in the manufacturing and retailing industries. To a lesser degree, the same held true in two of the three major financial industries (banking and insurance, but not brokerage), in some media industries (book publishing, radio, and television), yet clearly proved the opposite in others (most notably recorded music and just now starting in movies). The one general exception is that the cost of interacting with customers on myriad requests for information declined when companies used the Internet, regardless of the industry. Yet the cost of acquiring and holding customers remained high for two reasons: the normal costs of marketing, advertising, and processing customers had not changed, and second, an increased number of switches in vendors became easier to do thanks to computing and the Internet. Examples include rapid and easy changes in automobile and life insurance, all beginning to increase the churn in a firm's pool of customers. This pattern was also evident in the cell phone business, although it also proved expensive for users if they had to break a preexisting service contract.

Proposing general stages for any historical phenomenon has its limitations, of course, not the least of which is the fact that they are never absolutely comprehensive or normative. Nonetheless, defined patterns are helpful sufficiently in making sense of otherwise-disparate and numerous events that the limitations are worth enduring. Thus, while the periodization (stages) described above is generally true, one can overstate its reality. The process of moving from one state to another varied in speed and extent from one firm or industry to another. Surveys on the adoption of the Internet conducted by many industry associations and their trade magazines make these caveats explicit, albeit also supporting the generalizations made about the existence of these stages. To be sure, the trajectories and timetables varied, but enough proved similar to allow us to start generalizing about these stages. Finally, we need to recognize that not all firms went through all these stages; many never really made it past the first one even after years of using the Internet. They established a presence on the Web, offered information about their products and services, and usually also contact data (such as addresses and telephone numbers), but had no e-mail capability, access to databases, or the ability to conduct online transactions.

All that said, can we begin to identify the structural pattern in specific industries that made it possible for them to embrace the Internet quicker than others, much like Alfred Chandler Jr. had identified high-energy, continuous-process industries as the sources of multidivisional forms of corporations? Several features of an industry made the rapid use of the Internet possible, once awareness of the existence of the Net existed in an industry's firms and after technological constraints were overcome, such as the early lack of software to secure data from improper use—say, with credit card numbers. Having customers who were inclined to use technology, and indeed had the prerequisite hardware and software, were all essential elements. The brokerage industry comes to mind both for its employees and the day traders. For another, large enterprises with many locations and employees dispersed physically provided incentives for firms to use all manner of networks to stay in touch and conduct business. A third structural pattern evident in the early and substantive adopters of the Internet involved high-tech companies that could use information technology to move work and transactions back and forth, such as software developers, chip designers and manufacturers, and video game writers. They all worked with digital content to a large extent. One could argue that banks did too, but they did not meet the criteria of having large swaths of customers with the prerequisite information technology infrastructures in place (e.g., personal computers at home), secure software to protect files, or even for many years the regulatory permission to embrace fully the Net. Finally, as an unintended yet wonderful example of path dependency at work, those same firms and industries that Chandler pointed out as sources of the modern corporation, often were the early and rapid adopters of the Internet because of their prior knowledge and experience of telecommunications, possession of information technology skilled staffs, and the need to work on a continuous basis.

Role of Business-to-Business Commerce on the Internet

Business-to-business (B2B) commerce represented a class of work done over the Internet different from that done either internally within a firm or externally directly with consumers. Both in the early pages of this chapter and throughout this book, B2B is discussed often in confluence with business-to-customer, or even business-to-government, and even one government agency to another. Often they all merged and mixed, but there are some distinctions. In the B2B arena, companies either were selling to other firms, or buying products and services. For example, OfficeMax would sign a contract with a company to sell it office supplies at a discount off retail, provided that the firm doing the buying placed its orders over the Internet, thereby saving OfficeMax the labor cost of dealing with an order. The Internet could pull in the order into the supplying firm's order-processing software system, and populate other software-based processes, such as back orders, manufacturing, delivery, and billing, all with minimal or no use of expensive labor. Because many products could be sold that way by one company to another, the purchasing processes of customers and the order acceptance processes of providers could be highly automated or shared. In the chemical industry, one could bid, buy, and ship (or receive) bulk materials this way; the same was true in the petroleum industry, and for that matter, for most products and supplies used by corporations and small firms by the early years of the new century. Contracted-for services that involved the use of information also proved popular, such as accessing information offered by Gartner, Forrester, and other private providers of data—all examples of B2B.

In the case of B2B, historic precedence proved important. Prior to the availability of the Internet, companies had electronically done business with each other. They had thus worked out the necessary contractual terms and conditions, and had figured out the technical operational requirements so as to have one firm's software "talk" to that of another company's, and to integrate internal operating processes. Normally, however, the number of individuals with access to such systems in the 1970s and 1980s was restricted to those who had direct expertise in complex operations, such as scheduling the delivery of components for the manufacture of industrial equipment or, as at General Electric, consumer products. Much of how B2B functioned in the pre-Internet era remained the same after the arrival of the Net because these earlier networks and processes had high levels of data security—and they worked. The Internet provided a new channel for B2B, but until it became secure at the end of the century or less expensive, few had much incentive to migrate to the new network. In fact, despite enormous hyperbole about B2B, large corporations still used earlier forms of networking. B2B over the Internet became more attractive for small firms interacting with each other, often for more than just sales. Biotech firms, for example, often interact with each other and large pharmaceutical companies over the Internet or through shared intranets to conduct business. Estimates of how much business is done in a B2B environment over the Internet are unreliable and therefore not worth citing; nevertheless, it is already clear that it is a widespread practice that builds on the prior use of networks in an evolutionary manner.

Role of Customers on the Internet

The areas where the greatest effects of dealing with consumers over the Internet became evident involved services, particularly in reserving hotels and rental cars or buying airplane tickets. By the early 2000s, airline reservations became a highly visible model as all the major airlines began encouraging (or forcing) customers to query online flight databases, then reserve and pay for tickets online—a theme discussed in more detail by Jeffrey Yost in an earlier chapter. This allowed carriers to reduce their dependence on travel agencies, thereby avoiding the expense of paying commissions to these firms, while also reducing the number of airline employees who used to sell tickets and deal with the public. They were so successful in causing customers to start booking online that by the early 2000s, that task was fundamentally changed from what it had been even as late as 1998–2000. The industry enjoyed operating savings of hundreds of millions of dollars. Firms in other industries attempted to do the same thing, without as much success (so far), such as utility firms, telephone companies, and Internet service providers. Perhaps the longest-running attempt to get customers to use computing involved retail banking, which had tried to get customers to bank online since the 1980s, but with minimal success.[13] The reasons for the differences are not yet evident empirically, although one can suspect that airline customers were more computer savvy than most banking consumers since such a large portion of them were in business—and hence, more comfortable using computers—while we know that many groups of banking customers were low users of the Internet, let alone online banking.[14]

How the Internet Affected Internal Business Operations

Where the Internet did play a greater role in altering the work of firms involved internal operations hosted on intranet sites—that is, using the same technology as the Internet, but exclusively for use by employees and business partners. Access to such intranet sites could be done only by passwords and increasingly became highly secured. At first, these replaced existing internal telephone-based networks, initially through dial-up capabilities, and by the end of the century increasingly via high-speed links into existing local area networks or broadband networks. As it became more possible to perform secure access and transactions over intranet sites in the 1990s, and as the costs of doing so declined, large corporations in particular began moving transactions through the Internet that they previously did over privately run networks. Several examples illustrate the process at work. First, users moved their e-mail systems and databases that stored internal documents and forms over to intranets. In the case of all computer manufacturing firms, pre-Internet e-mail systems were in time converted from such software as the widely used PROFS to Lotus Notes (both internal networks), even adding access to the Internet and Microsoft's own e-mail tools. Large employers using e-mail internally linked to internal databases did the same. Examples include insurance companies, such as State Farm, and banks, such as Citicorp. Companies with large populations of office workers were the most likely to do this earliest. Contractors serving the military were also forced to move this way by the U.S. Department of Defense as early as the 1970s, using the predecessor of the modern Internet to communicate with the Pentagon and collaborate with academics.

Next, a variety of internal applications accessed often by employees on preexisting networks migrated over to intranets, such as filing for reimbursements of expenses, enrolling in training classes, updating contact information for directories and human resource systems, and accessing libraries of company publications, which then could be downloaded on to personal computers and be locally printed or published. While these functions did not change profoundly when moved to often less-expensive intranet sites, what they now made possible was the use of information in graphic format and linked to other data on the Internet, thereby providing a far richer body of material that one could work with day or night. By the end of the 1990s, at IBM almost all internal applications involving expense reimbursement, application for insurance and other services, publications, training, and information about products, organizations, telephone numbers, and so forth, could only be accessed or conducted by using intranets. The same applied to software firms, information technology companies, and increasingly federal agencies. But this general observation did not hold for all industries. For example, while insurance companies did more to use networking internally, as late as the early 2000s in dealing with customers outside the firm, they still were paper based, or required consumers to fax documents back and forth. The same proved true of many state and local government agencies, yet even here one has to be careful about generalizing. For instance, tax departments in state governments were still internally paper bound and in their external relations with citizens even in the early 2000s, yet license renewals for driving vehicles, or for professions, were highly automated and accessible through intranets or over the Internet.[15]

Even more significant, communities of practice could collaborate more closely by the end of the century than they had before, conducting their affairs remotely through on-line chat rooms and Web based conferencing, all using simultaneous telephone and presentation-based conversations. The technology made it possible and far easier than before to have a vastly enriched dialogue among employees located remotely. One consequence just emerging at the start of the new century was the integration of work teams around the world, such as scientists and engineers developing new products or solving problems. The extent to which this occurred has not been well explored, although extant case studies suggest a high level of use by Fortune 500 firms.[16] Still, this collaboration may ultimately represent some of the most important sets of changes in how organizations worked as a direct result of the introduction of the Internet into businesses, and thus will deserve extensive study by future students of the Net.

The outsourcing of work from one country to another also became increasingly a reality in the late 1990s. Most notably, American firms began shipping data entry and processing functions over intranets to less-expensive labor in India and Eastern Europe, building on prior practices of using internal telecommunications networks. Banks and insurance firms were some of the earliest to transfer this kind of work electronically in substantial quantities. American and West European firms frequently shipped a great deal of recoding of old programs to India, Eastern Europe, and Russia to address concerns of Y2K in the late 1990s, extending and demonstrating the ability of information technology and networks to support work around the world in an organized, practical, and cost-effective manner. By the early 2000s, this practice had become widespread for much back-office work, particularly in large American, British, and Japanese firms. By 2003, outsourced work—hence also redesigned to integrate internal and external interactions—spread to human resource functions, the processing of payments, finance, customer care, administrative tasks, and content development. The differences in labor costs were quite substantial. To do insurance claims work or run customer call centers in the United States cost ten times more per hour than in India; accountants were three times as expensive in the United States, and software engineers were six times more expensive.[17] With these kinds of deltas in costs, it became clear why outsourcing proved so attractive, especially as telecommunications became technically cheap and even easier to use than in earlier decades.

Management consulting firms also began using Indian labor to do increasing amounts of their research and the preparation of presentations for use in other countries so that work could, in effect, be done around the clock while lowering operating costs. In turn, however, that created new sources of competition for American firms as Indian companies sought to compete for higher-value consulting business precisely because they too could use intranet and Internet sites to deliver services by the early 2000s.[18] In any case, it is clear that intranet usage had already resulted in greater global collaboration and movement of work around the world to wherever it was least expensive to do, particularly data collection, analysis, and the preparation of reports and presentations.

Surveys from across many industries suggest a second-order effect caused by the use of both the Internet and intranets. The percent of employees in firms who had to interact with either of these two forms of telecommunications increased all through the 1990s and into the new century. No industry, large firm, government, or educational institution was immune from the trend.[19] While using internal networks via terminals and personal computers had long been evident, dating back to the 1960s for the former and the early 1980s for the latter, as access to both the Internet and the requirement to use intranet-based applications increased, two consequences became evident. First, the proverbial "everyone" had to become more personal computer or computer literate—that is to say, comfortable using such tools as Microsoft's products, accessing networks, sending and relying on e-mail, and even debugging minor software problems. Second, work done by less-expensive staffs increasingly shifted directly to higher-paid workers. For example, in the 1980s a middle manager in corporate America might still have dictated a letter to a secretary to write on a word processor, or relied on that secretary to manage his or her calendar and make flight arrangements for a business trip. By the end of the century, many of these functions were only being performed by secretaries for the most senior executives in a firm; the majority of others did their own. This resulted in the decline in the number of secretaries (by then more frequently called administrative assistants) at a time when the total population of workers in the American economy increased.[20] It is an ironic turn of events because it had been the advent of typewriters, telephones, and other "office appliances" in the last quarter of the nineteenth century that largely made possible the creation of jobs for women in clerical and secretarial positions. Now a century later, new information-processing office appliances were eliminating these positions.

Less studied, but quite evident, were two other developments facilitated primarily by networked personal computers, and secondarily by the availability of an ever-enriched Internet. The first concerned the growing opportunity to work at home. For example, mothers fixed software programs as part of the Y2K remediation initiatives of many companies, often receiving the software to modify over a network and returning the revised programs to a firm the same way. Marketing and other research work requiring the use of the Internet also became popular, while other home-based, information-dependent businesses emerged or expanded, such as writers and researchers for hire. The second development involved corporate employees now able to work remotely, yet stay linked to their employers, such as those in sales and services in manufacturing and retail firms.

Finally, as firms established Web sites and intranets, they had to hire experts in both forms of communications in order to create, maintain, and transform these classes of networks. While hard data on specific information technology professions are hard to come by, government employment data demonstrate that the number of employees in these professions actually went up, despite significant outsourcing of information technology support functions and software development to other countries. In 1995, there were several million employees in the United States working directly in information technology; by 2005, that number had increased nearly threefold.[21] They supported a growing number of employees who now used the Internet as part of the fabric of their work. The U.S. Census Bureau reported that by the end of 2001, some 71 percent of employees used the Internet. Nearly 83 percent of managerial, professional, and administrative staffs also used it. Most professions ranged in the 50 to 60 percentiles, while the lowest were those who did manual labor (ranging in the low 40s percentiles).[22]

This trend repeated one long evident in the field of information processing as every new class of computing required large and midsize firms to hire information technology experts—a practice dating back to the dawn of computing and even to earlier times for large data processing applications, such as those relying on punch-card tabulating between the 1890s and the end of the 1950s.

Before leaving this brief discussion of why firms used the Internet and intranets for internal operations, I should explain why they did this. First, these networks often proved less expensive to create and maintain than earlier forms of telecommunications. This proved especially so with regard to the Internet as a vehicle for supporting e-mail in small companies and an inexpensive yet rich source of information. Second, the Internet served as a way of reaching new customers through another channel of distribution and economic exchange that could potentially add revenue to that already extracted from existing physical markets. Third, intranets provided a far richer set of capabilities for internal operations than earlier networks, combining, for example, text, graphic data, and databases in convenient, interconnected forms. Intranets also proved reasonably easy to protect from external threats by the end of the 1990s. To be sure, not all companies enjoyed these various benefits, but these were their expectations for why they wanted to use the Internet and/or intranets.

Economic Effects

A debate ran continuously on whether the use of the Internet and intranets lowered the cost of labor, or the number of jobs, as firms continued automating the work of lower-skilled (hence lower-paid) employees, and also hired information technology experts to nurture these new systems.[23] The overall number of jobs increased in the American economy, including in those firms and industries that became extensive users of all manner of information technology and the Internet. Various lower-skilled positions evaporated, however, while the number of higher-level, better-paying positions increased. External economic and business issues affected the job mix, such as the role of wars and the price of oil. Competition on a global scale was made more intense by the emergence of worldwide processes and markets, and buttressed by an extensive network of telecommunications and transportation both effective and less expensive than in prior decades. Thus various events profoundly impacted the work of companies. In short, the Internet was not the only influence on business practices, and one can reasonably conclude that it was the least influential in the increase or decline of jobs in general.[24] That said, it is also too early to generalize with certainty about the effects of the Internet on what jobs were created and lost within specific sectors of the economy, and the effects on the quality of positions. Most discussions about the economic productivity of the 1990s in fact virtually ignore the role of the Internet.[25]

One should not completely ignore its economic presence, though. But as table 12.3 shows, it is also misleading to generalize too exuberantly about the business use of the Internet because as the evidence suggests, it varied widely between manufacturing, wholesale, and retail firms and industries. In addition, but not shown in the table, is the fact that the same agency that collected the data in table 12.3 also documented the continued use of EDI in manufacturing at the same time. The value of goods flowing through EDI channels to wholesalers and retailers approached nearly $238 million in 2000, and business conducted using EDI grew in each subsequent year with the latest data (2003) reflecting transactions worth $332 million. To be sure volumes were small when compared to what happened in the 1980s, for instance, but they were nonetheless still there. E-commerce varied by industry from negligible across the entire period for sales of general merchandise, to high for automobiles both in dollar terms and the percent of transactions. Approximately 75 percent of all e-tail sales came from electronic shopping and mail-order houses—those firms that essentially ran their internal and external operations over the Internet and intranets—while such sales were minimal to nonexistent for many small retail shops.[26]

The Brokerage Industry

Since new uses of technologies arrive at different speeds into firms and industries, are there bellwether uses of the Internet that can indicate the future effects of this new tool? At the risk of committing various sins of historiographical improprieties, including the use of the present to suggest past and future behavior, the answer is probably yes. A brief description of the experience of the brokerage industry is informative because here we can see the Internet's effects both on how work tasks change and how the construct of firms, even its industry, also can evolve with the role of the Internet visible.

The case of the brokerage industry is emblematic of many of the themes already discussed above apparent in many firms across numerous industries during the last several decades of the twentieth century and into the new one. For decades, all major and midsize firms in the brokerage industry had long used various forms of telecommunications with which to communicate with each other, clear transactions (the purchase and sale of stocks and bonds), and work with trading markets, such as NASDAQ. By the time the Internet became a commercially viable tool for these firms, they already had a set of "legacy" preexisting networks, experiences, and practices that worked. Through dial-up services using telephones, many had also made available to their customers their core services by the early 1980s. Charles Schwab and Company was one of the first; another widely known early entrant into this market was DLJdirect. In addition, various providers had emerged that made available online news about stocks through private networks (such as CompuServe and Dow Jones News Retrieval). Customers and brokers used personal computers, specialized software, and toll-free telephone numbers to conduct business, and in the process, firms had evolved organizations and practices to deal with electronic transactions.

Schwab is always cited as the poster child for how to conduct business online, since it was one of the first and most successful, and as a result compelled many of the largest firms in the industry to provide online brokerage services in the 1980s and 1990s in order to compete.[27] Yet another event was more influential than the Internet. In 1975, a nontechnical development stimulated the move to online services, which rapidly led to discounted brokerage fees. Federal regulations eliminated the long-standing practice of fixed commissions, creating an increased competitive environment that rapidly drove down the price of buying and selling securities. In turn, that led to lower costs for consumers, but in the process it dropped income for firms, which then had to make up the shortfalls in revenues and profits through increased volumes. So long before the Internet became a channel for conducting business, the industry's key firms had started the process of altering traditional practices, terms, and conditions, such that by the time they discovered the Internet, many—but not all—of the internal transformations required to operate in an electronic environment had been accomplished. One important by-product of these changes, and also as a result of alterations in tax laws and practices—such as the use of 401(k) savings plans—was the sharp increase in the number of customers for the services and products of this industry. Customers transitioned from working with full-service brokers in the 1970s to a combination of paper and telephone transactions in the 1980s, using personal computers and 800 numbers. By the late 1980s they used online query functions. Customers gained

direct access to increasing amounts of information; the number of consumers personally placing buy and sell orders grew steadily all through the 1980s and the early 1990s.

In 1995, the first online Internet-based services appeared, and by the end of 1997
there were over nine million online accounts conducting daily some five hundred
thousand trades. We now had a new *Homo economicus*, the day trader, who relied on
personal computers, specialized software, and the Internet to play the market all day
long, paying fees at a fraction of what they had been a decade earlier, and often interacting with a brokerage firm's computers without dealing with human brokers.[28] In the
1990s, in addition to existing firms using the Internet as an alternative channel
through which to conduct business (while keeping earlier telecommunications networks), new companies appeared that only operated over the Internet, such as E*Trade
and Datek. By the late 1990s, over two hundred firms provided Internet-based services
to over eight million customers.

One cannot overestimate the effects of the Internet on this industry because earlier
than in many other corners of the economy, the experience of its firms suggested
possible influences in the years to come across the economy in other sectors. One official in New York State explained in 1999 what was happening: "Online trading is rev-
olutionizing the securities industry in several critical ways" by causing relationships
between customers and brokers to change (such as was also happening in the automotive and retail industries), and making it possible for "individuals to manage their own
investments in a manner never before possible." The same commentator noted that
people could access massive quantities of financial data not available to them before
(the same as with automotive and retail sales), allowing them to "make an independent evaluation of stock performance" and "place their trades without the assistance
of a registered securities representative," thereby decreasing the requirement for so
many brokers.[29] As a consequence, as the volume of transactions increased in the late
1990s and even during the recessionary years of the early 2000s, the number of brokers
needed by firms declined and so they were laid off. This trend of a declining need for
brokers, caused in part by changing technological innovations, had started in the
1980s, and extended all through the 1990s. To be sure, less than ten firms dominated
online trading, reflecting a long-standing feature of the industry: it was always concentrated in the hands of a small number of companies.[30]

Enterprises trading on the Internet were able to improve the productivity of their
front offices substantially for the first time since before World War II. In the process
they displaced the increased costs for information technology, resulting from moving
away from a labor-intensive retail model to one relying more on technology. This was
all made possible by the fact that additional investors came into the market, largely
thanks to the convenience of the Internet along with changes in market and regulatory incentives to save, such as the availability of 401(k) and IRA accounts as well as

professionally managed mutual funds. The growth in 401(k) activity also illustrated the pattern of new use of technology emerging simultaneously with nontechnological events. In this case, many corporations were also shifting their defined benefit pension plans to 401(k) types in response either to the costs of doing business or changing accounting practices and tax laws, particularly in the United States.

Changes in the Information Balance of Power in the Marketplace

Fundamental changes took place in the market's balance of power, occurring simultaneously across many other industries as well, and facilitated directly by the use of the Internet and the growing availability of information online. This basic transformation involved those who had previously been constrained from influencing the practices of firms due to a lack of information now in the ascendancy. They obtained this new-found information through dial-up services in the 1980s and later through myriad sites on the Internet. Investors could know as much or more about a stock or industry than a broker, and act on that knowledge. Consumers buying automobiles, appliances, electronics, books, information, music, and health products did the same. Patients began challenging doctors and proposing therapies, often armed with more, and more current, medical information than their internists and general practitioners. Retailers could continuously transmit orders all day long to factories and wholesalers in a just-in-time manner, confident that they now had a better understanding of what consumers wanted than did manufacturers, which until at least the early 1980s, normally understood that part of the trading equation better than did retailers. The early use of inventory-tracking radio frequency identification (RFID) tags by wholesalers and retailers as this chapter was being written furthered the process.[31]

Effects of the Internet on the Public Sector

Finally, before discussing new uses by new companies, it is important to acknowledge briefly the effect of the Internet on the public sector because combined, all government agencies, elementary and secondary schools, and higher education accounted for over a quarter of the nation's gross domestic product. The public sector too made, bought, and sold a vast quantity of goods and services. It also employed millions of people. Hence, as both consumer and regulator, this sector influenced the work of the private sector. Much of the federal government's early involvement in the creation and use of the Internet is well understood. Less appreciated is what happened after the Internet became a widely known network by the mid-1990s. And like the private sector, local, state, and federal governments had long been extensive users of computing and telecommunications. Two of the central tenants of the Clinton administration's domestic policies of the 1990s were to leverage this technology to lower the costs of doing the federal government's work, and to make the Internet available to large segments of the American public, most notably schools. The patterns of usage mimicked those of the private sector, although deployed more slowly at the state and local levels. Also, the influence of the Internet on how work was done and the management of agencies proved less intense than what occurred in the private sector. Put less politely, the Internet had yet to budge American government agencies into new ways of working—a circumstance that stood in sharp contrast to the experiences of European governments, where consolidations of agencies providing services to citizens were being transformed and given new missions.[32]

New Economic Opportunities Made Possible by the Internet

So far in this chapter, I have discussed how firms bolted the Internet on to existing work processes and offerings. All major new technologies also create innovative economic opportunities, however, and the Internet proved to be no exception. Three patterns became evident by the end of the 1990s. First, new niche firms emerged that translated these new functions into business opportunities, dependent on the nature of the technology's capabilities in any particular year. Many of these firms emerged within the confines of an existing industry, such as E*Trade (brokerage), Amazon.com (book retailing), and e-Chemicals (chemical industry). In fact, most new Internet-based businesses came into existence this way. It made sense that this would be so as businesses could only recognize their new opportunities if led by individuals deeply knowledgeable about their industry and learning quickly about the Internet.

A brief example from the banking industry illustrates this point and a common pattern of emerging business opportunities. In October 1995, the Security First Network Bank became what the industry quickly realized was its first Internet bank, heralding a new era in banking that bankers recognized almost immediately. Customers could conduct their business over the Internet, using secure software, some of which had been developed for the U.S. military. This bank was part of a traditional brick-and-mortar one that had opened the prior year, and in that first year had acquired only 187 customers. Within two weeks of going online it had 750 online accounts. This bank did not pay interest on these online accounts, yet that did not matter because the convenience of banking online proved attractive enough to consumers to outweigh the competitive disadvantage of not paying interest on balances. Customers had personal computers and network access, saving the bank that cost, and regulators were satisfied with its business practices. Within one year of going on to the Net, this bank had over 1,000 accounts, all of which cost less to operate than traditional models of doing business. The industry as a whole took quick notice, and in 1996 other banks began to do the same—over 200 by spring 1998.[33] Services were routine: deposits, withdrawals, and payments.

The story of the Security First Network Bank is dramatic and telling because it illustrated how change could come to many firms, and again, how traditional services were first ported over to the new channel of distribution (the Net). In this instance it also required collaboration by various federal regulatory agencies, which in the mid-1990s were promoting the use of the Internet across the economy. The story also underscores that businesses in many industries adopted this new channel for distributing services and products after one or a few members of their industry had done so first, stimulating a rush to emulate that typically took an industry one to three years to do. Afterward, the use of the Internet had become relatively common, and companies were already moving to the new phases of adding the ability to conduct additional transactions on their Web sites.

A second pattern involved the creation of Internet-based businesses that went after opportunities in a nearby industry. Often these firms started out like the first type, offering a service relevant within their home industry and then additional services that impinged on other industries. For example, Amazon.com began as a bookseller and then offered other products, such as music, videos, furniture, and so forth. As federal regulators made it increasingly possible for banks, insurance companies, and brokerage firms to sell products from each other's industries in the late 1990s, all three industries enhanced their existing Internet-based offerings to provide new services. For instance, one could increasingly trade securities, deposit cash, write checks on securities accounts, and acquire insurance all on the same site within the same account, transferring funds from one service to another. In the early 2000s, the most dramatic occurrence of this pattern involved Apple Computer introducing iPods for portable music, forcing recorded music companies to supply the content on a song-by-song basis without forcing customers to buy a whole compact disc or paying a distributor, when all they wanted was one or a few songs. Apple was largely interested in selling iPods and was perfectly prepared to threaten long-standing sources of profits in the music business in order to do so (which is exactly what happened initially).[34]

The case of Apple is also important because the recorded music industry felt deeply threatened by the Internet, which millions of users went to in order to acquire free copies of music in violation of copyright laws beginning in the late 1990s, thereby denying recording firms and artists royalty payments for the "pirated" music. Apple's approach to licensing music for distribution through iTunes represented one of the first successful uses of the Internet as a source for music acquired in a legal manner respecting copyrights and compensating recording companies.[35]

All the while, service providers of either approach began collecting real-time information on how their consumers were behaving, which provided detailed insights influencing subsequent marketing, and about what new products and services to offer. General Electric along with Proctor and Gamble had long been the role models for marketing intelligence. Yet the new technology was of particular value now to vendors selling over the Internet. For example, Amazon.com used one's buying patterns to suggest other related products that might be of interest of a customer, such as specific books and music. By the early 2000s, this was a widespread application of the Internet by all major vendors selling over the Internet around the world. Google used a similar process for collating queries for information. The list of examples is endless, but the lesson is clear: as one used the Internet for any transaction, an interested party could track that use, and then leverage that information to target sales or improve responses to queries.

In short, the Internet provided a tangible way of disaggregating the boundaries of an industry, firm, products, and services. It did this in ways far different from what occurred with the use of EDI and other private networks in earlier decades. Now, work moved seamlessly in and out of one enterprise into another as part of the partnering relationships established to supply components, manufacture products, and sell goods.

A third approach involved creating a firm that did not have a strong identity or heritage within a particular industry. While discussed elsewhere in this book in more detail, suffice it to note here that these firms provided basic economic transactions, most notably online auctioneering and bidding for products. The most famous of all of these is of course eBay, which became the largest online trading operation for all manner of goods and which many have called the world's largest garage sale. Individuals and firms offered products for sale online, allowing consumers to bid on these up to a certain cutoff date, at which time the high bidder "won" and bought the product, using software provided by eBay to pay for these and communicate the consummation of the transaction to all concerned. Trust of suppliers surveys provided some capability for enforcing the execution of a sale within eBay. The success of this firm became emblematic of large changes in how the public evolved how it bought goods. Just citing the results of this one firm suggests the order of magnitude of what was happening. In 2005, eBay generated $4.552 billion in revenues while the total value of goods sold through its sites exceeded $12 billion; eBay's revenues had grown by 39 percent year over year, with 71.8 million active participants working through 212,000 "stores" online.[36] This firm was larger than many "old economy" major companies, as measured by revenues. In that same year of 2005, for instance, that included such household names as H&R Block ($4.2 billion), Mutual of Omaha Insurance ($4 billion), and Corning ($3.8 billion). So eBay must be recognized as a major force of transition in the American economy made explicitly possible by the availability of the Internet. Many commodity-bidding sites also came and went; some rooted in an industry or specific class of products (such as chemicals, agriculture, and ores), or as wholesalers for excess commodities, inventories, and government surplus. Such firms made money by charging a small transaction fee or taking a percentage of the value of a sale consummated through their Web sites.

In all three situations, common practices prevailed. Most obvious, they were established as largely (or only) Internet-based services, so their business models were designed to leverage the new technology. Brick-and-mortar cost structures were not involved, such as the cost of running stores and branch offices, but could be for warehouses and corporate offices. Expenses went more for marketing, advertising, and technology than for goods and staffs. Work was often outsourced to various other firms collaborating with the enterprise. Amazon.com might take an order, have a book publisher or bookstore fulfill the order, and use the U.S. Postal Service or some package delivery firm to get it to the consumer. This business model made it possible to synchronize the whole fulfillment process with everyone involved using the Internet within their individual tasks in the larger process. Profit margins and line items in a firm's financial chart of accounts were thus different from those of companies operating in either an Internet-free market or one that mixed old forms of business with Internet-based activities. Specific differences in such things as profit margins, managerial practices, and consumer behavior all have yet to be studied by historians, but have been the subject of most of the contemporary discussions about the business of the Internet.

In all three instances, the circumstances were of recent origin, mostly from the late 1990s onward. As others in this book have pointed out, many enterprises experimented with different business models and offerings, some with spectacular success (most notably eBay, Yahoo!, Google, and Amazon) and many thousands with dramatic failure (the dot-com meltdown). Participation occurred all over the economy; in other words, these new firms did not concentrate in just a few industries. As one result, a substantial debate began about the effects of the Internet on the productivity of the nation's economy.[37] To be sure, this pattern of success and failure was well-known to economic historians. For instance, in the United States scores of railroad companies came into existence in the 1840s to 1860s, only to fail, be consolidated into larger ones, and settle down into a dozen firms that then thrived for decades. The same happened with pharmaceuticals, beginning in the 1870s and 1880s, and then settling down by the 1920s into a dozen or so firms, not to be disturbed again until new technologies and science (genomics and DNA) began upsetting historical patterns of behavior, opportunities, and profits in the 1990s and early 2000s. So what happened to the Internet-only firms is familiar and probably, as of this writing, not concluded.

Convergence of Old and New Business Practices

As with the earlier examples, what constitutes new business models or even Internet businesses (and its attendant markets, firms, and industries) is still subject to debate. In fact, this ambiguity is an important component of any discussion about the new wine in new bottles phenomenon of the business of the Internet. Is it a combination of infrastructures (such as Google or Yahoo!) and content providers (such as those offering up news or business data)? Is Cisco, a leading manufacturer of telecommunications switches used in the hardware makeup of the Internet, an Internet business? What about IBM, which provides large mainframes that house and distribute data over the Internet? Is this nearly one-hundred-year-old enterprise an Internet company, as its marketing suggested in the mid-1990s? A group of Internet watchers at the University of Texas introduced a model of different levels of players in the Internet world that could lead one to believe that many industries were crucial components of this new Internet business. They divided the world of the "Internet Economy" into four clusters of participants: providers of infrastructure (such as Cisco, Dell, and America Online), applications (Microsoft, IBM, and Oracle), intermediaries (online travel agencies, portal providers, and online advertisers), and commerce (e-tailers, content providers, and Amazon.com).[38] The problem with this model is that it would be as if one were to argue that paint companies providing paint used on Cisco's telecommunications switches were part of the mix.

More sensibly, the U.S. government recently redefined its categorization of industries and segments, creating the Information Segment as part of its replacement of Standard Industrial Classification codes with the North American Industry Classification, and subsumed in that new typology elements of industries, firms, content and service providers, and so forth, that participated in the expanding world of Internet business. That process is beginning to provide a more useful approach for understanding the activities of today's American economy.

We already know clearly some things. First, the number of consumers using the Internet to gather information and make purchases has been increasing each year both in the United States and around the world since the late 1990s. So potential markets and traffic are growing. These volumes are counted on a continuous basis by governments, businesses, and professors, including by some of the contributors to this volume. Second, there is hardly a business offering a product or financial or information-based service in the United States that does not conduct transactions over the Internet. That pattern is rapidly becoming the case in Europe and parts of Asia, albeit a bit later than in the United States. Third, governments all over the industrialized world are rapidly shifting the distribution of information, correspondence, and services to the Internet from national to local agencies. Fourth, there is much about Internet businesses that mimic old economy practices, such as that a line of business or a firm on the Net must be profitable sooner rather than later to survive.[39] It must also have a way to generate profitable revenue, although often different from in a brick-and-mortar world, such as by selling advertising displayed to users every time they click on to a site.

This last point may seem pedantic and obvious. But it was not during most of the 1990s. In a book published in 1999, two economists, Carl Shapiro and Hal R. Varian, reminded managers that Internet-based business had to generate revenues and spin off profits at some point to be viable, and then explained how to accomplish that task. Their message came as a surprise to so many that the book became a best seller that year, and has now become an important resource for those contemplating running a business over the Internet and indeed a minor classic in the area of Internet studies.[40] One would have thought that such a message did not have to be delivered, but typical of what many business executives and observers noted, a member of the chemical industry in 1999 publicly stated that "I'll bet that at this time last year, no one had heard of E-commerce."[41] The dot-com fiasco was not so much a "fiasco" as it was another collection of events demonstrating that business tied to the Internet really was new to its participants, and therefore called for drawing from proven prior wisdom and experience in creating new practices and skills while adhering to some immutable laws of economic behavior.

Analysis of Patterns and Practices

Several observations can be extracted from the experiences of old and new companies using the Internet. These observations reaffirm broad patterns evident in the adoption of other technologies, including computing and telecommunications, which prevailed again with the Internet. Perhaps the most obvious is that companies took their time to understand the characteristics of this new technology, asking the same questions about it as they had for earlier forms of information technology, reflecting a practice in existence since the Gilded Age of the nineteenth century. Managers asked questions about applicability, economic value, risks of deployment (or not), costs, and effects on operations, management, revenue streams, and customer service. They did this while the technology underwent crucial transformations in the 1990s and beyond, with the introduction of more content, functions that improved the ease of use, security features, and the emergence of new rivals from within, without, and based on the Internet as primary channels for sharing information, partnering with other firms, and going after new business, often with new offerings, or combinations of old and new products and services. The technological churn is not over. Open and closed standards, new software platforms, redesigned intranet protocols to reduce the chaos and insecurity so prevalent on the Net today, and new search tools all remain as unclear as did earlier issues related to the commercial use of the Internet. So it should come as no surprise that as with many prior forms of information technology, companies almost universally moved in a slow and deliberate manner to embrace the technology.

Their initial forays focused on limited uses, such as the posting of information for consumers, employees, and others, and only later in an evolutionary way adding functions. But as the data on EDI showed, most companies that were in existence before the arrival of the Internet, added Net-based functions to preexisting networks and work processes in ways intended to minimize the risk to revenues and work. Management proved cautious due to their lack of knowledge about the potential benefits and risks of using the Internet. Other chapters in this book describe these activities and consequences in considerable detail, leading in hindsight to the conclusion that a cautious tact to adoption of the Internet was the prudent approach. Before the dot-com bubble happened, thousands of American companies and thousands of public-sector agencies and local governments had chosen to use the Internet more slowly than the Net enthusiasts would lead us to believe.

It may seem like an odd comment to make at a time when the Internet is such a topic of discussion and focus, but for all businesses there are other issues besides the Internet in which sits today's latest form of information technology. The most obvious technical one concerns the ongoing use of proven non-Internet-based technologies, such as private networks and EDI, and centralized or decentralized computing. All of these were integrated into the operations and organizational structures of companies over the past half century, and few companies were about to dismantle them in favor of a new, hardly understood, poorly managed information technology, otherwise known as the Internet. In short, legacy systems, worth trillions of dollars in prior investments and currently supporting tens of trillions of dollars of commerce around the world, continued to influence the work of companies far more so than the Internet. In fact, the Internet is so new to business, and its technical form is still morphing at a substantial rate, that it remains difficult to even project out one decade about the role of the Internet in American life, let alone in business.

External gating factors did (and do) influence profoundly the interaction of this new technology with business practices. Examples abound. For one, businesses using the Internet to sell products and services need to have customers who access and use the Internet, and are willing to conduct business using this medium. That is why firms contemplating leveraging the Internet for sales always closely watched who used the Internet and earlier forms of telecommunications. It is no accident, for instance, that brokerage firms became extensive providers of transaction-based offerings over the Internet because they had a set of customers more familiar with buying and selling via telecommunications than many other sets of customers in other industries. Internally within firms, a gating factor concerned the degree to which there were employees knowledgeable about the Internet and willing to use it. The health industry, and more specifically doctors and hospital administrators, were notorious for their ignorance of computing of all types, and thus it was not until nearly a decade after its debut on the business stage that even hospitals began using the technology. Few doctors did, even as late as 2006. Other examples could be cited, but the key observation is that technology itself is not always the dominant influencer of how and when it is used. Often more important are those externalities that can come from market realities, the nature of a firm's workforce, prior experience with and reliance on earlier technologies, or the economic threats and opportunities made evident by the existence of the new technology. That last point is the subject of several chapters in this book for good reason because it is central to any discussion of the effects of the Internet on American business. And we know the effects varied widely from one firm to another and across industries.[42]

Finally, what are we to conclude about the construct of business organizations? A basic premise of much of the business community's discussion about the role of the Internet is the collection of effects it has, or could have, on how organizations are structured and operate. There is hardly a business book today talking about the Internet that does not elaborate on how it is changing everything, and most notably companies. Alfred D. Chandler Jr.'s construct of how corporations were formed and Michael Porter's descriptions of how they responded to competition are sometimes prematurely dismissed as portraits of bygone eras. Corporate executives are quoted frequently in their trade press and at conferences extolling the future of their companies, in which work is dispersed geographically through mobile workers, the use of business partners, and across continents.

However much this may be the case now or in the future, memories can be short or nonexistent, as these trends have been emerging for some time, often lubricated by the use of telecommunications. Charles Handy spoke of such organizational configurations in the 1980s, years before the Internet was an important part of business life, contending that

80 percent of the value [of products and services] was actually carried out by people not inside their organization. These 20/80 organizations do not always realize how large the contractual fringe has grown because it has become a way of their life. It is only recently that more individual professionals, more small businesses, more hived-off management buyouts have shown a spotlight on a way of organizing which has, in fact, always existed.[43]

And he was not alone. Even earlier than Handy, Peter F. Drucker commented on the same theme. In 1980 he wrote, "Electronics are becoming the main channel for the transmission of graphic, printed information," predicting the imminent emergence of organizations similar to what Handy later saw in place.[44] Still earlier in 1973, Drucker devoted many pages in his seminal work *Management: Tasks, Responsibilities, Practices* to the role of technologies of all kinds, warning his readers that any "technology in which a company has distinction must be central rather than incidental to the product or service into which a company diversifies. To disregard this rule invites frustration."[45] Arguing for effective communications (not just verbal), he pleaded the case that communication "is not a means of organization. It is the mode of organization."[46] This statement could have easily been made in the early 2000s by Bill Gates of Microsoft, Andy Grove of Intel, or Eric Schmidt of Google.

Because businesses began using telecommunications networks long before the wide availability of the Internet, much structural change had already occurred that the Internet has, even now, yet to alter. Using IBM as one example, workers in sales and services began to operate in a mobile fashion at the start of the 1990s, while the use of intranets did not start until the mid-1990s and were not widely deployed until the end of the century. Before the Internet for e-mail they used PROFS (developed in the 1960s), VM mainframe operating systems (also circa 1960s), and an internal network that has yet to be rivaled by either intranets or the Internet in speed, reliability, or security. As Professor Martin Campbell-Kelly pointed out in his study of the software industry, one of the most widely installed telecommunications transaction-based software tools in the world was a pre-Internet software package called CICS. In 2006, it still remained the most widely used transaction software facilitator among businesses, and also one of the least publicized software packages in the history of computing.[47]

We can conclude, however, that with the enormous variety and indeed richness of the Internet's ability to mix and match all manner of digital media in ways not possible with prior software and telecommunications, we have yet to see companies either exploit extensively this technology or transform into some new construct as a result. As the technology's capabilities become more sophisticated and its use more manageable, there is little doubt that it will play a far more prominent role than it has so far. To a large extent, that will not be the case just because the technology improves but also because prior experience and sound business practices will motivate management to more fully embrace it.

Perhaps the least visible, but ultimately the most important consequence of the Internet's effects on American business may be in conjunction with all prior use of information technologies. The historical record is quite clear that for over a half century, managers preferred to make incremental decisions about the adoption of some new technology or use of information technology, for this approach ensured that the daily operations of the firm continued, and minimized risk of some technological or operational disruption to business. But because of millions of incremental changes in how companies functioned, the consequences were cumulative—that is, they were in response to the existing circumstance at the time when a new action was taken. Invariably, when a manager looks back—say, in a decade or so—at what happened in their firm regarding the use of information technology, that person concludes, "We made a revolution." Yet with few minor exceptions, a close study of what happened in that decade in any individual firm or industry almost always shows that dozens, even hundreds of small changes took place to resolve immediate issues that in turn, cumulatively, in nondramatic form, caused changes to occur. The Internet played the same role in this process as all other forms of technology, regulatory behavior, and globalized business practices.

Thus, like new wines poured into old bottles, and new wines into new bottles, the Internet has given companies the opportunity to harvest yet a new crop of ways of working, and given new enterprises new ways to generate profitable revenues.

Notes

1. In no way do I intend to imply that these studies are hysterical or incompetent; rather, many are thoughtful and innovative. See, for example, Don Tapscott, *The Digital Economy: Promise and Peril in the Age of Networked Intelligence* (New York: McGraw-Hill, 1996); David Ticoll and Alex Lowy, *Digital Capital: Harnessing the Power of Business Webs* (Boston: Harvard Business School Press, 2000); Ted G. Lewis, *The Friction-Free Economy* (New York: HarperBusiness, 1997); Philip Evans and Thomas S. Wurster, *Blown to Bits: How the New Economics of Information Transforms Strategy* (Boston: Harvard Business School Press, 1999).

2. The major sources for ongoing monitoring are the U.S. Department of Commerce and the Pew Foundation, although many academics and other organizations do similar work as well.

3. Use the Department of Commerce for industry usage, and the Pew Foundation for understanding the role of individuals.

4. An important finding of a study of over forty U.S. industries; James W. Cortada, *The Digital Hand*, 3 vols. (New York: Oxford University Press, 2004–2008).

5. The point-of-sale story is particularly instructive on this point; Stephen A. Brown, *Revolution at the Checkout Counter* (Cambridge, MA: Harvard University Press, 1997).

6. The two major exceptions were universities and defense industry firms, both of which worked closely with the Pentagon on developing weapons and scientific projects that required them to use the Internet to collaborate.

7. Students of the process brought much attention to the subject through their publications, and none more so than Carl Shapiro and Hal R. Varian, *Information Rules: A Strategic Guide to the Network Economy* (Boston: Harvard Business School Press, 1999).

8. Charles H. Fine and Daniel M. G. Raff, "Automotive Industry: Internet-Driven Innovation and Economic Performance," in *The Economic Payoff from the Internet Revolution*, ed. Robert E. Litan and Alice M. Rivlin (Washington, DC: Brookings Institution Press, 2001), 62–86; Cortada, *The Digital Hand*, 1:140–143.

9. By the late 1990s, consultants and business professors began adopting language reflecting types of Internet-based interaction. These included B2B, which means business-to-business dialogue, such as a manufacturer might have with a supplier of components or a distributor; and B2C, which means business-to-consumer, which is about interactions between a retailer, for instance, and customers, or what otherwise might also be called e-tailing or e-commerce. There were other permutations of these kinds of phrases that kept appearing in the late 1990s and the early 2000s, such as e-CFO, e-chemicals, or G2G (government-agency-to-government-agency).

10. The most reliable tracking is done by the U.S. Census Bureau, with the results reported every quarter and published as the "Quarterly Retail E-Commerce Sales" in a press release.

11. A subject only recently seriously studied, such as by Graham Tanaka, *Digital Deflation: The Productivity Revolution and How It Will Ignite the Economy* (New York: McGraw-Hill, 2004).

12. Of all these vendors, Amazon.com has been the most studied; see, for example, Rebecca Saunders, *Business the Amazon.com Way* (Dover, NH: Capstone, 1999).

13. Jude W. Fowler, "The Branch Is Dead!" *ABA Banking Journal* 87 (April 1995): 40.

14. The Pew Foundation's dozens of studies of Internet users supports this conclusion. For these studies, see Pew Internet and American Life Project, available at ⟨http://www.pewinternet.org⟩.

15. I deal with these issues more fully in volume 3 of *The Digital Hand* (forthcoming).

16. See Rob Cross, Andrew Parker, and Lisa Sasson, eds., *Networks in the Knowledge Economy* (New York: Oxford University Press, 2003); and Don Cohen and Laurence Prusak, *In Good Company: How Social Capital Makes Organizations Work* (Boston: Harvard Business School Press, 2001).

17. "The Specter of Outsourcing," *Washington Post*, January 14, 2004; "Big-Bank Perspectives on Offshore Outsourcing," *American Banker*, March 8, 2004; "Bank of America Expands India Outsourcing," Associated Press, May 9, 2004; "Near-Term Growth of Offshore Accelerating," Forrester Research, May 17, 2004; Unpublished, IBM Institute for Business Value analysis, Somers, N.Y.

18. A subject that has generated more passion and emotional outbursts than serious study; see, for example, Lou Dobbs, *Exporting America: Why Corporate Greed Is Shipping American Jobs Overseas* (New York: Warner Business Books, 2004).

19. Monitored continuously by the Pew Foundation and the U.S. Department of Commerce. By late 2001, strong evidence had accumulated that nearly 80 percent of all managerial and professional workers in the United States used a computer at work, and 66 percent did the same with the Internet. For details, see Steve Hipple and Karen Kosanovich, "Computer and Internet Use at Work in 2001," *Monthly Labor Review* (February 2003): 26–35.

20. U.S. Department of Labor tracks these trends, particularly through the Bureau of Labor Statistics, which also develops projections of future occupations.

21. U.S. Census Bureau, *Statistical Abstract of the United States: 2006* (Washington, DC: U.S. Government Printing Office, 2005), 414, 418.

22. Ibid., 423.

23. Recently discussed well by Amy Sue Bix, *Inventing Ourselves Out of Jobs? America's Debate over Technological Unemployment, 1929–1981* (Baltimore, MD: Johns Hopkins University Press, 2000).

24. Documented in the numerous raw statistics of the U.S. government; U.S. Census Bureau, *Statistical Abstract of the United States: 2006* (Washington, DC: U.S. Government Printing Office, 2005).

25. For example, Dale W. Jorgenson, Mun S. Ho, and Kevin J. Stiroh, *Information Technology and the American Growth Resurgence*, vol. 3, *Productivity* (Cambridge, MA: MIT Press, 2005); an important yet limited exception is Litan and Rivlin, *The Economic Payoff from the Internet Revolution*.

26. U.S. Census Bureau, available at ⟨http://www.census.gov/eos/www/historical/2003ht.pdf⟩ (accessed February 15, 2006).

27. John Kador, *Charles Schwab: How One Company Beat Wall Street and Reinvented the Brokerage Industry* (New York: John Wiley and Sons, 2002): 273–277.

28. Described by Gregory J. Millman, *The Day Traders: The Untold Story of the Extreme Investors and How They Changed Wall Street Forever* (New York: Times Business Press, 1999).

29. Office of New York State Attorney General Eliot Spitzer, "Online Brokerage Industry Report," available at ⟨http://www.oag.staste.ny.us/investors/1999_online_brokers/execsum.html⟩ (accessed February 15, 2001).

30. One can also speculate reasonably that part of why the stock bubble occurred could be attributed to a wave of new and inexperienced stock investors moving into the market using the Internet, attracted by the fast-rising market, especially the NASDAQ. Fortunately, this speculation is increasingly being substantiated by serious research on how stock markets behave. See, for example, the crucial analysis of Didiet Sornette, *Why Stock Markets Crash: Critical Events in Complex Financial Systems* (Princeton, NJ: Princeton University Press, 2003), especially on the "herding" effect, 88, 91–111.

31. RFIDs are small devices, such as computer chips with antennae, that serve as tags attached to an object, such as a piece of inventory, that can store and receive data remotely, and thus serve as a tool for tracking inventory or the location of an object, animal, or person. Wal-Mart and the U.S. Department of Defense have become early and aggressive adopters of this technology.

32. IBM and the Economist Intelligence Unit have conducted a global annual survey on this issue since 2000. Surveys from 2005 and 2006 reported on best practices that had been developing over time; Economist Intelligence Unit, *The 2005 E-Readiness Rankings* (London: Economist Intelligence Unit, 2005), and the 2006 updated edition of the report.

33. For the story in more detail, see Cortada, *The Digital Hand*, 2:104–105.

34. Jeffrey S. Young and William L. Simon, *iCon Steve Jobs: The Greatest Second Act in the History of Business* (New York: John Wiley and Sons, 2005), 275–330.

35. The subject of pirated music and movies had become one of the loudest issues concerning the Internet in the early 2000s. For an introduction to the issue, see William W. Fisher, *Promises to Keep: Technology, Law, and the Future of Entertainment* (Stanford, CA: Stanford University Press, 2004).

36. eBay, press release, January 18, 2006.

37. The two most currently influential collections of debates are Litan and Rivin, *The Economic Payoff from the Internet Revolution*; Erik Brynjolfsson and Brian Kahin, eds., *Understanding the Digital Economy: Data, Tools, and Research* (Cambridge, MA: MIT Press, 2000).

38. A. Barua, J. Pinnell, J. Shutter, and A. B. Winston, "Measuring the Internet Economy," 1999, available at ⟨http://crec.mccombs.utexas.edu⟩ (accessed February 13, 2006).

39. One of the problems so many of the dot-coms faced was the fact that they and their investors had frequently embraced the notion that getting a high market share first was more important than making a profit early in a venture. That thinking proved incorrect.

40. Shapiro and Varian, *Information Rules*.

41. Rick Witting, "Information Week 500: Chemicals: IT Opens Doors for Chemical Makers," *InformationWeek*, September 27, 1999, 113. While conducting research for all three volumes of *The Digital Hand*, I came across similar comments for every industry studied (forty plus) from the same period.

42. This is one of the central findings of my three-volume study, *The Digital Hand*.

43. Charles Handy, *The Age of Unreason* (Boston: Harvard Business School Press, 1989), 92.

44. Peter F. Drucker, *Managing in Turbulent Times* (New York: Harper and Row, 1980), 53.

45. Peter F. Drucker, *Management: Tasks, Responsibilities, Practices* (New York: Harper and Row, 1973), 701; on technology, see ibid., 698–710.

46. Ibid., 493.

47. Martin Campbell-Kelly, *From Airline Reservations to Sonic the Hedgehog: A History of the Software Industry* (Cambridge, MA: MIT Press, 2003), 149–152.

13 Communities and Specialized Information Businesses

Atsushi Akera

This chapter focuses on the opportunities afforded by the new Internet-based businesses and services that cater to communities. On the one hand, this domain has been occupied and studied by researchers and practitioners who comprise the emerging field of *community informatics*, which defines itself as a group concerned with "the use of Information and Communications Technologies (ICT) for personal, social, cultural or economic development within communities, for enabling the achievement of collaboratively determined community goals, and for invigorating and empowering communities in relation to their larger social, economic, cultural and political environments."[1] Yet a quick look at how the notion of "community" has been mobilized by various commercial Internet-based entities—eBay, America Online (AOL), MySpace, and Match.com among them—suggests that the term has had substantial versatility and utility beyond what amounts to a more narrow academic definition. Despite this apparent difference, however, I hope to rely on historical perspectives to draw out the common threads that unite these two different approaches to the use of community. Indeed, by the end of this chapter, it should be clear that all those who want to launch a successful community-oriented information service, either as a public service or for commercial gain, have to do so by understanding the specific information needs and motivations of the community they wish to serve.

This chapter is divided into five sections, beginning with a brief history of the origins of online communities. The subsequent four sections, rather than tracing a straight chronology, then describe subsequent developments as classified according to a dominant business model. They consist of *converted services*, such as AOL and LexisNexis, which successfully carried their early dial-up subscribers to their current Web-based services; *brokerage services* such as eBay, Napster, and Amazon Marketplace, where the primary revenue stream came to lie with transaction fees; *social network services*, such as Match.com and Meetup.com, which are based substantially or exclusively on subscriptions revenue; and *community networking services*, which as we will see has been dominated so far by public-sector initiatives. Different notions of community appear under these different modes of operation, and rather than engaging in the futile and largely academic exercise of defining what a "real" community is, this chapter sets out to document more precisely how different notions of community have been utilized for different commercial and public venues. Where possible, I also try to draw the connections—both historical and lateral—that existed across these different kinds of information services.

Although I will rely partly on the conventions of historical writing, my primary intent is to make this history useful to those who wish to create new community-oriented information services. Nevertheless, let me reiterate the major thesis here: those who seek to profit from this market must be willing to develop a real understanding of the social and informational dynamics of a given community and application in a way that goes beyond a traditional market-research model. In particular, historical evidence suggests that it is necessary for entrepreneurs working in this highly "social" segment of the Internet to understand how, exactly, technology interacts with and *mediates* the social interactions within specific communities. Without exception, the most "successful" services, including eBay, Napster, and Meetup.com, achieved their standing by designing a popular interface, and providing the necessary social services and infrastructure that generated enthusiasm within a specific community. This understanding of the relationship between technology, communications, and social organization is to be found not in the field (or at least not exclusively in the field) of community informatics but in the broader one of *social informatics*, of which community informatics is a part. This is discussed below. In any event, in a manner quite different from the "strictly business"–oriented strategies presented elsewhere in this volume, it should become clear that the most successful services described in this chapter earned their position by advancing not only a valid business model but also a genuine interest in the needs and interests of different communities. Having said this, it should be apparent that different strategies were and are possible.

Origins of Online Communities and Information Services

While there are many who would attribute the origins of online communities to Howard Rheingold's 1993 publication of *Virtual Communities: Homesteading on the Electronic Frontier*, online communities predated Rheingold's work by many years. As already described in Greenstein's chapter on Internet service providers (ISPs) herein, many early online communities had their origins with computer bulletin board and general dial-up services such as CompuServe. While commercial services entered the field quite early on, there were a number of more explicitly community-minded networks—the Cleveland Free-Net, the Blacksburg Electronic Village, and Seattle Community Network, among others—that were set up explicitly to strengthen real-life communities and advance a specific vision of deliberative democracy. Many of these sites gained a substantial audience and remained viable so long as the cost of Internet access remained high. But the declining cost of access along with the general commercialization of the Internet undermined the basic rationale for providing "free" Internet access. Some did make the transition to more commercially oriented Web-based services.

Still, this commercialization of the early Free-Nets should not obscure the extent to which the communitarian ideals from this period in the history of online communities continue to affect the conduct of many users. Indeed, these values remain embedded in the design of mainstream communications software, as documented by Stanford University media studies scholar Fred Turner. Specifically, Turner describes how liberal and countercultural norms were embedded into the design of The WELL, the very online, text-based community that Rheingold popularized through his notion of a virtual community.[2]

The WELL, whose full title was The Whole Earth 'Lectronic Link, was a direct outcome of the *Whole Earth Catalog* and other publications produced by the Whole Earth organization, which was established by the progressive entrepreneur Stuart Brand. For those no longer familiar with it, the *Whole Earth Catalog* was an independently produced consumer products catalog launched in 1968 that offered highly valued reviews of natural and technological products designed for alternative living. As depicted by Turner, Brand explicitly created the *Catalog* in the name of restoring community in an era marked by both political and technological alienation.[3]

Many regard the *Whole Earth Catalog* as a document that promoted a countercultural alternative to materialism and mainstream consumer culture. As Turner points out, however, the *Catalog*, along with the *Whole Earth Review* and other Whole Earth publications, offered a much more direct precedent for The WELL by providing a discursive space for conversations among a geographically distended community. The product reviews published in the *Catalog* were often written by trusted members of the community. In fact, the *Catalog* did not sell any products directly but aimed simply to introduce its readers to alternative products manufacturers and their resellers. What "made" the catalog was then the mode of self-presentation whereby the mostly voluntary, amateur authors of the reviews came to generate a community of shared interest around alternative products and a vision of alternative living more generally. The dialogue within the *Whole Earth Catalog* and the *Whole Earth Review* helped define alternative attitudes toward consumption as well as broader countercultural values about a wide range of subjects from mysticism and religion, to technology and community.[4]

One can see in the *Whole Earth Catalog* an important precedent for the practice of customer reviews found on sites such as Amazon.com. Even in its diluted form, this practice owes itself to a certain set of populist, if no longer fully democratic, ideals about consumer empowerment. Those who have tried to manipulate customer reviews to their advantage have often discovered the underlying norms that remain widely diffused among online customers. Although not the direct point of his study, Turner's account makes it clear that these values can be traced back to the *Whole Earth Catalog* and other print-based forums for product reviews. (The culture of book reviews being the other major precedent, especially for Amazon.com.)

Turner's main focus was on the broader influence of the *Catalog* on online communities and community networks. Working from the thesis that technologies can be designed to support a specific set of values, Turner explained how The WELL's creators—Brand, along with former members of the Tennessee-based commune The Farm who served as The WELL's initial staff—developed a vision that called for The WELL to be free, or at least inexpensive to access; that it nevertheless was "profitable" enough to survive as a going concern; that it be open and self-governing; and that it would be a "self-designing experiment" just like the *Whole Earth Catalog*. Moreover, the mode of self-presentation found in the *Catalog* became a distinctive feature of The WELL, along with other community networks. Having a core body of active and visible contributors who regularly write to a public forum, which is then read by a silent but larger body of readers who give the venue legitimacy, became an important hallmark for the social interactions found in nearly all virtual communities. It should be noted that communitarian ideals were embedded into other early community networks, such as the Blacksburg Electronic Village, that influenced subsequent developments. Whether by plan or habit, contemporary e-commerce sites have indeed made use of such social dynamics; for Amazon.com, this can be found in its system of "spotlight reviews" and the "badges" it awards to its top reviewers.[5]

Rheingold's *Virtual Communities* was in fact a major watershed in the popularity of online communities. Up until then, most of the scholarly interest lay with computer-mediated communication—generally using computers to improve productivity in the workplace—while prevailing community networks remained tied to a physically local, geographically bounded community. His book, and subsequent career as the first executive editor of *HotWired*, also helped to carry forward the democratic and communitarian ideals that remained integral to The WELL and other existing online communities. This is not to say that there were transcendent, normative criteria for all new community-oriented information services. Sites such as AOL, eBay, and Match.com all represent varying degrees of departure from earlier norms. Nevertheless, even as the Internet became substantially more commercialized, certain value systems remained crucial to a fairly broad population of computer users, as documented below. So long as community networks operate under the general logic of a network economy, there remained substantial incentives to be attuned to the more "vocal," tech-savvy users, who as early adopters could help select successful ventures within a highly competitive field.

Converted Services

Chronologically, the first set of services to capitalize on the growing interest in online communities was the converted services—namely, information services that were able to carry an established clientele over from an earlier dial-up format to their present-day Web-based platform. The WELL itself was one of the sites that made such a transition, but a better example for demonstrating how a strong interest in community can be instrumental to the success of a more mainstream, commercial enterprise can be found in AOL. As described by Christine Ogan and Randal A. Beam in this volume, AOL was one of the early ISPs that made a successful transition to an informational environment dominated by the Web. Without doubt, there were many different reasons for AOL's success; nevertheless, in this section I wish to focus on how AOL successfully marketed its online services by appealing to a projected sense of community, especially as observed by the social historian and labor analyst Hector Postigo.

Unlike its competitor, CompuServe, which concentrated on offering general Internet access, AOL marketed itself as a medium of communications that aimed to build a vast online community. Beginning with CEO Steve Case as AOL's self-appointed mayor, AOL chose to devote considerable energy to building community and establishing content internal to the AOL domain. This material was built and maintained using an extensive network of volunteers—some fifteen thousand volunteers by 1999—who helped to constitute AOL into something more than a simple ISP. Relevant content, in the form of chat rooms and discussion forums on topics ranging from education, to computers, to sports and other leisure activities, was an important part of AOL's identity early on. AOL was able to surpass CompuServe in part by offering meaningful content at a time when the Internet itself contained relatively little material of interest to all but expert users.[6]

Once the Web became a familiar and ubiquitous entity, AOL itself made a transition into being more of a portal to the Internet at large; dedicated content is less of an attraction for many AOL users today. Still, AOL's volunteer network continued for a while to help novice users enter the era of digital literacy in drawing on the traditions of mentoring and voluntary assistance that have long been a feature of both academic and commercial computing facilities. Whether or not a majority of users needed or benefited from this service, the presence of a volunteer community inside the AOL domain made it possible for AOL to market itself as a user-friendly service. In all of these respects, the notion of community played into AOL's strategy of expansion and market domination, which combined ease of access and relevant content with an aggressive pricing scheme. (In 1996, AOL pioneered the shift from an hourly connection charge to its original fixed-price, $19.99 per month fee.)

Another well-known information service that made its transition by retaining close ties, here with two preexisting professional communities, is LexisNexis. Initiated as an experiment by the Ohio State Bar Association, Lexis was originally a dedicated, dial-up legal information service. The service was offered to the general legal community in 1973 by Mead Data Central. Starting with full-text search capabilities for all Ohio and New York legal cases, by 1980 Lexis had assembled a hand-keyed archive containing all federal and state legal cases in the United States. Also in 1980, Mead Data Central added the Nexis information service, which provided journalists with access to a searchable database of published news articles. Since then, LexisNexis has continued to diversify its data offerings by expanding into other legal jurisdictions, offering unpublished case opinions, and compiling ancillary information such as a record of deeds and mortgages as well as a mailing list with the addresses of all U.S. residents. It has also considerably expanded the scope of coverage within the Nexis portion of the database.[7]

Without a doubt, LexisNexis' success was based on being the first mover in a market that was of considerable value to the members of these two professional communities. A searchable database of legal case opinions was an invaluable asset to lawyers since this fell squarely within the existing work routine of the profession: every case had to be built up through rigorous references to past precedents. Likewise, Nexis provided journalists with a vast store of information about past stories and developments. It gave independent journalists and reporters working at smaller newspapers access to the same kind of news archive or "morgue" maintained by the largest metropolitan papers. LexisNexis has had competition from other firms—the West Publishing Company (now Thomson West) in the legal information services sector—and various online services in the news and general information sector. Yet the firm has enjoyed an oligopolistic market position as a result of a loyal client base that typically chose to remain with a familiar service to which they had adapted their professional practice. As such, neither LexisNexis nor its major competitors found much difficulty getting their users to move over to a Web-based service, although in moving to the Web they have all adopted a strategy of serving also as an information portal to the more diverse sources of information available on the Web.[8]

For the most part, the changes at LexisNexis occurred through a customer-driven, evolutionary model where the service was gradually expanded to include information resources and features that were either identified by or thought to be of value to the firm's major clients. There was little direct invocation of the notion of community, except to the extent to which the designers, developers, and executives at LexisNexis sought to understand the professional communities they wished to serve. A different approach, in which a more intricate knowledge of the conduct within a given professional community is explicitly built into the design of an information service, can be found in the example of Counsel Connect. Counsel Connect, as described by well-known cyberlaw scholar Lawrence Lessig, was created by the legal entrepreneur David Johnson. From the outset, Johnson envisioned his service as a kind of lawyer's cooperative, where the idea was to "give subscribers access to each other; let them engage in conversations with each other," and that through this access, lawyers would "give and take work; they would contribute ideas as they found ideas in this space. A different kind of law practice would emerge—less insular, less exclusive, more broadly based."[9] With this as a goal, Johnson created Counsel Connect with several prespecified features: individuals were expected to identify themselves through their real-world identity; discussion groups were to be facilitated by a leader (but not a moderator authorized to cancel a post); all posts, moreover, were retained in an archive and organized into threads so that each contributor had the ability as well as implied obligation to work through an existing sequence of exchanges before adding their own post. Membership was also limited to those belonging to the legal profession.

As noted by Lessig, each of these rules was established to strike an appropriate balance between the free flow of ideas and accountability for the ideas expressed, all in the interest of enhancing the quality of the legal discourse on this site. The anonymity allowed at other online forums would have undercut personal and professional accountability, even as open membership could have undermined professional dialogue. The site design also enabled personal reputations that were built up through posts to emerge as a valuable asset (similar to the way reputations accrued in The WELL), because law was a profession where it was common for individuals to give and take work in their respective specialties. Far from being an abstract design exercise, Johnson's understanding of the customary conduct of the legal profession led him to the features he wanted to see in Counsel Connect. Counsel Connect was in fact one of the diversified services offered for a while through LexisNexis before it was sold to American Lawyer Media, which converted it to a Web-based service in 1997. It has continued to operate there since.[10]

As mentioned above, this practice of designing and adapting information technologies to be carefully integrated with their social and institutional contexts (or at least retrospectively assessing their efficacy in such terms) can be associated with the emerging field of social informatics. A term brought into general use by Rob Kling at Indiana University, social informatics is a field that grew out of early studies of computer-mediated communication and especially computer-supported cooperative work. Rooted in academic inquiry, this line of research drew on the broad interest in a study of labor processes inaugurated during the 1970s by labor historians and those studying labor and industrial relations. This focus on labor processes by computer scientists and systems analysts revealed that new technologies could as often be disruptive as helpful when introduced into a tightly organized workplace, such as the payroll operations of a large company, the course registration procedures of a university, and even the checkout lanes of a grocery store. The problem was frequently not just that there was a lack of an adequate understanding of the existing operations but a failure to appreciate the more subtle and often unstated social dynamics that exist within any workplace.

All workplaces function, in effect, as their own kind of community. It is also important to acknowledge how the politics of a workplace can serve either to hinder or facilitate the introduction of new technology.

Drawing on the language of the broader "constructivist" movement in the social sciences, Kling targeted the largely situation independent, or "context-free" guidelines that many computer science departments promoted as a sound approach to the design of new information systems. His decision to create the Center for Social Informatics at Indiana University also coincided with the popular concern that emerged during the mid-1990s around the "productivity paradox." During the dot-com boom, firms invested heavily in information technology infrastructure. Aggregate spending on information technology resources by U.S. firms rose to become as much as one-half of all current capital expenditures. Though the trend has been reversed, indiscriminate spending on information technology resources, especially when paired with the lack of attention to the context surrounding each application, produced what in retrospect was an unsurprisingly low return on investment.[11]

Kling's classic essay "What Is Social Informatics and Why Does It Matter?" is still a fresh description of many of the common errors associated with information systems design and deployment during this period. In it are portraits of the failed introduction of Lotus Notes into Price Waterhouse (as originally studied by Wanda Orlikowski), where technological enthusiasm prevented the firm from considering how the prevailing incentive structures within a consulting firm prohibited the system from being embraced by the firm's junior analysts, who were the main intended audience; IBM's initial attempt to design a new civilian air traffic control system (in 1993) that would have required controllers to complete a sixty-five-field database entry in clear violation of this application's requirement for rapid, real-time responses; and the different fates of two new electronic journals in artificial intelligence and cognitive science research, only one of which took into account the social mechanisms of peer review and academic reputation so as to ensure its success (similar to Counsel Connect).[12] Social informatics remains an active and growing field, and much of the work can be viewed as supporting Lessig's assertion that *code* can be structured so as to support a specific pattern of social interaction, and that it ought to be designed with the interests of users in mind.[13]

Brokerage Services

Next to arrive on the scene were a series of brokerage services such as eBay and Napster—services that were integral to the expansion of e-commerce as discussed more generally by Kirsch in this volume. The reference here is not to general online brokerage services such as E*Trade Financial but rather different kinds of brokerage services that operate primarily within the leisure and entertainment sector of the economy. Similar to the *Whole Earth Catalog*, these brokerage services operated not through the direct sale of goods but through their circulation. Like the *Catalog*, they served as vehicles for cultural production. In addition, these brokerages services drew more directly on the various elements of early online communities, although I do also consider the more limited case of Amazon Marketplace below.

Most readers will undoubtedly be familiar with the basic history of eBay. It was founded in 1995 as an online auction site by San Francisco Bay Area computer programmer Pierre Omidyar. The company has since admitted that the story that Omidyar created eBay to help his fiancée collect and exchange PEZ dispensers was a fabrication by one of the company's public relations managers. Nevertheless, the site clearly tapped into a widespread collectors' culture—one that constituted a significant underground economy that could be brought to the surface and developed to its full potential via the Internet. eBay has since diversified into other areas of operation, including the popular online bills payment service PayPal. Its principal revenue stream as well as business model, however, remains that of charging transaction fees from users who list and sell items on eBay ($0.20 to $80 per listing, plus a 2 to 8 percent commission).[14]

eBay did quickly develop into a diverse, international marketplace, extending well beyond the collectors' culture in which it had its origins. By 2004, the firm was estimating that some 430,000 people around the world were making all or most of their income selling items on eBay or its affiliated sites, making eBay, in effect, the second-largest Fortune 500 "employer" after Wal-Mart. Although a good deal of this expansion occurred through eBay's transformation into a general resale market, the firm drew on diverse forms of subcultures to do so. This included the U.S. culture of garage sales, the more general obsession with "bargains," and an established culture of both urban and rural auctioneering. During its early growth, eBay proceeded to tap into various subcultures surrounding distinct hobbies and markets, from antiques and vintage clothing, to computers, gaming, and photography. It also appealed to more uniquely definable market segments, such as environmentalists committed to purchasing secondhand goods rather than promoting the manufacture of new products. Another key element was the cultural dynamics of gambling. The sustained interactions and time-bounded auction period that Omidyar chose as the format for online auctions tapped into a dynamic of addiction that remained latent within the more general auctioneering culture. In other words, eBay successfully built itself up by structuring the interactions among users so as to mobilize and augment the desire for exchange that existed within, and at the juncture of, multiple subcultures.[15]

Having more or less fortuitously designed a new commercial engine for exchange, Omidyar was more or less able to preside over eBay's growth, which occurred largely through the efforts of its clientele. The system he developed basically required the users to design and maintain the multitude of Web pages that constituted eBay's online products catalog. Still, it is worth considering what the implications of this specific business model (more specific, that is, than the general model of a brokerage service) were from the standpoint of corporate management and strategy. It was in March 1998 that Omidyar brought in the Harvard MBA and former Disney executive Meg Whitman as the president and CEO of eBay. As described in a 2004 cover story in *Fortune* magazine by journalist Patricia Sellers, Whitman quickly developed a management style that was consistent with eBay's mode of operation. Building on her background in consumer products marketing, Whitman placed her initial emphasis on cultivating eBay's existing clientele, and indeed allowing the machinery of eBay to meld into the traditional channels of communication that already tied together distinct consumer subcultures. None too different from the *Whole Earth Catalog*, and drawing on both the mechanism and habits of self-presentation found in online communities such as The WELL, eBay created the means for many intersecting subcultures to augment their cultures of consumption. This was, albeit, without the progressive vision, and with the direct sales model absent in the *Whole Earth Catalog*.

Whitman accurately judged that the best way to accomplish outreach and growth was through eBay's ever-expanding network of loyal buyers and sellers. Thus, unlike the flashy national advertising campaigns launched by the then competitive entities such as Priceline.com, eBay did not partake in television advertising until 2004. Unlike Amazon.com's CEO, Jeff Bezos, Whitman chose to keep eBay tightly focused on its core business. As the U.S. market growth began to plateau, the firm did move cautiously into the realm of online superstores (via its eBay Stores and the acquisition of Shopping.com), financial services (via PayPal), and telecommunications (via Skype). Nevertheless, the firm's primary business strategy to this day remains that of exporting its tightly integrated and multifaceted formula to international markets. This has allowed eBay to outperform other e-commerce sites and weather as well the general collapse of the dot-com boom.[16]

eBay has also attempted, more explicitly, to cultivate its buyers and sellers into a community of its own. It has done so using not only chat rooms, discussion boards, and other techniques (such as blogs) that derive from online communities but through formal training sessions, mentoring programs, and an annual gathering, eBay Live! All of these latter methods draw on quite traditional practices for cultivating a sales force. But the most important element of community that eBay has incorporated into its business model has been a social network of trust—something that eBay has consistently emphasized as being at the top of its list of "community values." This is expressed through their phrase, "We believe people are basically good." This normative position remains a necessary stance for a business where every purchaser must trust that a seller will deliver the goods one has paid for. eBay places the verified cases of fraud at less than 0.01 percent of all transactions.[17] The social trust is backed in part by dispute resolution procedures, enforcement mechanisms, and financial guarantees offered through services such as the PayPal Buyer Protection Plan, which can be used to guarantee purchases up to $1,000. Yet the most essential mechanism for establishing trust has been eBay's system of customer feedback, where both a rating system and a series of customer-generated comments create, over time, a reliable composite of each seller's identity. This too mirrors the mechanisms used to construct identities in early online communities; it also transcends them in some ways. For instance, the conduct associated with eBay's feedback mechanism has drawn some managerial interest from the standpoint of game theory. But the overall effect of this system has been to foster a general sense of trust in an environment where face-to-face evaluations of veracity are not possible.[18] It too is a specific instance of how code can be used to structure social interactions within an online environment.

Another classic example of an information service that has, or at least had, the promise to profit from being a culture broker is the music file-sharing service Napster. The general phenomenon of musical file sharing is described by William Aspray later in this volume, and therefore will not be discussed extensively here. Yet I wish to draw specific attention to how the notion of community, as found both within and across various musical subcultures, contributed to the rise and fall of Napster, and has shaped file-sharing practices since.

It is worth reiterating that although Napster was shut down in July 2001 through a court order that ruled that the service continued to facilitate massive copyright violations, musical file sharing continued, and still occurs through more fully decentralized (and hence less prone to regulation) peer-to-peer file-sharing programs such as Kazaa and LimeWire. While all of these services may seem in clear violation of the law to many—which they are—the musical industry has long operated with the tacit acceptance of (and some would argue, active reliance on) illicit exchanges of copyrighted material. Thus, although Fanning is generally credited with having launched the widespread interest in music file sharing, Napster, like eBay, was ultimately a service that built on a robust, preexisting exchange network. Those belonging to various musical subcultures had long participated in the exchange of musical recordings, first in analog form. Perhaps the most well-known example is the circulation of tapes of live concert recordings among Grateful Dead fans. And while the culture surrounding the Dead Heads is quite unique, a good deal of the interest and excitement in any musical genre involves cultural activities and exchanges that happen between musicians and across a musical audience—again similar to the way countercultural identities were constructed through the *Whole Earth Catalog*.

In his chapter, Aspray provides a strong set of reasons for the rapid ascent of musical file sharing. But in this chapter, let me again emphasize how aspects of community contributed to this ascent. In fact, although there are those who like to regard the recording industry's lawsuit against Napster, filed in 1999, as itself having fueled the popular interest in music file sharing, the outcry that emerged against the recording industry should be seen as something already deeply encoded into the culture of the musical listening audience. As documented by Aspray, there were well-established patterns of discontent with the recording industry, ranging from the price of compact discs, to insufficient royalties paid to artists, to the negative consequences various recording industry practices had on musical creativity. Also significant was the fact that this was an industry that was already founded on mechanisms of communications among members of the community. There were both online and off-line mediums, including publications such as *Rolling Stone* and even *Wired* magazine, broadcast media such as MTV, and the massive informal networks among high school students, college students, and various musical audiences that provided immediate venues for expressions of dissatisfaction with the recording industry.

From this point of view, the various arguments that emerged to justify music file sharing in the wake of the lawsuit—claims, for instance, about the legitimacy of acquiring digital tracks for recordings already purchased on vinyl, or predictions about the demise of copyright in the age of digital information, or even the "inevitability" of musical file exchange amid the "paradigm-shattering things called Napster"[19]—can best be understood within the framework of the sociology of deviance. All of these assertions constituted predictable efforts by a diverse and yet substantially organized community to construct an alternative normative system that lay external to mainstream legal and commercial institutions. That many of the file-sharing systems sprung up in academic environments draws not only on the facts mentioned by Aspray—the arrival of the MP3 compressed audio format, and the general availability of high-speed Internet access on university campuses by the mid-1990s—but the broader ways in which high schools and colleges support a liminal period of adolescence when students are able to create an environment with a sense of culpability and accountability different from that found elsewhere in society. That schools and colleges have long served as important sites for the illicit exchange of music, and that there was a rampant culture of software piracy that served as a direct precedent for musical file sharing, should no doubt be considered as well.

It is also interesting that despite the recording industry's legal victory, collective attitudes formed during the lawsuit have continued to influence business within this market. For instance, it is useful to compare the recent success of the iTunes Music Store, which is not a free service, against the relative failure of the repackaged Napster service (whose name and trademarks were acquired by the digital media software firm Roxio), even though Napster 2.0 attempts to provide users with some access to free music. The iTunes Music Store did undoubtedly enjoy a first-mover advantage in an industry where image and reputation play a major part in determining success. Still, it seems equally important that Roxio/Napster chose to promote an exclusive player and file format in an attempt to retain what it felt was the musical audience's demand for a "free" music service.[20] By contrast, Apple chose to stick to a publicly supported audio compression format, AAC (again, see Apray's chapter herein), which it integrated into its popular iPods. This in effect permitted iPod users to continue to engage in local, licit and illicit exchanges of musical tracks—a practice that in the end proved to be more valuable to the target community than a so-called free music service. Here again, success was determined by how well a service was structured to support the underlying interests of a community. The current business model implicitly includes an extralegal component, but one that unlike Kazaa and the old Napster, the recording industry has been willing to again grant its implied consent.

I also consider here, more briefly, Amazon Marketplace. Unlike the larger operations of Amazon.com, Amazon Marketplace operates as a brokerage service, competing in this respect with eBay. Its more specific business model involves allowing sellers of both new and used merchandise to list their products alongside Amazon's regular listings, charging sellers a 6 to 15 percent commission plus a transaction and closing fee that varies from $1.43 to $2.22 or more. Amazon is able to do this because these fees compare favorably with the margins Amazon earns on its routine book sales, and because this kind of business incurs no inventory or warehousing costs. As noted by Gary Rivlin of the *New York Times* in his recent review of Amazon's operations, the existence of Amazon Marketplace has also served to reassure customers who held doubts about Amazon's deep discount policies. Amazon Marketplace has been a critical part of the firm's revenue stream since its introduction in 2001.[21]

Amazon Marketplace draws on and operates with much less of a sense of community when compared to eBay. For the most part, users view it as a comparison-shopping engine that allows them, in most instances, to purchase a used item in lieu of a new product offered by Amazon. Nevertheless, Amazon Marketplace is sustained through the same mechanism of social trust—one upheld through a similar vehicle for customer ratings and feedback. (The only significant difference is that Amazon Marketplace uses a five-point rating scale rather than the positive/neutral/negative rating system employed by eBay, in keeping with the scheme for customer product reviews found on its larger site.) On the other hand, given that the products listed on Amazon Marketplace are restricted to a relatively narrow range of commodities—mostly books, compact discs, videos, DVDs, computer games, and software—this system of trust has unfolded with a subtle difference. On Amazon Marketplace, customer reviews and ratings have allowed sellers who are able to offer a good product selection and reliable service to surface to the top of the market. This has had a significant impact, especially in the used book market. This too is a market with a well-established history and subculture, and unlike the effects of Amazon on independent booksellers more generally, many independents specializing in used and remaindered books have benefited from the new national outlet created by Amazon Marketplace.

Social Networking Services

A different kind of community-oriented information service that has enjoyed more recent growth is the social networking service, such as Meetup.com or Match.com. Clearly sites such as Match.com, AmericanSingles, and eHarmony are online versions of the personals column found in local newspapers and entertainment weeklies. The general availability of graphic browsers and the low cost of online publishing more generally, however, have provided users with an opportunity to employ a much greater bandwidth in presentations of their "selves." Online dating services have been quick to carefully package this capability, drawing simultaneously on earlier elements of online services as well as the traditional personals column in creating a uniform format by which individuals can compose a personal "profile" through a combination of images and text.

The basic business model here revolves around subscription revenues (Match.com's current rate varies from $12.99 to $39.98 per month, depending primarily on the contract duration). This is an industry that draws most explicitly on the network effect, and this tends to give commercial services that advertise heavily a distinct advantage— although there are a number of "free" online dating sites that continue to operate through advertising revenue. Beyond this basic choice, the specific business models advanced by the firms in this industry remain somewhat fluid. Many of the sites have recently augmented their "coaching" services and other customer relations functions in emulating more traditional matchmaking services. Meanwhile, a key component for all of these services has been their proprietary survey and matchmaking algorithm, which are based increasingly on scientific claims about efficacy. As of 2004, online dating accounted for 43 percent of the $991 million dating services market, and was estimated to have been tried by approximately 20 percent of all singles in the United States.[22]

Personals ads as a whole work on the basis of a presentation of self. Yet as some cultural analysts working in the "symbolic interactionist" tradition in sociology have observed, the increased bandwidth of the personal profiles presented via online dating sites has provided individuals with a more extensive opportunity to experiment with their identity. Based on detailed interviews with thirty-three active users of these services, Jennifer Yurchisin, Kittichai Watchravesringkan, and Deborah McCabe found that individuals chose quite different strategies for negotiating through their different "possible selves," ranging from those who chose to construct stable, "honest" profiles, to those who used the medium to try on different identities (for instance, by drawing on traits found in other profiles they themselves found to be attractive). This ability to experiment with projected identities has proved to be of special value to distinct segments of the dating population, such as divorcees seeking to regain self-confidence and refashion an identity they feel would be conducive to a better relationship.[23]

With regard to the proprietary surveys and matchmaking algorithms, the scientific basis of these instruments has attracted some criticism both in the popular and academic literature.[24] There are those who continue to doubt whether matchmaking algorithms produce results any better than a blind date, and a client's manual perusal through personality profiles, including those generated by a matchmaking algorithm, remain an important part of the overall online dating experience. But irrespective of these algorithms' validity, the specific body of scientific theory employed by each of the major services—a similarities-based model of compatibility at Match.com, and the biochemistry of physical attraction at Chemistry.com—has functioned as part of an integrated business plan for differentiating each service. Broader strategies for differentiation include how a site structures its profiles (Match.com, for example, places greater emphasis on the subjective comments that clients attach in the "In My Words" section of each profile), screening policies, and targeted advertising campaigns designed to develop specific subpopulations within its membership. This has allowed each site to constitute different communities of courtship, from sites such as Match.com and Chemistry.com that encourage casual dating, to eHarmony's focus on marriage and long-term relationships, to True.com's stress on personal safety and security.[25]

Viewed from a broader cultural perspective, it is clear that each of these services have adapted themselves to the different practices of courtship that can be found in U.S. society at large. Overall, courtship in this country has evolved into a regime that emphasizes individual initiative and responsibility, even as demographic changes in divorce rates, the age of marriage, and women's participation in the workforce, among other things, have introduced complexity and uncertainty to the process. Changes in sexual mores have also ensured that the purpose of dating has not exclusively been that of long-term relationships and marriage. In this respect, the variation that can be found on current online dating sites, which now includes many specialized sites that cater to specific religions, ethnicities, and other social groupings, can all be seen as explicit attempts to structure interactions in a manner consistent with the courtship rituals of different dating populations.

A more general form of social network service can be found in the now highly popular sites such as MySpace, Facebook, Friendster, and Orkut. *Time* magazine recently reported on such sites as having "become, almost overnight, booming teen magnets exerting an almost irresistible pull on kids' time and attention." Less than three years old, MySpace and Facebook, as the leading two sites, have attracted well over twenty million subscribers, and consistently rank within the top ten Web sites accessed on the Internet. Much more so than the online dating sites, these sites have explicitly adopted elements of early community networking software, augmenting them with new and distinctive features designed to facilitate specific patterns of social network formation. Thus, in addition to the traditional use of discussion boards and threads, Facebook, for instance, makes it possible for an individual to "poke" a friend (to signal that you're around), or to check the "pulse" at your high school or college by finding out what music or motion picture is currently the most talked about at your school. Orkut restricts membership to a referral network, thereby encouraging circles of social exclusivity. Friendster allows members to offer "testimonials" of other members, and re-presents this through a system of ranking that fuels popularity contests and contests for visibility.[26]

These sites have uniformly adopted the text and image-based presentation of the personal profiles found on online dating sites. And although courtship occurs through these sites, they have served more generally as a place for creating conversational communities among individuals with shared interests and especially within self-designated friendship circles. Although their membership base has begun to extend beyond the high school and college populations, these sites remain highly popular with youth. As far as their specific business model goes, because the efficacy of these sites depends on the size of the membership that actively contribute to ongoing conversations, they generally operate as free services, where income is generated primarily through advertising revenues.

Meanwhile, the idea that specific design features within a social networking site can structure the nature of the social interactions that occur within the site can be seen in the recent uproar over certain design changes made to Facebook. On September 5, 2006, Facebook installed a couple of controversial new features. The most contentious of these was the "Newsfeed," which provided each member with an automatically generated, minute-by-minute log of every action taken by those in a member's friendship circle. It would report everything from the most mundane actions of a friend, to breaking news that someone you admired had broken up with their partner. Moreover, the designers provided no features that would either allow a user to turn off the Newsfeed or block others from viewing their actions through this utility. Facebook members were quick to decry the "'Big Brother' nature of the new layout." Within two days, there were multiple protest groups organized as discussion groups inside Facebook itself, with the largest group, "Students Against Facebook News Feed," attracting over seven hundred thousand members in this two-day period. This again demonstrates the extent to which specific values and expectations about online communities remain embedded within a large segment of users.[27]

This incident was more or less a lesson in poor design—a design that failed to take into account some basic principles in social informatics. The company was in fact quick to add new privacy features. Yet it should also be noted that online dating services and other social networking sites have attracted broader scrutiny and criticism. The underlying architecture of existing systems, which requires users to construct a single profile, can significantly constrain the range of possible interactions because each user is required to reveal at the outset an extensive, "authentic" identity. This might offset the value and personal freedom that individuals can hope to gain from the identity play that can occur over time. Especially with social networking services such as Friendster, the investment required to design an interesting profile (which often includes an original Web site), and the "assets" accrued through the feedback received from other members in the community, tends to promote a fixity of identity rather than routine reconstruction and experimentation with identity. These sites have also been described as producing highly insular, "decontextualized" communities for promoting an egotistical culture of "personalism," and for the general commodification of identity that remains anathema to a more progressive vision of community.[28]

This progressive vision may be found, by contrast, on the social networking site Meetup.com. Though a for-profit entity, Meetup was set up in response to a well-known essay, titled "Bowling Alone: America's Declining Social Capital," written in 1995 by Harvard political scientist Robert Putnam.[29] While Putnam's use of the metaphor of social capital has itself drawn criticism for contributing to the commodification of culture, Putnam nevertheless presented a broad-based argument that the general decline of community and voluntary associations in the United States was undermining the active civic engagement required for a strong democracy. Entrepreneurs Scott Heiferman, Matt Meeker, and Peter Kamali created Meetup.com in an explicit attempt to cultivate communities of interest through the use of the Internet.

Meetup basically provides its members with an opportunity to identify other people with shared interests and then organize a local gathering of these members. Although Meetup's publicists insist that it is not a social networking site, the service clearly operates through social networking. The only difference is that instead of enticing users to create social networks that "thrive on virtual online relationships and anonymity," the social networks assembled through Meetup are built exclusively on preexisting hobbies and interests. This serves to work against the insularity of virtual communities by both relying on and promoting natural connections back to the family and broader elements of community.[30] Meetup events are also real-world gatherings and therefore work under the model of local, geographically situated communities.

As far as its specific business model, Meetup operates by combining several different sources of income including advertising revenues, fees paid by restaurants and other businesses that host a Meetup event, and a modest monthly fee paid by the organizer of each Meetup group. Meetup has drawn considerable interest from venture philanthropists, including eBay founder Pierre Omidyar. Omidyar had specifically created his Omidyar Network as an alternative to traditional philanthropy after realizing that a for-profit entity such as eBay could have as great, if not greater, an impact in terms of social and economic development and transformation. (Meetup also gained a $2 million investment directly from eBay in 2006.)[31]

While many initially regarded Meetup to be an eclectic idea born of the dot-com boom, the site gained considerable notoriety when staff members from the 2004 Howard Dean presidential campaign took note of the grassroots political organizing for Dean that was occurring on Meetup and made use of it to augment Dean's support base. Meetup soon garnered national media attention, as the "topic" representing Dean's presidential bid, "Democracy for America," attracted over 140,000 members. Meetup gave Dean's campaign a considerable boost, especially by amassing a highly loyal following at an early stage in the campaign. Regardless of this happenstance, Meetup has maintained itself as a nonpartisan operation in keeping with the broader principles of a strong democracy. Its 2.5 million members have meanwhile continued to form an endless variety of communities of interest—knitting groups, autism groups, pug owners' groups, and groups of single parents and stay-at-home moms—in advancing the vision of voluntary association that Putnam argued was essential for strong democracy.[32]

Community Networking Services

Finally, I would like to turn to a range of services that have upheld the ideals associated with early online communities such as The WELL, in carrying their progressive vision forward into the new infrastructure and bandwidth offered by the Internet. Meetup is in fact one instance of a site that has pursued such an end. Any business seeking to enter this arena, however, should be aware that most other sites that operate in this domain are nonprofit entities supported primarily through public funding and philanthropic activity. Indeed, much of the activity has been sustained through academic and community advocacy organizations dedicated to social action and intervention. This work is mostly carried out under the general banner of community informatics, as mentioned at the beginning of this chapter.

In fairness to the reader, I should be up-front about my own inclinations about the work in this arena. While I am quite content to have businesses draw on different notions and elements of community in designing a successful commercial entity, my hope is that firms operating in this particular domain will aim to earnestly understand and respect the needs of the communities they wish to serve. Especially for economically disadvantaged communities that want to use online services as a means of fostering economic redevelopment, it is important that the overall business model is not an extractive one that deprives communities of the resources that are essential for their well-being.

But there should be plenty of opportunities for synergy rather than tension. First, the most successful initiatives within this sector have been those that were attuned to and sought to support the real needs of the communities they hoped to serve. A purely extractive approach is likely to alienate a community whose close cooperation is necessary for success within this arena. Second, given that most of the work in this sector has been supported through governmental and philanthropic resources, a purely commercial entity is less likely to be competitive within this sector. Lastly, unlike the strength of social services funding in Europe and elsewhere, resources for the civic sector in the United States are quite limited. The discontinuous nature of philanthropic grants also introduces considerable uncertainty and instability. This suggests that there ought to be opportunities for greater public-private partnerships, where more sustainable operating incomes might be gained through a commercial component designed to augment existing nonprofit and public-sector initiatives.

Having said this, there have been few successful public-private collaborations to date. Confirming the limitations of purely commercial efforts, private initiatives such as TownOnline (a site that serves as the online version of the *Boston Herald* and other community newspapers owned by the Herald Media syndicate) have evolved into little more than a means of accessing newspapers online. Such efforts to add to the community, without first forging strong ties with established community institutions, have had difficulty generating the interest and relevant content needed to sustain a lively discursive community in the manner of social networking sites such as MySpace and Facebook. This is consistent with the findings of the academic literature on community informatics as well as the practical how-to manuals produced by community networking advocacy organizations. Both emphasize that the most effective forms of community networks begin with active community involvement and planning prior to the launch of a new community-oriented site on the Internet.[33]

From the standpoint of business, probably the most promising point of entry would be two different kinds of community networking services that have grassroots origins, but offer attractive opportunities for partnering with for-profit entities. The first of these would be Community Technology Centers (CTC): multiuse facilities that typically offer computer training and digital literacy programs as well as provide Internet access and other complementary services to members of a specific, geographically localized community.

Many CTCs had their origins in earlier, community-based computer training programs and facilities, many of which were located in inner-city schools, public housing projects, and other such locations. From this start, CTCs have come to offer a broad range of digital literacy and jobs/skills transition services. This can be discerned by examining any of the more successful centers, such as the East Palo Alto organization Plugged In.[34] In the United States, the CTC movement gained substantial momentum as a result of the digital divide initiatives launched by the Clinton administration. Community networking advocates have been somewhat ambivalent about the notion of a digital divide. Some see the term as all too easily misconstrued as a matter of access, which fails to address the underlying issues about the skills people need to fully benefit from a digital economy. This construal, along with the declining cost of computers, has allowed conservative pundits to portray the digital divide as a "digital lag." Moreover, there has been considerable retrenchment at the federal level with the change in administration. On the other hand, local and regional CTCs gained considerable strength during this era. Many remain affiliated with the nonprofit Community Technology Centers' Network, more commonly known as the CTCNet created by former Manhattan public school teacher Antonia Stone. This network has helped local CTCs create sustainable programs that are already substantially integrated into their community.[35]

As viewed from a broader socioeconomic and structural perspective, CTCs are but a stopgap measure within the major social transformations associated with economic globalization and the shift to a service-oriented, postindustrial economy. Both from the standpoint of reaching displaced individuals and the curriculum that CTCs can present, there remains a considerable gap between these centers and the junior colleges and technical institutes that can offer the sufficiently broad retraining program necessary for entering the "new economy." While the dot-com bust has relieved some of the labor-market pressures for more extensive worker training in the field of information technology, as this sector resumes its growth, there will be renewed calls for workforce development. Public-sector funding in the United States is unlikely to be able to meet such a demand. This is where there ought to be opportunities for greater public-private partnership, where new revenue streams, possibly as collected from private-sector employers who have to meet staffing requirements, can contribute toward complementing or augmenting the capabilities of present-day CTCs. This could take various forms, such as the development of more unified curricula, more standard software packages that assist with CTCs' operations, or internship and placement programs that make it possible for those who use CTCs to obtain a career as a member of a growing information technology–based workforce.

A second strategy, and one where there is an even more direct synergy with commercial interests, would be Assets-Based Community Development (ABCD). Based on the work of John Kretzmann and John McKnight at the Institute for Policy Research at Northwestern University, ABCD operates under the basic philosophy that economically strained communities are not poor in every respect. It encourages communities and aid organizations to begin by surveying the socioeconomic assets that every community has—assets that are absolutely vital in providing a foundation for social and economic redevelopment. For instance, rather than considering unemployment in an exclusively negative light, ABCD initiatives would begin by considering this as an important part of a community's assets in personnel: these residents can bring new income into the community, given the appropriate training and retraining programs.[36]

When implemented specifically as one part of a broader urban redevelopment initiative, one of the common practices in ABCD has been that of mapping out the various commercial and public institutions and assets that already belong to a given community. This offers a way to promote local commerce, thereby improving the community's cash flow and jobs situation. Access to computers remains a significant barrier to this approach, and there will always be a place for publicly funded components (such as public kiosks). Still, this is clearly a realm where there is an opportunity for public-private partnerships, with the more progressive programs allowing cross subsidies of various community-oriented activities through advertising and other revenues. So far, the most successful programs of this sort, such as the Neighborhood Knowledge Los Angeles' asset-mapping program, remain university-based public-service initiatives.[37]

All of these efforts can be properly classified within the emerging field of community informatics. The term, while in use for some time, was made commonplace by Michael Gurstein, currently a visiting professor at the New Jersey Institute of Technology. Gurstein is known for having edited the compilation *Community Informatics: Enabling Communities with Information and Communications Technologies*.[38] As suggested by the definition offered at the start of this chapter, those engaged in this work have not focused exclusively on economic development but rather the broader well-being of a community and its members. The researchers and practitioners who make up this field have constituted themselves into an active research network, and their published findings can be found in the online *Journal of Community Informatics*.[39] The academic component of this field remains highly interdisciplinary, and while much of the work to date has centered on physical communities, those in the field generally see their work as extensible to virtual communities built around shared interests. For-profit entities that wish to work in this arena should know in advance that there are some tensions, for the most part productive, between academicians and those with a grassroots, bottom-up orientation. The benefits and risks of public-private partnerships have also been discussed explicitly among members of this community. This is not to say that "advocacy"-based initiatives are the only viable approach to community networking. Efforts to define what constitutes a real community, which have surfaced in debates among the researchers and practitioners in this field, have done as much to foreclose certain approaches as to open up viable options. Nevertheless, to the extent to which current initiatives have been carried by those belonging to this field, commercial entities seeking to enter this arena should be aware of the values and assumptions that exist in this domain.

Conclusions

Without question, the Internet is a medium for business communication that has substantially altered the prevailing patterns of commerce by facilitating rapid interactions across diverse, international markets. As a medium of both personal and cultural communication, the Internet has also redrawn traditional lines of community, making it possible to create new associations and reinforce existing ones. It is at the intersection of these two trends that new enterprises and initiatives—from eBay, to Meetup, to Plugged In—emerged.

Given that my own training is in history, and not business or management, I am less able to offer astute observations about the current business prospects within this market. It can be said that community-oriented information services have generally operated under the principles of network economies, and as such, nearly all of the services, regardless of the specific sector in which they operate, have benefited substantially from first-mover advantages. Given the way in which the notion of community is mobilized to cultivate loyal users and customers, it is unlikely that a new site can be created to unseat an entity like eBay—something that eBay itself learned through its delayed entry into certain international markets, most notably Japan.[40] Significant opportunities remain, though, in many niche markets. In fact, this has been a trend within certain segments of the community-based information services market. Much of the most recent growth in the dating and matchmaking services market, for instance, has occurred through highly specialized sites—from Islamic matrimonial sites, to gay and lesbian dating services—where the distinct cultural pattern within a specific subculture creates a demand for a system different from that offered by mainstream services. Community networks that are tied to a specific geographic location, or specific interests, also present many opportunities for interventions and creative initiatives.

Regardless of the specific application or approach, successful enterprises have forged a unique synthesis between a viable business model and the interests of the specific community they have come to serve. Insofar as all community-oriented information services operate with distinct cultures and within the overall realm of cultural production, it has been necessary to design information systems that support the social interaction that uphold and strengthen existing communities. There are undoubtedly many other opportunities to do so.

Notes

1. This definition is provided by Larry Stillman, one of the active researchers in this community, as posted under "Community Informatics" on Wikipedia, available at ⟨http://en.wikipedia.org/wiki/Community_informatics⟩ (accessed July 19, 2006).

2. Fred Turner, "Where the Counterculture Met the New Economy: The WELL and the Origins of Virtual Community," *Technology and Culture* 46 (2005): 485–512.

3. Ibid.

4. Ibid.

5. Ibid., 498.

6. Hector Postigo, "Emerging Sources of Labor on the Internet: The Case of America Online Volunteers," *International Review of Social History* 48 (2003): 205–223. See also Tiziana Terranova, "Free Labor: Producing Culture for the Digital Economy," *Social Text* 63 (2000): 33–58. The

primary focus of both Postigo and Terranova's articles are on voluntary labor as characteristic of the kind of employment relations to be found in the new economy.

7. LexisNexis, "Company History," available at ⟨http://www.lexisnexis.com/presscenter/mediakit/history.asp⟩ (accessed July 19, 2006); "LexisNexis," available at ⟨http://en.wikipedia.org/wiki/Lexis-Nexis#History⟩ (accessed July 19, 2006). While I generally draw on published sources in this chapter, I also provide references to Wikipedia entries, where available, to facilitate direct access to current information about each service.

8. "LexisNexis," ⟨http://en.wikipedia.org/wiki/Lexis-Nexis#History⟩.

9. Lawrence Lessig, *Code and Other Laws of Cyberspace* (New York: Basic Books, 2000), 71–72.

10. Ibid.

11. John Leslie King, "Rob Kling and the Irvine School," *Information Society* 20 (2004): 97–99.

12. Wanda Orlikowski, "Learning from Notes: Organizational Issues in Groupware Implementation," *Information Society* 9, no. 3 (1993): 237–250.

13. Rob Kling, "What Is Social Informatics and Why Does It Matter?" *D-Lib Magazine* 5 (January 1999), available at ⟨http://www.dlib.org/dlib/january99/kling/01kling.html⟩ (accessed July 19, 2006).

14. Steven Levy, Anjali Arora, and Esther Pan, "Wired for the Bottom Line," *Newsweek*, September 20, 1999, 42–49; see also "EBay," available at ⟨http://en.wikipedia.org/wiki/EBay⟩ (accessed July 19, 2006).

15. Patricia Sellers, "eBay's Secret," *Fortune*, October 18, 2004, 160–178. On Internet addiction, see Kimberly Young, "Internet Addiction: A New Clinical Phenomenon and Its Consequences," *American Behavioral Scientist* 48 (2004): 402–415.

16. Sellers, "eBay's Secret." On Bezos and Amazon.com, see Gary Rivlin, "A Retail Revolution Turns 10," *New York Times*, July 10, 2005, sec. 3, 1.

17. Given unreported cases of fraud, the actual figure is generally regarded to be somewhat higher.

18. For a study of trust on eBay, see, for instance, Bob Rietjens, "Trust and Reputation on eBay: Towards a Legal Framework for Feedback Intermediaries," *Communications Technology Law* 15 (2006): 55–78.

19. John Barlow, "The Next Economy of Ideas," *Wired*, October 2000, available at ⟨http://www.wired.com/wired/archive/8.10/download.html⟩ (accessed July 19, 2006). Statements of this kind continued to proliferate in serious publications even after the injunction was upheld. See, for instance, Alan Karp, "Making Money Selling Content That Others Are Giving Away," *Communications of the ACM* 46, no. 1 (2003): 21–22.

20. As described by Aspray, under its current business model, all Napster users are allowed to listen to every track in the firm's musical library five times for free, via a negotiated arrangement with various labels within the recording industry. Beyond this, listeners have to subscribe to

Napster. They do also have the option to subscribe to a premium service that allows them to download musical files to compatible MP3 players.

21. Rivlin, "A Retail Revolution Turns 10."

22. Andrea Orr, *Meeting, Mating, and Cheating: Sex, Love, and the New World of Online Dating* (Old Tappan, NJ: Reuters Prentice Hall, 2003); Alan Smith, "Exploring Online Dating and Customer Relationship Management," *Online Information Review* 29 (2005): 18–33; James Houran, Rense Lange, P. Jason Rentfrow, and Karin Bruckner, "Do Online Matchmaking Tests Work? An Assessment of Preliminary Evidence for a Publicized 'Predictive Model of Marital Success,'" *North American Journal of Psychology* 6 (2004): 507–526.

23. Jennifer Yurchisin, Kittichai Watchravesringkan, and Deborah McCabe, "An Exploration of Identity Re-Creation in the Context of Internet Dating," *Social Behavior and Personality* 33 (2005): 735–750.

24. Lori Gottlieb, "How Do I Love Thee?" *Atlantic* 297 (2006): 58–67; Houran et al., "Do Online Matchmaking Tests Work?"

25. Gottlieb, "How Do I Love Thee?"

26. Michael Duffy, "A Dad's Encounter with the Vortex of Facebook," *Time*, March 19, 2006, available at ⟨http://www.time.com/time/magazine/article/0,9171,1174704,00.html⟩ (accessed September 9, 2006).

27. Justin Hesser and Dave Studinski, "Feed Angers Facebook Users," *Ball State Daily News Online*, September 7, 2006, available at ⟨http://www.bsudailynews.com/media/storage/paper849/news/2006/09/07/News/Feed-Angers.Facebook.Users-2260799.shtml?norewrite200609222001&sourcedomain=www.bsudailynews.com⟩ (accessed September 9, 2006); Andrew Cheong to Atsushi Akera, September 5, 2006, letter in author's possession.

28. Given the recent introduction of these services, critical studies of this nature are themselves a relatively recent development, and may be found at current conferences such as those held by the Association of Internet Researchers. See, for instance, Alice Marwick, "'I'm a Lot More Interesting Than a Friendster Profile': Identity Presentation, Authenticity, and Power in Social Networking Services," and Felicia Song, "From Hanging in The WELL to Going to Meetups: An Analysis of the Evolution of Online Communities and Their Democratic Potential" (papers presented at Internet Research 6.0, Chicago), available at ⟨http://conferences.aoir.org/papers.php?cf=3⟩ (accessed July 19, 2006).

29. Robert Putnam, "Bowling Alone: America's Declining Social Capital," *Journal of Democracy* 6 (1995): 65–78.

30. Meetup.com, "Over 2.5 Million People," press kit, available at ⟨http://press.meetup.com/pdfs/mediakit.pdf⟩ (accessed July 19, 2006); see also "Meetup.com," available at ⟨http://en.wikipedia.org/wiki/Meetup⟩ (accessed July 19, 2006).

31. "Pierre Omidyar," available at ⟨http://en.wikipedia.org/wiki/Pierre_Omidyar⟩ (accessed July 19, 2006).

32. "Meetup.com."

33. TownOnline, available at ⟨http://www.townonline.com⟩ (accessed July 19, 2006).

34. Plugged In, available at ⟨http://www.pluggedin.org/⟩ (accessed July 19, 2006).

35. Community Technology Centers' Network, "About the Network," available at ⟨http://ctcnet
.org/who/network.htm⟩. See also its "CTC Startup Manual" (Washington, DC: Community Technology Centers' Network, 1997), available at ⟨http://ctcnet.org/what/resources/startup_manual
.htm⟩ (accessed July 19, 2006).

36. John Kretzmann and John McKnight, *Building Communities from the Inside Out* (Chicago:
ACTA Publications, 1993); Alison Mathie and Gord Cunningham, "From Clients to Citizens:
Asset-Based Community Development as a Strategy for Community-Driven Development," *Devel-
opment in Practice* 13 (2003): 474–486.

37. Deborah Page-Adams and Michael Sherraden, "Asset Building as a Community Revitalization
Strategy," *Social Work* 42 (1997): 423–434.

38. Michael Gurstein, ed., *Community Informatics: Enabling Communities with Information and
Communications Technologies* (Hershey, PA: Idea Group, 2000).

39. Available at ⟨http://ci-journal.net/⟩ (accessed July 19, 2006).

40. Sellers, "eBay's Secret."

VI Newly Created or Amplified Problems

14 File Sharing and the Music Industry

William Aspray

Stealing artists' music without paying for it fairly is absolutely piracy, and I'm talking about major-label recording contracts, not Napster.
Courtney Love, Digital Hollywood Conference, 2002

What's new is this amazingly effective distribution system for stolen property called the Internet—and no one's gonna shut down the Internet.
Steve Jobs, *Rolling Stone*, December 3, 2003

This chapter tells the story of the use of the Internet to share music.[1] It is a story of business interests, legal prohibitions, artistic expressions, and consumer desires. The most famous episode in this story is the use of Napster between 1999 and 2002 by millions of people to share music freely with one another—freely in both the sense of sharing with anyone connected to the Internet and also in the sense of sharing at no cost—and the ultimate destruction of the original Napster through the efforts of the music establishment using legal means. However, the story is much broader. The technology of centralized file sharing as embodied in Napster is only one of several new technologies, such as digital audio tape and decentralized file sharing, that have both created new opportunities to share music and presented new challenges to the music establishment.

In order to keep this chapter to a manageable size some closely related topics have not been covered. One is recent technologies for sharing music that are alternatives to peer-to-peer file sharing. These include webcasting, CD burning, and personal video recording such as Tivo. A second topic that receives only passing coverage, where necessary for understanding music file sharing, is the use of peer-to-peer networks to share films over the Internet. Because audio files are much smaller than video files and thus easier to transfer over the Internet, and because of some differences between the established distribution systems for music and film, file sharing of music over the Internet has come first. There is no question, however, that as storage, processor speed, and broadband connectivity improve, the film industry will face the same challenges as the music industry of having their intellectual products shared over the Internet. The story of digital film sharing needs to play out further before it is ready for historical examination. A third topic not covered here concerns copyright infringement (or is it fair use because of the exception for commentary and criticism in the copyright laws?) through the blending of copyrighted materials into new audio and video works. There are long traditions in both painting and American jazz where artists appropriate, adapt, and comment on previous works of art, but there is less acceptance of audios and videos that follow this same pattern of appropriation and customization.[2] A fourth topic that only receives brief mention is the notion of semiotic democracy, of using the Web as a place for individual expression instead of having the Web dominated by large commercial interests such as the major news, movie, and record companies. Finally, there are many different peer-to-peer systems for sharing music. Although the chapter covers a number of these systems, it does not cover two fairly popular ones, eDonkey and BitTorrent.

The Traditional Music Distribution System

From the end of the Second World War until around 1990, the system for distributing records was well established and relatively stable.[3] It was an oligopoly, dominated by six record companies: Capitol-EMI, CBS, MCA, Polygram, RCA, and Warner.[4] The system was tremendously profitable, but the profits were distributed in a highly uneven fashion, with the record companies generating large revenue and profits, but with the composers and recording artists generally gaining little financially from the distribution of their music. Among the recording artists, the distribution of income was also highly uneven. A few stars made a great deal of money, but it was not uncommon for other recorded artists—even well-known ones—to make little or no income from the distribution of their records once the record companies and various intermediaries took their cuts. Indeed, most recorded musicians made their income not from royalties on their recordings but instead from proceeds on their concerts.

This music distribution system was complex and economically inefficient.[5] Composers of a song would typically assign the rights to a music publisher. The music publisher would license companies to print and sell sheet music, sell the foreign rights, sell synchronization licenses to use the song in a movie, sell mechanical licenses (through an intermediary, The Harry Fox Agency) to use the song on an album, and use other intermediaries (the performing rights organizations ASCAP, BMI, and SESAC) to sell the licenses that enabled the song to be legally played as background in a restaurant or performed by radio or television stations.

During this period of time it was expensive to record a song. It required the use of a professional recording studio that cost many thousands of dollars to furnish. This has changed since 1990, with increasingly powerful recording equipment within reach of individuals through the use of their personal computer and some powerful software. It was also difficult before 1990 for an individual artist to promote and distribute his or her songs. The rise of the Internet changed the asymmetry, giving the individual musician a viable way to publicize and distribute songs, independent of the record establishment. Until the 1990s, a musician had little choice but to sign on with one of the record companies, and the record companies took advantage of their oligopolist position to exact a high price from the artist. From the 10 to 20 percent of proceeds credited to the artist from each record sold, the artist had to pay for labor costs of the producer and backup musicians; production costs such as studio and equipment rental, and mixing and editing costs; the "packaging cost" of putting the song onto the media (phonograph record, audio tape, or compact disc); promotional costs such as the cost of free records to distribute for promotional purposes, part of the cost of promotional videos, and part of the cost of independent promoters who persuade the radio stations to "air" the song; a reserve fund for recordings returned by stores; and even a "breakage fee" for damage to recordings during shipping (held over from the day of fragile vinyl LP phonograph records). Debits from any one recording were charged against profits on future recordings. As a result, in most cases the recording artist never saw much, if any profit from the record sales.

After 1990 the record companies and their intermediaries worked assiduously to maintain their favored economic position. Their position was threatened not by companies trying to enter their market and compete head-to-head with them at their own game, but instead by waves of technological innovation that enabled musicians and consumers to consider alternatives to the established system. The big record companies fought back through business means by using their tight lock on the distribution and airing systems for music and their large public relations and marketing arms, through technological means such as encrypting recordings on compact discs, and through legal means such as lobbying Congress for industry-friendly laws such as the Digital Millenium Copyright Act and aggressive use of these laws against alternative, Internet-based music distribution companies and eventually against musicians and individual listeners themselves.

Underlying Technologies

Until the 1980s, all of the means for recording and distributing music were analog. These included, for example, phonographic records and traditional analog tape recorders. Since then, however, a series of alternative digital technologies, such as digital audio tape recorders, compact disc burners, and MP3 file sharing technologies, have become available. Digital recordings have largely but not entirely taken over the market from analog recordings. It is still possible to buy recordings in phonographic or analog tape format, but there is a much wider selection available and many more copies are sold in digital form. The compact disc, which was developed jointly by Sony and Philips in 1981, rapidly became the preferred form of sale by the record companies, passing sales of phonograph records in 1988.[6] Breakage and shelf space in retail outlets was smaller for CDs than for phonograph records, and the initial higher sales price for CDs compared to phonograph records—presumably because of the added fidelity (disputed by record connoisseurs) and the initial high cost of building CD production facilities—was maintained even after the CD production system matured and it became much less expensive to manufacture and ship a CD than it did a phonograph record. The record companies used their business muscle with the large retail chains to get them to switch quickly to selling more of the digital CDs than the analog phonograph records.[7]

Interestingly, the record companies considered mechanisms, such as watermarking, for protecting their recordings from being copied when they first brought out CDs, but they decided against doing so.[8] Digital technologies have some differences from analog technologies for recording music. The most important difference for music file sharing is that analog recordings lose fidelity each time they are copied, whereas digital recordings do not. If one took a piece of music and copied it using an analog tape recorder, the copy would sound worse than the original and a copy of the copy would sound worse than the original copy. So by the time a recording had been passed around and copied and recopied multiple times, the fidelity would be much worse than the original. With a digital recording medium, however, no matter how many times it is copied and recopied, the copies would have the same fidelity as the original. Thus the digital copies would be a competitive product to the originals produced by the record company, whereas the analog copies would not have the same quality as the recordings the record companies sold and did not create a particularly strong reaction from the record industry. The reason the record companies did not worry about competition from digital reproductions at the time CDs were introduced, and thus take protective measures such as watermarking, was that they did not envision individual consumers having either a means for making digital copies or for distributing them widely.

Several technological innovations changed the situation entirely, creating a serious problem for the record companies in the form of digital reproduction:

• The Internet became a significant distribution channel for music files. Although, as discussed in the introductory chapter, work on the Internet began in the late 1960s with funding from DARPA, there was limited public access to the Internet until the 1990s, changed mostly by the rise of AOL, which reached a million subscribers in the mid-1990s. In fact, until 1991 there were strong norms against using the Internet for commercial or personal purposes; it was intended only for military and scientific uses.
• The creation of the MP3 standard, which is a compression method that enables audio files to be reduced in size to a point that they can be stored on a personal computer

and transmitted over the Internet in a reasonable amount of time but without a notice-able loss in sound quality to humans. The MP3 standard was the result of a European Union project on digital audio associated with the Moving Picture Experts Group MPEG-1 (audio level 3). An earlier version, MP2, was released in 1993 and was used on the Internet to share music files, using the Xing MPEG Audio Player, often to play legal recordings from the Internet Underground Music Archive. The MP3 format, which achieved technical improvements over MP2 (the same sound quality was achieved on MP3 at 128Kbits/s as MP2 achieved at 192Kbits/s). The Fraunhofer Soci-ety, a German applied research organization, introduced the first software encoder for MP3 in 1994 and the first MP3 player, Winplay3, in 1995. This player decompressed the file as it played it and did not require manual decompression, which had been nec-essary before this player was developed. The development in 1997 of the Winamp MP3 player by Justin Frankel (discussed later in this chapter) provided many additional features over the Winplay3 and helped lead to the wide adoption of the MP3 format. Winamp remains in 2007 one of the most popular media players, available as both freeware and for a modest cost a pro version.[9]

• The development of machinery available to the general public that enabled people to record as well as play digital recordings. Digital Audiotape (DAT) recorders were intro-duced in the United States by Sony and Philips in 1986 and are discussed later in this chapter.

• The development of powerful processors for personal computers. Processor capability followed Moore's Law, doubling about every 18 months. It was only with the introduc-tion of the Intel 486 chip in 1989 that it became possible to decompress MP3 files in real time without the music sounding as though it stops and starts (no "jumpiness").

• The development of large and inexpensive storage devices for personal computers. The most common storage medium today for personal computers is the hard drive. Hard drives were invented by IBM in 1955, but for the next twenty-five years these were large, cumbersome devices more often found in a computing center than on a personal computer. The technical breakthrough for personal computer hard drives was made by Seagate in 1980 when it introduced a $5\frac{1}{4}$ inch floppy hard drive that could hold 5 MB of data. Through a series of technical innovations over more than a quarter century the amount of storage on a personal computer has grown exponentially. In 2007 it is typical for a hard drive made for a desktop PC to have a capacity of 160 GB (32,000 times the capacity of the 1980 Seagate drive), and hard drives were available with as much as 750 GB capacity (150,000 times the capacity of the 1980 Seagate). Also important to the rise of downloadable music was the introduction during the sec-ond half of the 1990s of fast external interfaces (USB and FireWire) to connect to the drives.[10]

• During the dot-com years of the late 1990s, 2000, and early 2001, there was a tremen-dous growth in broadband telecommunications capacity. This translated into lower prices for the amount of service offered. Broadband capacity available in most American homes, such as Digital Subscriber Line (DSL), cable modem, and satellite, increases the speed of the Internet for consumers typically by a factor of more than 100 over dial-up modems or over even the IDSN technology that was traditionally available on the voice telephone line. This is important when transferring large music files, and the number of U.S. households that had broadband access had increased to about one-third by 2003. As of March 2006, 42% of all American adults had a high-speed Internet connection at home.[11]

Early Legal Foundations: The Sony Betamax Case and the AHRA

Two new technologies, video cassette recorders (VCR) in the mid-1970s and digital audio tape (DAT) recorders in the 1980s, led to legal and political responses that set the stage for the later battles over music sharing on the Internet. With VCRs it was a precedent-setting legal case (*Sony* v. *Universal City Studios*), and with DAT it was the passage of the Audio Home Recording Act.

In 1976, a year after Sony introduced its Betamax videotape recording system, the company was sued by Universal Studios for contributory copyright infringement based on the fact that this equipment could and was being used to make unauthorized copies of copyrighted material.[12] The U.S. District Court ruled in Sony's favor, deciding that noncommercial home recording fell under the fair use provisions of the copyright law. However, the decision was overturned by the appellate court, which recommended damages, injunctions, and forced licensing for future use. The decision was reversed again in 1984 by the U.S. Supreme Court in a 5–4 vote, and thus the final ruling was in favor of Sony. The Supreme Court argued that "the sale of copying equipment, like the sale of other articles of commerce, does not constitute contributory infringement if the product is widely used for legitimate, unobjectionable purposes. Indeed, it need merely be capable of substantial noninfringing uses...."[13] In this case, the legitimate use was private, noncommercial "time shifting" of television programming—recording a football game on home television, for example, to watch at a later time. This ruling was one of the most important rulings in intellectual property law in the twentieth century, and it was a foundation for many of the legal battles over copying of music.[14]

In 1986 Sony and Philips completed the development of the first digital audio tape recorders and had them ready for the market, including the large U.S. market. The new DAT recorders had several advantages over existing recording technologies. They could record up to 180 minutes on a tape, which was better than the 120-minute recording limits on traditional analog tapes. They also used a digital recording technology that produced very high quality—higher than the quality of a compact disc recording.

The record companies had tolerated the use of traditional analog tape recording because the quality was low and degraded quickly with repeated copying, but they were very concerned about the DAT technology—both about individuals making high-fidelity copies for their friends and families and about pirates using DAT recorders to make copies for sale on the black market. The record companies threatened to sue Sony and Philips for vicarious and contributory copyright infringement, causing Sony and Philips to delay introduction of DAT recorders into the U.S. market. (Vicarious copyright infringement occurs when a person or organization has the right and ability to control copyright infringement but chooses not to do so for personal gain. Contributory copyright infringement occurs when the person or organization is aware of the copyright infringement and provides assistance or inducement to it.)

The record companies, working through their trade association, the Record Industry Association of America (RIAA), pressed Congress to legislate on this matter.[15] A DAT bill was introduced in the Senate Commerce Committee that required DAT recorder makers to include a serial copy management system (SCMS). This system would allow copies of originals but not copies of copies, as a way to protect the record industry. However, the bill died in committee because of testimony that it would be easy to circumvent the copy management system and also because the record companies and music companies wanted the makers of DAT recorders and DAT tapes to pay taxes on their sales that would be redistributed by the government to the record industry to compensate for lost sales.[16] In 1991 Dennis DeConcini (D-AZ) introduced the Digital Audio Tape Recorder Act in the U.S. Senate with both the SCMS and the tax provisions, and Congress easily passed this bill into law in 1992.[17]

Once the original bill died in committee, Sony began selling DAT recorders in the U.S. market without any copy management protection system. In response, three music publishers and the songwriter for Frank Sinatra, Sammy Cahn, filed a class-action lawsuit against Sony.[18] An out-of-court agreement was reached that included a provision that both sides would abide by new legislation regularizing the transfer of royalty payments from the DAT recorder manufacturers to music writers and music publishers. Less than a month after the Cahn suit was settled, both the U.S. House and Senate introduced bills that regulated DAT recorders, called the Audio Home Recording Act.[19] The bill directed that all DAT recorders sold in the United States should include a serial copy management system, which provided that 2 percent of the wholesale price of DAT recorders and 3 percent of the wholesale price of DAT recording media would be held back and placed in a federal fund for distribution: one-third to song writers and two-thirds to music publishers.[20] With these protections, the law specified that no copyright infringement suits could be brought against the makers of DAT recorders or DAT recording media. The bill was written broadly enough that it applied not only to DAT recorders but also to some recording devices and media developed later, including digital minidisc players.[21]

This episode has to be seen as favoring the record companies. They delayed the introduction of DAT technology for about five years through the threat of lawsuits and pending legislation; and when legislation was passed, there was protection for the record companies in the form of required copy management systems and taxation to recover lost revenue. Perhaps because of the delay, few DAT recorders were ever sold in the United States.

Early Uses of the Internet and Related Technology for Marketing and Sharing Music

Until 1991 there was no commercial use of the Internet, and only limited personal use. During the first five or six years after the rules were relaxed about uses of the Internet, various bands and a few record companies experimented with using the Internet to publicize and distribute music. This was not the first time technology had been used to share music outside the establishment record distribution system. An important early example was the rock band, The Grateful Dead. Beginning in 1968, the band encouraged people to record their live concerts and trade copies with other "Deadheads." Out of that experience, the band's fans became some of the earliest and most active users of networked bulletin boards, such as The WELL, where they traded information about the band and many other topics.[22] They later became early users of the Internet for this same purpose. Another technology that reshaped the music industry was MTV, which was created by Bob Pittman for the Warner American Express Satellite Company in 1981. The success of MTV forced the record industry to assume the added expense of making and distributing videos of albums they released.[23] This additional cost and partial loss of control angered the big record companies, but the videos helped to sell more records. Michael Jackson's album *Thriller* sold more than 20 million copies.

One of the musicians who most actively experimented with the Internet in the early and mid-1990s was David Bowie, who created an Internet Service Provider (ISP) called BowieNet. The ISP failed but the Web site—still active today—provides a discussion forum for fans, gives information about road tours as well as exclusive photos and video, provides a portal to published reviews, background information about Bowie's career, fan polls, and an online store to buy music, art, apparel, and other items.[24]

In 1993 three students at the University of California-Santa Cruz (Jeff Patterson, Rob Lord, and Jon Luini) created the Internet Underground Music Archive. It provided a forum for bands that were not receiving commercial attention to tell about themselves on a free Web page and offer their music for download, both directly from the Web and by using the file transfer protocol (FTP), which required less computing power. This pioneering site, which was one of the principal places to find legal music for download in the 1990s, ceased operation in 2006.

In 1994 Capitol Records decided as an experiment to use the Internet as the main marketing tool for the hip-hop group the Beastie Boys. The band liked the results and on their own began to place MP3 versions of songs from their live concerts on the Web. They were probably the first major band to use the MP3 format to promote their music. Without authorization and to the annoyance of the Beastie Boys, Michael Robertson, the founder of the company MP3.com, posted the Beastie Boys songs on his Web site. After a while, the band decided that placing live concert versions of their songs on the Internet was harmful to both sales and control over authorized versions of their songs, so they decided to remove the live concert versions from their Web site. Fans, however, saw this action as the band bowing to pressure from Capitol Records and denounced both the company and the band.[25]

In 1993 Rob Glaser, a former Microsoft executive, formed Progressive Networks as a means to distribute progressive political content on the Internet. In 1995 the company developed a technology called RealAudio for use by radio stations to rebroadcast programs, especially talk shows where high fidelity was not much of an issue. The company had some success, being used by CNET and HotWired, among others. Glaser took the company public in 1997 and renamed it RealNetworks. The company provided a system for highly compressing music (with relatively low fidelity) so that people who had only (slow) modem access could stream music—listen to the music online but not download a copy to their computer.[26]

The songwriter and musician Todd Rundgren was also an early user of the Internet for the distribution of his music. He had experimented with interactive CD-ROMs in the late 1980s. A rudimentary system was built into the CD-ROM so that the listeners could mix the songs as they wished. Although this experiment did not generate much fan interest, Rundgren was nevertheless inspired to try out the new Internet technology. Between 1996 and 1998 he developed PatroNet, which was one of the earliest music subscription services.[27] Subscribers could download songs in an unprotected MP3 format from his Web site or use the site to order CD recordings of his music. One thing that Rundgren liked about this distribution model was its flexibility, both with respect to when he released new works and how long they were. This distribution model freed him from making every work fit the standard album-format length. At first, Rundgren was not concerned about people sharing his unprotected MP3 files, but when the Napster online file sharing system came into widespread use in 2000, he became alarmed and threatened to terminate subscriptions to PatroNet for anyone he found on Napster sharing his songs.

In 1996 Gerry Kearby—who had been a founder in the mid-1980s of the company Integrated Media Systems that had built a digital recording studio for movie pioneer George Lucas—formed a new company, Liquid Audio. It produced a music player and a CD burner based on a new compression system known as Advanced Audio Coding (AAC), which was developed by Dolby, Fraunhofer, AT&T, Sony, and Nokia. AAC was technically superior to the MP3 compression system in several respects, but AAC did not catch on like MP3 did, and Liquid Audio had to add MP3 capability to its products in 1999 because this had become the industry standard. (AAC has gotten a new life in recent years since Apple began using the format in its iTunes Music Store and Sony, Nokia, Motorola, Siemens, and other mobile phone companies have introduced music phones based on the AAC format.)

Thus by 1997 a number of experiments had been tried in using technology to share music. Over the next three years, events moved at a furious pace, with multiple technical issues being played out simultaneously. The next three sections discuss several of the most important of these technical developments as they pertained to music sharing over the Internet: encryption for serial copy management and circumvention of that encryption, music lockers and subscription services, and centralized and noncentralized file sharing systems.

Streaming, Encryption Circumvention, and the DMCA

The most powerful legal tool in the arsenal of the record companies, the Digital Millennium Copyright Act, took effect in 1998. The record and movie industries got behind this sweeping legislation in response to threats they perceived in the practice of streaming. This technology, developed by RealNetworks, enabled users to listen to music or watch a movie that was being continuously streamed on the Internet but which could not be downloaded. Entertainment companies liked streaming because it enabled a pay-per-use pricing model and did not leave permanent copies in the hands of users who might share them with others. However, the concern in the record and film industries was that they would build a business model around streaming and then someone would break the security codes of RealNetworks, enabling users to download large portions of their music and film backlists free of charge.

Prior legislation prohibiting the circumvention of security technologies on entertainment media had been passed in a sectoral fashion. The Cable Act of 1992 prohibited unscrambling of cable television signals. The Audio Home Recording Act (AHRA) made it illegal to produce and distribute devices that would undermine the serial copy management systems for musical recordings. The Motion Picture Association of America, the trade association representing the major film companies, had been lobbying for a bill like AHRA that mandated serial copy protection on DVDs and other devices that could play and copy movies, but the association had been opposed by the computer and personal electronics industries and had not succeeded in getting a bill passed.

Both the motion picture and record industries were supportive of a broad bill that covered all acts of circumvention. This was particularly desirable because the courts had ruled in several cases in ways that limited anticircumvention laws. In *Vault Corporation* v. *Quaid Software*, circumvention of the copy protection mechanism on computer discs was judged to be legal.[28] In *Sega Enterprises* v. *Accolade*, the court ruled that circumvention of a lock-out mechanism on the Sega game console was legal.[29] In *Lasercomb* v. *Reynolds*, circumvention of software product protection was allowed.[30]

President Clinton and Vice President Gore—the first high-tech presidential team— had a vision that the private sector would build a new national information infra- structure with leadership from the federal government. The president created the Information Infrastructure Task Force (IITF) to create the federal view on this new infrastructure.[31] Lobbyists for the entertainment industry received support from an IITF white paper that argued anticircumvention legislation was in the public interest and pushed for legislation that would codify the position in this white paper. However, the American Library Association and several consumer interest groups opposed broad anticircumvention language, and there was no traction in Congress for an anticircum- vention bill.

The passage of anticircumvention legislation came through indirect means, led by Bruce Lehman, the chair of the IITF's Working Group on Intellectual Property Rights. Lehman, who was appointed by President Clinton as a U.S. representative to the World Information Property Organization (WIPO) in 1996, convinced the WIPO to adopt a policy containing anticircumvention language that was modeled in many respects after the IITF white paper. Nations that were members of the WIPO were expected to bring national law into conformance with the WIPO decree, so a mandate was created to introduce legislation in the United States that would track the WIPO's anticircum- vention rules.

The politicking over this bill, the Digital Millennium Copyright Act, was intense; and the legislation that resulted was lengthy and complex, addressing the interests of many different constituencies. The entertainment industry lobbied hard for broad legislation that ruled against all kinds of circumvention. The computer industry was concerned that DMCA would outlaw reverse engineering as well as security testing and research, so it favored a more limited bill narrowly addressing circumvention of copyright. Internet service providers and companies that maintained and repaired computers were concerned about pass-through liability issues. Would an ISP be liable, for example, if someone used its service to provide information about how to circum- vent a scrambling system for films?

The political maneuvering took place at a time when the Congress was distracted by the Monica Lewinski scandal, and the legislation passed by voice vote in the fall of 1998. In the end, the entertainment industry had prevailed. The DMCA made it illegal to circumvent technology measures that control access to a copyrighted work.[32] The law also made it illegal to produce or distribute technologies that could be used in cir- cumvention, much the way that it is illegal to possess or distribute lock-picking tools (not only to use them to break into a house or business). Entertainment companies injured by circumvention could seek civil damages under the law. There were also provisions for criminal penalties against willful circumvention for commercial or personal gain. The law did contain a few narrowly crafted exceptions, such as for libraries to circumvent to see if they wanted to purchase a copyrighted work, for reverse engineering to achieve interoperability between computer programs, for encryption research, and to control minors' access to pornography.

The DMCA was very quickly applied in a series of court cases, most of them brought by the entertainment industry. In the majority of cases, the entertainment industry prevailed.[33] RealNetworks successfully sued Streambox, which had developed a technology that circumvented the encryption on RealNetwork's streaming technology and enabled the audio files to be turned into MP3 files.[34] A more widely publicized case involved DeCSS. In 1995 a standardized format for DVDs was agreed upon, and a trade association, the DVD Forum, was formed by more than 200 companies, including both movie makers and DVD player manufacturers, to create a standard encryption system for DVDs known as the Content Scramble System (CSS). This standard was widely licensed, and by 1997 DVD players were being produced that incorporated the CSS technology and movie companies began selling scrambled discs. However, there were no CSS DVD players that could operate in a computer using the Linux operating system. In 1999 Jon Johansen of Norway developed a program to circumvent CSS in order to build a Linux-based DVD player. Johansen's DeCSS program, as it was known, was widely accessed on the Web, and many people used it to illegally copy DVDs. The Motion Picture Association of America filed suit against Web sites that refused to take down the DeCSS software.[35] Particularly defiant was Eric Corley, the editor of *2600: The Hacker's Quarterly*. The courts ruled in favor of the movie industry, and the Web sites were sued not only to remove DeCSS from their sites but also to remove links to other sites where a person could find a copy of DeCSS. While Corley was forced to comply with the court ruling, there were nevertheless places on the Web where the determined user could find the DeCSS software.

In 1998 the music industry founded the Secure Digital Music Initiative (SDMI), modeled after the motion picture industry's DVD Forum.[36] SDMI included 120 member companies from the record industry and also including personal electronics manufacturers, personal computer manufacturers, and Internet service providers. Their goals were first to set standards for portable MP3 players and then to develop detection and watermarking technologies that could control the number of times an individual compact disc could be played or copied. SDMI was led by Leonardo Chiariglione, who had worked on the MPEG format that produced MP3. The standard for portable players was released in 1999. However, there were technical problems and contradictory objectives among the members concerning the encryption process.[37] Chiariglione resigned in frustration in 2001. Concerned about both the widespread use of Napster and the problems in achieving a technological fix to illegal copying, the major record companies began to rely more on legal remedies—for example, aggressively challenging Diamond Multimedia over its Rio MP3 player.[38]

The main example of anticircumvention in the music industry involved the watermarking standards of the SDMI. In 2000, as a way to test the vulnerability of and possibly improve their music copy protection system, SDMI announced an open competition. Any group that wanted to enter the competition was given a piece of music protected by the digital watermarking technology that SDMI had developed and had three weeks to break the watermark and return the music file in playable form without the watermark. Princeton University professor Edward Felten and some of his students and colleagues entered the contest and were able to break the watermark. Felten had chosen not to sign the nondisclosure agreement associated with the contest, even though it meant his group was not eligible for the prize associated with the contest. He planned on presenting a lecture on the methods for breaking the watermark at a professional computer science conference, and he was threatened by SDMI and RIAA with a lawsuit under the DMCA for disclosing methods to circumvent an encryption technology for protecting copyrighted materials. Felten decided against making that presentation, and SDMI issued a public statement that they did not intend to sue Felten. Nevertheless, Felten, with assistance from the Electronic Frontier Foundation, sued the groups that had threatened him with lawsuits, asking for a declaratory judgment that the paper was not a violation of the DMCA, given that it was for the purposes of research and security testing. The judge dismissed the lawsuit, noting that the music industry had already stated it would not sue Felten. Felten subsequently presented the results at a professional conference in 2001 and received some assurances from the Department of Justice that there was protection for this kind of research under the DMCA.[39]

By and large, the DMCA has been highly effective for the entertainment industry as a tool for controlling access to music and films. Many legal experts believe that the law is too broad. For example, in December 2002 it was invoked by the computer printer manufacturer Lexmark in the Kentucky courts against a company (Static Control) that had circumvented Lexmark's authentication sequence and made unauthorized replacement toner cartridges for Lexmark printers. The courts provided a preliminary injunction against Static Control in 2003, but in 2004 the appeals court overruled the lower court, arguing that the toner loading program was not subject to copyright protection, and the Supreme Court refused to hear the case.[40] In 2003 Chamberlain, a manufacturer of garage door openers in Illinois, sued the Canadian company Skylink Technologies for selling remote control devices that work with Chamberlain garage door openers. Chamberlain argued that the DMCA applied because Skylink had circumvented the access codes in the program that made the garage door opener functional. The case was ruled in favor of Skylink in the District Court of Illinois in 2003, and that ruling was upheld by the Federal Circuit Court in 2004.[41]

Creating an Internet Music Business From MP3 Technology

It was not until 1996 that there was a serious effort to create an Internet music business from the MP3 technology. Between 1996 and 2000 a number of business models were attempted. The world began to change in late 1999 and 2000, with the meteoric rise and fall of Napster; thereafter, all Internet music businesses were shaped by the Napster experience. The story of Napster and other file sharing companies is told in the next section. This section examines the early efforts to create an Internet music business.

Until 1996 the only MP3 players available in the United States were made by German and Korean firms, and these were not all that user friendly and were not widely marketed. The void was filled by a highly creative teenager, named Justin Frankel, from Arizona, who later made major contributions to Web broadcasting with his SHOUTcast MP3 streaming media server and to file sharing through his creation of Gnutella. In 1996 Frankel developed a new MP3 player called Winamp. It attracted a following because it had a good graphical user interface, which made it seem to the user more like a radio than a piece of software, and it had some useful features. Frankel formed a company, called Nullsoft (a play on Microsoft since null is even smaller than micro), to market Winamp. Features included the ability to organize the music in a certain order and plug-ins that enabled it to play various secure formats. At first, Frankel was uncertain how to make money off of Winamp. He formed an alliance with Rob Lord, formerly of the Internet Underground Music Archive and now with the Internet music company N2K, to market his MP3 player. In particular, Lord helped Frankel reach a deal with the newly formed music merchandising company Artist Direct, which brought in $300,000 for the placement of advertisements. Frankel sold Nullsoft to AOL in 1999 for $400 million. It was an uneasy alliance, however. There were regular conflicts between the technology workers at Nullsoft, who were eager to get new technology ideas into the hands of users, and the managers at AOL, whose first concern was looking after AOL's business interests. In particular, there were battles over Nullsoft's unsanctioned release of two file sharing systems: Gnutella in 1999 (discussed later in this chapter) and Waste in 2003. Nullsoft's San Francisco offices were closed by AOL in December 2003. Frankel resigned in January 2004, and most of the rest of the staff left that same year. Winamp version 5 continues to be marketed by AOL.[42]

In 1997, using the open source MP3 technology, Michael Robertson founded MP3.com. The company hosted Web sites for bands and became the principal site to visit for people who were interested in online music. Because of the strong lock that the traditional record companies had on the big-name musicians, MP3.com struggled to find both attractive music content for its site and a profitable business strategy. One approach it tried was to give stock in the company to the artists Ice-T and Alanis Morissette to place songs on the MP3.com site. They also tried other approaches that carefully avoided direct competition with the big record companies. For example, they formed Digital Audio Music, which would press and mail a CD on demand of any band on the MP3.com site, with 50 percent of the revenue returning to MP3.com. However, the service was not well advertised, and the stable of MP3.com artists was not in heavy demand, so the service had limited success and attracted no serious attention as competition from the big record companies. In 2000, MP3.com tried two subscription service models. For popular music, it allowed users to subscribe to individual artists or individual record labels, with fees set by the artists and MP3.com retaining 50 percent of the proceeds. For classical music, there was a monthly fee that gave access to everything classical in MP3.com's list.

MP3.com went public in 1999 and in the frenzy of the dot-com boom took on a stock value of $6.9 billion, which was higher than one of the big record companies, EMI. MP3.com's downfall came through its entry into the music locker business. In 1998 and 1999 a number of companies had started music locker services (e.g., RioPort, MyPlay, and Driveway) in which the company would store MP3 files either for free or for a nominal cost. Music lockers were not all that popular because it took time and some technical knowledge for an individual user to "rip" their CD (i.e., copy it) into an MP3 format and post it on the Web.

MP3.com believed it had an answer to this obstacle in its new service, called Beam-It. The company built up a large library of music, including music from the major record companies, and stored it on its own Web site. A Beam-It customer would prove ownership of a particular CD either by placing it in the CD drive on his or her computer and letting MP3.com read the title or by purchasing the CD through one of MP3.com's authorized sellers, which kept a registry of CDs sold. Once the customer proved ownership of a copy of a particular work, he or she could download that work in MP3 format from MP3.com's library for personal use. This avoided the individual user from having to rip the CD and post it on the Web. Over 4 million songs were downloaded during the first three weeks of Beam-It's operation. Whereas the record industry had not been bothered by the other music locker services, they filed suit against MP3.com only nine days after Beam-It started operation and received a summary judgment that shut Beam-It down. When the trial came to court in 2000, the judge ruled that the DMCA allowed individuals themselves to make copies of CDs they owned, but they could not have a commercial firm provide that service for them.

MP3.com tried various strategies to preserve its locker service and the company's viability. It lobbied Congress to except locker services from the DMCA. Representative Rick Boucher (D-VA) introduced the Music Owners' Licensing Rights Act, which died in committee under strong music industry opposition. MP3.com also tried to negotiate individual deals with the big record companies that were represented in the suit. MP3.com was able to negotiate deals with all these companies with the exception of Universal Music Group, paying each one $20 million in a one-time up-front fee and $10 million dollars per year in licensing fees. Because Universal would not settle, the case went to court.[43] In September 2000 the court awarded the largest damages in copyright history, requiring MP3.com to pay Universal statutory damages for willful copyright violation in the amount of $25,000 per song downloaded illegally—a total of $118 million. MP3.com could not generate enough income from its subscription services to meet the court-ordered payment terms and eventually sold out to Universal, which eliminated the locker service. MP3.com was later resold to CNET but did not reenter the locker service business. Locker services were not a viable business in light of DMCA and the attitude of the big record companies, and all of the locker services either went out of business or moved into another business soon after MP3.com lost its court case.

There were a number of other attempts to build Internet music companies, especially in 1999 when venture capital money became widely available. These companies followed various business models, as a few examples illustrate:

• Diamond Multimedia believed there was a market for a portable MP3 player and introduced the Rio in 1998, a portable player with long battery life and no moving parts so that it would not skip while a user was listening while jogging. The music industry sued under the AHRA because the player did not have serial copy protection software.[44] The music industry won the case, but it was reversed on appeal by a judge who ruled that the Rio would fall under the scope of the AHRA only if it could accept input from a consumer electronics device such as a stereo. This case soured many in the Internet music community about the record industry, whereas before this time there was optimism in the Internet music companies that they could find a way to coexist with the major record companies.

• One problem for the Internet music sites was finding good content, given that the Internet companies were essentially locked out from the musicians who were signed with a major record label. Garageband.com handled this issue by having a competition for undiscovered talent, not unlike the *American Idol* television competition today. Music in MP3 format from bands was made available anonymously, people could sample the music at the site and vote, and the winner won a recording contract. (The site continues to be active as of the time this book was in production in 2007.)

• In 1998 Goodmusic (later renamed eMusic) set up as a music distribution company, charging 99 cents for a song and $8.99 for an album, downloaded as MP3 files. In order to build up a music base, eMusic used its venture capital to provide one-time signing bonuses of up to $40,000 to small record labels and independent artists. Most of the artists were unknown, but the company eventually signed up the well-known singer Tom Waits. Later eMusic introduced a subscription service, under which people could download music files if they signed up for a one-year subscription at $19.95 a month. Emusic then arranged a deal with Hewlett-Packard to bundle an eMusic subscription into the sale of new Hewlett-Packard personal computers.[45]

- In 1999 Riffage used its venture capital to buy The Great American Music Hall in San Francisco, with the plan to stream audio from the live concerts held there. The company did not have the capital to continue operating and folded in 2000.[46]
- RealNetworks entered the music distribution business in 1999 with its Real Jukebox. They had a good product, good marketing, and a user-friendly CD ripper built into the jukebox. One could acquire, play, and manage an entire digital music collection with this software. In 2001 the company integrated the jukebox software into its simple player for streaming audio and video, known as RealPlayer, to create a new, more powerful media player called RealOne. This was done in response to Microsoft adding new features to its media player.[47]
- Gracenote was formed as a market research firm, to determine what online users were listening to and selling that information for marketing purposes. Over time it added a number of other services, including a service it licenses to makers of CD and MP3 players that enables them to build into their machines software so that the player will recognize a CD and display to the user the artist, title, track list, and other information.[48]

At the same time the music establishment was also beginning in tentative ways to explore the opportunities that the Internet presented for their business:

- Listen.com had strong financial backing from the big record companies and Madonna's Maverick Records. Listen.com offered what was intended to be a portal to all online music, with editorials about content.[49]
- In 2000 EMI was the first of the major record labels to place albums online. However, it was a highly tentative effort, limiting the number of albums to 100 and not providing strong marketing for the effort. The albums were made available using Windows Media format because of both the wide availability of the format and the digital rights management system incorporated in it.[50]
- The other major record companies all followed, again in somewhat tentative ways, but increasingly rapidly in response to the popularity of Napster. For example, a typical early response was to place a single, promotional song on Liquid Audio or Mjuice. (In 2002, a little later than the main focus of this section, EMI's Christian Music Group entered into a partnership with Liquid Audio, called BurnITFIRST.com, which enabled users to join a subscription service for $9.95 per month, burn a CD of all tracks of up to 20 high-quality recordings per month, download the recordings to a portable player, and have no expiration date on these recordings.[51])
- By far, the most active of the major record companies in the Internet area has been Bertelsmann. In 1999 Bertelsmann and Universal created GetMusic.com in order to distribute music online. GetMusic.com included several online music channels such as Peeps Republic (www.peeps.com) for hip-hop, Bug Juice (www.bugjuice.com) for alternative artists, and Twang This! (www.twangthis.com) for country music.[52] In 2001

The Bertelsmann Music Group (BMG) collaborated with the Israeli firm Midbar to release two albums using Midbar's Cactus Data Shield technology for copy protection, but they abandoned the project when there were technical problems.[53] The company's music division, BMG, formed partnerships with ARTISTdirect, Riffage.com, Egreetings Network, FanGlobe, Listen.com, and Eritmo.com.[54] The parent company invested heavily in online companies including large minority positions in Lycos Europe and BN.com (the online Barnes and Noble business), invested in AOL Europe, and acquired the online CD distributor CDNow.[55]

As one can readily see, between 1996 and 2000 there were no clearly successful business models in the digital music business. Music locker services were found to be illegal. There were small business opportunities for independent companies in subscription music services, streaming technologies, and media players. The established record companies were beginning to experiment with online distribution, but only in a very cautious way.

Napster and File Sharing

The sharing of files over the Internet has a long history.[56] The file transfer protocol (FTP) was one of the original applications of the Internet, and it was used extensively and legally, for example, by scientific researchers in the 1980s and 1990s to share data. FTP is still used for many legal purposes today. There was also sharing, generally by teenagers, of (often pirated) software and images in the late 1980s and early 1990s on bulletin board systems. Most of the bulletin boards were not used for sharing music or videos because the files were too large until compression systems such as MP3 were developed in the 1990s. In the mid-1990s, before the peer-to-peer file sharing networks such as Napster were invented, people uploaded MP3 files to Web sites built by companies such as GeoCities, where individuals were offered free Web sites.[57] When GeoCities learned about the sharing of copyrighted music in this way, it became concerned about legal liabilities and the large amount of space these files could take up, so it banned illegal MP3 files and deleted those it found. Users began to use alternative file extension names to avoid detection, but GeoCities nevertheless found means to detect these files and continued its fight against illegal MP3 files.[58]

Relatively few people were using services such as those offered by GeoCities to share music. That all changed when the easy-to-use, peer-to-peer file sharing system Napster was created in 1999. It was created by Shawn Fanning, a student at Northeastern University in Boston, and two friends in order to trade MP3 files easily. In their system, a central server held the names of all MP3 files of everyone connected to Napster but did not hold any of the MP3 files themselves. Someone on the Napster network who wanted to download a particular song would check the central file sharer to see who had a copy and directly download the file from that person. There was no charge for the software to participate in Napster or for access to the centralized list of MP3 file names. Napster existed for its first year on funds from family and large numbers of donated hours of programming time. In May 2000 it received its first venture capital—$15 million from Hummer Windblad, which brought in a professional CEO as well as provided funds.

Napster was launched in June 1999 and received immediate popularity, especially with high school and college students. By mid-2000 there were a half million users of Napster every night and an estimated 80 million users overall in the first sixteen months. Quite naturally the record industry was concerned, and the RIAA filed suit under the DMCA against Napster. Because the central server held no music files, Fanning had assumed he would not be liable for copyright infringement. The RIAA agreed that Fanning was not liable for direct copyright infringement but sued him for contributory copyright violation for abetting others in their copyright violation.

Napster brought in as their chief counsel the big-name lawyer David Boies, who had been primarily responsible for winning the U.S. antitrust case against Microsoft.[59] Boies made three major arguments: according to the AHRA, sharing music is legal for consumers, so long as they are not trying to profit from it; as in the Sony Betamax case, *Sony* v. *Universal City Studios*, there are legitimate uses for Napster, not just ones that infringe copyright; and, as an Internet service provider, Napster could not be held liable for the listings in its centralized directory of MP3 files. Judge Patel dismissed all of Boies's arguments and ruled in favor of the RIAA. During the appeal process, in October 2000, Bertelsmann's Music Group dropped out of the suit against Napster and invested $50 million in the company. In February 2001 the appellate court supported Patel's ruling, holding Napster liable for copyright infringement and overturning a court-brokered deal that would have limited Napster's financial liability to the record companies. The court required the record companies to compile a list of records for Napster to block from being downloaded. In March 2001, the record companies submitted a list of 135,000 records. Napster made a good faith effort to comply, but the judge ruled that there had to be 100 percent blockage. When Napster was unable to do so, Judge Patel closed Napster down, and this ruling was upheld in the appellate court. In June 2002, Napster filed for bankruptcy after the courts blocked a sale to Bertelsmann. The digital media company Roxio bought Napster's brand and logos in a bankruptcy auction and in 2003 bought Pressplay, an online music store that had been formed as a joint venture by the record companies Universal and Sony, to form a purchase and subscription music service known as Napster 2. Roxio subsequently renamed itself Napster, Inc. and is a competitor of Apple's iTunes today.[60]

Napster was not the only centralized file sharing system. Scour Exchange (SX) was formed in 1997 by UCLA students and financially backed by a group of investors led by former Walt Disney executive Michael Ovitz. When SX was threatened with a lawsuit from the RIAA and the MPAA in 2000 (SX could download movies and images as well as music), the investors pulled their money, and SX shut down later that year.[61] Aimster (later known as Madster after objections from AOL, whose instant messaging system was named AIM) was a file sharing system built in 2000 by a group of Rensselaer Polytechnic Institute students on top of AOL's instant messaging system. The site was popular, growing to several million users. Aimster was sued in May 2001 by the RIAA. Aimster tried two tactics in its efforts to avoid the restrictions of the DMCA: breaking up the company into several pieces, each of which provided a piece of the service but none of which could operate in a way to infringe copyright, and adding encryption so that it could argue in court that it was illegal under the DMCA for the company to decrypt in order to monitor their users' behaviors. Both tactics proved futile, and the courts shut Aimster down in 2002, a decision upheld in appellate court in 2003.[62]

The greatest legal vulnerability of the peer-to-peer file sharing systems such as Napster was the central server. If a file sharing system could be designed that did not involve this feature, then the legal risk seemed significantly lessened. Moreover, a peer-to-peer file sharing system without a central server would have a more distributed architecture; so even if one part was put out of business by the courts, the system could continue to operate. This was the motivation behind a number of the decentralized peer-to-peer file sharing systems that were developed.

The earliest important decentralized peer-to-peer file sharing system was Gnutella, built by Justin Frankel (the creator of the WinAmp MP3 player, discussed earlier in this chapter). Frankel's company, Nullsoft, had been purchased by AOL in 1999, and Frankel created Gnutella in 2000 as a bootleg project, without permission from the AOL management. (Gnutella's name supposedly came from a blending of the GNU General Public License, which is popular with the open source community, and Nutella, a commercial nut spread that Frankel was fond of eating.) When AOL, now merged with Time Warner—a company that ran one of the most conservative, anti-Internet music record labels—found out about Gnutella, they immediately forced Frankel to remove it from the AOL Web site. However, several thousand people had already downloaded the files, several individuals reverse engineered Frankel's source code, and the Gnutella protocol was continued as open source software.

Gnutella was a distributed network with no centralized server. A requesting computer wishing to download an MP3 file could make contact with another computer having the Gnutella software. That computer was connected to others, which were connected to others, and so on for a few levels of path. If the song was found on this network, it would be sent through all the requesters and back to the computer making the original request.

There were some technical shortcomings of Gnutella. It did not scale well. For example, it could not search more than 4,000 other computers for a song. Moreover, performance could be slow, especially if some of the computers checked along the way were using dial-up connections. It was also vulnerable to denial of service attacks.

Although some technical improvements were made to Gnutella, resulting in new file sharing systems such as LimeWare, the greatest promise came from a new proprietary software system in 2001 offered by the Dutch company Consumer Empowerment. The principal designers for the new peer-to-peer protocol, called FastTrack, were Niklas Zennstrom from Sweden, Janus Friis from Denmark, and Jaan Tallinn from Estonia. More recently, these three inventors created the Skype Internet telephone software. Kazaa and Grokster are two of the most important file sharing systems based on FastTrack, but they are incompatible with one another. The main improvement in FastTrack over Gnutella is the use of "supernodes" to improve scalability of the system. Supernodes are fast computers with fast Internet connections that are automatically enlisted by the FastTrack software to help in rapidly identifying which computers have the MP3 file in question by talking with other currently active supernodes. Once the appropriate file is located, the requesting computer downloads directly from the peer computer having the file, without the file being sent through a chain of other computers, as with Gnutella. One technical difficulty with FastTrack is that the hashing algorithm used in checking the correctness of files is built for speed and is vulnerable to not catching massive corruption in files. The RIAA has taken advantage of this feature to spread corrupt files on FastTrack networks.

The file sharing system offered by Consumer Empowerment that was based on Fast-Track is Kazaa (originally KaZaA). Kazaa has been used widely to download both music and films, and it has a somewhat controversial history because of both its dealings with various national legal systems and the fact that the free download of the software includes various kinds of malware, including (reportedly) software that collects information on the computer's Internet surfing activities and passes on that information to a private company, installs popup advertising, and monitors Web sites visited when using certain Web browsers and provides links to competing Web sites.

Consumer Empowerment was taken to court in the Netherlands by the Dutch music publisher Buma/Stemra. The court directed Consumer Empowerment to stop its Kazaa users from infringing copyright or it would levy a heavy fine on the company, but instead of complying, the company sold Kazaa to Sharman Networks, a company headquartered in Australia but incorporated in the South Pacific island nation of Vanuatu, where the legal system does not provide strong copyright protection. After this sale took place, the Dutch court of appeals reversed the ruling, claiming that Consumer Empowerment was not responsible if users infringed copyright. In 2002 the RIAA and the MPAA sued Sharman in the U.S. legal system.

In 2003, the RIAA filed suit in civil court against a number of individuals who transferred large numbers of copyrighted files using Kazaa. The RIAA won settlements against most of these individuals, gaining penalty payments typically ranging from

$1,000 to $20,000. This was part of a more general strategy by the record companies to pursue legal cases for copyright infringement against individuals who traded large numbers of copyrighted files and not only sue the companies providing file sharing services. For a long time, the RIAA had been reluctant to sue individuals, concerned about the potential backlash to the record companies from their customers, but illegal file sharing had become so prevalent and new technologies that could be used for file sharing were springing up so quickly that the RIAA had trouble keeping up the attack on the service providers by either legal or technical means. The RIAA was able to collect the IP numbers of individuals who were illegally trading music files, and they used a provision in the DMCA to force Internet service providers to provide names and addresses of the individuals holding these IP numbers. Verizon resisted giving out the names of their customers but were eventually compelled by the courts to do so.[63] One intended effect of this strategy was to get the attention of college students and the university computer service administrators, since many of the downloaders were students and the network facilities for file sharing were unusually good at many universities. The strategy was successful in doing so, and a number of colleges took measures to prohibit using campus facilities for file sharing beginning with the 2003–04 academic year.[64]

In 2004 the Australian Record Industry Association (ARIA) filed suit against Sharman. The courts decided in 2005 that Sharman had not itself infringed copyright but that its executives had knowingly allowed Kazaa users to share copyrighted songs. Sharman faced millions of dollars in civil damages from the record companies, and the courts shut down access to Kazaa for anyone with an Australian IP address after the company did not make the changes in the software mandated by the court. The case is under appeal, but meanwhile Sharman has opened offices in the Netherlands, presumably because of the favorable ruling from the Dutch appeals court.

Morpheus is an interesting story of a file sharing system that has had a series of technology problems over time. It is offered by StreamCast (formerly MusicCity). At first Morpheus used an open source version of Napster's software, but with Napster's legal problems it switched to FastTrack in 2001. StreamCast and the owners of FastTrack had a business dispute, and FastTrack was updated in 2002 in a way that prevented all Morpheus users from logging in to the network. The owners of Morpheus then turned to the Gnutella network, which was slower, but they also had business disputes with the people running Gnutella, so they switched to an older and even less powerful Jtella technology, which they used until they were able to switch back to Gnutella. Finally, in 2004, the company moved to its own NeoNetwork. The legal status of Morpheus is closely linked to the legal status of Grokster, which is discussed below.

Grokster was another of the file sharing programs based on FastTrack and available to all computers using the Microsoft Windows operating system. It was developed by Grokster, Ltd., a private company headquartered in the West Indies. The RIAA and the MPAA filed suit against both Grokster and StreamCast (makers of Morpheus), but in 2003 the federal court in Los Angeles ruled in favor of the defendants, finding that their file sharing software is not illegal. The RIAA and the MPAA appealed the decision, and in 2004 the appellate court ruled that Grokster was not liable for contributory or vicarious copyright infringement by the users of the software. The case was then appealed to the U.S. Supreme Court, and a ruling came down in 2005 unanimously declaring that Grokster could be sued for contributory copyright infringement for the uses of the software. The case hinged on how to apply the ruling from the Sony Betamax case, in which the court had ruled that the sale of copying equipment does not constitute contributory infringement if the equipment is capable of "substantial" noninfringing uses. Grokster argued that it could fill this substantiality criterion by demonstrating there were reasonable actual or potential noninfringing uses. The RIAA and the MPAA argued that it is not sufficient to have an incidental noninfringing use but that the noninfringing use had to be the principal one. With the unfavorable Supreme Court decision, Grokster shut down its operations in November 2005.

The most recent major developments in peer-to-peer file sharing systems are those that allow users to send and retrieve files anonymously. There are several of these systems, including Freenet, Earth Station Five, and Entropy. Perhaps the best known of these is Freenet. It was developed by Iam Clarke in Ireland in 2000 as a tool to enable people to speak freely and anonymously, and to avoid state censorship. The system operates more slowly than most other peer-to-peer file sharing systems and is difficult to use; hence it has attracted few users. However, these systems do a good job at providing privacy and anonymity. Freenet is not dependent on a centralized server, as Napster is, and is therefore harder to shut down through court actions than Napster.

The iTunes Music Store

The original business models for selling music legally had not worked very well. Some approaches, such as music lockers, were found illegal under the Digital Millennium Copyright Act. Subscription services had not really caught on. As of 2003, there were only about 50,000 people signed up for subscription services, such as Rhapsody and PressPlay.[65] Part of the reason was that the big record companies did not trust Internet delivery systems and thus limited the offerings available in this way, and users regarded the offerings as too limited to justify the high subscription rates. Another reason was that people were used to buying rather than leasing; they had always bought their phonograph records and CDs. Subscribers made an increasingly large investment as they paid their $10 to $20 per month to have access; and all that investment was completely lost if they elected to stop paying the monthly fee, the subscription company went bankrupt, or the company decided to get out of the subscription business.

Because of the limited success of the subscription service, Steve Jobs, the CEO of Apple Computer, decided to develop a digital music purchase business. Apple already had in place some of the components for this business. The iTunes software had originally been bundled with all Apple computers as a way to store and play digital music on the computer, but it now became the software that enabled access to Apple's music store of digital offerings (also called iTunes).[66] The iPod, Apple's portable digital music player, which was already widely successful, was the only player that could access music from Apple's music store. The only missing element was access to a wide selection of music at a reasonable cost. This required Apple to make deals with the record companies.

Throughout 2002, Jobs made repeated visits to the big record companies. They were highly skeptical at first. Many of the record industry executives were technology averse and unknowledgeable and simply did not want to enter the digital music business. Others believed that a subscription service was the most viable business plan. But the strategy of suing individuals for illegal downloading, which had become an increasing practice of the recording industry, was not particularly effective at limiting illegal downloads and was badly eroding consumer attitudes toward the record companies. Jobs hammered away at the record executives about the practical futility of digital copy management software. He also made the pitch that Apple shared the record companies' predicament in that it too had to be concerned about illegal copying of its intellectual property, which for Apple was software. Apple got Warner and then Universal to sign on, and then he got the other major record companies and numerous smaller record companies to agree to participate.

In April 2003 Apple launched its music store, where Mac users could download individual songs for 99 cents and albums for $9.99, with no subscription fee. There was a wide selection of music available (more than 200,000 songs originally, and the offerings rapidly doubled), including those of many—but not all—artists from the five major record companies and more than two hundred smaller record companies. The record companies insisted on digital rights management, but Jobs, who had never been a fan of it, negotiated one of the milder forms. Songs can be sampled for up to 30 seconds free of charge in the music store. When purchased, the songs can be downloaded to as many as three Apple computers or an iPod, and unlimited numbers of CDs can be burned.

Part of the reason the record companies were willing to sign deals with Jobs was that Apple held only 5 percent of the personal computer market in the United States and only 3 percent worldwide. Thus some of the record company executives saw this agreement as only another in a long line of experiments with digital music. None of them anticipated the success that Apple's Music Store would have. In the first week alone, over a million songs were purchased. But these numbers paled in comparison to the numbers sold after Apple introduced innovations making the Music Store available to

PC users in October 2003. The PC version had a new feature that enabled parents to send money to the music store, which was paid out each month as an allowance that gave their children credit toward buying digital downloads from the music store. This enabled parents to keep their children's music downloading habits legal without having to give them a credit card. By December 2003, there had been 20 million downloads. Part of the success is attributable to how easy it is to use the system. Partly it is attributable to the pricing. At 99 cents for an individual song, this was one-quarter the price of some of the previous digital music sales. Apple had also entered the largest music promotion ever with Pepsi Cola, launched at the SuperBowl in early 2004, to give away a potential of 100 million free songs from the Apple Music Store based on messages in Pepsi bottle caps.[67]

As of this writing, the record companies are not entirely happy with the deal they cut with Jobs. They are making money—65 cents on every song sold by the iTunes Music Store. However, with Apple's success, which Jobs claims is over 80 percent of all digital music sold, the record companies have lost some of the control they have traditionally had. Jobs wants to keep the business model simple, so he has insisted that all songs be sold for 99 cents and that the amount not increase at least in the near future. The big record companies would like to increase revenues by using differential pricing—lower amounts for older songs and higher amounts for recent songs by popular artists. Jobs refuses, and the record companies cannot change the agreement or afford not to participate in Apple's venture. However, they are hoping to outmaneuver Apple by making their own deals with the mobile phone companies. There are many more cell phones in the world than iPods, and cell phone users have shown that they are willing to pay as much as $2.95 for a few seconds of a song to use as a ring tone. Moreover, the record companies feel more akin to the cell phone companies because both have business models that are intended to make money by selling things to play on the hardware devices, whereas Apple is mainly in the music store business in order to sell more iPods.

One of the things that Jobs likes about the music store business is that it is resistant to competition. Of the 99 cents revenue on the sale of each song, 65 cents goes immediately to the record companies as royalties, and at least 25 cents goes to credit card fees, digitizing songs, hosting the digitized files on servers, and writing the software that runs these activities. Estimates vary on the profit made by Apple on each 99-cent sale at less than 10 cents to negative profit. The reason Apple is interested in the music store is that it generates enormous sales of iPods—for example, over one billion dollars of sales in the first quarter of 2005.[68] Although Jobs claims that "there are sneakers that cost more than an iPod," Apple makes substantial profits on these hardware sales.[69] One estimate is that Apple profits $175 on the sale of one of its more expensive models.[70] Apple is protecting this business by not licensing its FairPlay antipiracy technology and not accepting other music formats, in particular those of Microsoft.[71] Other companies that are selling music online are not selling this music hardware. So the short-term future, at least, looks bright for Apple.

Attitudes Toward and Impacts of the Internet on the Music Business

How did the Internet affect the various players in the music business—the big record companies, the Internet startup music firms, the computer and personal electronics industries, the musicians, and the fans of music—and what were their attitudes toward these developments?

There is disagreement about the impact that the Internet had on sales of the big record companies.[72] The RIAA has argued that the harm has been significant. In particular, they have argued that the shipment of CD singles dropped 39 percent in 2000, but some industry observers argue that the time of the single was waning anyway, and so the decrease was not due to music pirating.[73] Album sales dropped 3 percent in the first half of 2001, but some argue that it was because of the shutdown of Napster, which had enabled people to sample music before buying it.[74] A study by SoundScan, which tracks sales in the record industry, showed a two-year decline in sales of 4 percent at record stores near university campuses, and 7 percent decline at record stores near universities that had to take action to stop file sharing because of the heavy usage of computer facilities for these purposes.[75]

The data from the academic community is also mixed. A study by University of Texas-Dallas management professor Stan Liebowitz shows strong negative impact of file sharing on record sales.[76] Another academic study of music file sharing by college students, conducted by economics professor Rafael Rob and business school professor Joel Waldfogel at the University of Pennsylvania, showed that per capita expenditure on hit albums released between 1999 and 2003 dropped from $126 to $100.[77] However, a study by economist Koleman Strumpf of the University of North Carolina and Felix Oberholzer-Gee of the Harvard Business School that looked at 0.01 percent of all downloads of a large number of albums showed no statistically significant impact on sales.[78]

There is at least anecdotal evidence of file sharing helping the sales of a record:

A leak of Radiohead's album Kid A on Napster may have actually stimulated sales. Tracks from Kid A were released on Napster 3 months before the CD's release and millions had downloaded the music by the time it hit record stores. The album was not expected to do that well to begin with as it was an artsy endeavor by a band that had only hit the Top 20 in the US once before with the album Ok Computer. There was very little marketing employed and few radio stations played it so Napster was expected to kill off whatever market was left. Instead, when the CD was released Radiohead zoomed to the top of the charts. Having put the music in the hands of so many people, Napster appears to be the force that drove this success. Nonetheless, the record industry was reluctant to credit a company it was suing.[79]

Whatever the actual effect, the attitude of the record establishment has been clearly hostile to this new threat to their established business model, whether it was from illegal sharing of music files or a legal Internet music business. At an MP3 summit organized in 1997 by MP3.com's founder Michael Robertson, for example, the new Internet music businesses showed good will and an interest in working with the big record companies, but they received a frosty reception from the representatives of the big record companies. At a conference of the Los Angeles chapter of The Recording Academy a few weeks later, the view of the record industry was unequivocally hostile. The record industry worked hard to get the DMCA passed, and they used it with considerable vigor against companies that threatened their revenue stream. It was not only the record companies but other parts of the record establishment as well that shared this attitude. For example, the Harry Fox Agency and the National Music Publisher's Association used the legal system to stop the On Line Guitar Archive from offering free on their Internet site note transcriptions of 33,000 songs prepared by volunteers—many of the songs copyrighted. It is interesting that the record establishment did not get together to form a subscription service to fill the vacuum left after they had demolished Napster, but the major record companies were busy fighting one another while the RIAA was moving on to do battle with the newest file sharing technological threats.

The Internet music companies were of course dependent on the file sharing technologies for their livelihood. However, it is possible that the illegal file sharing, such as on Napster, hurt these companies more than it hurt the big record companies. Companies such as Mjuice and Emusic had great difficulty selling any music in an environment in which listeners could obtain the same music for free. At one point, Emusic was giving away a portable MP3 player worth $150 to anyone who signed up to buy $25 worth of music. Although there is little evidence the Internet music companies were supportive of the tactics of the record establishment, as embodied by the RIAA, they could not help but be glad that Napster was put out of business and young people began to be more concerned about illegal file sharing. A subscription model seemed to be a viable business model to attract young people who previously had used Napster.

The computer and personal electronics industries were by and large supporters of the Internet music boom. As providers of the devices, whether they were portable MP3 players or entertainment enhancements for a personal computer, the movement generated sales without threatening their intellectual property. Apple found a business model for enhancing the sales of its iPod players through its iTunes music service, which itself is largely a loss leader. The computer industry was concerned about the broad sweep of the DMCA, which threatened the ability to do security testing and research, as in the case of Professor Felton at Princeton. At least one of the companies (Sony) in the personal electronics industry was also in the music industry, and this at times presented internal business tensions, for example, over the breadth of the DMCA legislation.

The musicians were deeply divided about Napster and illegal file sharing. The vast majority of musicians did not fare well under the traditional record distribution system. Most musicians could not become part of the establishment by getting recording contracts with the major record labels and getting air time on the radio, and even the majority of those who did found themselves earning little from their record sale royalties. The stars who could make significant income from their record sales often chafed under the one-sided rules that protected the interests of the record companies, music publishers, and licensing companies at the expense of the songwriters and performers. Thus some musicians welcomed the Internet developments as the first real alternative to the record establishment while others felt they had worked hard to make it to the top of the heap in the record establishment and wanted to protect their interests there. It must be noted, however, that few artists gained fame solely through use of the Web. One of the few successful examples is the Chicago-based Puerto Rican rap group La Junta. It actively participated in AOL chat rooms and provided individual Web sites with logos and information about the group. This strategy helped significantly to get their name out.[80]

Reporting on a study by the Pew Internet & American Life Project, the *Washington Post* reported that:

about half of the artists it surveyed think unauthorized file-sharing should be illegal, it also concluded that "the vast majority do not see online file-sharing as a big threat to creative industries. Across the board, artists and musicians are more likely to say that the internet has made it possible for them to make more money from their art than they are to say it has made it harder to protect their work from piracy or unlawful use," according to the study, which also found that "two-thirds of artists say peer-to-peer file sharing poses a minor threat or no threat at all to them."[81]

Some of the performers who supported the Internet development included Motley Crue, Limp Bizkit, Courtney Love, Dave Matthews, Jewel, Moby, Public Enemy, and the David Crowder Band.[82] Some artists, such as David Bowie and Madonna, tried to remain part of the establishment but incorporate the opportunities of the Internet into their plans.[83] Those opposed included Dr. Dre, Eminem, Busta Rhymes, Metallica, Loudon Wainwright III, Hootie and the Blowfish, and Oasis.[84]

The story of the heavy metal band Metallica is of particular interest. Metallica had encouraged its fans to trade tapes of its live concerts, although the band was careful to keep control of the master copies. When Napster became active and a great number of Metallica songs were being downloaded that way, Metallica became concerned. The tipping point for action occurred when several versions of not-yet-released Metallica songs started being traded over Napster, and the band believed that it had lost artistic control as well as lost revenue. In April 2000, Metallica used RICO racketeering laws to sue Napster and three universities (Indiana, USC, and Yale) whose facilities were being used extensively for Napster downloads of Metallica songs. Metallica asked Napster to remove all Metallica songs from its servers, but Napster replied that, since the individuals using Napster named the files, Napster could not tell reliably which file names corresponded to Metallica songs. Napster invited Metallica to supply them with the names of people illegally trading Metallica songs and they would take them off Napster. Perhaps surprising to Napster, Metallica came up with a list of names, and Napster followed through by removing them. Many of the removed individuals were unhappy about this, and 30,000 petitioned to be reinstated. According to the procedures spelled out in the DMCA, Metallica had only ten days to respond to the 30,000 listeners who asked to be reinstated on Napster, and the band did not have the resources to do so, so all 30,000 of these individuals were reinstated. The fan reaction to Metallica varied widely, but there were some angry fans, including those who formed the Web sites MetallicaSucks.com and BoycottMetallica.com.

The relation between bands and their fans was often a complicated one, and never more so with the fan sites that were created on the Internet. On the one hand, the bands were happy to have the community building around their work and the free publicity, but it could be a problem when the bands could not control what happened on the sites. In 1997 the band Oasis, with support from its record company, Sony Music, informed all of its fan sites that they had to remove all copyrighted material, such as photos, sound clips, and lyrics. The fans reacted with anger toward the band.[85]

Individual music listeners have by and large found music file sharing an unmitigated boon. They seem relatively untroubled by the argument that what they are doing is illegal. Many of the users of Napster, especially early adaptors, were college students in part because of their access to high-speed connections. In this community there was a "music should be free" ideology and a desire to punish the greedy record companies for excessive profits, high record prices, and not sharing the revenue fairly with the performers and songwriters.[86] We have seen this same sense of what constitutes ethical behavior on the Internet more recently in the widespread sense that it is a fair business practice for Apple to charge 99 cents for a song, and in Google's concern about how to treat fairly the repression of information in its service to China and not become an "evil" company. *New York Times* writer Kelly Alexander described how the use of Napster was not restricted to young people but also included "respectable, mortgage-paying geezers in our 30's and up. We are over-aged teenagers for whom Napster is not just about freedom, but regression. The freedom to turn on a virtual radio that is always playing your favorite song."[87] UCLA sociologist Peter Kollock has described how Napster users have a hard time believing their one act of downloading can have a serious impact on the record companies, and UC Berkeley legal scholar Robert MacCoun has argued:

This is a kind of classic situation of what social psychologists call "pluralistic ignorance." The idea is that, often, social situations are ambiguous and we look to other people to help us define what's appropriate in a given situation. And we often infer from the fact that no one else is acting alarmed that there's nothing alarming going on. Everyone is agreeing tacitly not to ask the hard questions.... It's a little reminiscent of the 1970s period when marijuana use became so prevalent that people acted as though it had been de facto legalized.[88]

It may be that online music sharing habits are beginning to change. In a report from the Pew Internet & American Life project in 2005, the point is driven home that Americans are not giving up on file sharing, but they are moving away from traditional peer-to-peer networks and online subscription services. This will present a new set of challenges for the music establishment and the musicians:

About 36 million Americans—or 27% of internet users—say they download either music or video files and about half of them have found ways outside of traditional peer-to-peer networks or paid online services to swap their files.[89] Some 19% of current music and video downloaders, about 7 million adults, say they have downloaded files from someone else's iPod or MP3 player. About 28%, or 10 million people, say they get music and video files via e-mail and instant messages. However, there is some overlap between these two groups; 9% of downloaders say they have used both of these sources. In all, 48% of current downloaders have used sources other than peer-to-peer networks or paid music and movie services to get music or video files. Beyond MP3 players, e-mail and instant messaging, these alternative sources include music and movie websites, blogs and online review sites.

This "privatization" of file-sharing is taking place as the number of Americans using paid online music services is growing and the total number of downloaders is increasing, though not nearly to the level that existed before the recording industry began to file lawsuits against suspected music file sharers in mid-2003.[90]

The digital sharing of music over the Internet that has occurred over the past decade introduced a set of knotty business, social, and legal issues, but it is really just the beginning of a larger story about individuals becoming increasingly digital in their means of entertainment and expression. This story involves the *convergence* of access to film, television, radio, games, and news as well as music. All of these media are being accessed in a digital format, often using the same digital rights management systems and sometimes the same hardware devices to receive and play content. It is also a story of *customization*, of people blending together many sources to create new collections of music or even new works of art themselves. Some thinkers, such as Harvard Law professor William Fisher, talk of a semiotic democracy in which individuals can use the power of the Web to express themselves and are not limited to the content provided by big business.[91] However, these stories of privatization, convergence, and customization are left for future historians to consider.

Notes

1. The material in this chapter, and particularly the next section, is heavily influenced by several lectures the author has attended by Professor William Fisher of the Harvard Law School. This

material is brilliantly codified and argued in Fisher's book *Promises to Keep: Technology, Law, and the Future of Entertainment* (Stanford University Press, 2004).

2. This issue received one of its first investigations in the exhibit Illegal Art: Freedom of Expression in the Corporate Age, San Francisco Museum of Modern Art, Artists Gallery, San Francisco, CA in 2003.

3. For a discussion of this industry, see Andre Millard, *America on Record: A History of Recorded Sound* (Cambridge University Press, 1995).

4. These were later consolidated into four companies (Universal Music Group, Sony BMG Music Entertainment, EMI Group, and Warner Music Group), each of which receives between one and ten billion dollars a year in revenue. For more details about the record labels in each group, see "List of Record Labels" on Wikipedia.

5. However, it was not as inefficient as it was between the two world wars, when graft was rampant. For exposés of the music industry, see, for example, Frederick Dannen, *Hit Men* (Crown, 1990) or Peter M. Thall, *What They'll Never Tell You About the Music Business* (Watson-Guptill Publications, 2002).

6. Kees Immink, "The Compact Disc Story," *Journal of the Audio Engineering Society*, 46 (May 1998): 458–465. See http://www.exp-math.uni-essen.de/~immink/pdf/cdstory.pdf (accessed May 26, 2006).

7. On the move from phonograph records to CDs, see Andre Millard, *America on Record* (Cambridge University Press, 2005); Pekka Gronow and Ilpo Saunio, *An International History of the Recording Industry* (Continuum, 1998).

8. *Watermarking* is the process of embedding information into an object such as a music recording, video recording, or piece of software used for the purpose of copy protection.

9. See the Wikipedia articles on MP3, Winamp, and Winplay3 (accessed May 26, 2006).

10. See the Wikipedia article on Hard Drive (accessed May 25, 2006).

11. See Wikipedia article on Broadband Internet Access; Mary Madden and Lee Rainie, *America's Online Pursuits* Pew Internet and American Life Project, 12/22/2003, http://www.pewinternet.org/PPF/r/106/report_display.asp.

12. For context on the Betamax and video recording, see Edward A. Jeffords, "Home Audio Recording After Betamax: Taking a Fresh Look," *Baylor Law Review* 36 (1984): 855, 857; Margaret Graham, *RCA and the VideoDisc: The Business of Research* (Cambridge University Press, 1989).

13. *Sony Corp. of America* v. *Universal City Studios, Inc.*, 464 US 417 (1984).

14. On this case, see, for example: the Wikipedia article on Pamela Samuelson (2006); Pamela Samuelson, "The Generativity of *Sony V. Universal*: The Intellectual Property Legacy of Justice Stevens" *Fordham L Rev* 74: 831 *Sony Corp* v. *Universal City Studios*; and Paul Goldstein, *Copyright's Highway: From Gutenberg to the Celestial Jukebox* (Stanford University Press, 2003).

15. The RIAA was formed by the major record companies in 1952 and is responsible for seeing after their common interests, especially in lobbying Congress and pursuing legal interests of the companies. While the industry has been marked by intense competition among the major players, with the exception of the occasional straying of Bertelsmann, the major record companies have been highly cooperative concerning the threat of the Internet and related information technologies to their business models.

16. Consumers at the time and later complained about serial copy management, arguing that they have a right to make as many copies as they wish for personal use. However, these concerns did not appear to have much of an effect in Congress in the politics of the DAT bill—or all that much influence on later bills either.

17. For more information from Philip Greenspun, one of the people who testified on these two bills, see "Tape Tax History," http://philip.greenspun.com/politics/tape-tax-history.html (accessed May 27, 2006). For additional context, see "Chronology of Events in Home Recording Rights," http://www.andrew.cmu.edu/course/19-102/timeline.txt (accessed May 27, 2006).

18. *Cahn* v. *Sony*, 90-CIV-4537 (S.D.N.Y. filed Jul. 10, 1991).

19. U.S. Public Law 102-563, Audio Home Recording Act of 1992, an amendment to The Copyright Law of 1976.

20. A similar legal remedy was proposed more recently for CD burners. Some people see this as part of a tax on the Internet. For a discussion on Internet taxation, see William Aspray, "Internet Use," in William Aspray, ed. *Chasing Moore's Law* (SciTech Publishing, 2004).

21. For discussion of the AHRA, see Saba Elkman and Andrew F. Christie, "Regulating Private Copying of Musical Works: Lessons from the U.S. Audio Home Recording Act of 1992," Intellectual Property Research Institute of Australia, Working Paper No. 12/04, ISSN 1447–2317. http://www.law.unimelb.edu.au/ipria/publications/workingpapers/2004/IPRIA%20WP%2012.04 .pdf. Also Gillian Davies and Michele E. Hung, *Music and Video Private Copying: An International Survey of the Problem and the Law* (Sweet and Maxwell, 1993); Joel L. McKuin, "Home Audio Taping of Copyrighted Works and the Home Audio Recording Act of 1992: A Critical Analysis," *Hastings Comercial Entertainment Law Journal* 16 (1994); Christine Carlisle, "The Audio Home Recording Act of 1992," *Journal of Intellectual Property Law* 1994; Gary S. Lutzger, "DAT's All Folks: Cahn v Sony and the Audio Home Recording Act of 1991—Merrie Melodies or Looney Tunes," *Cardozo Arts and Entertainment Law Journal* 145 (1992); Lewis Kurlantzick and Jacqueline E. Pennino, "The Audio Home Recording Act of 1992 and the Formation of Copyright Policy," *Journal of the Copyright Society of the USA* 45 (1998); Jessica Litman, "Copyright Legislation and Technological Change," *Oregon Law Review* 68 (1989); Jessica Litman, "Copyright, Compromise, and Legislative History," *Cornell Law Review* 72 (1987).

22. There is a rich and related history, which it would take us too far afield to explore in this chapter, concerning the origins of cyberculture out of counterculture. This story involves John Perry Barlow, a lyricist for The Grateful Dead, who became one of the earliest and most outspoken and influential voices about the value of the Internet to the individual; *The Whole Earth Catalog* and The WELL online community, and the Electronic Frontier Foundation. For more information

on this topic, see Fred Turner, *From Counterculture to Cyberculture* (University of Chicago Press, 2006); Katie Hafner, *The WELL* (Carroll & Graf, 2001); or the Wikipedia articles on The WELL (Whole Earth 'Lectronic Link) and John Perry Barlow.

23. For more on MTV, see a good communication study by Jack Banks, *Monopoly Television: MTV's Quest to Control the Music* (Westview, 1996); and two journalistic works by Tom McGrath: *MTV: The Making of a Revolution* (Running Press, 1996); and *Video Killed the Radio Star: How MTV Rocked the World* (Random House, 1996). There are also a number of cultural and gender studies and fan books about MTV that are less useful for understanding the business impact of MTV on the record distribution system.

24. See http://www.davidbowie.com for the current version of Bowienet.

25. The discussion of this episode and many others in this chapter concerning individual artists and bands draws heavily on John Alderman, *Sonic Boom: Napster, MP3, and the New Pioneers of Music* (Basic Books, 2001).

26. See the Wikipedia article on RealNetworks, http://en.wikipedia.org/wiki/RealNetworks.

27. See Todd Rundgren, "The History of PatroNet," http://www.lightos.com/TRmessage.html, June 29, 1998 (accessed May 28, 2006); this subscription service for music, books, and films is still available (as of May 28, 2006) at http://www.patronet.com.

28. *Vault Corp.* v. *Quaid Software, Ltd.*, 847 F.2d 255 (5th Cir. 1988). For a discussion of the case, see http://cyber.law.harvard.edu/ilaw/Contract/vault.htm.

29. *Sega Enterprises Ltd.* v. *Accolade Inc.*, U.S. Court of Appeals, Ninth Circuit October 20, 1992, 977 F.2d 1510, 24 USPQ2d 1561. For a discussion of the case, see http://cyber.law.harvard.edu/openlaw/DVD/cases/Sega_v_Accolade.html or http://digital-law-online.info/cases/24PQ2D1561.htm.

30. *Lasercomb America Inc.* v. *Reynolds*, U.S. Court of Appeals, Fourth Circuit August 16, 1990, 911 F.2d 970, 15 USPQ2d 1846. For a discussion of the case, see http://digital-law-online.info/cases/15PQ2D1846.htm.

31. For more information on IITF, see http://www.ibiblio.org/nii/NII-Task-Force.html.

32. http://www.copyright.gov/legislation/dmca.pdf provides a summary of the DMCA prepared by the U.S. Copyright Office in 1998.

33. For a discussion of these and other anticircumvention cases, see Fisher, *Promises to Keep*; Lorraine Woellert, "Intellectual Property," in William Aspray, ed. *Chasing Moore's Law* (SciTech Publishing, 2004).

34. RealNetworks, Inc. v. Streambox, Inc., United States District Court for the Western District of Washington, *2000 U.S. Dist. LEXIS 1889*, January 18, 2000. For a discussion of the case, see http://www.law.harvard.edu/faculty/tfisher/iLaw/realnetworks.html.

35. *Universal City Studios, Inc.* v. *Reimerdes*, 82 F. Supp.2d 211 (SDNY 2000). For a discussion, see http://cyber.law.harvard.edu/openlaw/DVD/NY/trial/op.html.

36. Although SDMI has not been active since 2001, it still maintains its official Web site at www.sdmi.org. For a discussion of some of the legal issues associated with copy protection, see Andre Lucas, "Copyright Law and Technical Protection Devices," 21 *Columbia-VLA Journal of Law & Arts* 225 (1997).

37. For more about these problems, see Eric Scheirer's editorial for MP3.com, "The End of SDMI," http://www.eff.org/cafe/scheirer1.html.

38. For a discussion of the legal challenge against Rio, see Brenden M. Schulman, "The Song Heard iRound the World: The Copyright Implications of MP3s and the Future of Digital Music," *Harvard Journal of Law & Technology* 12, no. 3 (1999).

39. The paper as presented was Scott A. Craver, Min Wu, Bede Liu, Adam Stubblefield, Ben Swartzlander, Dan W. Wallach, Drew Dean, and Edward W. Felten, "Reading Between the Lines: Lessons from the SDMI Challenge," *Proceedings of 10th USENIX Security Symposium*, August 2001. On the SDMI-Felten case, see, for example, the Wikipedia article on Ed Felten; the Electronic Freedom Foundation's site on *Felten, et al.,* v. *RIAA, et al.* http://www.eff.org/IP/DMCA/Felten_v_RIAA/; and Edward W. Felten, "The Digital Millennium Copyright Act and Its Legacy: A View from the Trenches," *Illinois Journal of Law, Technology and Policy*, Fall 2002.

40. For more on this case, see Ken "Caesar" Fisher, "Lexmark's DMCA Aspirations All But Dead," *Ars Technica* Web site, February 21, 2005, http://arstechnica.com/news.ars/post/20050221-4636 .html; also see the Electronic Frontier Foundation Web page, "Lexmark v. Static Control Case Archive, http://www.eff.org/legal/cases/Lexmark_v_Static_Control/ (accessed 5 June 2006).

41. For more on *Chamberlain* v. *Skylink*, see the Electronic Frontier Foundation Web page, *Chamberlain Group, Inc.* v. *Skylink Technologies, Inc.*, http://www.eff.org/legal/cases/Chamberlain _v_Skylink/.

42. See Nate Mook, "Death Knell Sounds for Nullsoft, Winamp," *BetaNews*, November 10, 2004, http://www.betanews.com/article/Death_Knell_Sounds_for_Nullsoft_Winamp/1100111204 (accessed June 5, 2006).

43. *UMG Recordings, Inc.* v. *MP3.com, Inc.*, 92 F. Supp. 2d 349 (S.D.N.Y. 2000). For an analysis of the case, see for example http://www.law.uh.edu/faculty/cjoyce/copyright/release10/UGM.html.

44. *Recording Industry Association of America* v. *Diamond Multimedia Systems*, 1998. For an analysis of the case, see http://cyber.law.harvard.edu/property00/MP3/rio.html.

45. For a discussion of the more recent business model and business environment for eMusic in the face of iTunes, see Nate Anderson, "Making Money Selling Music without DRM: The Rise of eMusic," *ArsTechnica*, May 22, 2006, http://arstechnica.com/articles/culture/emusic.ars (accessed June 5, 2006).

46. Brad King, "Riffage.com Pulls the Plug," *Wired News*, December 8, 2000, http://www.wired .com/news/business/0,1367,40561,00.html; John Borland, "Riffage Site Pulls its Own Plug," CNET News.com, December 8, 2000, http://news.com.com/2100-1023-249653.html.

47. See "RealNetwork Mixes Player, Jukebox Software," ITWorld.com, September 24, 2001, http://www.itworld.com/AppDev/1470/IDG010924real/ (accessed August 3, 2006).

48. See Gracenote's Web site for more information about its services, in particular http://www.gracenote.com/music/corporate/FAQs.html/faqset=what/page=all.

49. For more information about Listen.com, see Courtney Macavinta, "All 'Big Five' Labels, Madonna Back Listen.com," CNET News.com, February 4, 2000, http://news.com.com/All+Big+Five+labels,+Madonna+back+Listen.com/2100-1023_3-236507.html.

50. More recently, EMI has taken a major step into the digital world. Between 2003 and 2008 it plans to spend between $130 million and $175 million to digitize its more than 300,000 master copies recorded over the past century and to develop an infrastructure that will support sales of digital content. See Laurie Sullivan, "EMI Takes Its Music Digital," *Information Week*, January 20, 2006, http://www.informationweek.com/news/showArticle.jhtml?articleID=177102644 (accessed June 5, 2006).

51. See "EMI CMG Partners with Liquid Audio to Offer Burnitfirst.com," EMI Press Release, April 29, 2002, http://www.emigroup.com/Press/2002/PRESS18.htm.

52. BMG Entertainment and Universal Music Group Form GetMusic—An Unprecedented Internet Content and Commerce Alliance, Universal Music Group Press Release, April 7, 1999, http://new.umusic.com/News.aspx?NewsId=28. In 2001, Universal bought out Bertelsmann's interest in GetMusic. Lori Enos, "Universal to Put All Eggs in GetMusic.com," *E-Commerce Times*, April 26, 2001, http://www.ecommercetimes.com/story/9264.html (accessed June 5, 2006).

53. It is interesting to see what was going on in Europe at the same time. Australian pop star Natalie Imbruglia's album *White Lilies Island* was released throughout Europe with the Cactus Data Shield in late 2001. When played in a PC's CD-ROM drive, the Cactus Data Shield corrupts the disc so it is unreadable, but if played on an ordinary CD player the player's error-correction system can make the disc readable with no noticeable loss in audio quality. Bertelsmann had not informed Virgin MegaStore, its largest retailer, that this album was being released only with the anticopy software and had not labeled the individual discs or their packaging to that effect. Following numerous complaints from purchasers, Bertelsmann set up a free return-envelope system to replace the copy-protected CDs with non-copy-protected ones. However, only one year later Bertelsmann was planning copy protection on all its new releases in Europe. See Tony Smith, "BMG to Replace Anti-Rip Natalie Imbruglia CDs," *The Register*, November 19, 2001, http://www.theregister.co.uk/2001/11/19/bmg_to_replace_antirip_natalie/; John Lettice, "'No More Music CDs Without Copy Protection' Claims BMG Unit," *The Register* November 6, 2002, http://www.theregister.co.uk/2002/11/06/no_more_music_cds_without/.

54. See, for example, Paul Kushner, "Streaming Media Company Spotlight: ARTISTdirect," November 26, 2001, Streaming Media.com, http://www.streamingmedia.com/article.asp?id=8071; Ellie Kieskowski, "Riffage's Strategy Combines Online and Offline Resources," October 20, 2000, Streaming Media.com, http://www.streamingmedia.com/article.asp?id=6356; "BMG and Microsoft to Advance Secure Distribution of Digital Music," Microsoft, PressPass—Information for Journalists," April 6, 2000, http://www.microsoft.com/Presspass/press/2000/apr00/bmgpr.mspx;

Pamela Parker, "BMG Entertainment, Egreetings Ally to Promote Artists," April 27, 2000, ClickZNews, http://72.14.203.104/search?q=cache:W4zRFtZGrxwJ:clickz.com/news/article.php/349791+bmg+egreetings&hl=en&gl=us&ct=clnk&cd=1; "Eritmo.com Unveils B2B Strategy," *The South Florida Business Journal*, October 16, 2000, http://www.bizjournals.com/southflorida/stories/2000/10/16/daily6.html.

55. Andy Wang, "Riffage.com Hits High Notes with AOL, BMG Funding," *Ecommerce Time*, December 15, 1999, http://www.ecommercetimes.com/story/2006.html; Knowledge@Wharton, "CDNow Founders Profit as Investors Take a Beating," CNET News.com, August 10, 2000, http://news.com.com/2100-1017-244218.html.

56. For an overview of antecedents as well as peer-to-peer systems for sharing music, see Shuman Ghosemajumder, "Advanced Peer-Based Technology Business Models," masters thesis, Sloan School of Management, MIT, 2002, http://shumans.com/p2p-business-models.pdf.

57. GeoCities, formed in 1994 as Beverly Hills Internet, began offering free Web hosting in 1995 and had a million users at its heyday in 1997. See the Wikipedia article on GeoCities, http://en.wikipedia.org/wiki/GeoCities.

58. Pete Brush, "The MP3 Eating Spider," APBNews.com, March 12, 1999, http://www.namesuppressed.com/ns/press-19990312.shtml.

59. While Napster was not concerned about whether its actions damaged the record industry's intellectual property, Napster did act to protect its own intellectual property. When another dot-com company (Angry Coffee) implemented a system called Percolator that enabled users to use Napster to download music without directly connecting to Napster, access from Angry Coffee to Napster was blocked and legal warnings were issued.

60. See the Wikipedia articles on Napster, Napster (pay service), Pressplay, and Roxio.

61. See the Wikipedia article on Scour, http://en.wikipedia.org/wiki/Scour.

62. See the Wikipedia article on Madster, http://en.wikipedia.org/wiki/Aimster.

63. This decision was eventually reversed on appeal. See *Recording Industry Association of American v. Verizon Internet Services, Inc.*, F.3d, No. 03-7053, 2003 WL 22970995 (D.C. Cir., Dec. 19, 2003).

64. See, for example, all the discussion of file sharing on the site of Educause, the organization that serves university chief information officers: "P2P (Peer to Peer)/File Sharing" at http://www.educause.edu/645?PARENT_ID=608 (accessed June 6, 2006).

65. Jeff Goodell, "Steve Jobs: The Rolling Stone Interview" Rolling Stone, December 3, 2003. http://www.rollingstone.com/news/story/5939600/steve_jobs_the_rolling_stone_interview/.

66. All music in the Apple Music Store is encoded in the AAC format, which is superior to the MP3 format in its compression of file size and which gives high music quality, similar to that of an uncompressed CD. See John Borland "Apple Unveils Music Store," CNET News.com, April 28, 2003. http://news.com.com/Apple+unveils+music+store/2100-1027_3-998590.html.

67. Leander Kahney, "iTunes, Now for the Rest of Us," *Wired News*, October 16, 2003.

68. John Borland, "Music Moguls Trumped by Steve Jobs?" CNET News.com, April 15, 2005. http://news.com.com/Music+moguls+trumped+by+Steve+Jobs/2100-1027_3-5671705.html.

69. *Newsweek*, October 27, 2003, as quoted in *Wired News*, "Steve Jobs' Best Quotes Ever" http://www.wired.com/news/culture/mac/0,70512-1.html?tw=wn_story_page_next1. Also see the partisan account in Andrew Orlowski, "Your 99¢ Belong to the RIAA—Steve Jobs," *The Register* November 7, 2003. http://www.theregister.co.uk/2003/11/07/your_99¢_belong/.

70. Chris Taylor, "The 99¢ Solution," *Time Magazine* 2003, http://www.time.com/2003/inventions/invmusic/html (accessed 31 July 2006).

71. Borland, "Music Moguls."

72. One reviewer of an early draft of this chapter indicated that not only changes in the technology but also in the nature of the music that is published and recorded could have an effect on record sales. This may be true, but it is beyond the scope of this chapter to make an analysis of the effect of record content on sales.

73. "Napster Said to Hurt CD Sales," *New York Times*, February 25, 2001, http://msl1.mit.edu/ESD10/nytimes_nap_cdsales.pdf.

74. Joel Selvin, "Did Napster Help Boost Record Sales?," Lively Arts, *San Francisco Chronicle*, August 5, 2001, http://www.sfgate.com/cgi-bin/article.cgi?file=/chronicle/archive/2001/08/05/PK220163.DTL.

75. John Borland and Rachel Konrad, "Study Finds Napster Use May Cut Into Record Sales," CNET News.com, May 25, 2000, http://news.com.com/2100-1023-241065.html.

76. Stan J. Liebowitz, "Testing File-Sharing's Impact By Examining Record Sales in Cities," School of Management, University of Texas at Dallas, September 2005, http://www.utdallas.edu/~liebowit/intprop/cities.pdf. He found: "Using a data set for 99 American cities containing information on Internet use, record sales, and other demographic variables, an econometric analysis is undertaken to explain the change in record sales before and after file-sharing. The results imply that file-sharing has caused the entire decline in record sales that has occurred and also appears to have vitiated what otherwise would have been fairly robust growth in the industry. Looking at sales in individual musical genres reinforces the primary conclusions."

77. Rafael Rob and Joel Waldfogel, "Piracy on the High C's: Music Downloading, Sales Displacement, and Social Welfare in a Sample of College Students," NBER Working Papers, 10874, National Bureau of Economic Research, Inc. (2004).

78. For a discussion of this study, see Beth Potier, "File Sharing May Boost CD Sales," *Harvard University Gazette*, April 15, 2004. The paper by Oberholzer-Gee and Strumpf, "The Effect of File Sharing on Record Sales: An Empirical Analysis," June 2005, can be found at http://www.unc.edu/~cigar/papers/FileSharing_June2005_final.pdf.

79. http://en.wikipedia.org/wiki/File_sharing.

80. See, for example, Neil Gladstone, "Virtual Conference," March 25–April 1, 1999, Philadelphia CityPaper.net, http://citypaper.net/articles/032599/mus.sxsw.shtml.

81. Cynthia L. Webb, "Musicians Sing Different Tune on File Sharing," *Washington Post*, December 6, 2004, http://www.washingtonpost.com/wp-dyn/articles/A39155-2004Dec6.html. The study cited by the *Washington Post* is Mary Madden, "Artists, Musicians, and the Internet," Pew Internet & American Life Project, December 5, 2004, http://www.pewinternet.org/pdfs/PIP_Artists.Musicians_Report.pdf.

82. Neil Strauss, "File-Sharing Battle Leaves Musicians Caught in the Middle," *New York Times*, September 14, 2003, http://home.att.net/~mrmorse/nytimes20030914musi.html.

83. Some other examples include Tom Petty, Dionne Warwick, Willie Nelson, and the Blues Brothers. See, for example, Jon Pareles, "Musicians Want a Revolution Waged on the Internet, *New York Times*, March 8, 1999; Michael Robertson, "Top Tier Artists Do MP3," MP3.com (as cited in Schulman, "Song Heard iRound the World" note 50, but with a link no longer active).

84. "MP3 Revolution Splitting Music Industry Along Cyber Lines," CNN.com, December 16, 1998, http://www.cnn.com/SHOWBIZ/Music/9812/16/digital.music/index.html.

85. A legal discussion for bands about protecting intellectual property on fan sites can be found at Erika S. Koster and Jim Shatz-Akin, "Set Phasers on Stun: Handling Internet Fan Sites," Oppenheimer Wolff & Donnelly LLP, copyright 2002, http://www.oppenheimer.com/news/content/fan_sites.htm (accessed June 6, 2006).

86. See chapter 2, "Why Napster Spread Like Wildfire," in Trevor Merriden, *Irresistible Forces: The Business Legacy of Napster & the Growth of the Underground Internet* (Capstone Publishing, 2001).

87. "The Day My Free Computer Music Died," *New York Times* February 18, 2001.

88. As quoted in Merriden, *Irresistible Forces*, p. 25.

89. One example is the increasing number of people who are legally swapping music from jam bands. This phenomenon is in the tradition of the Deadheads, mentioned earlier in this chapter, who were swapping audiotapes of live concerts with the permission of the Grateful Dead, and more recently fans who traded MP3 recordings of live concerts of the (now disbanded) rock band Phish, which seldom performed a song twice in the same way. There are more than 25,000 live recordings from more than a thousand artists available, for example, on Brewster Kahle's Live Music Archive. The RIAA is not opposed to this kind of legal trading. For more on this legal trading, see "Live Music Trading Goes on Net," *Los Angeles Times*, latimes.com, August 8, 2005, http://hypebot.typepad.com/hypebot/2005/08/live_music_trad.html (accessed 2 August 2006).

90. Mary Madden and Lee Rainie, "Music and Video Downloading Moves Beyond P2P," Pew Internet Project Data Memo, March 2005, Pew Internet & American Life Project, http://www.pewinternet.org/pdfs/PIP_Filesharing_March05.pdf.

91. The term *semiotic democracy* was introduced by the media critic John Fiske in his book *Television Culture* (Routledge 1988), in which he showed that viewers were not passive but give their own interpretations of television programming. More recent examples have included the so-called Phantom Edit of Jar Jar Binks out of the movie *Star Wars: The Phantom Menace* and the rewriting of J. K. Rawling's Harry Potter stories into homosexual romances. See Lorraine Woellert, "Intellectual Property," in William Aspray, ed., *Chasing Moore's Law* (SciTech Publishing, 2004), and the Wikipedia articles on fan fiction, slash fiction, and semiotic democracy (accessed 25 May 2006), respectively.

15 Eros Unbound: Pornography and the Internet

Blaise Cronin

A business history of cybersex is badly needed.
Jonathan Coopersmith, "Does Your Mother Know What You *Really* Do?" 2006

Noël Coward thought pornography was "terribly, terribly boring." Tens of millions of Americans disagree. At least it would be difficult to conclude otherwise given the billions of dollars spent annually on legal sex goods and services in this country—a point made, I believe, by Nadine Strossen, president of the American Civil Liberties Union: "Of the $10 billion sex industry, it's not ten perverts spending $1 billion a year." And she might have added, it is not just the United States; nor is it just what might be termed the Internet effect. By way of illustration, in Finland, a small country with a population of roughly five million, the annual market for pornographic videos alone is estimated to be in the $300 to $500 million range. In Sweden, even in the mid-1970s, more than 650,000 pornographic magazines were sold weekly, and that in a country with a population of just over eight million.[1] Sex, it is safe to say, sells; it always has. In Renaissance Venice, 10 percent of the city's population earned a living from prostitution.[2] Clearly, the "sin market," whether epitomized by whorish Venice, contemporary Thailand's sex industry, or the triple X-rated corners of world of the Internet, is neither a trifling nor a peculiarly modern affair.[3]

Definitions of the legal sex industry typically include the following activities, products, and services: adult videos and digital video discs (DVDs), magazines, branded merchandise, sex aids and games, pay-per-view television, Web-based subscription services, sex clubs, phone sex, and escort services; excluded are prostitution, sex trafficking, and child pornography. Of course as technology evolves, new distribution media, products, and services will come onstream, expanding the spectrum of pornographic possibilities and in turn refining established product categories. In this chapter I focus predominantly on the legal sex industry, concentrating on the output of what, for want of a better term, might be called the respectable or at least licit part of the pornography business. I readily acknowledge the existence of, but do not dwell on the seamier side, unceremoniously referred to by an anonymous industry insider as the world of "dogs, horses, 12-year old girls, all this crazed Third-World s—." Additionally, I provide some historical background on the production and consumption of pornography so that the significance of the Internet to the euphemistically labeled adult entertainment industry can be better conceptualized as well as appreciated.

If sex sells, then technology-mediated sex sells spectacularly. Historians, sociologists, sexologists, and cultural and communications theorists have sedulously analyzed the symbiotic relationship between pornography and new media. These days, it is scarcely necessary to point out that representational technology and Eros have from time immemorial performed an inventive pas de deux. Depending on your perspective, it is a marriage made in heaven or hell—in either case, an indissoluble marriage.

We have come a long way since the making of the first pornographic daguerreotype in the mid-nineteenth century and the first public showing of a pornographic film in the United States in the early part of the twentieth. The invention of photography, more particularly cinematography, brought visual pornography to the masses for the first time; it allowed for a class of pornography, moreover, that was distinguished by previously unimagined levels of objectification, amplification, directness, and verisimilitude. As the painter David Hockney observed, "Photography moved in closer so your eye sensed the flesh more and you got closer to the figure."[4] The big screen close-up effortlessly trumped the gunmetal gray lithograph; mediated action eclipsed the sepia postcard. With the invention of the Internet and wireless communications, we are witnessing another major transition in the evolution and expansion of populist pornography, and the correlative emergence of altogether new forms of "pornographic spectatorship."[5]

This chapter does not pretend to be a strict, chronological analysis of the Internet and pornography. That history is still in its infancy and much remains emergent. Although the Internet is the binding thread of my narrative, I have chosen to analyze the historical evolution of pornography production indirectly, by exploring a number of major themes (e.g., media substitution, domestication, and corporatization) that illustrate the relationship between information and communication technologies, consumer markets, and forms of sexual representation and expression. In one sense, the Internet is just the latest in a succession of technological advances that have taken pornographic products and practices to the next level of sophistication. In another sense, it is a continuously evolving sociotechnical innovation that supports genuinely novel modalities of pornographic expression and confession. To appreciate why both these viewpoints have merit, we need to understand something of the prehistory of the Internet.

Representations and Definitions

Visual representations of sex, ranging from the mildly erotic to the troublingly transgressive, span the documented history of humankind and are everywhere to be found:

from prehistoric caves, through Attic pottery and seventeenth-century political pamphlets, to present-day mass media advertising. Despite cycles of repression and prohibition, pornography is a more or less accepted facet of contemporary life—a recognized "cultural category"—albeit one that continues to be forensically elusive, with Supreme Court justice Potter Stewart's famously visceral "I know it when I see it" approach as prevalent as ever.[6] Pornography is also, as historians are wont to observe, a construct of fairly recent origin, first appearing in the *Oxford English Dictionary* in the mid-nineteenth century. Maturity, stability, and consensus are not terms one readily associates with "pornography."

Whose definition of "erotic," "arousing," "indecent," "obscene," "prurient," or "deviant" holds sway at any given moment? What is the juridical standard for material that is "patently offensive"? How do we establish "beyond reasonable doubt" that pornography causes actual harm, "cultural pollution," or leads to an increase in sex crimes? Who gets to define "the average man" or "reasonable person" when community standards are being invoked? What constitutes the community when the focus is Web-based pornography and virtual addresses, rather than state lines and national borders? Who is best qualified to decide whether a putatively pornographic piece of work or object possesses "any redeeming, literary, artistic, political, or scientific value"? Whose reading or semiotic analysis of pornographic images is deemed to be correct? How important are intention, context, and packaging in determining what should be labeled as "pornographic"? What are the pros and cons of "disgust-based legislation"[7]? These, it should be noted, are not just legal abstractions but matters that have a material bearing on the activities and operations of the adult entertainment industry, and also the lifestyle choices of millions of Americans.

In the United States, the litmus test for obscenity continues to be the three-pronged Miller test, named after the *Miller v. California* case in 1973. In his ruling, Justice Warren Burger defined as pornographic a work that when taken as a whole, appeals to a prurient interest, depicts sexual conduct in a patently offensive way, and lacks serious literary, artistic, political, or scientific value. If a work conforms to this definition, then it may, according to prevailing community standards, be considered as obscene. Of course, the problem of achieving consensus on any of the key terms embedded in this admirably concise ruling remains: What precisely is "prurient interest," and how is it demonstrated? Who, to use the words of John Milton, is blessed with "the grace of infallibility" necessary to answer such unyielding questions? Probably not constitutional lawyers and social scientists, who together have assembled a mass of conflicting evidence and contested opinions over the decades, often parceled in voluminous reports issued by august government commissions (more on that later). A pragmatic and highly parsimonious solution is to define pornography as that which is produced and distributed by self-confessed pornographers. Thus, one looks to the market rather than the Supreme Court or the social and behavioral sciences to settle definitional debates.[8]

Having said that, a brief philological excursus is probably in order here. The word pornography, which trips so easily off the contemporary tongue despite a striking lack of consensus as to what it denotes, stems from the Greek word *porne*, itself derived from the verb for selling/being sold (*pernanai*). In the "sexual economy" of ancient Athens—so vividly described in James Davidson's *Courtesans and Fishcakes*—the word porne was commonly used to mean "prostitute" and gave rise to an associated lexicon (e.g., *porneion* meaning "brothel," and *pornoboskos* meaning "pimp").[9] The word pornography from the late Greek *pornographos*, "writing about prostitutes," only surfaced more than two millennia later, seemingly for the first time in the late eighteenth century.[10] In short, the social construction of pornography is a really rather recent phenomenon.

In some Western industrialized nations (e.g., Denmark and Germany), the liberalization of censorship laws in the 1960s and 1970s profoundly changed public practices and social values with respect to pornography. In the United States, however, the world's largest producer and consumer of pornography, public attitudes continue to be schizophrenic, with a perplexing variation in community standards across state lines and an equally perplexing inconsistency in the application of existing antiobscenity statutes. Nonetheless, access to pornography in the United States has been relaxed to a considerable extent—the "feminist rhetoric of abhorrence" as well as the growth of religiosity and neoconservative values notwithstanding.[11] The subject has been progressively decriminalized and destigmatized, partly due to changing sexual mores, partly to the effects of technology, and partly as a result of growing media and scholarly coverage, providing further evidence for some of the "modern incitement to sexuality" and the coarsening of society, daily revealed in the pages of the tabloid press, reality television programming, and the proliferation of Web-based voyeur sites (see, for example, Voyeurweb.com).[12]

Table 15.1 shows the frequency of newspaper and magazine mentions for two terms, "adult entertainment" and "porn*" (the * signifies truncation) based on searches of the LexisNexis news databases over a seven-year period (1999–2005). The subject is evidently newsworthy, and indeed has been for several decades: equivalent data relating to media coverage for the period 1984–1994 can be found in Brian McNair's clear-headed study of pornography and postmodern culture.[13] Judging by the available evidence and indicators, it would be hard to avoid the conclusion that we live not only in a "public culture of sex" but in a "culture of public sex," where frankness and the absence of privacy are taken for granted.[14]

Since the invention of photography, each successive wave of technological innovation—Polaroid, videocassette recorder (VCR), camcorder, CD-ROM, digital camera, Webcam, cell phone, and iPod—has generated a new battery of products, formats, and delivery possibilities for enterprising individuals to extend the boundaries of representational fantasy. This phenomenon is most strikingly apparent in the case of the

Table 15.1

LexisNexis search in general news for *adult entertainment* and *porn*

| Year | Adult entertainment | | Porn* | |
	Limited to magazines and journals	Limited to major papers	Limited to magazines and journals	Limited to major papers
1999	6	236	151	Over 1,000*
2000	19	227	165	Over 1,000
2001	20	263	229	Over 1,000
2002	30	211	285	Over 1,000
2003	34	237	299	Over 1,000
2004	15	189	242	Over 1,000
2005	19	199	264	Over 1,000

*Does not record actual number above one thousand.

Internet, an enabling platform of enormous channel capacity and potentially huge market reach for pornographers of all kinds, be they publicly quoted corporations, small enterprises, or sole traders. Significantly, as documented in both the scholarly and popular press, pornographers and others have not only exploited the reproductive and distributional potential of the Internet (and before it, albeit on a more modest scale, Minitel in France) to great effect; in many instances they have been, and continue to be, either directly or indirectly responsible for major technical innovations, or at the very least, for taking such developments to the next level of sophistication (e.g., subscription services, exit consoles/pop-up windows, streaming video, and affiliate marketing schemes). As the futurist Paul Saffo put it: "The simple fact is porn is an early adopter of new media. If you're trying to get something established ...you're going to privately and secretly hope and pray that the porn industry likes your medium."[15] This, it has been argued by more than a few amateur historians, was illustrated in the 1980s when the industry's preference for VHS over Betamax format helped shape the final market outcome. Something similar is happening (at the time of writing) with respect to the competition between Blu-ray and HD-DVD for dominance in the high-definition DVD format market, and the debate has been reignited.[16]

Many of these technical and marketing innovations have subsequently become, for better or worse, standard practice in the wider world of electronic commerce. As Jonathan Coopersmith, a historian, observed: "Pornographic products have served to stimulate initial interest on these new technologies despite their higher initial costs.... As each of these technologies matures and prices drop, the role of pornographic products diminishes relatively, but not absolutely."[17] Pornography may credibly claim, along with gambling, to be e-commerce's much-heralded "killer application." It is no exaggeration to say that the Internet is taking us, in the words of Michel Foucault, to "a different economy of bodies and pleasure," one in which the viewer/consumer has increasing control over both the specification of carnal content (be it live or canned, passive or interactive) and the conditions (social, temporal, or locational) under which it is consumed or experienced.[18]

The "sexualization of the public sphere" began in earnest in the "Swinging Sixties."[19] Since then, public attitudes regarding sex and the public tolerance of pornography have changed measurably and irrevocably. It is not that the Internet has somehow fashioned entirely new sexual tastes and tropes, or given rise to entirely new forms of sexual representation and expression, as anyone with even a fleeting familiarity with the history of human sexuality and its representation knows all too well. Rather, in addition to an unprecedented scaling up of production, distribution, marketing, and consumption possibilities, it has created a multiplicity of meticulously categorized spaces, both commercial *and* free, in which individuals of like mind and predilection can find customized content, experience expressive echo, and experiment with novel forms of engagement, ranging from chat rooms through virtual sex games and interactive DVDs to technology-mediated, spatially distributed sex (the burgeoning realm of teledildonics, cybersex, and haptic interfaces). And consumers are able to take advantage of Internetworking features and functionalities because the means of production (cameras, computers, etc.) have become progressively smarter, cheaper, and easier to use by nonexperts—a general trend in the history of information and communication technologies referred to as "blackboxing."[20]

Thus, along with the growth of the market for professionally produced pornographic goods, we are witnessing the rapid maturation of relatively low-cost, amateur modes of production and exchange, often collective in character, resulting in the creation of a parallel not-for-profit trading zone predicated on barter and gift exchange—one that scarcely existed previously. In fact, it would be more accurate to say that galaxies of micromarkets and virtual sexual communities, often fiercely libertarian in character—reflecting the early Internet culture—are being created, some legal, some not, alongside the more immediately visible and traditionally structured commercial sector. In these virtual (liminal) spaces, social norms and conversational etiquette are routinely suspended. Foucault believed that the "moral code" (that which is forbidden, that which is allowed) had changed little over the centuries, but he was writing in pre-Internet times.[21] It is hard to gainsay the challenge to the prevailing moral order provided by the extraordinary confessional power of the Internet. It allows the little man—millions of little men, more accurately—to cock a snoot at the "code," in the process achieving a kind of postmodern purification through unfettered libidinous self-expression on a scale heretofore unknown. And it's not just men: women, too, have been adept at exploiting the presentational power of the Internet, be it "risqué...self-portraiture" on blogs and Web sites such as Flickr.com, or willing participation in the production and sharing of "reality pornography."[22]

Nor did the Internet in any sense "invent" some of the more bizarre and violent forms of pornography that have attracted so much media comment in recent years—saliromania, bestiality, pedophilia, and other practices that are commonly deemed to be *contra naturam* have a long, if less than edifying, pedigree—but it certainly has increased awareness of as well as opportunities for creating, experiencing, and sharing nonmainstream forms of pornography.[23] Sean Thomas reflects popularly held views on the protean nature of online pornography when he speaks of its "infinite and remarkable variety," and its "voluptuously and seductively diverse" character.[24] Specifically, more children and young adults, boys in particular, are being exposed to pornography, and more varieties of pornography, than ever before—and at a significantly earlier age, thanks to the Internet.[25] It should be noted in passing that the same phenomenon can be seen in relation to online gambling. It is unclear, however, whether the consumption ratio of mainstream to marginal or bizarre pornographic content has changed with the arrival of the digital age. But it is not unreasonable to assume that in a competitive industry characterized by oversupply—ten to fifteen thousand pornographic movies are made annually in the United States according to various estimates—that one effect is a ratcheting up of the production of hard-core output targeted at increasingly demanding and specialized markets. As the tolerance threshold rises, producers will do whatever it takes to boost sales and differentiate their wares from those of the competition. This is, of course, a feature of competitive, capitalist markets generally, not a phenomenon peculiar to pornography.

Yet the central question remains: does the easy availability of pornography on the Internet change individuals' dispositions and actions? There is a body of empirical evidence, reviewed by William Fisher and Azy Barak, that suggests "normal range individuals will ordinarily choose sexually explicit Internet materials which are not antisocial in nature." Most people have "a lifetime learning history and set of expectancies about acceptable and unacceptable sexual behavior" that are sufficiently robust to constrain their search and consumption behaviors. The experimental evidence does not seem to support the presumptive "'monkey see, monkey do' assumption" when it comes to Internet pornography. In sum, Fisher and Barak argue, wide availability and easy access to online pornography may not in themselves be sufficient to alter individuals' tolerances and tastes for the simple reason that our behavior is "self-regulated," based as it is on "preexisting response tendencies."[26]

Social Construction of Technology

The early history of the Internet, beginning more than four decades ago with the experimental ARPANET, is a stirring tale of engineering inventiveness and inspired collective action.[27] In a relatively short time, though, a system that was designed primarily to transport computer files and facilitate reliable communication between research scientists mutated into a massively distributed, global network with an installed user base of billions. The extraordinary growth and rapid diffusion of the Internet, and subsequently the Web, cannot be explained satisfactorily in engineering terms alone; a sociotechnical analysis is required. How many engineers would have predicted the emergence of chat rooms, or envisaged developments such as blogs or folksonomies? Such innovations testify eloquently to individual users' (who are often amateurs rather than experts) willingness and ability to exploit the potentialities and affordances of the Internet in ways that traditional systems developers would not have imagined. In some respects, the Internet is a compelling instantiation of Eric von Hippel's notion of the "democratization of innovation."[28] Here, social forces and values matter as much as technical expertise and professional status. Nowhere perhaps is this general point more tellingly exemplified than in the world of online pornography.

As Elisabeth Davenport and I have shown, a social construction of technology perspective provides an effective theoretical framework for understanding developments relating to digital pornography in both the public and private spheres.[29] It can, in the words of the late Rob Kling, offer "a terrific illustration of the ways that IT can be used . . . that are well outside the intentions of original designers."[30] Sociotechnically speaking, the Internet and the pornography business are co-constitutive: technology alone does not determine the means and modes of production. In the vernacular, the Internet is shaped as much by pornography as pornographic products and practices are shaped by the Internet: human values, personal choices, and individual inventiveness all come into play.

To be sure, other information and communication technologies have had a profound influence on the nature of the adult education business, but as we shall see, none has resulted in remotely comparable economies of scale and scope. The Internet allows pornographers to extend their reach to larger and more geodemographically diverse markets. It also enables them to distribute their otherwise-perishable product portfolio to a multiplicity of consumer segments at close-to-zero marginal cost, generating for the lucky ones highly attractive rates of return. But there are challenges: credit card fraud and unpaid charges (consumers frequently deny having made a subscription or purchase) are persistent problems for the online pornography industry, with producers and merchants being held responsible for all charge backs by credit cards companies.[31]

Disintermediation has created a direct producer-to-consumer market space, which has stimulated the evolution of new business models and approaches, and greatly enhanced customer choice. A good example are Web sites such as Candida Royalle's pioneering Femme (⟨http://www.royalle.com⟩) that are owned, designed, and operated by women with products fashioned specifically for the female market. Continuous product refreshing is a *conditio* sine qua non for a business whose paying customer base has a notoriously short attention span; the traditional pornographic product life cycle curve is priapic in character. As is well-known, customer loyalty is extremely fickle when it comes to representations of sex; the "endless slippage of desire," as Zabet Patterson so delicately puts it, drives the consumer to the next available source of supply.[32]

This is a market propelled by an insatiable appetite for freshness and novelty, though not necessarily premium quality—one that at first glance would seem to instantiate what economists refer to as the "law of diminishing marginal utility." Happily, from the pornographer's perspective, the Internet supports multiple product life cycles. The same digitized product can be sold, cross sold, up sold, recycled, syndicated, and bundled almost ad nauseam, although in fairness it should be noted that earlier generations of print pornographers had long shown an aptitude for creative, if sometimes shady, product reuse and bundling (the same models, the same photographs, but different magazine covers). Furthermore, the Internet facilitates innovative forms of customization and interactivity (e.g., chat rooms or live sex); the consumer now has the possibility of shaping real-time audiovisual interactions with both professional sex performers and fellow amateurs—a radical shift in the historically asymmetrical take-it-or-leave-it relationship that has existed between producer and consumer in this industry. Lastly, the Internet has resulted in a reconfigured geography of pornography production and distribution, with new locations (e.g., Budapest) assuming importance alongside established centers such as Los Angeles and Amsterdam. Domain name analysis shows Hungry, Russia, and Thailand to be emerging centers for the production of pornographic content.[33] Brazil has also emerged as both an important production site and also an outsourcing center for American companies; the attractions include the relatively low production costs and an exotic talent pool.

Media Substitution

The adult entertainment sector provides a textbook illustration of technological diffusion and media substitution. With the rise of video technology, the number of cinemas nationwide showing sexually explicit movies—variously referred to over the years as sexploitation films, nudies, hard-core features, pornos, skin flicks, art studies, white coaters, hard-core loops, beavers, or split beavers, depending on the period, equipment, subject matter, and degree of sexual frankness—swiftly plummeted.[34] Interestingly, general cinema attendance does not seem to have been similarly affected by the rise in video sales and rentals.[35] Video production was not only much cheaper and more efficient than traditional filmmaking but it allowed pornography consumption to be removed from the public gaze. The VCR revolutionized the pornography industry in the late 1980s and the early 1990s, allowing mass-produced goods (still generating high margins) to be consumed in the remote-controlled comfort and safety of the home. This shift in the locus of consumption from public to private spaces was crucial to the industry's subsequent growth trajectory.

As the VCR-installed base grew nationally, a mass consumer market emerged, powered in no small measure by the sale of premium-priced adult movies; consumers will happily pay more to see or rent adult movies than mainstream, commercial releases, just as they once did for pornographic magazines. By the 1990s, according to various video industry trade statistics, millions of pornographic movies were being viewed each year in American households. An insight into the nature and significance of this sector of the adult entertainment business is provided regularly by the trade magazine *Adult Video News* (⟨http://www.avn.com/⟩). Today, in addition to the estimated eight hundred million adult tapes and DVDs that are either sold or rented annually—roughly a third of the total video sales, according to various industry and independent estimates—millions of pornographic images and video clips, many of them in all likelihood obscene as defined by prevailing community standards and thus technically illegal, are being viewed daily in U.S. homes, offices, and public spaces such as libraries via the Internet.

Film was not the only medium to suffer in the video revolution; pornographic magazine sales went into a free fall. Predictably, with the advent of the Internet, Web, and on-demand DVDs, traditional video sales are themselves suffering—a trend that will accelerate. By way of illustration, Private Media Group, a major adult entertainment company (profiled later in this chapter), reported an 83 percent decrease in video sales for the first half of 2005 compared with the same period in 2004. But it is not just a case of one medium or format displacing an earlier convention or standard; the shift from one genre or generation of technology to the next can also have the effect of boosting overall levels of consumption. At the peak of their pre-video popularity around about 1980, U.S. adult movie theaters were attracting two and a half million paying customers per week. Yet that figure palls alongside the fifteen million hardcore movies that were being rented weekly less than a decade later.[36] Business not only migrated from one delivery medium to another but also grew as a consequence—a "substitution plus" phenomenon that is occurring once again, this time thanks to the Internet.

Pornography consumption is no longer the preserve of an elite, male minority, as had been the case until the end of the nineteenth century and the advent of the "dirty" postcard, the prototypical, mass-market, pornographic product.[37] Nor is it something that happens at the periphery of polite society. Pervasive computing has spawned polymorphous pornography on television screens, desktops, laptops, cellular phones, and iPods. Producers are now crafting minimovies specifically for personal digital assistants and cell phones, extending the reach of their X-rated products and in the process making fresh use of existing video stock. The statistical reality is that millions of Americans view millions of pornographic images daily, often paying for the pleasure. We live in what Walter Kendrick terms a "postpornographic era," in which a growing slice of the nation's disposable income (and leisure time) is allocated to the acquisition of highly diverse visual sexual representations, mediated sexual experiences, and sexual fantasy goods.[38]

Each successive wave of technological innovation—telephony, photography, and computing—seems to take the adult entertainment business to a new level of economic significance and, in the case of the Internet, public awareness. Coopersmith has this to say on the trend: "As the purveyors of pornography have shifted from the black (illegal) to grey (legal and low profile) and white (legal and 'normal' profile) markets, its profits and prominence have increased."[39] Pornography has long had a shadowy presence in the public sphere, tucked behind frosted glass on magazine stands or sequestered in adult video stores. As a general rule, pornographic goods and sex services are quarantined from mainstream retail and residential spaces (due to the fear of negative secondary effects on the local community) and concentrated instead in "'zones of commodification,' magic spaces that turn people into products," to borrow Davidson's felicitous phrase.[40] But while little has changed in the physical realm, the Internet has enabled pornography to achieve near ubiquity, and greatly heightened public salience by circumventing the "time, place and manner regulations" long associated with brick-and-mortar sex businesses.[41]

One should be careful to avoid ahistoricism, though. The adult entertainment industry first exhibited dramatic growth in the latter decades of the twentieth century, helped by the relaxation or abolition of antipornography legislation. Even a quarter of a century ago, the pornography business in West Germany was estimated to have a turnover of roughly $250 million; for the United States, the corresponding figure was in the $4 to $7 billion range.[42] At its zenith, the print-based adult industry was producing literally hundreds of different pornographic magazines for newsstands and unmarked mail distribution. In 1975, sexually explicit magazines had a monthly circulation of more than fourteen million is in the United States, peaking at sixteen million five years later, according to Department of Justice statistics.[43] And to take a specific example, it was in the 1980s in Japan that one saw the emergence of hard-core pornographic comics targeted at twenty- to thirty-year-old females.[44] This pre-Internet genre has since grown in popularity, migrated across media, and begun to attract a broad international following. But again, one should be mindful of historical antecedents. Depression era America produced the intriguingly named Tijuana bibles (aka "eight-pagers"), ribald, satiric, pornographic comic books, which though illegal, remained extremely popular until the emergence of mainstream adult magazines in the 1950s. They have since acquired near folk art status among collectors of Americana.

In addition, the first generation of professionally run sex and pornography retail businesses was established during the second half of the twentieth century. In the early

1960s, the legendary Beate Uhse opened the first of her many sex emporiums in Germany. Two decades later her pioneering business had achieved a turnover of $55 million. Today, the company (⟨http://www.beateuhse.com⟩), still trading under her name and now managed by her children, is listed on the Frankfurt Stock Exchange and generates revenues of approximately $300 million. Like some other first-generation pornography/sex businesses, Beate Uhse Ag. is developing a strong Web presence in order to capitalize on its impressive brand recognition (survey research shows that almost all adult Germans know the company's name).

While recognizing that the Internet has spawned numerous start-ups, one should not overlook the fact that it has also provided established pornographers with a powerful platform for expansion and diversification. The amplification (of product range and customer base) attributable to the Internet may indeed be exceptional (even revolutionary), but the phenomena of technology-induced market growth and aggressive retailing are certainly not without precedent in the history of pornography.

Domestication

With the invention of the VCR, the domestication of pornography was set in motion. During the 1990s, the trend was intensified by the increased availability of pay-per-view adult content, via both cable and satellite television. Most recently, DVD by mail has been added to the delivery mix, and in the near future video on demand seems likely to become the preferred option. In addition to the home, major hotel chains quickly established themselves as highly profitable sites for discrete pornography consumption in both domestic and foreign markets. Today, roughly half of all American hotel rooms offer adult movies to their guests (the most extreme material, however, will be available in offshore locations), and according to billing records this is a highly lucrative business, especially for the distributors, who can take the lion's share of the revenues.[45] Typically, the video producer signs a multiyear license (which may or may not require an exclusivity agreement) with the operator (e.g., Playboy Enterprises). Some hotel chains (e.g., Hilton, Marriott, and Sheraton) as well as distribution companies (e.g., DirecTV, EchoStar, and Time Warner) are blue-chip corporations, and not surprisingly, they have shown reluctance to discuss on the record the matter of their place in the pornography supply chain. A few hoteliers (e.g., the Omni Group) believe it makes good public relations sense to discontinue offering adult pay-per-view movies, allowing them to differentiate themselves from near-market competitors.

Truth be told, the public face and private life of the nation tell two quite different stories. In contemporary America, motherhood and apple pie coexist, albeit awkwardly, with round-the-clock erotica and hard-core pornography. The country's historically ambivalent attitude to matters sexual was revealed most tellingly in the public's fascination with the "discourse of semen" that flowed through the *Starr Report*.[46]

Modern America's contradictory attitudes to sex and pornography have also been summarized deftly by Walter Kendrick in his book *The Secret Museum: Pornography in Modern Culture*: "The fact is, Americans concern themselves much less with sex itself than with representations of sex, any words or pictures that convey or may arouse sexual feelings. The American obsession with sex reflects a deep ambivalence about the power of representations; it is the by-product of a passion for image making, uneasily yoked to a passionate fear of image.... America vacillates hysterically between controlling sexual images and letting them run free."[47] Roland Barthes has also observed Americans' dependence on "a generalized image-repertoire," noting wryly that in a New York porn store "you will not find vice, but only its *tableaux vivants*."[48] One is tempted to say that the United States is a floating point on the erotophobia-erotophilia scale. In the months and years to come, we can expect to see much more jousting between representatives of the Department of Justice and the American Civil Liberties Union, as the battle lines between the various antipornography and free speech lobbies are drawn even more starkly.

At the time of this writing, mammon and morality seem set on a collision course—something that always seemed unlikely during the Clinton presidency. The Bush administration, its probusiness stance notwithstanding, is actively attempting to curb the excesses of the adult entertainment industry through the following:

• Legislation (most recently by requiring pornography promoters and producers to maintain records—something proposed much earlier by the Meese Commission—certifying that their models and actors are over the age of eighteen, a demand (18 U.S.C. 2257) that has caused some companies to either pull products from the marketplace or withdraw from the market altogether)
• Criminal investigation (the establishment in 2005 of an antiobscenity squad within the Federal Bureau of Investigation)
• Prosecution (such as the high-profile legal action against Extreme Associates, the first federal obscenity prosecution against a pornographic video manufacturer for over a decade)

Currently, the conservative Right is mobilizing support against the possible introduction of a new top-level domain name (.xxx) by the Internet Corporation for Assigned Names and Numbers for adult sites on the grounds that such a move would in effect create a virtual red-light zone for pornographers, and lead to an increase in the trafficking and consumption of adult materials, including, as the federal government is wont to point out, child pornography. In contemporary America, pornography is inconveniently located at the intersection of free speech and free enterprise.

This, it need hardly be said, is an industry whose top line is especially vulnerable to shifts in prevailing ideology and public opinion, as pioneer pornographers such as Larry Flynt (founder of the determinedly misogynist *Hustler* magazine and related ventures) have occasionally learned at their expense. Surprisingly perhaps, law enforcement has tended to concentrate its interdiction efforts on the distribution rather than the production end of the business, though that could conceivably change. Unsurprisingly, some U.S.-based companies have contingency plans in place to take their operations offshore in the event of either a sustained crackdown on the industry or a serious refocusing of existing law enforcement efforts. It seems clear in any case that pornography will not easily be put back in Pandora's box. As Linda Williams has noted, "Mainstream or margin, pornography is emphatically part of American culture, and it is time for the criticism of it to recognize this fact." Her remarks should not be construed as either an active endorsement or jaded acceptance of pornography but as a pragmatic and value-neutral acknowledgment of its inextricable presence as a "cultural form" in the Internet Age—one that merits systematic, scholarly investigation.[49]

Corporatization

Pornography is no longer an economically peripheral activity but a moderately significant—albeit for many, still unpalatable—component of the larger entertainment economy. The phenomenon is most visible in the state of California, where some ten thousand or so individuals are employed, directly or indirectly, in the adult video business alone. The San Fernando Valley region in Southern California is a textbook example of a dynamic industry cluster—a dense network of mutually supportive and complementary micro- and midsize enterprises catering to the needs of a specific sector.[50] It was also, as it happens, the subject matter of HBO's documentary series *Going Down in the Valley*.

As with other facets of the larger entertainment industry, mainstream pornography now has its own lobby groups, trade magazines, heath education programs, worker alliances, and business-to-business expositions.[51] It is adroitly surrounding itself with the paraphernalia of propriety and the clichés of marketing communications. The sleazy strip joints, tiny sex shops, dingy backstreet video stores, and other such outlets may not yet have disappeared, but along with the Web-driven mainstreaming of pornography has come—almost inevitably, one has to say—full-blown corporatization and cosmeticization. The small topless bar can still be found, yet it is being squeezed out by the appearance of upscale nightclubs. Some of these so-called gentlemen's clubs (e.g., Million Dollar Saloon, ⟨http://www.milliondollar.com/⟩; Rick's Cabaret International, Inc., ⟨http://www.rickscabaret.com/⟩) are multimillion dollar, publicly quoted companies: both the Million Dollar Saloon and Rick's trade on the NASDAQ. This, it is worth noting, is a growth segment of the overall adult entertainment market: there are almost four thousand such clubs in the United States, employing more than five hundred thousand people.[52]

Generally speaking, the archetypal mom-and-pop business is being replaced by a raft of companies with business school–trained accountants, marketing managers, and investment analysts at the helm—an acceleration of a trend that began at the tail end of the twentieth century. As the pariah industry strives to smarten itself up, the language used by some of the leading companies has become indistinguishable from that of Silicon Valley or Martha Stewart. It is a normalizing discourse designed to resonate with the industry's largely affluent, middle-class customer base.

This extract from New Frontier Media's Web page (⟨http://www.noof.com⟩) captures the tone: "New Frontier Media, Inc. is a technology driven content distribution company specializing in adult entertainment. Our corporate culture is built on a foundation of quality, integrity and commitment and our work environment is an extension of this. . . . The Company offers diversity of cultures and ethnic groups. Dress is casual and holiday and summer parties are normal course. We support team and community activities." Discursive reframing is very much à la mode in an effort to reposition pornography as a lifestyle proposition. But a cautionary note is in order: one should not assume that a company like New Frontier is representative of even the production side of the industry. The reality is that porn actors are not unionized, and do not have access to medical insurance or pension benefits. A fortunate few may have contracts with companies such as Vivid (⟨http://www.vivid.com⟩) or Digital Playground (⟨http://www.digitalplayground.com⟩), and a fortunate few—females in the main—may earn significant sums for each movie they make, but most are commodified bodies for hire, paid off the books. They have little security or leverage in a business blessed with an endless supply of willing, and sometimes desperate, labor.

The silent, grainy, black-and-white sixteen millimeter stag films and the accompanying whiff of organized crime so redolent of the mid-1900s have given way to high-quality digital productions (the first $1 million adult video, *Pirates*, a coproduction of Digital Playground and Adam and Eve, was released in 2005 with initial sales of a hundred thousand being claimed), sometimes featuring contracted studio porn stars, real-time streaming video, and vast libraries of digitized images catering to every conceivable sexual taste and deviance, from anal sex to zoophilia. What was once underground and the preserve of all-male audiences—roughly two thousand examples of the stag genre were produced between 1915 and 1968—is now aboveboard, Internet-accessible, and readily available to a mixed gender market.[53] Indeed, with the demise of the self-consciously primitive stag has come the end of "a ritual folk tradition of the American male."[54] Postmodernity is no respecter of medium.

Web-based interactive sex and product customization have taken the adult entertainment industry to new levels of technical sophistication, even if, in the words of Fenton Bailey, "endless repetition is the beat of pornography."[55] Merchandising, brand extension, creative product development, and pricing innovations of a kind that would have been inconceivable in the predigital age also have been introduced and progressively refined. What better illustration of the trend to corporatization than the efforts of some adult entertainment companies—whose adoption of this innocuous-sounding label is itself a revealing example of discursive reframing—to seek a public offering on Wall Street?

Despite the popular stereotypes—think of feature movies such as *Boogie Nights*, *Showgirls*, and *Striptease*—and social stigma long associated with the penumbral world of pornography and sex businesses, some companies (examples follow) have floated on the stock market, and continue to trade more or less successfully. Moreover, media interest in the subject is as widespread as it is persistent. Not only does sex sell; stories about sex sell too. The subject of pornography has been frequently discussed and occasionally analyzed in depth in a range of broadcast and cable television network documentaries in recent years (e.g., on ABC, Fox News, HBO, PBS, and in the United Kingdom, BBC's Channel 4). Hundreds of newspaper and magazine articles, including many in reputable business publications such as the *Wall Street Journal*, *New York Times*, *Forbes*, and *Business Week*, have been written on the nature, evolution, and most recently personalities of the adult entertainment industry (see table 15.1 for trend data): the world of pornography now has its very own celebrity culture, just like the rest of the entertainment industry, yet further evidence of the mainstreaming trend and the fervid quest for respectability that drives sections of the business.

Nonetheless, one should not lose sight of the fact that even as the industry seeks to soften its public image, consumer demand for hard-core products and new types of mediated sex is growing apace. A convenient barometer of public appetites are the free (intermediary) Web sites such as Mark's Bookmark (⟨http://book-mark.net/mark.html⟩) or the Hun's Yellow Pages (⟨http://www.thehun.net/index.html⟩), which carry complimentary samples, including video clips, of the materials offered by commercial sites. Whether this hardening trend is a result of gradual consumer desensitization and addiction, expanded technological possibilities, growing social tolerance, greater supply-side risk taking, or the interaction of these and other factors is difficult to say. The reality, however, is unmistakable. Whether shaping or simply responding to the market, pornographers have demonstrated a heady willingness to take their product to the next level of explicitness and fashion a bewildering array of "alternative pornographies."[56]

Adult movies are sometimes referred to as either "features" (these usually have a minimalist narrative structure and are used to supply the pay-per-view market) or "gonzos" (plotless, wall-to-wall, uninhibited sexual action). As customer tastes harden, progressing from erotic through soft- and hard-core to extreme forms of pornography, producers are demanding more in terms of performance from their actors—the ultimate commodified product—whose careers suffer as a result. The actors burn out more rapidly, both physically and psychologically, become overexposed, and gradually

lose their market appeal. There are, of course, a few notable and enterprising exceptions to the general rule, such as Danni Ashe (⟨http://www.danni.com/⟩), the world's
most downloaded porn star; Jenna Jameson (⟨http://www.clubjenna.com/⟩), a self-
proclaimed "cultural icon"; and Jill Kelly (⟨http://www.jillkelly.com/⟩), founder of
JKP, a company with ambitions to be traded publicly.

The market and other forces that drive this superficially glamorous industry have
been graphically captured by photographer Stefano de Luigi and writer Martin Amis
on their world tour of production sites, wittily deconstructed by the journalist and
onetime pornographic scriptwriter A. A. Gill, and clinically revealed in a number of
autobiographical accounts by industry insiders and "star" performers such as Jenna
Jameson.[57] The above-mentioned exceptions to the general rule, Ashe, Jameson, and
Kelly, are just that: these performers-turned-entrepreneurs stand out from the commodity crowd by virtue of their tenacity, entrepreneurial intelligence, and good fortune. They have reversed the dominant model (male ownership) by taking over
control of the business, the core product of which is their own objectified, corporeal
selves, and in no small measure by astutely exploiting the branding possibilities offered
by the Internet. But in general, the pornography business remains a compelling illustration of Karl Marx's notion of alienation.

Democratization

Although my primary focus here is the distribution and consumption of professionally
produced pornography via the Internet, it is important to remember that alongside the
much-vaunted formal sex economy, there exists an informal online gift economy. As
already mentioned, Internetworking has resulted in both the rapid domestication and
progressive corporatization of pornography. But that is not the whole story. It has also
created a multiplicity of online spaces, such as Second Life (⟨http://secondlife.com/⟩)
or AdultFriendFinder (⟨http://www.adultfriendfinder.com/⟩), on which amateurs and
exhibitionists can post, annotate, or trade images; act out fantasies; create or cocreate
sexual products and online interactive experiences; chat, blog, and cruise; and shop,
stare, and share. In this "heterotopic space"—to appropriate Foucault's term—we are
witnessing the democratization of the processes of sexual representation and the emergence of "do-it-yourself (DIY) pornography."[58] Sites such as Flickr.com or MySpace
.com, though proscribing pornography, carry material that might reasonably be
deemed bawdy, lewd, or mildly erotic. In that regard, they reflect the growing tolerance
of sexual exhibitionism in everyday life, and the role played by the Internet in both
facilitating and legitimating such practices.

In fact, in many cases there is a blurring of the traditional lines between professional
and amateur, producer and consumer; the age of the pornographic "prosumer"—
Alvin Toffler's perspicacious neologism of almost three decades ago—is at hand.[59] The

proliferation of amateur and home videos available on the Internet testifies to both the domestication of pornography and the "porning" of the domestic.[60] Homegrown, a longtime market leader in the production of amateur pornography established in the early 1980s in the era of the VCR and camcorder, has been described by *U.S. News and World Report* thus: "Homegrown, made by the people for the people, represents the democracy of porn" (see ⟨http://www.HomegrownVideo.com⟩). At the risk of oversimplifying, (genuine) amateur pornography equates with, or at least aspires to, authenticity, realism, and spontaneity—attributes that are not usually associated with the highly formulaic productions made by professional directors and performers. In the language of critical theory, commercially produced pornography displaces "the real by the simulacral."[61] Many consumers have a strong fondness for genuine amateur content, but in actuality much so-called amateur pornography is faux, consciously engineered or staged by commercial producers to have an amateur look and feel.

Caveat Lector

Before offering some vital statistics on the adult entertainment industry and its component parts, I would like to offer a cautionary tale—one told at my own expense. In our 2001 article in the *Information Society*, Davenport and I provided estimated global sales figures for the legal sex/pornography industry. The legend accompanying the relevant table read as follows: "This table is reproduced from Morais (1999, p. 218), who sourced the data from Private Media Group Inc. The figures are consistent with those quoted by Rose (1997, p. 221), who sourced his from a range of trade and media organizations."[62] Private Media Group is a Barcelona-based, publicly traded adult entertainment company (discussed below). Following the publication of our article, I was contacted by Private's investor relations department wondering if *I* could furnish the company with growth projections and other related information for the adult entertainment industry—I, who had sourced some of my data from *its* Web site. Revealingly, Private states in its 10–K filed in spring 2005 with the Securities and Exchange Commission (available from EDGAR Online) that "the total worldwide adult entertainment market exceeds $56 billion annually," the very number we cited in 2001. A case of the blind leading the blind? To compound the confusion, our essentially derivative 2001 table resurfaced in 2004, reproduced as a footnote (minus the original legend) in an article in the online journal *First Monday*.[63] What this means is that decade-old data continue to circulate in the public domain and are treated by some readers as received wisdom.

Such circularity tells one a great deal about the difficulty of obtaining credible and consistent statistics on this shadowy sector of the larger entertainment economy. The inherent nature of the business is such that individuals and companies are understand-

ably reluctant to speak openly; pornographers, in the main, prefer a culture of anonymity or pseudonymity, striving to stay out of the limelight and away from the
unwelcome attention of the Internal Revenue Service, the Immigration and Naturaliza-
tion Service, or Department of Justice. Furthermore, this is a highly fragmented industry, one with relatively few large or midsize companies but many small, fly-by-night
enterprises. And since few businesses are publicly quoted, reliable data are hard
to come by. Many of the numbers bandied about by journalists, pundits, industry
insiders, and market research organizations are therefore lazily recycled, as in the case
of our aforementioned table, moving effortlessly from one story and one reporting context to the next. What seem to be original data and primary sources may actually be
secondary or tertiary in character. When it comes to pornography, "accounting commonly takes the form of dubious quantitative impressions and largely notional statistics," as Gordon Hawkins and Franklin Zimring wryly observed more than fifteen years
ago.[64]

Some of the startling revenue estimates and growth forecasts produced over the years
by reputable market research firms such as Datamonitor, Forrester Research, and the
Gartner Group have been viewed all too often with awe rather than healthy skepticism. Even pornography's own trade association, the high-mindedly named Free
Speech Coalition (⟨http://www.freespeechcoalition.com/⟩), in its 2005 white paper on
the adult entertainment industry in the United States, acknowledges that reliable figures are simply not available. Finally, it should be noted that it is difficult, for a variety
of well-documented technical reasons, to measure and interpret Internet-based traffic
and usage patterns with precision, whether we are talking about pornography or any
other e-commerce domain.

Running the Numbers

It can be instructive to trace the migration paths of trade and market research statistics
from their originating source over time (see tables 15.2–15.8 for some idea of the
many sources and estimates in circulation). Upper- and lower-bound revenue figures
for the pornography business diverge greatly, but the consensus seems to be that
this is an industry whose top line can be reckoned in billions (yet is it $16 billion or
$12 billion?). In the United States alone, the Free Speech Coalition (which cites, but
doesn't comment on the trustworthiness of the various sources used in its statistical
compilation) estimates that adult video/DVD sales and rentals amount to at least
$4 billion annually, while revenues from phone sex are thought to be approximately
$1 billion.

There is huge variation in the statistics pertaining to, for instance, the online content market (more or less than $1 billion?), the number of Web pages devoted to sex

Table 15.2

Aggregate porn revenues

Revenues/remarks	Cited source	Retrieved source
$10–14 billion	*Fortune/Forbes* (No dates given)	J. M. LaRue, Concerned Women for America 2005
$8 billion	*U.S. News and World Report* 1997	Free Speech Coalition 2005
$10–11 billion	*Adult Video News* 1997	Free Speech Coalition 2005
$10–14 billion	Forrester Research 1998	F. Rich, *New York Times Magazine* 2001
$16.2 billion (as of 2000)		L. Perdue, *EroticaBiz: How Sex Shaped the Internet* 2002
$2.6–3.9 billion	Adams Media Research, Forrester Research, Veronis Suhler Communications Industry Report (No dates given)	D. Ackman, Forbes.com 2001
$10–14 billion	Forrester Research 2001	M. A. Zook, *Environment and Planning* 2003
$11 billion	D. Dukcevich, Forbes.com 2001	Anonymous, *History and Technology* 2005
$20 billion (as of 2001)		D. Campbell, Guardian.com 2001
$4–10 billion	National Research Council Report 2002	ProtectKids.com 2005

(billions or hundreds of millions?), the percentage of the Web's total bandwidth consumed by pornographic traffic (more than half?), the number of commercial sex sites (thousands?), the value of the video market segment (is it about half the size of Hollywood's box office receipts, as claimed by *Adult Video News*?), the number of sites hosting child pornography (several thousand or more?), the number of free sites (tens or hundreds of thousands?), the annual churn rate for subscription sites (20 percent, 40 percent, or more?), the number of unique visitors per day to pornographic Web sites (millions or tens of millions?), and the potential market for third-generation mobile phone pornography (more than $10 billion by the end of the decade?). These, by the way, are just some of the numerous estimates (and guesstimates) to be found in the open literature—a few of which are invoked with suspicious regularity.

Table 15.3

Porn revenues: Internet

Revenues/remarks	Original source	Retrieved source
Less than $200 million (as of 1969)	U.S. Congress, Commission on Pornography and Obscenity Report 1970	W. Kendrick, *The Secret Museum: Pornography in Modern Culture* 1996
$40 million	A. Edmond, CEO of Sextracker 2000 (for 1996)	E. Rampell, *Adult Video News*, AVN Online.com 2000
$50 million	Forrester Research 1996	M. A. Zook, *Environment and Planning* 2003
$200 million	A. Edmond, CEO of Sextracker 2000 (for 1997)	E. Rampell, *Adult Video News*, AVN Online.com 2000
$550–800 million	A. Edmond, CEO of Sextracker 2000 (for 1998)	E. Rampell, *Adult Video News*, AVN Online.com 2000
$970 million– $1.4 billion	Datamonitor 1998	Industry Standard 2000 M. A. Zook, *Environment and Planning* 2003
$750 million– $1 billion	Forrester Research 1998	E. Rampell, *Adult Video News*, AVN Online.com 2000 D. Ackman, Forbes.com 2001 Caslon Analytics 2004 M. A. Zook, *Environment and Planning* 2003
$176 million	Jupiter 1999	Screen Digest 2000
$900 million	Forrester Research 1999	Screen Digest 2000
$1.39 billion	Datamonitor 1999	Screen Digest 2000
$4–10 million monthly ("The largest pay site")	A. Edmond, CEO of Sextracker 2000	E. Rampell, *Adult Video News*, AVN Online.com 2000

Table 15.3

(continued)

Revenues/remarks	Original source	Retrieved source
$2 billion	F. Lane, *Obscene Profits: The Entrepreneurs of Pornography in the Cyber Age* 2000	Caslon Analytics 2004
$1.78 billion (projection for 2000)	Datamonitor 1999	Industry Standard 2000 M. A. Zook, *Environment and Planning* 2003
$2.5 billion (as of 2000)		L. Perdue, *EroticaBiz: How Sex Shaped the Internet* 2002
$1–1.8 billion (business-to-consumer)	A. Edmond, CEO of Sextracker 2000	E. Rampell, *Adult Video News*, AVN Online.com 2000
$3–6 billion (business-to-business transactions)	A. Edmond, CEO of Sextracker 2000	E. Rampell, *Adult Video News*, AVN Online.com 2000
$185 million	D. Card, Jupiter Media Metrix 2001	ABC News abcnews.com 2001
$1 billion	Adams Media Research, Forrester Research, Veronis Suhler Communications Industry Report (no dates given)	D. Ackman, Forbes.com 2001
$2.26 billion (projection for 2001)	Datamonitor 1999	Industry Standard 2000
$2.3 billion (projection for 2001)	Datamonitor 1999	M. A. Zook, *Environment and Planning* 2003
$10–12 billion	W. Lyon, Free Speech Coalition 2001	Caslon Analytics 2004
$1 billion	National Research Council Report 2002	ProtectKids.com Recent Statistics on Internet Dangers 2005
$2 billion	*Adult Video News* 2002	J. Swartz, *USA Today* 2004

Table 15.3

(continued)

Revenues/remarks	Original source	Retrieved source
$2.4 billion	D. Thornburgh and H. S. Lin, *Youth, Pornography, and the Internet* 2002	Caslon Analytics 2004
$2.7 billion (projection for 2002)	Datamonitor 1999	Industry Standard 2000
$3–4 billion	Financial Times Limited 2002	J. M. LaRue, Concerned Women for America 2005
$3.12 billion (projection for 2003)	Datamonitor 2000	Industry Standard 2000
$2.5 billion	TopTenReviews.com Internet Pornography Statistics 2005	TopTenReviews.com Internet Filter Review 2005 W. Hungerford, About.com 2005
$5–7 billion (projection for 2005)	A. Edmond, CEO of Sextracker 2000	E. Rampell, *Adult Video News*, AVN Online.com 2000
$7 billion (projection for 2006)	VisionGain 2003	Caslon Analytics 2004

Reliable supply- and demand-side statistics on the pornography market are frustratingly elusive, and market research firms may not be averse to a little boosterism. As Hawkins and Zimring have shown in *Pornography in a Free Society*, even seemingly credible data from ostensibly trustworthy sources do not always stand up to rigorous scrutiny. Lewis Perdue's *EroticaBiz*, its jaunty tone notwithstanding, provides a dogged review of North American adult industry trade figures, broken down by category, along with an assessment of their sources, while *Porn Gold* by David Hebditch and Nick Anning, though somewhat dated, is a solid piece of investigative journalism covering both Europe and the United States.[65]

Some of the oft-quoted numbers acquired additional perceived legitimacy when, in 2001, the *New York Times* published an article by Frank Rich on the U.S. pornography business.[66] Rich took as his baseline data purportedly cited in a 1998 study by Forrester Research. He was subsequently taken to task by another journalist, Forbes.com's Dan Ackman, for failing to exercise sufficient skepticism in his analysis of the industry's revenue claims. Ackman robustly challenged many of the figures cited by Rich, including

Table 15.4
Porn sites: Number

Statistics/remarks	Cited source	Retrieved source
Sites offering free sex content: 22,000 in 1997 to 280,300 in 2000	*American Demographics Magazine* 2001	M. A. Zook, *Environment and Planning* 2003
E-commerce sex Web sites: 230 in 1997 to 1,100 in 2000	*American Demographics Magazine* 2001	M. A. Zook, *Environment and Planning* 2003
In 1997–2002, the percentage of pornography on the Web dropped from about 20% to 5% as commercial and informational Web use grew	J. Spink, *Web Search: Public Searching of the Web* 2004	Anonymous, *History and Technology* 2005
20,000–40,000 adult sites	R. Morais, *Forbes* 1998	M. A. Zook, *Environment and Planning* 2003
90% of free porn sites, and nearly all pay porn sites, buy their material rather than create it themselves	M. Rosoff, cnet.com 1999	M. A. Zook, *Environment and Planning* 2003
Pornographic-oriented Web sites account for less than 1.5% of Internet Web sites	G. Lawrence, *Nature* 1999	M. A. Zook, *Environment and Planning* 2003
500,000 pornographic sites	L. Carr, *Grok*, a publication of the Industry Standard 2000	M. A. Zook, *Environment and Planning* 2003
In mid-2000, New Frontier Media owned over 1,300 cybersex domains that sent traffic to its 27 paid sites, and Cyber Entertainment Network claimed approximately 3,000 and 14, respectively	T. C. Doyle, *VARBusiness* 2000	Anonymous, *History and Technology* 2005
In 2001, Yahoo! offered 469 categories of sex on 3,407 sites		D. Campbell, Guardian.com 2001
Internet porn makes up barely a fifth of American porn consumption		F. Rich, *New York Times Magazine* 2001
Over 100,000 subscription sites in the United States, and 400,000 subscription sites worldwide	D. Thornburgh and H. S. Lin, *Youth, Pornography, and the Internet* 2002	Caslon Analytics 2004

Table 15.4

(continued)

Statistics/remarks	Cited source	Retrieved source
The two largest individual buyers of bandwidth are U.S. firms in the adult online industry	National Research Council Report 2002	ProtectKids.com 2005
40,000 expired domain names were "porn-napped"	National Research Council Report 2002	ProtectKids.com 2005
Over 260 million pornographic Web pages (twentyfold increase in 5 years)	N2H2, Inc. 2003	ProtectKids.com 2005
1.3 million porn Web sites	N2H2, Inc. 2003	ProtectKids.com 2005
Pornographic-oriented Web sites account for less than 0.8% of Internet Web sites		M. A. Zook, *Environment and Planning* 2003
Around 74,000 commercial sites of adult content	OCLC researchers 2004	Caslon Analytics 2004
Around 200,000 commercial sites of adult content	UAS/IFA 2004	Caslon Analytics 2004
1.3 million porn sites	National Research Council Report 2004	J. Swartz, *USA Today* 2004
Over 4 million pornography Web sites, housing 372 million Web pages (12% of total Web sites)	TopTenReviews.com Internet Filter Review 2005	W. Hungerford, About.com 2005

those attributed to Forrester and others who had written on the finances of the adult entertainment industry, and duly proffered a set of considerably more conservative estimates. His conclusion deflates some of the hyperbole: "When one really examines the numbers, the porn industry—while a subject of fascination—is every bit as marginal as it seems at first glance."[67] This journalistic spat itself became the stuff of a story penned by Emmanuelle Richard, all of which provides yet another cautionary tale for those who think that the dimensions and structural dynamics of the pornography business can be easily grasped.[68]

Mainstream Ambitions

What one *can* say with some assurance is that every month many millions of Americans, with greater or lesser regularity, access commercial and/or freely available

Table 15.5
Porn sites: Traffic volume

Statistics/remarks	Cited source	Retrieved source
Sextracker tracks 13 million actual visitors on to porn sites	A. Edmond, CEO of Sextracker 2000	E. Rampell, *Adult Video News*, AVN Online.com 2000
About 15–20 million Internet users visit porn sites monthly	Media Metrix Research 2001	ABC News, abcnews.com 2001
A study carried out in September 2003 found that the adult industry had the second-highest percentage of 18–34-year-old male visitors, with 24.95% of its audience, or 19,159 unique individuals, ranking just below gaming sites		J. Shermack, comScore Networks 2003
Of 400,000 queries from February 6–23, 2003, on the Gnutella file-sharing network, 42% were for pornography compared with 38% for music	D. C. Chmielewski, siliconvalley.com 2003	Anonymous, *History and Technology* 2005
More than 32 million individuals visited a porn site in September 2003 (71% were male, and 29% female)	Nielson/Net Ratings 2003	ProtectKids.com 2005 Free Speech Coalition 2005
7–9 of the top 10 terms in image and video searches are pornographic compared with 2–3 in general searches	J. Spink, *Web Search: Public Searching of the Web* 2004	Anonymous, *History and Technology* 2005
25 million Americans visit adult sites 1–10 hours per week	MSNBC/Stanford/Duquesne Study 2005	ProtectKids.com 2005
There are 72 million worldwide visitors to pornographic Web sites annually		TopTenReviews.com Internet Filter Review 2005
In August 2005, Internet users viewed over 15 billion pages of adult content	comScore Media Metrix 2005	ProtectKids.com 2005
71.9 million people visited adult sites in August 2005, 42.7% of the Internet audience	comScore Media Metrix 2005	ProtectKids.com 2005
There are 1.5 billion monthly pornographic downloads (peer-to-peer) (35% of all downloads)		TopTenReviews.com Internet Filter Review 2005
There are 68 million (25% of total) daily pornographic search engine requests	TopTenReviews.com Internet Filter Review 2005	TopTenReviews.com Internet Filter Review 2005 W. Hungerford, About.com 2005

Table 15.6

Porn revenues: Videos

Revenues/remarks	Cited source	Retrieved source
$400 million (mail-order video sales)	FSC and Video Software Dealers Association 1999	Caslon Analytics 2004
$4.1 billion	FSC and Video Software Dealers Association 1999	Caslon Analytics 2004
$4 billion (as of 2000)		L. Perdue, *EroticaBiz: How Sex Shaped the Internet* 2002
$4 billion	*Adult Video News* 2000	Caslon Analytics 2004
$500 million–$1.8 billion	Adams Media Research, Forrester Research, Veronis Suhler Communications Industry Report (no dates given)	D. Ackman, Forbes.com 2001
$4.2 billion	*Adult Video News* 2001	F. Rich, *New York Times Magazine* 2001
$3.95 billion plus (as of 2005)		Free Speech Coalition 2005
Rentals of hard-core videos rose from 79 million in 1985 to 759 million in 2001, an increase of almost 1000%	*Adult Video News* (no date given)	P. Kloer, *Atlanta Journal-Constitution* 2003
Adult movie rentals are available in about 1.5 million hotel rooms, and account for 80% of in-room movie profits	*Nightline*, ABC News 2002	P. Kloer, *Atlanta Journal-Constitution* 2003

pornographic material on the Web, buy millions of print copies of pornographic magazines (*Playboy*, for instance, has an audited, albeit declining, paid readership of more than three million with a median age of thirty-three; the magazine's paid readership peaked in 1972 at just over seven million), and view tens of millions of pornographic videos/DVDs in their own homes. Market and survey research data show that age, disposable income, and gender correlate positively with pornography consumption, though some of the popular assumptions (e.g., that only young males are interested in pornographic goods and experiences, or that females do not purchase or rent adult materials) are challenged by the point-of-sale, subscription, and audience/consumer

Table 15.7
Porn revenues: U.S. television/cable

Revenues/remarks	Cited source	Retrieved source
$54 million (adult pay-per-view)	Showtime Event Television 1993	T. Egan, *New York Times* 2000
$367 million (adult pay-per-view)	Showtime Event Television 1999	T. Egan, *New York Times* 2000
$200 million (GM's DirecTV) (as of 2000)		T. Egan, *New York Times* 2000
$310 million (U.S. adult cable and satellite industries)	*Forbes* 2000	J. M. LaRue, Concerned Women for America 2005
$500 million (as of 2000)		L. Perdue, *EroticaBiz: How Sex Shaped the Internet* 2002
$465 million	E. Schlosser, *Reefer Madness: Sex, Drugs, and Cheap Labor* 2001	P. Kloer, *Atlanta Journal-Constitution* 2003, August
$128 million (adult pay-per-view)	Adams Media Research, Forrester Research, Veronis Suhler Communications Industry Report (no dates given)	D. Ackman, Forbes.com 2001
$800 million (movie subscriptions and pay-per-view)	Kagan Research 2005	Free Speech Coalition 2005

research data. As Linda Williams notes, there is a "move towards the feminine domestication of pornography"—a trend that is not warmly welcomed by the feminist anti-pornography movement.[69]

Many Americans, *pace* the distinguished thespian mentioned in the introduction to this chapter, do not find pornography boring, an incontrovertible fact that entrepreneurs and large corporations alike have exploited to great effect of late through the medium of the Internet. In the last decade, the subject of pornography has moved from the periphery to the center of public discourse, and in its softer manifestations, become a media and advertising staple. In short, pornography is being maneuvered into the cultural mainstream—a phenomenon variously referred to as "the pornofication of public space" or "porn chic."[70]

Table 15.8

PMG Inc. product sales (EUR in thousands)

	Years ended December 31		
	2002	2003	2004
Net sales by product group:			
Magazine	5,777	5,316	5,121
Video	7,986	7,206	2,130
DVDs	15,482	18,309	19,488
Internet	6,059	4,819	4,859
Broadcasting	4,106	2,841	4,014
Total	39,410	38,491	35,612

Source: Form 10–K, 2004, 77.

Based on statistical data from a range of reports and surveys (e.g., by the U.S. Government Accounting Office and the National Academy Press) we know that access to pornography, both hard- and soft-core, has never been easier or more widespread, that more minors than previously are consumers, and that there is a growing risk of inadvertent exposure of juvenile users of chat rooms and peer-to-peer networks to all kinds of pornography—and also to pedophiles. This in turn raises the age-old question of whether "the immature will be precociously excited into sexuality" as a result of the proliferation of pornography in contemporary society and the ease with which it can be accessed.[71] One seemingly related consequence of these trends has been a significant rise in child pornography crime, up by 1,500 percent in the United Kingdom from 1988 to 2001, according to a study by NCH, the Children's Charity.[72] These and related matters were discussed in a recent issue of the journal *Pediatrics*.[73]

Such concerns, whether valid or not, have provided particular impetus to the anti-pornography movement in the United States, and help explain the increasingly interventionist stance taken by some state governments and also the federal government—in the last decade witness the contentious passing of the Communications Decency Act, Child Online Protection Act, and the Children's Internet Protection Act, all three of which were ultimately challenged in cases heard by the Supreme Court. It is not hard to see why large commercial producers, working under the umbrella of the Free Speech Coalition, go to great lengths to distance themselves from extralegal pornographers as well as all forms of illegal material (child pornography) and practices (use of minors and coercion). Changes in the legal and regulatory environments in which pornographers operate could, in theory, have a potentially catastrophic effect on the commercial viability and creeping social acceptability of their activities. It is therefore hardly surprising that the maturing industry, keen to leave behind the "outlaw tradition," is endeavoring to reframe the public rhetoric surrounding its activities by repositioning its wares as lifestyle goods and engaging in some degree of voluntary self-regulation, an example of which is Adult Sites against Child Pornography.[74] The organization is now known as the Association of Sites Advocating Child Protection (⟨http://www.asacp.org/⟩) and its list of sponsors features many well-known companies from the adult sector.

The establishment of the U.S. Commission on Obscenity and Pornography (aka the Johnson Commission) in the late 1960s and the U.S. Department of Justice's Commission on Pornography (aka the Meese Commission) in the 1980s, along with various more or less successful legislative and policy initiatives, are all part of what Hawkins and Zimring refer to as recurring "ceremonies of adjustment" to the reality of pervasive pornography in contemporary society.[75] Those ceremonies of adjustment are likely to continue, in one guise or another, but without significantly reshaping either public policy or the public's behavior, if recent history is any guide. At the same time, we can expect to see the larger companies intensify their preemptive commitment to image rehabilitation and collective self-regulation.

Industry Structure

As mentioned earlier, the adult entertainment industry comprises many minnows and a small number of relatively large fish. With the rapid growth of the Internet, the number of minnows has grown spectacularly; the number of large fish has grown too, although less dramatically. Factors such as low entry barriers, network externalities, access to niche markets, and high innovation potential have attracted waves of would-be entrepreneurs, experimentalists, and established businesses, many of whom in all probability make little or no money. In the last decade numerous adult entertainment companies have been founded, with wildly varying degrees of success. Some have launched and failed, while others have grown and prospered. Some successful enterprises have been "born digital," while others have extended or migrated their activities to cyberspace. As in more conventional sectors of the entertainment industry (e.g., movies and music), there is a rich mix of players involved across the entire production chain. These include content producers, investors, performers, distributors, marketers, carriers, retailers, and service providers. Among the infrastructure providers—telecommunications companies (e.g., DirecTV), online portals (e.g., AOL Time Warner), server hosts (e.g., MCI), and search engines (e.g., Alta Vista)—there are many household names, which though occasionally embarrassed by their arm's length association with professional pornographers, are loath to spurn the high profit margins. On occasion, however, the level of embarrassment is such that a quick exit from the business may be the best course of action, as when Yahoo! discontinued selling adult content in response to mounting public pressure.

Some firms have forward integrated from production into distribution, some have backward integrated, some have diversified from print-based products into online content delivery, and others have launched state-of-the-art interactive services or used the Web to achieve brand extension for existing lines of business. There are some highly successful companies with multimillion dollar revenues and several large firms that are publicly quoted, a few of which I profile later. But size is not all that matters in this business. Digital Playground has established itself as a leader in interactive formats, including high-definition DVDs that offer behind-the-scenes looks, multiple viewing angles, and Dolby Digital surround sound. It was also granted official trademark registration for the term "Virtual Sex" in 2006. In a different vein, RetroRaunch (⟨http://www.retroraunch.com⟩) is a cleverly conceived and tightly developed business that exploits to the full the potential of the Web. The company, established in 1997 by Hester Nash, offers vintage erotic photographs (some forty thousand, many of which are contributed by its members) and claims to have one of the highest subscriber retention rates in the industry. Having found its niche, RetroRaunch recently acquired a complementary site, Martha's Girls (a combined six-month subscription costs $99.99), and is now planning to launch retro movies, a textbook case of "sticking to the knitting." And as in the music industry, several independent producers—indies, in popular parlance—such as Suicidegirls (⟨http://suicidegirls.com⟩) have emerged whose product line aims to be more natural and authentic than that offered by mainstream producers; in effect, online pornography now has its own subversive equivalent of punk rock and grunge that seeks the "abolition of the spectacular" in favor of the quotidian.[76]

The adult entertainment industry is an economic and social reality, whatever one may think of its products, personnel, and practices. Yet despite all the recent media coverage, the contours of the business are still difficult to discern. As we have seen, there is constant speculation, and not a little hyperbole, regarding both aggregate and sector-specific earnings and forecasts. One thing is clear, however: the trend to corporatization is irreversible, which is not quite the same as saying that the sector is integrating with mainstream publishing, moviemaking, or retailing corporations via mergers, acquisitions, or joint ventures of one kind or another. To date, the adult entertainment industry has few articulated links to other business sectors; it is highly self-contained, though its existence necessarily contributes to the profitability of Internet service providers and a raft of technology companies.

This is an industry nonetheless in the throes of a major makeover—one determined to exploit maximally the affordances and functionalities of the expanding digital environment. In short, pornography is being repositioned and repackaged in the age of electronic commerce. At the same time this is an industry that is anathema to many politicians, lawmakers, religious leaders, and concerned citizens. It is an industry whose degrees of operating freedom could be swiftly and severely curtailed by the introduction of new regulations and laws that affect trading practices both on the ground and in cyberspace.

Structurally, the content production side of the adult entertainment industry resembles a pyramid, with a small number of large companies at the apex and a large number of small companies at the base. In revenue terms, though, it resembles an inverted pyramid: a small number of companies accounts for a significant share of the gross revenues. This is not an altogether new phenomenon, as Hebditch and Anning have shown: "Although newcomers with sufficient capital can still get started in the business, few become as prosperous as the dozen or so key figures who dominate the industry world-wide and probably control more than 50 per cent of the trade by value."[77] Of course, innovation is not the prerogative of the corporate elite, and new entrants will continue to introduce novel business methods, products, and technical features, in the process challenging the mainline players. This will be particularly true in the digital environment, where the barriers to entry and the transaction costs are relatively low, which helps explain the remarkable growth in the number of content producers.

But if we are interested in the likely future shape of the industry, we would be well advised to explore the strategic mind-sets, financial makeup, and business models associated with the leaders, those few companies with access to large markets, a dynamically updated asset base, advertising expenditures, and sources of investment capital. This, unsurprisingly, is a view shared by some industry leaders. Private Media Group, Inc. maintains that technological and other factors will cause a shakeout, as this extract from its 2005 K–10 filing makes clear: "While the adult entertainment industry is currently characterized by a large number of relatively small producers and distributors, we believe that the factors discussed above will cause smaller, thinly capitalized producers to seek partners or exit the adult entertainment business, leading to a consolidation." (available from EDGAR Online) It is hard to disagree with this assessment.

Some Major Players

With this background in place, I want now to bring the preceding generalizations to life by profiling in greater detail a few commercial enterprises, publicly quoted and privately held, that together represent the content production, distribution, and retailing sides of the business: Private Media Group, Inc., Playboy Enterprises, Inc., New Frontier Media, Inc., Club Jenna, Inc., and Vivid Entertainment Group, Inc. All are American or have a presence in U.S. markets. In the case of publicly traded companies, we have easy access to their annual 10–K forms, filing of which is required by the Securities and Exchange Commission. These documents offer a good insight into the finances, core business activities, and strategic objectives of the major players in the industry, and thus provide a bellwether of future trends. Additional information is available from corporate Web sites, news releases, and general media coverage. All such sources, it need hardly be said, should be treated with caution. Together, these companies illustrate the plenitude of pornographic products, services, and delivery options available to consumers, and demonstrate the growing importance of electronic commerce to would-be global leaders in the expanding and highly competitive adult entertainment industry. In each case, I highlight these companies' views on, and strategic approaches to, the Internet and online content delivery.

Private Media Group

Private Media Group's (PMG) corporate profile is available on the Web (⟨http://www.PRVT.com⟩). It is a polished document, professional in its presentation of the company's history, hard-core product portfolio, business highlights, and ambitions. The Barcelona-based company, a self-described "multinational powerhouse," was originally founded in Sweden in 1965 by the present chairperson's father with the launch of its flagship publication *Private*—the world's first full-color adult magazine. PMG's vision is "to be the world's preferred content provider of adult entertainment to all consumers, anywhere, anytime, and across all distribution platforms." PMG was also the first adult entertainment company to be traded on the NASDAQ National Market, with a public listing in 1999 (PMG is also listed on the German DAX). At the time of this writing, the company's stock was trading at $2.60 and its market capitalization was $138.19 million. Its global net sales in 2004 were more than $48 million with a net income of roughly a third of a million. Fidelity Investments, the blue-chip mutual fund, held 12 percent of PMG's stock, making it the largest institutional investor in the company—a fact that raised some eyebrows in the mainstream financial press.

PMG acquires worldwide rights to still images and motion pictures from independent directors, tailoring and processing these into a variety of media formats (magazines, videos, DVDs, etc.) for distribution via multiple channels, including retail outlets, mail-order catalogs, cable/television, wireless, and the Internet. The company has built up a digital archive of three million photographs and has more than eight hundred titles in its movie library, which is currently being digitized to take advantage of Internet distribution possibilities. These photographic and video assets are valued at $20 million on the company's 2004 balance sheet. PMG also sells two million magazines (*Private*, *Pirate*, etc.) annually, has produced hundreds of hard-core videos and DVDs, and since 2000 has launched three television channels (Private Gold, Private Blue, and Private Fantasy) to deliver its content to a global market. After more than forty years in the business, PMG has a well-established distribution network and claims to have the potential to access almost two hundred and fifty thousand points of sale in forty countries.

The brand-conscious company engages in selective, upscale sports sponsorship and actively seeks cross-business branding opportunities, particularly in Europe, where in 2006 it launched a duty-free adult retail outlet in partnership with Beate Uhse in

Munich airport. The company created its flagship Web site (⟨http://www.private.com⟩) in 1997 to take advantage of the growth in Internet access and new methods of product distribution. In 2003 it outsourced all Internet operations (as it had done earlier with its printing operations) to reduce expenses. PMG currently offers its subscribers full-length Private movies, photo sets, and sex stories. Magazines are available in image and downloadable formats, and broadband video-on-demand is deliverable from PrivateSpeed.com. In addition to proprietary materials, the company's Web site offers third-party content. Subscribers also have access to a live-sex chat service, adult personals, and Private's range of licensed lifestyle products. PMG claims that its prizewinning Web site is one of the most successful of its kind, attracting over 2.5 million visits and 60 million page views per month. According to PMG's 10–K for fiscal year 2004, the company "sold over 72,000 memberships and had approx. 10,000 active members at any time." These figures reinforce the widely held view that subscription churn is a major challenge for the adult entertainment sector; subscribers are fickle, and customer loyalty is hard-won. Elsewhere in the report, net sales are broken down by product group (magazine, video, DVDs, Internet, and broadcasting). The total sales for 2004 were $35.6 million, of which the Internet accounted for $4.8 million, a relatively modest slice of PMG's overall business—one that has yet to achieve straight-line growth (see table 15.8).

In an August 2005 press release, the company was bullish about the future growth of its online division, noting that the Internet was "a high-margin business" (available from EDGAR Online). At the same time, PMG cautiously acknowledged in its 10–K filing that "new laws or regulations relating to the Internet, or more aggressive application of existing laws, could decrease the growth of our websites, prevent us from making our content available in various jurisdictions or otherwise have a material adverse effect on our business, financial condition and operating results." In any event, PMG is a highly diversified company with a good understanding of the dynamics of the adult entertainment industry and the potential of new media to boost its bottom line.

Playboy Enterprises

Playboy, the iconic soft-porn magazine, was founded by Hugh Hefner in 1953. More than half a century later it remains the best-selling, general-interest men's magazine, albeit with considerably fewer readers than in its heyday.[78] According to the parent company's (Playboy Enterprises Inc.) 10–K for fiscal 2004, *Playboy*'s circulation was 3.1 million copies monthly, with the 17 licensed international editions adding a further 1.1 million to that number. Market research shows that *Playboy* is read by one in every seven men in the United States between the ages of eighteen and thirty-four. The strength of the Playboy brand name is unrivaled in the adult entertainment industry, and the company has exploited both its name and distinctive logo (the rabbit head) to great effect domestically and internationally. It also has a reputation for reacting aggressively to copyright infringements and product counterfeiting. According to the company's 2004 annual report, Playboy-branded products account for more than 120 license holders in more than 130 territories. Although *Playboy* magazine continues to be termed soft, the company's product range now contains harder-core material for delivery via its television networks. In fact, Playboy is facing competition at both end of the soft-hard spectrum, given the number of existing players (so-called lads' magazines, such as *Maxim* or *Loaded*) and new, Internet-enabled entrants (such as Nerve, which aims to be "more graphic, forthright, and topical than 'erotica,' but less blockheadedly masculine than 'pornography'" [⟨http://www.nerve.com/aboutus/what/⟩]).

Today, the company, in which the founder remains the dominant shareholder and his MBA-holding daughter is CEO, employs more than 650 people. It generated net revenues of $329 million in fiscal 2004; the net income was almost $10 million, following several years of operating losses. Playboy Enterprise's stock has been quoted on the New York Stock Exchange since 1971. At the time of this writing, it was trading at $14.90 and the company's market capitalization was $420.97 million.

The company describes itself as "a worldwide leader in the development and distribution of multi-media lifestyle entertainment for adult audiences." It comprises three divisions: entertainment, publishing, and licensing. The entertainment and online businesses were consolidated for reporting purposes in 2004. The operations of its entertainment group include the production and marketing of "lifestyle adult entertainment television programming for our domestic and international TV networks, web-based entertainment experiences, wireless content distribution, e-commerce, worldwide DVD products and online gaming under the Playboy and Spice brand names" (available from EDGAR Online). Playboy's networked programming portfolio is the backbone of it overall business, one that has been augmented by strategic acquisition (e.g., the Vivid TV network). In fiscal 2004, the combined revenues for Playboy's domestic and international television networks were $142 million (see table 15.9).

Unlike PMG, which either buys in or commissions its stock of photographs and movies, Playboy produces original adult content using its own centralized studio facilities in Los Angeles that offer playback, production control, and origination services for the company's networks; Andrita Studios is a wholly owned subsidiary of Playboy Enterprises Inc. and the facilities are available for commercial hire. Annually, the company spends almost $40 million on producing proprietary programming and acquiring licensed content for its stable of movie networks (e.g., Playboy TV, Playboy Hot HD, Spice, Spice Ultimate, and Hot Net). Over time, Playboy has built up a domestic library of primarily exclusive, Playboy-branded programming totaling almost 2,700 hours—an asset base that gives it real competitive advantage in the fast-growing, content-hungry online sector.

In 1994, *Playboy* was the first national U.S. magazine to establish an online presence, an initially loss-making move that has since proved to have been inspired. Today, the

Table 15.9
Playboy Enterprises, Inc. revenues results of operations (in millions, except per share amounts)

	Net revenue for fiscal year ended		
	12/31/04	12/31/03	12/31/02
Entertainment			
Domestic television networks	$96.9	$95.3	$94.4
International	45.3	37.9	19.1
Online subscriptions	21.5	18.2	11.0
E-commerce	18.7	16.8	14.4
Other	6.8	7.5	13.7
Total entertainment	189.2	175.7	152.6
Publishing			
Playboy magazine	101.5	102.0	94.7
Other domestic publishing	11.9	13.0	11.7
International publishing	6.4	5.7	5.4
Total publishing	119.8	120.7	111.8
Licensing			
International licensing	12.1	8.0	5.6
Domestic licensing	3.0	3.2	3.3
Entertainment licensing	2.0	1.4	0.5
Marketing events	3.0	2.9	2.8
Other	0.3	3.9	1.0
Total licensing	20.4	19.4	13.2
Total net revenues	$329.4	$315.8	$277.6

Source: Form 10–K, 2004, 30.

Web site attracts more than 150 million page views per month. Playboy Cyber Club offers access to more than one hundred thousand proprietary photographs. The company's Internet activities generate a relatively modest share of the overall revenues, but Playboy.com is the fastest-growing facet of the business, and the company licenses sites in over a dozen countries, thereby allowing for customization of materials to reflect local market conditions and demands. In fiscal 2002, online subscriptions amounted to $11 million; by 2004 the figure had risen to $21.5 million (see table 15.9). Online subscription revenues increased 18 percent ($3.3 million) during 2004, reflecting increases in the average monthly subscriber count and per subscriber revenues. Online subscribers can choose from six premium clubs with monthly rates ranging from $19.95 to $29.95 (standard pricing for the industry); annual rates are in the range $95.40 to $189.96.

Although the rates of return from its Internet businesses are impressive, Playboy acknowledges that continuous growth requires significant investment in technology and marketing. Additionally, the company is aware of the fact that the Internet business is both highly competitive and potentially vulnerable to regulatory developments. In its 2004 10–K, the company makes the following observations, very similar indeed to those expressed by PMG in its filing to the Securities and Exchange Commission:

Various governmental agencies are considering a number of legislative and regulatory proposals which may lead to laws or regulations concerning various aspects of the Internet, including online content, intellectual property rights, user privacy, taxation, access charges, liability for third party activities and jurisdiction. Regulation of the Internet could materially affect our business, financial condition or results of operations by reducing the overall use of the Internet, reducing the demand for our services or increasing the cost of doing business. (available from EDGAR Online)

Playboy has exceptional brand salience and market reach. In addition, it is a primary producer of high-quality adult content, with dedicated production facilities. Its expanding digital asset base will help propel the growth of its Internet-based businesses in the United States and overseas, barring any major regulatory interventions from national governments.

Both Playboy and Private are, by the standards of the adult entertainment industry, long-established companies. Both are professionally organized and managed, publicly traded, technologically astute, and proven innovators. The two have evolved from being print-based publishers to diversified multimedia producers of adult materials with a presence in multiple sectors of the adult entertainment and lifestyle markets. In addition, both have amassed considerable digital resources that position them for future growth in the Internet and wireless sectors, and in the broader adult lifestyle marketplace. Not surprisingly, both these adult heavyweights understand the value of strategic partnerships when the conditions are right; in 2005, Playboy and PMG announced the merger of their European pay-television channels, Private Gold and Spice Platinum, to form Private Spice. Under the terms of the agreement, Playboy TV International will take responsibility for the management, marketing, and distribution with Private providing the premium content. Further collaboration between industry leaders across all parts of the value chain seems likely.

New Frontier Media

New Frontier Media (⟨http://www.noof.com⟩) is a technology-driven distributor of branded adult entertainment content with 121 employees, based in Boulder, Colorado. Its Pay TV Group encodes, originates, distributes, and manages seven networks (TEN, TEN*Clips, TEN*Blue, etc.) through a fully digital broadcast center that delivers around-the-clock programming. The Pay TV Group acquires all of its feature-length

Table 15.10

NFM Inc. product sales (in millions)

	Twelve months ended March 31		
	2005	2004	2003
Net revenue			
Pay-per-view/cable/direct broadcast satellite	$24.0	$21.3	$19.2
Video on demand/cable/hotel	15.6	12.6	2.2
C-band	3.9	5.7	7.5
Total	$43.5	$39.6	$28.9
Net membership	$2.4	$2.7	$5.3
Sale of content	0.3	0.6	1.2
Sale of traffic	—	—	1.3
Total	$2.7	$3.3	$7.8

Source: Form 10–K, 2005, 31, 35.

content and stock of still images (more than half a million) through a mix of exclusive and nonexclusive licensing agreements with almost fifty third-party studios and independent producers, typically for a five-year term. Some of these deals are large, such as the company's 1999 licensing (in exchange for a large block of New Frontier Media stock) of three thousand film titles and other adult material from Metro Global Media, Inc.

New Frontier Media does not create original content nor does it possess proprietary archives. Nonetheless, it does have a track record of successfully distributing its licensed content on a range of platforms (cable television, direct broadcast satellite, Internet, and wireless). The great bulk of its sales is generated in the North American market. Revenues for New Frontier Media's Pay TV Group in 2005 were $43.5 million, up from $28.8 million in 2003 (see table 15.10). The company is traded on the NASDAQ, with a share price, at the time of this writing, of $6.63 (down from $7.40 four months earlier) and a market capitalization of $147.91 million (compared with $138.19 million and $420.97 million for PMG and Playboy, respectively).

According to New Frontier Media's 10–K for the fiscal year ended March 31, 2005, the company's main competition in the subscription, pay-per-view, video-on-demand markets is Playboy Enterprises, with its established presence, brand power, size, and market position. In addition, Playboy has an extensive, continuously expanding, proprietary digital asset base that can be leveraged across both established and emerging technologies to give it a long-term competitive advantage over a pure distribution company—something that New Frontier Media lacks. On its Web site, New Frontier Media describes its online ambitions in the context of the presumed "$1 billion, and growing, adult Internet market." In 1993, New Frontier Media absorbed restructuring costs of more than $3 million with respect to the Internet Group's in-house data center, which was subsequently relocated from California to the company's Boulder headquarters. Despite these changes, and despite the features and functionality (e.g., searching for clips by the scene, type of action, and name of actor) offered by its flagship site, TEN.com, revenues from Internet membership have declined by almost 50 percent from $5.3 million in 2003 to $2.4 million in 2005 (the reverse of Playboy's experience). Income from the sale of Internet content has also fallen from $1.2 million to $0.3 million over the same period (see table 15.10). The company has now abandoned its incentive-based traffic-generation programs that rewarded affiliated Web masters for referrals (the affiliate would keep between 40 and 67 percent of the first month's $29.95 membership). This income stream, which amounted to $1.3 million in 2003, has since been discontinued. The following extract from the company's 10–K provides a cautionary note regarding the marketing of digital pornography: "The adult entertainment industry is highly competitive and highly fragmented given the relatively low barriers to entry. The leading adult Internet companies are constantly vying for more members while also seeking to hold down member acquisition costs paid to webmasters." New Frontier Media is hoping to exploit its existing distributor relationships and the emergence of the wireless market to boost the annual revenues of its Internet Group in the years ahead.

Club Jenna

Both PMG and Playboy Enterprises have carefully nurtured their respective brands over the years, building up and out from their eponymous founding magazines. A different approach to branding is exhibited by Club Jenna (⟨http://www.clubjenna.com⟩), the corporate vehicle for promoting Jenna Jameson, one of the most successful and popular performers in the adult entertainment business. Jameson, an industry veteran— sometimes referred to as "the reigning queen of porn," with numerous movie titles, magazine covers, and awards to her credit—made a strategic career decision to capitalize on her name recognition by taking personal control (along with her husband and professional co-performer) of all aspects of the business, from content production through Internet delivery to branded merchandizing. Her company has also moved into Web site management for other performers in the adult business, and according to Matthew Miller, who profiled the holding company on Forbes.com, combined hosting and membership subscription revenues amounted to $12 million.[79] Jameson's autobiography, *How to Make Love Like a Porn Star*, spent weeks on the *New York Times* best-seller list, reinforcing her proto-celebrity status and suggesting that she might be the first adult industry performer with the potential to cross over successfully into the mainstream entertainment sector.

Clubjenna.com showcases a diverse range of products and services designed, customized, licensed, or authorized by Jenna (the company's sex toy inventory includes a mold of the star's pelvic region, which retails for $179.95). The company has also moved into movie production with a stable of actresses who appear under the Club Jenna label and feature on the Web site. Jameson's movies are distributed and marketed by Vivid (⟨http://www.vivid.com⟩), the adult industry's largest film company, which takes a 30 percent commission. If Miller's reported figures are accurate, a typical Jameson release sells fifty thousand copies at $50 each, well above the industry norm, which helps make her one of, if not the, top-grossing performer in the business.[80]

It remains to be seen whether Jameson's efforts to further monetize her name by moving into the operation of strip clubs and the marketing of nonsexual goods, for example, will pay off in the longer term. Her strategy—and in this general regard she is no different from, say, Donald Trump—of leveraging name recognition in both the core and peripheral adult market segments sets her apart from almost all other individually owned pornographic enterprises. Jameson has, additionally, capitalized on her appeal to gay men by launching an interactive Web site (⟨http://www.clubthrust.com⟩) for her homosexual fan base—the first lateral move of its kind by a female porn star. She also recognizes the value of cobranding: Jameson currently hosts *Jenna's American Sex Star* on Playboy TV.

Club Jenna is not a publicly quoted company, and reliable figures are thus hard to come by. Jameson claims in her autobiography that Club Jenna, which offers a wide variety of features along with much exclusive content, was "hugely profitable in its first month," but she provides no actual financial data.[81] Miller, however, estimates revenues of $30 million in 2005, with profits amounting to almost $15 million.[82] If correct, these are impressive numbers for a relatively new entrant to the pornography business; recall that PMG, a longtime presence in multiple markets, posted revenues of $48 million in 2004.

A major challenge for any personality-branded adult entertainment company is transitioning from what might be termed an ego-centered start-up to a mature business that allows for sustainable corporate growth once the star's shelf life has come to its inevitable end. Club Jenna is largely synonymous with Jameson; she is its unique selling proposition. For now, the company's success is linked, inextricably, to the salience and market appeal of its principal, and the digital asset base that she has diligently constructed around her own stage persona. It remains to be seen whether she can take the business to the next level, at once reinforcing brand identity while at the same time reducing the company's dependence on her anatomical assets. What can be said with assurance is that the Web enabled Jameson to conceive, launch, and extend a business in a way, on a scale, and at a pace that would simply not have been possible in the pre-Internet days.

Vivid Entertainment Group

Vivid, a privately held company with a hundred employees, is one of the world's largest adult film producers, with estimated revenues, according to various sources, of $100 million. It was cofounded by David James and Steven Hirsch (yet another second-generation pornographer) in 1984, and despite recurrent speculation that it would launch an initial public offering, has continued to remain in private hands. Vivid has built up a strong reputation as a producer of high-quality, high-cost adult content, releasing sixty or so films per year, many of which are produced at the company's studio in the San Fernando Valley. The production costs range from $40,000 to $200,000 per movie.[83] Vivid sells its videos to retail outlets, hotels, and cable systems, and is making its extensive movie library (fifteen hundred titles) available via the Internet to members. In 2000, it was estimated that 25 percent of Vivid's pretax income was profit.[84]

Vivid has been a successful presence in the competitive adult entertainment industry for more than twenty years, during which time it has developed strong brand awareness and demonstrated a capacity to innovate. The company pioneered the contract system for its top female stars (known as the Vivid Girls), who purportedly earn from $80,000 to $750,000 annually, and it has an exclusive distribution deal with the aforementioned Jameson, who coproduces some of her videos with Vivid. As with the other companies profiled here, Vivid has moved into branded merchandise, such as condoms, snowboards, herbal supplements, and fine cigars. In 2003 it launched Vivid Comix—a sexually explicit comic series featuring the Vivid Girls—in conjunction with Avatar Press, a specialty publisher. Most recently, the company announced plans to sell downloadable porn movies that consumers can burn to copy-protected DVDs and watch on their televisions.

Vivid.com allows for the digital distribution of the company's extensive stock of proprietary programming; unlike, for example, PMG, Vivid does not rely on third-party suppliers for its Web content. The pay-per-view site (comprising images, movies, and live sex) is structured around the Vivid Girls concept, which gives the company a distinctive product offering, rather like Playboy in some respects. Vivid's CEO, Hirsch, is on record as saying that about $25 million comes from online sales—a number that he expects to increase significantly as the markets for wireless and on-demand video grow. It is not clear what proportion of the company's revenues are generated by monthly online subscriptions.

Vivid may seek a public listing or stay independent; or alternatively, the current owners may decide to sell the company, though there are few other adult entertainment businesses with the financial wherewithal to acquire it outright. Vivid has diversified and grown considerably in recent years—both in terms of revenues and product portfolio—and given its large Internet-accessible archive, brand presence, technical capabilities, and industry know-how, may choose to continue as a strong independent presence in the market, seek a public flotation, or position itself as an acquisition target.

Conclusions

The pornographic Internet comprises amateur and professional producers, conglomerates and microenterprises, mass and niche markets, expensive and free goods, and legal and illegal products. The pyramidal Internet-based adult entertainment business is increasingly dominated, in revenue terms, by a small number of resource-rich multimedia enterprises (such as PMG and Playboy Enterprises) along with a handful of major content producers (e.g., Vivid Entertainment and Wicked Pictures), but these technological pacesetters coexist with a raft of ego-centered, Web-based companies (e.g., Danni Ash and Jenna Jameson) whose competitive edge, initially at least, resides in the reputation and brand value of their eponymous principals. In addition, there are some clearly defined and astutely differentiated niche market suppliers (e.g., Retro-Raunch and Candida Royale) along with countless smaller enterprises offering either general or themed adult content (e.g., ⟨http://www.kink.com⟩), not to mention the many sites offering masses of free material made available by commercial producers seeking visibility and referrals. As in any other business sector, it is reasonable to assume that some of the more enterprising but weakly capitalized companies will be taken over by larger ones.

Quite apart from the widely acknowledged difficulty of measuring the dimensions of the Internet (size, traffic, and demographics), a number of assertions can be made with reasonable confidence:

- The "sexual socialization of children" is happening earlier than in pre-Internet times[85]
- Many more young people are exposed more often to a greater variety of pornographic material than ever before
- Many more people are spending more of their disposable income on pornographic goods and services than ever before
- A greater mass and variety of sexual representations are available on the Web than was the case in the pre-Internet world
- Interactivity and customization have become defining features of the contemporary pornographer's portfolio
- Web content may not be a billion dollar market in the United States, but depending on how it is measured, it likely generates hundreds of millions of dollars
- Many more individuals and organizations are involved in the production and distribution of pornographic content than previously

- The Internet has greatly augmented the commercial distribution of pornographic content, and resulted in media displacement and substitution
- It has democratized the production and consumption of pornographic goods and experiences

The Internet facilitates the uninhibited sharing of sexual feelings and representations across time, space, and class, creating a sociosexual realm that did not previously exist. It is not an overstatement to say that the Internet is sui generis in the sociotechnical annals of pornography.

Acknowledgments

I am grateful to William Aspray, Paul Ceruzzi, Elisabeth Davenport, Yvonne Rogers, Howard Rosenbaum, and an anonymous reviewer for their comments and suggestions. I also acknowledge useful feedback from seminar participants at Napier University in Edinburgh and Indiana University in Bloomington. Sara Franks helped with data collection.

Notes

1. Hannamari Hänninen, "The Money Streams in Retail Business of Pornography in Finland," 2005, available at ⟨http://www.nikk.uio.no/arrangementer/referat/tallin03/hanninen.html⟩.

2. Quoted in Lynn Hunt, *The Invention of Pornography: Obscenity and the Origins of Modernity, 1500–1800* (New York: Zone Books, 1993), 357.

3. George Gordon, *Erotic Communications: Studies in Sex, Sin, and Censorship* (New York: Hastings House, 1980), 181.

4. David Hockney, *That's the Way I See It* (London: Thames and Hudson, 1993), 130.

5. Linda Williams, *Hard Core: Power, Pleasure, and the "Frenzy of the Visible"* (Berkeley: University of California Press, 1999), 299.

6. Brian McNair, *Mediated Sex: Pornography and Postmodern Culture* (Oxford: Oxford University Press, 1996), vii.

7. Martha Nussbaum, *Hiding from Humanity: Disgust, Shame, and the Law* (Princeton, NJ: Princeton University Press, 2006), 152.

8. See, for example, Edward Donnerstein, Daniel Linz, and Steven Penrod, *The Question of Pornography: Research Findings and Policy Implications* (New York: Free Press, 1987); William A. Fisher and Azy Barak, "Internet Pornography: A Social Psychological Perspective on Internet Sexuality," *Journal of Sex Research* 38, no. 4 (2001): 312–323.

9. James Davidson, *Courtesans and Fishcakes: The Consuming Passions of Classical Athens* (New York: Harper Perennial, 1999).

10. Hunt, *The Invention of Pornography*; Walter Kendrick, *The Secret Museum: Pornography in Modern Culture* (Berkeley: University of California Press, 1996).

11. Williams, *Hard Core*, 4.

12. Ibid., 283.

13. McNair, *Mediated Sex*, 24–25, 35–36.

14. Michael Warner, quoted in Rich Cante and Angelo Restivo, "The Cultural-Aesthetic Specificities of All-Male Moving Image Pornography," in *Porn Studies*, ed. Linda Williams (Durham, NC: Duke University Press, 2004), 145.

15. Paul Saffo, quoted in Dawn C. Chmielewski and Claire Hoffman, "Porn Industry Again at the Tech Forefront," *Los Angeles Times*, April 20, 2006, available at ⟨http://www.latimes.com/business/la-fi-porn19apr19,1,5662527.story?coll=la-headlines-business⟩.

16. Lucas Mearian, "Porn Industry May Be Decider in Blu-ray, HD-DVD Battle," *Computerworld*, May 2, 2006, available at ⟨http://www.computerworld.com/hardwaretopics/storage/story/0,10801,111087,00.html⟩.

17. Jonathan Coopersmith, "Pornography, Technology, and Progress," *ICON* 4 (1998): 94.

18. Michel Foucault, *The History of Sexuality: An Introduction, Volume 1*, trans. Robert Hurley (New York: Vintage Books, 1990), 159.

19. McNair, *Mediated Sex*, 31.

20. Coopersmith, "Pornography, Technology, and Progress," 106.

21. Hubert L. Dreyfus and Paul Rabinow, *Michel Foucault: Beyond Structuralism and Hermeneutics*, 2nd ed. (Chicago: University of Chicago Press, 1983), 237–240.

22. Tony Allen-Mills, "Web Gallery Hit by 'Nipplegate,'" *Sunday Times*, July 23, 2006, 27; Giles Hattersley, "The Blog That Turned My Life Upside Down," *Sunday Times*, July 23, 2006, 7; Ruth Barcan, "In the Raw: 'Home-made' Porn and Reality Genres," *Journal of Mundane Behavior* 3, no. 1 (2002), available at ⟨http://www.mundanebehavior.org⟩.

23. See, for example, Kurt Eichenwald, "Through His Webcam, a Boy Joins a Sordid Online World," *New York Times*, December 19, 2005, A1; Roger Shattuck, *Forbidden Knowledge: From Prometheus to Pornography* (New York: St. Martin's Press, 1996); Yaron Svoray with Thomas Hughes, *Gods of Death* (New York: Simon and Schuster, 1997).

24. Sean Thomas, *Millions of Women Are Waiting to Meet You* (London: Bloomsbury, 2006), 157.

25. Ragnhild T. Bjornebekk, "Accessibility of Violent Pornography on the Internet: Revisiting a Study of Accessibility—Prevalence and Effects," 2005, available at ⟨http://www.nikk.uio.no/arrangementer/konferens/tallin03/bjornebekk_e.html⟩.

26. Fisher and Barak, "Internet Pornography," 312, 320.

27. Stephen Segaller, *Nerds 2.0.1.: A Brief History of the Internet* (New York: TV Books, 1999).

28. Eric von Hippel, *Democratizing Innovation* (Cambridge, MA: MIT Press, 2005).

29. Blaise Cronin and Elisabeth Davenport, "E-rogenous Zones: Positioning Pornography in the Digital Marketplace," *Information Society* 17, no. 1 (2001): 33–48.

30. Rob Kling, "Letter from the Editor-in-Chief," *Information Society* 17, no. 1 (2001): 1–3.

31. For more on the nature and scale of the revenue losses suffered by large and small businesses, see Lewis Perdue, *EroticaBiz: How Sex Shaped the Internet* (New York: Writers Club Press, 2002).

32. Zabet Patterson, "Going On-line: Consuming Pornography in the Digital Era," in *Porn Studies*, ed. Linda Williams (Durham, NC: Duke University Press, 2004), 110.

33. Matthew Zook, "Underground Globalization: Mapping the Space of Flows of the Internet Adult Industry," *Environment and Planning* 35 (2003): 1261–1286.

34. For an explication of the various genres and appellations, see Eric Schaefer, "Gauging a Revolution: 16 mm Film and the Rise of the Pornographic Feature," in *Porn Studies*, ed. Linda Williams (Durham, NC: Duke University Press, 2004), 370–400.

35. Michelle Pautz, "The Decline in Average Weekly Cinema Attendance," *Issues in Political Economy* 11 (2002), available at ⟨http://org.elon.edu/ipe/Vol%2011%202002.htm⟩.

36. David Hebditch and Nick Anning, *Porn Gold: Inside the Pornography Business* (London: Faber and Faber, 1988), 228.

37. Lisa Z. Sigel, "Filth in the Wrong People's Hands: Postcards and the Expansion of Pornography in Britain and the Atlantic World, 1880–1914," *Journal of Social History* 33, no. 4 (2000): 859–885.

38. Kendrick, *The Secret Museum*, 95.

39. Jonathan Coopersmith, "Does Your Mother Know What You *Really* Do? The Changing Nature and Image of Computer-Based Pornography," *History and Technology* 22, no. 1 (2006), 1.

40. Davidson, *Courtesans and Fishcakes*, 112.

41. Gordon Hawkins and Franklin E. Zimring, *Pornography in a Free Society* (Cambridge: Cambridge University Press, 1991), 212.

42. Hebditch and Anning, *Porn Gold*, 70–71.

43. Hawkins and Zimring, *Pornography in a Free Society*, 36.

44. Deborah Shamoon, "Office Sluts and Rebel Flowers: The Pleasures of Japanese Pornographic Comics for Women," in *Porn Studies*, ed. Linda Williams (Durham, NC: Duke University Press, 2004), 78.

45. Michael Kirk, "American Porn," *Frontline*, Public Broadcasting Service, 2002.

46. Fedwa Malti-Douglas, *The Starr Report Disrobed* (New York: Columbia University Press, 2000), xix.

47. Kendrick, *The Secret Museum*, 242.

48. Roland Barthes, *Camera Lucida*, trans. Richard Howard (London: Vintage Books, 2000), 118.

49. Williams, *Hard Core*, 2, 5.

50. On the theory of industry clusters, see Michael E. Porter, "Clusters: The New Economics of Competition," *Harvard Business Review* (November–December 1998): 77–90.

51. See Cronin and Davenport, "E-rogenous Zones."

52. Angelina Spencer, *The Erotic Economy*, available at ⟨http://www.google.com/search?hl=en&ie =ISO-8859-1&q=%22Angelina+Spencer%22+%22Erotic+economy%22⟩.

53. Thomas Waugh, "Homosociality in the Classical American Stag Film: Off-screen, On-screen," in *Porn Studies*, ed. Linda Williams (Durham, NC: Duke University Press, 2004), 127.

54. Joseph H. Slade, "Eroticism and Technological Regression: The Stag Film," *History and Technology* 22, no. 1 (2006): 27–52; Williams, *Hard Core*, 58.

55. Fenton Baily, introduction to *Pornography: The Secret History of Civilization*, by Isabel Tang (London: Channel Four Books, 1999), 19.

56. McNair, *Mediated Sex*, 128.

57. Stefano de Luigi and Martin Amis, *Pornoland* (London: Thames and Hudson, 2004); A. A. Gill, "When DD Met AA: US Pornography, November 1999," in *AA Gill Is Away* (London: Cassell, 2002), 134–145; Jenna Jameson, *How to Make Love Like a Porn Star: A Cautionary Tale* (New York: Regan Books, 2004).

58. Michel Foucault, "Of Other Species," *Diacritics* 16 (1986): 22–27; Jonathan Coopersmith, "Pornography, Videotape, and the Internet," *IEEE Technology and Society Magazine* (Spring 2000): 29.

59. Alvin Toffler, *The Third Wave* (London: Pan, 1980).

60. Minette Hillyer, "Sex in the Suburban: Porn, Home Movies, and the Live Action Performance of Love in Pam and Tommy Lee: Hardcored and Uncensored," in *Porn Studies*, ed. Linda Williams (Durham, NC: Duke University Press, 2004), 51.

61. Franklin Melendez, "Video Pornography, Visual Pleasure, and the Return of the Sublime," in *Porn Studies*, ed. Linda Williams (Durham, NC: Duke University Press, 2004), 401.

62. Cronin and Davenport, "E-rogenous Zones," 38.

63. Indhu Rajagopal with Nis Bojin, "Globalization of Prurience: The Internet and the Degradation of Women and Children," *First Monday* 9, no. 1 (2004), available at ⟨http://firstmonday.org/ issues/issue9_1rajagopal/index.html⟩.

64. Hawkins and Zimring, *Pornography in a Free Society*, 30.

65. Hawkins and Zimring, *Pornography in a Free Society*; Perdue, *EroticaBiz*; Hebditch and Anning, *Porn Gold*.

66. Frank Rich, "Naked Capitalists: There's No Business Like Porn Business," *New York Times*, May 20, 2001, Section 6, 51.

67. Dan Ackman, "How Big Is Porn?" May 25, 2001, available at ⟨http://www.forbes.com/2001/05/25/0524porn.html⟩.

68. Emmanuelle Richard, "The Perils of Covering Porn," *Online Journalism Review*, June 5, 2002, available at ⟨http://www.ojr.org/ojr/business/1017866651.php⟩.

69. Williams, *Hard Core*, 283.

70. Hänninen, "The Money Streams in Retail Business of Pornography in Finland."

71. Geoffrey Gorer, "Chapter 3," in *Does Pornography Matter?* ed. C. H. Rolph (London: Routledge and Kegan Paul, 1961), 39.

72. NCH, the Children's Charity, *Child Abuse, Child Pornography, and the Internet*, 2004, available at ⟨http://www.nch.org.uk/information/index.php?I=94#internet⟩.

73. S. Liliana Escobar-Chaves, Susan R. Tortolero, Christine M. Markham, Barbara J. Low, Patricia Eitel, and Patricia Thickstun, "Impact of the Media on Adolescent Sexual Attitudes and Behaviors," *Pediatrics* 116, no. 1 (2005): 303–326.

74. Hawkins and Zimring, *Pornography in a Free Society*, 206.

75. U.S. Commission on Obscenity and Pornography, *Report* (Washington, DC: U.S. Government Printing Office, 1970); U.S. Department of Justice, *Attorney General's Commission on Pornography Final Report* (Washington, DC: U.S. Government Printing Office, 1986); Hawkins and Zimring, *Pornography in a Free Society*, xi.

76. Patterson, "Going On-line," 112.

77. Hawkins and Zimring, *Pornography in a Free Society*, 2.

78. Joan Acocella, "The Girls Next Door: Life in the Centerfold," *New Yorker*, March 20, 2006, 144–148.

79. Matthew Miller, "The (Porn) Players," July 4, 2005, available at ⟨http://www.forbes.com/forbes/2005/0704/124_print.html⟩.

80. Ibid.

81. Jameson, *How to Make Love Like a Porn Star*, 554.

82. Miller, "The (Porn) Players."

83. Brett Pulley, "The Porn King," April 5, 2005, available at ⟨http://www.forbes.com/business/2005/03/07/cz_bp_0307vivid.html⟩.

84. William Li, "Porn Goes Public," *Industry Standard*, November 6, 2000, 94.

85. James B. Weaver III, *Prepared Testimony, U.S. Senate Committee on Commerce, Science, and Transportation Hearing on "Protecting Children on the Internet,"* January 19, 2006.

VII   Lessons Learned, Future Opportunities

16 Market and Agora: Community Building by Internet

Wolfgang Coy

An agora (αγορά), translatable as marketplace, was an essential part of an ancient Greek polis or city-state. An agora acted as a marketplace and a forum to the citizens of the polis. The agora arose along with the poleis after the fall of Mycenaean civilization, and were well established as a part of a city by the time of Homer (probably the 8th century BC). The most well-known agora is the Ancient Agora of Athens.
Wikipedia, as of November 5, 2006

AGORA, originally, in primitive times, the assembly of the Greek people, convoked by the king or one of his nobles. The right of speech and vote was restricted to the nobles, the people being permitted to express their opinion only by signs of applause or disapproval. The word then came to be used for the place where assemblies were held, and thus from its convenience as a meeting-place the agora became in most of the cities of Greece the general resort for public and especially commercial intercourse, corresponding in general with the Roman forum. At Athens, with the increase of commerce and political interest, it was found advisable to call public meetings at the Pnyx or the temple of Dionysus; but the important assemblies, such as meetings for ostracism, were held in the agora. In the best days of Greece the agora was the place where nearly all public traffic was conducted. It was most frequented in the forenoon, and then only by men. Slaves did the greater part of the purchasing, though even the noblest citizens of Athens did not scruple to buy and sell there. Citizens were allowed a free market.
Encyclopedia Britannica, 1911

In the beginning for academia, the Internet could be seen as an extension of the public library, enhanced by data files, access to distant computers, and a new communication path called e-mail. Though the use of distant computing services to access prevailed, the use of e-mail and file transfer protocol demonstrated the usefulness of the Net also to scientists and students outside computer science or particle physics. The Net was initially just another, albeit new, example of scientific communism, the liberal sharing of resources among "those that had access" as part of academia.[1] This idea of a sharing community is deeply rooted in scientific research, a special economic and political regime deeply based on free access to published results, and an open exchange of valuable ideas (including some not so valuable ones too). As the academic networks were substantially supported by governments, predominantly via military projects or National Science Foundation funding, and many locally financed computing centers, they were perceived as a free and available service—free of both charge and content regulations. (In fact, neither was the actual case, once one understood their workings.)

Usenet, a term that stands for user network, differed from other Unix services in that it built communities—from net.unix.compiler and mod.newprod to distant extensions like net.rumor and mod.ki.[2] Usenet was a bulletin board software, not developed as part of the ARPANET. It was an independent "hack," stemming from the bulletin board system (BBS) scene, although it was a Unix service based on Unix-to-Unix CoPy and the initial Unix System V7. Basically Usenet newsgroups were started before the public introduction of the IBM personal computer in 1979 at Duke University when two students, Tom Truscott and Jim Ellis, began to connect several Unix computers. Steve Bellovin, another student at the University of North Carolina, wrote the first version of the news software using shell scripts and installed it on two sites: "unc" and "duke."[3] Usenet was described at the January 1980 Usenix conference, after which a C version of the underlying programs was published. The oldest newsgroup article that can be found via Google (where Usenet news may be found under Google Groups) is dated Monday, May 11, 1981, 10:09 a.m., but there were groups well before that date.[4] Since 1980, some newsgroups from ARPANET became available via a network gateway at the University of California at Berkeley; these groups were collected under the header "FA" (abbreviated for "from ARPANET"). In 1982, a rewrite was publicly distributed with the Berkeley Software Distribution enhancements of Unix, but Usenet fully started in 1986, after it was transformed to the meanwhile-developed Network News Transfer Protocol, shifting from Unix-to-Unix CoPy to the Transmission-Control Protocol/Internet Protocol as specified in Request for Comments 977, and from external network backbones to the Internet. The original newsgroup structure was changed in the "Great Renaming" from net.*, mod.*, and fa.* hierarchies to comp.*, misc.*, news.*, rec.*, soc.*, sci.*, and talk.*.

As these groups were regulated by a majority voting system, advocates of free speech opposed this step. Brian Reid, whose "recipe group" was lost during the renaming, started the alt.* hierarchy to initially circumvent the backbones. As part of the zeitgeist, he formally included groups like alt.sex, alt.drugs, and for symmetry reasons alt.rock-n-roll, transforming Usenet into a community where nearly everything could be discussed. Usenet may be seen as the first global interactive community outside ARPA research (with the exception of point-to-point e-mail, of course.) Newsreaders, programs to subscribe selected newsgroups, were developed for practically all operating systems, thereby opening much of the closed ARPANET to a broad number of BBS users connected via modem to their local BBS provider. The elementary power of networks became immediately visible: connecting different networks via gateways extends both simultaneously. Local BBSs became part of the global Internet, and the Internet reached new groups of supporters. With Usenet, it finally became obvious that computers are digital media machines, not only computing machinery—and not only multimedia workstations but full-fledged intercommunication devices with all the signs of communicative and distributive media. Even certain kinds of audio or visual communication were imagined quite early—from the Internet to telephony equivalents, such as videoconferencing, Internet radio, and on-demand video. It all would depend on the bandwidth, the appropriate political regulation, and sufficient financial support.

Some understood the Internet as a supplement or successor to the still-dominant mass media—that is, television, radio, and newspapers. The corresponding media industries started quite a number of experiments in the direction of media convergence, usually missing the point: namely, the highly interconnected structure of the Internet, where user participation was obvious. Nevertheless, these experiments pushed the Internet into commerce, including the vision of a highly asymmetrical Net where a few senders opposed large numbers of receivers or consumers. The user perception of Usenet was quite different, even if many users were only readers, or leechers as they were sometimes called. It was the sheer possibility of being active that made Usenet users feel a part of communities—much stronger than traditional mass media ever could.

This collective experience of global users, many of them technically quite knowledgeable and experienced, raised the awareness that many network resources were externally financed (even though the university computing center is considered external for many academic users). This resulted in a broad understanding of Internet resources as something shared but also valuable that ought to be treated with care—an attitude generally described as the Acceptable Use Policy or netiquette. Netiquette often included the notions that any commercial use of the Internet was considered inappropriate independent of any formal rules, that Net services should be free of charge for end users, and especially that any assumed inappropriate use of the network demanded counteraction—or at least support of such action.

It was on the Usenet where the (mis)use of the Internet as a commercial platform was discussed broadly, especially after the first case of ubiquitous spam. In April 1994, the term spam was not common except for on Hormel cans and in *Monty Python*'s restaurant sketch, but in a single act the problem of unwanted commercial use became visible to all Usenet news readers when a Phoenix-based lawyer couple named Laurence A. Canter and Martha S. Siegel placed a message using a Perl script on thousands of Usenet newsgroups, advertising their rather-peripheral services in the U.S. green card lottery.[5] The name spam was adopted widely for these practices.

Though Canter and Siegel's abuse of Net resources was not the first case of mass advertising on the Net (chain letters, undirected questions, cries for help, or list mail postings running astray were common to some extent), it was arguably the first commercial mass advertising on Usenet news and a point of common self-reflection. While most users felt disturbed by the prospect of reduced bandwidth because of mass mailing, marketing specialists made this obvious observation: electronic mail marketing is much cheaper than letters or phone calls, and it can be easily done by existing e-mail list software. The relaxation of the academic Internet for commercial purposes demonstrated a new kind of Internet beyond academic netiquette.

Later that same year, 1994, Pizza Hut and the SCO Group announced the PizzaNet, "a long-awaited service" for the geek community, allowing one to order pizza over the Net: "In a revolutionary spin on business use of the Information Superhighway, The Santa Cruz Operation, Inc. (SCO) and Pizza Hut, Inc. today announced PizzaNet, a pilot program that enables computer users, for the first time, to electronically order pizza delivery from their local Pizza Hut restaurant via the worldwide Internet. Pizza Hut will launch the PizzaNet pilot in the Santa Cruz area on August 22."[6]

As the pizza service was restricted initially to the Santa Cruz, California, area, Pizza-Net did not become a "killer application." Other mail-order services demonstrated a better-aimed base for commercial Internet exploits. Company Web sites became the ubiquitous number one choice for the initial foray on the information superhighway, and electronic catalogs were among the first digital goods to be offered. Mail-order services were well suited for such a commercial step as they were delivered free of cost, a plus in the tradition of the academic Internet, and also because payment services were then not a strength of the Net technology.

Again in 1994, the newly founded Netscape Communications Corporation announced its Web browser Navigator and offered it as a giveaway to all interested users. Netscape was derived from the National Center for Supercomputing Applications' Mosaic, the first Web browser with a wide distribution.[7] The company offered its successful initial public offering on the NASDAQ, promising that its income would come from the sale of server software. This set the tone for further commercial Web content: to sell something, something had to be given away "for free" as an add-on. The hope was to turn a profit because of the sheer number of customers, but the result was that nearly all Web browsers are now freely distributed since Netscape's experiment. Though this seemed to be a clever move, the company was somewhat unfortunate, as shortly after its initial public offering, successful Web server software like Apache, another National Center for Supercomputing Applications derivative, was also offered "free of charge." Apache was well received and now serves as the basic software for more than two-thirds of all Internet servers, while Netscape's Enterprise Server became part of Sun's Java System Web Server.[8] The once-dominant Netscape Navigator browser became the open-source foundation for the Mozilla group of browsers in early 1998, viewed as an economic and cultural counterstrike to Microsoft's Internet Explorer, which was bundled with Microsoft Windows. Later in 1998, Netscape was bought by America Online (AOL), and it was closed in 2003 after AOL switched to Microsoft's Internet Explorer as part of a larger, juridical settlement between Microsoft and AOL. Netscape Corporation may be seen as an archetypal case of early dot-coms from its glamorous initial public offering to the end.

While online payment is possible, though neither secure nor reliable via credit card or other banking services, the delivery of material goods is obviously beyond the capacity of the Net. Electronic goods in contrast are well suited for network delivery, and it became rapidly clear to the public that shrink-wrap packages were a rather accidental way of software distribution. Software and updates are by their nature digital artifacts—as is money. Hence the sale of software is a perfect fit for the Internet. The same holds for written material that can be read on a screen or printed for a more convenient use. E-books and audio books along with music and movies are all perfect for digital networks of sufficient bandwidth, while audio CDs, CD-ROMs, and DVDs are only coincidental package forms.

Strangely, the cultural industry was initially reluctant about or even ignorant of the new electronic possibilities of distribution. Others had to clear the path, and they did. File transfer protocol servers and Usenet were already used in the early 1990s for the propagation of software and music. This centralized approach changed with the advent of peer-to-peer networks, where every download was technically combined with an upload of the already-downloaded data. Peer-to-peer technology resolves the asymmetry between server and client computers, distributing bandwidth demands between all engaged network nodes. It is perfectly suited for the spread of files when time and actual bandwidth is noncritical. Peer-to-peer networks may be differentiated by the degree of centralized support for the decentralized peers, and by their scalability, which depends on the lack of central services.

Peer-to-peer technology became more widely recognized by the public when Napster, one of the first such networks, was drawn into a complex court case with the Recording Industry Association of America in 1999, later partially settled in 2001. As a consequence Napster changed to a subscription system, losing many of the users that it had gained during the court case. Napster, a centralized peer-to-peer network, had specialized in the exchange of MP3-compressed music. It was probably the first music distribution system that published illegal prereleases of recorded music—among others from Metallica, Madonna, and Dr. Dre—but it was not the first peer-to-peer channel. Yet Napster was easier to use than, say, Hotline, a system combining chat, message boards, and file sharing. Hotline was also used for MP3 distribution, a decent number of which were legally deposited on Web sites like Live365.com, MP3.com, and the Internet Underground Music Archive (⟨http://www.iuma.com⟩).[9] While the number of Napster users reached a high in 2001, similar services like KaZaA, Gnutella, or Limewire with different degrees of centralized directories were started. These are now large communities; Gnutella claims to serve more than two million users, with one-third being active concurrently. The dominant peer-to-peer distribution protocol is BitTorrent, invented and maintained by Bram Cohen of BitTorrent, Inc.

With the advent of home broadband access, the content distribution shifted from highly compressed short titles to albums, lossless compression, and whole DVDs. This made copyright violation a major issue—with intense arguments centering around "piracy" and "free use." The obvious way out in the common interest of the copyright holders as well as the music and movie industries—namely, inexpensive legalized downloads—was only halfheartedly discussed until 2001, when Apple announced its iTunes Store as a commercial distribution channel via the Internet. Though the initial price fixed at $0.99 per music track was compatible with the music industry's usual pricing scheme, this industry entered this distribution channel late, leaving Apple a major share of the market until now. It is interesting to compare this to the behavior of the producers and distributors of software, computer game, music, e-books, and movies. Though they all were victims of a certain amount of illegal copying, they reacted quite differently. All tried some technical schemes of copy protection, some tried to influence legislation and law enforcement, and others attempted to influence legal and illegal users. E-book publishing houses seemed to accept the new situation to some extent, as did most software producers, despite some lobbying by the Business Software Alliance and attempts by dominant market players like Microsoft and Adobe to install some complex registration processes. The personal computer gaming industry relies mainly on technical copy protection schemes, including a technique borrowed from Sony that allows for the unannounced intrusion into operating systems with viral software (root kits)—considered illegal in some countries. The music and movie industries also take legal actions more or less at random against suspects, found sometimes by illegal actions. The movie industry uses rude advertising methods, including making legal buyers watch long, noninterruptible trailers with each film. Though digital content industries are victims of product piracy worldwide ranging from more or less perfect industrial copies to openly sold illegal ones in some countries, the music and movie industries choose to engage in open legal fights with underage or adolescent end users predominantly in the United States and Europe—a fight they probably cannot win. Despite all these measures a "Darknet," a term coined in a paper by Microsoft engineers, seems to be here to stay.[10]

As a positive aspect of these fights, a digital content trade is emerging despite the reluctant attitudes of the music and media giants—featuring software, games, audio and e-books, music, movies, and now videos of television series. This supports an enormous growth of digital content available in and outside the Net. And we see communities sharing—sometimes illegal or questionable ideas, but also many new and untested concepts as well as intellectual and cultural stimuli. BitTorrent servers are by no means restricted to the distribution of audio CDs, CD-ROMs, and DVDs. They use a resource-protecting protocol that allows for the distribution of large files in a reasonable amount of time. As an example, a server like chomskytorrent.org is a multiplier of political discussions far beyond the exchange of Hollywood's entertainment sedatives. While peer-to-peer networks usually focus on the exchange of files, they generally also have forum spaces, initially thought of as support mechanisms. Some sites build living communities with their forums, while others restrict the verbal exchange to a bare minimum—often controlled by a designated administrator(s).

Discursive Communities

Slashdot was started in 1998 as a news provider for the field of "digital technology and society" in a broad sense, with an accent on the open-source software movement, financed by advertising.[11] Slashdot represents a bottom-up journalism relying strongly on its readers: "News for Nerds. Stuff That Matters." While anyone with Net access can read Slashdot, only registered users can write a comment, and only paying subscribers can get an ad-free version. As of 2006, Slashdot had more than a million sub-scribers paying a small fee for extra services. It has been an important active user com-munity on the Net over the years, in the tradition of the early BBS communities. Slashdot serves as a model for other community-building processes. It elegantly solved the question of a commercial, ad-supported Internet by proposing a different model: an ad-free Web site for paying subscribers along with a free, accessible version paid for by advertising. The costs of this Internet site are not hidden to its users.

In November 2006, the number of comments on Slashdot exceeded 16.77 million, or to be precise, 2^{24} in binary notation, taking up all 24-bit addresses of Slashdot's data-base, and leading to a temporary shutdown of several hours for user comments. An-other unexpected side effect of an active Web site like Slashdot became visible in its first years: URLs mentioned by Slashdot (or other active discursive sites) could bring down the referred site by the sheer volume of hits by interested users. This phenome-non is sometimes called the "Slashdot effect"—an Internet prize of fame and glory.

Free Internet: Free Speech or Free Beer?

In the early 1990s, Internet users generally agreed that the Internet was and should be free. But the free use of Internet resources was translated quite differently to different users with the two perspectives of "free speech" and some vague notion of "free of charge." As Richard Stallman of the Free Software Foundation clarified on many occa-sions, articulating a political statement for Net users grouped around the GNU license and other open-source activism: "Think of free as in free speech, not free as in free beer."[12] This did not necessarily reflect the attitude of young student users or the view-point of companies interested in a commercial Internet, however, and it was certainly beyond the political understanding of a regulated Net held by many governments worldwide.

Opening the Internet to nonacademic users included different modes of financing. As with many BBS-accessing points, Internet service providers took money, but the driving force for commerce was the "vision to do business" over the Net. During the first decade of the commercial use of the Internet, many business visionaries found that Internet users were not eager to pay for what they thought "ought to be free," and even if people were willing to spend money they did not necessarily feel secure about sending in their credit card numbers. There was some trouble with Internet fraud, and print and television media were quick to report the "new form of crime." The security issue was solved to some extent by the development of the "secure" https protocol—defined in the Request for Comments 2818 in May 2000—that demands no additional cryptographic software as well as the further development of payment services by banks, but also by new services like PayPal, founded in 1998 and bought by eBay in 2002. Still, the reservations about having to pay anything for Net goods are visible even today.

But independent of the fraud issues, direct payment is not the only way to finance a global Net. The modern economy offers many ways to spend money, some with long-term profit perspectives, and others with different attitudes, like those of learning, exploring, amateur engagements, or sometimes sheer fun.

Advertising and public relations work is a solid source of the first type. This includes giveaways and add-ons, as client retention becomes increasingly important compared to immediate sales. But despite many deregulation efforts, public services still exist, allowing some financial support for goods and services offered on the Net—and the same holds for the by-products and services of academia research, at least as long as external funding works. And then there is still an enormous amount of hobbyists and others spending many unpaid hours on the Internet as a public service. The Internet becomes a perfect amplifier for such community work, as a single endeavor may result in a thousand- and millionfold uses worldwide—reward enough for most engaged amateurs. There is no question that services on the Internet must be financed, but the commercial market model of direct payment transactions is not the only solution.

Perhaps the most elegant way to finance a Net service was demonstrated by Google, the search engine giant founded in 1998. Since 2000, Google has used text-based advertising related to the search entry and is paid for by the number of hits the advertised link receives. The relatively unobtrusive text ads were accepted by users, and the payment mode based on actual hits was well received by the clients. This user-query-directed Internet advertising is also relatively cheap, so that small companies can easily afford it. As of 2006, Google was selling about $10 billion worth of advertising per year, with a profit of one-quarter of that amount.

With the commercial penetration of the Internet, netiquette as developed in the early 1990s has changed dramatically. One of the first technical steps that heralded new rules can be found in the asymmetrical digital subscriber line protocol. With this basic technical decision, Net nodes were divided into servers and clients, with a much higher download than upload capacity. This holds true despite the observation that by conservative estimates, one-third to one-half of the traffic is done via peer-to-peer-networking that would perfectly fit a symmetrical line load. An asymmetrical digital subscriber line is deeply rooted in a model of mass media and trade, demonstrating "Telecom Rules."

The academic understanding of fair use for intellectual property differs in some aspects from the commercial understanding. For example, under certain circumstances the right of citation allows for the partial use of otherwise-protected content, while the academic prohibition against plagiarism disallows some otherwise-acceptable noncited transfer of ideas. Copyright, patent, brand name, and trademark disputes arose unavoidably while the Internet expanded its commercial wing. A long debate over licenses started with the first open-source programs, and now covers all kinds of digital content. Open access and the creative commons are only two perspectives on the political and legal transformation from material trade to electronic commerce.

Besides all these changes, security and privacy issues remain unsolved problems in terms of the Internet. One of the issues focuses on the basic commercial transaction of payment. Secure payment is still an ongoing concern. Internet banking procedures, including credit card payments, are ubiquitous on the Net, but so is fraud and theft. Even if the number of criminal acts may be not higher than outside the Net, the general perception may be different. This is not unfounded. As a recent FBI study states, "Internet auction fraud was by far the most reported offense, comprising 62.7 percent of referred complaints. Non-delivered merchandise and/or payment accounted for 15.7 percent of complaints. Credit/debit card fraud made up 6.8 percent of complaints."[13] This is especially unfortunate given that technical solutions like secure cryptographed payments, micropayments, anonymous payments, or digital signatures have been available for several years now, but are usually not implemented for various reasons. Early work done by David Chaum (DigiCash), Dennis Charter (PaySafe), Phil Zimmerman, or Bruce Schneier was up until now widely ignored in the mainstream of Internet commerce.

Other issues are related to data mining, privacy, spam, anonymity, or censorship. None of these problems have been solved, and every technological twist of the Net may generate new aspects to these questions. Obviously, the Internet constitutes a global community that is only partially regulated by national or transnational law. The long-term influence of Net technologies on national laws, including not only trade but even constitutional balances, is not yet settled. The specific U.S. understanding of free speech is, for example, by no means a global fact, even if the United Nations Declaration of Human Rights includes such a right—but not necessarily the specifics of U.S. legal interpretation.[14]

Games and Their Communities

Computer games were initially driven by fantasies of artificial intelligence, and artificial intelligence had some early triumphs with computer games. Alan M. Turing, Konrad Zuse, Claude Shannon, and many others among the pioneers of electronic computing were fascinated by the prospect of a computer program that could play chess. Zuse even stated that he learned chess in order to program it.[15] In 1958, IBM researcher Arthur Samuel built a program that defeated a local Tennessee master player some years later in a game of checkers. The program was able to "self-learn" by adjusting its own parameters and playing repeatedly against a copy of itself, but it did not improve over a certain level and never defeated a world champion in checkers. Chess programming was more successful, culminating in the 1994 defeat of the human world champion It should be noted that the inner workings of IBM's special hardware "Deep Blue" and its program were never disclosed to the public, and that its astounding victory has never been repeated. These early board games were based on logical rules, appropriate for the restricted output of early computers. Story-based "adventure games" were a completely different kind of game, sometimes called interactive literature. But there was even an early visual game, Lunar Landing, that used the restricted twenty-four by eighty black-and-green alphanumeric terminal screen successfully, generating a minimal visual experience. Later Pong and other games copied that minimalistic visual experience and brought it to the arcade gaming places.

Gaming had a strong push with advanced computer graphics, and computer graphics hardware and software experienced a strong push with the specialized gaming hardware used in Nintendo machines, Sony's Play Station, and Microsoft's Xbox. Strangely enough, early computer games were in the tradition of the solitaire paradigm, playable by a single person at a time. Only the arcade games allowed for sometimes-consecutive rounds of playing for different players. Even graphic-based screen games were typically first-person shooter games. This is an obvious contrast to the human gaming experience of centuries, where gaming was well established as a common experience between groups of players.

The Net changed that cultural anomaly, especially after broadband access became more common. Some retro-movement computer networks now serve as media between players in different places worldwide. There are interconnected chess communities as well as communities for every board or card game. Online gambling also became a commercial enterprise, much to the chagrin of the tax authorities.

Network games do more than merely replicate traditional game circles, however. Commercial network games like World of Warcraft or Second Life assemble large communities. WoW claims to have more than six million paying players, at least half a million of which are present in the game at any time.

Wikis and Open Access

While the original Web structure concentrated on the hyperlinked presentation of texts, later augmented by graphics, sound, and music formats, it is not well suited for interaction beyond formulary input. This changed with server-based services dependent on databases, PHP, Python, Perl, Javascript, Java, and other programming aids. A decent step toward the cooperative, interactive use of the Web—instead of using it only as stored, albeit linked documents—was taken with the introduction of wikis, server-based browser windows that allowed all readers to write and edit directly into some text fields of that page. In addition to regained writing capabilities, the server allows users to step back in time, ignoring changed entries starting with most recently changed ones.

While wikis are of value in and of themselves, their principle is widely recognized through Wikipedia, an encyclopedia written by all interested participants willing to share their knowledge—and errors. Though Wikipedia claims its roots in the Library of Alexandria and the encyclopedias of Denis Diderot and Jean d'Alembert, it is probably more closely related to the Nupedia Internet project, an edited free encyclopedia that failed after Wikipedia's successful launch in 2001.[16] The aim of Wikipedia is the open and free access to written knowledge worldwide: "Imagine a world in which every person has free access to the sum of all human knowledge. That's what we're doing." Financially, Wikipedia relies on donations and the unpaid work of subscribers —without advertising, and with only a small sum from merchandising. This has to cover a budget of around $700,000 as of 2005, with two-thirds of that going to hardware and network access.

Wikipedia is a large community with over one million registered users already for the English edition, and there are editions in thirty-three other languages with their respective communities. Besides the encyclopedia, there are other projects under the umbrella of the Wikimedia Foundation: Wiktionary, Wikiquote, Wikibooks (including Wikijunior), Wikisource, Wikimedia Commons, Wikispecies, Wikinews, Wikiversity, and Meta-Wiki collaborative projects.

The English Wikipedia now contains nearly 1.5 million articles; as of 2006, according to Alexa Radar, Wikipedia was number four among the fastest-growing Web sites (measured by page hits).[17]

Wikipedia is rooted in a much broader environment: namely, that of open-source programming, and open access to scientific and cultural heritage. There are other projects of this type that may be less visible—among them Open Courseware, Creative Commons, or Open Archive—but are certainly as far-reaching as the Wikimedia projects.

All these approaches are not, or not primarily, commercial enterprises. People searched for and found ways to finance their projects, because they provide services to the public, and they are kept as open and free of charge as possible.

Social Software, Social Networking

Direct communication is perhaps the closest form of collaboration, and the Internet has extended the idea of letter writing and telephony with e-mail, discussion lists, and relay chats to video- and audio-enhanced messaging services like voice-over Internet protocol telephony and video services like AIM, Skype, and iChatAV. Even if these services are in part commercially driven, they allow for a decent price reduction for the users. And they allow for services like videoconferencing that were quite expensive before and depended on special equipment that is now in the reach of the average personal computer user (especially if the personal computer is a Mac). Skype now has over fifty million users, eight million of which are concurrently online. Its technology includes video telephony, which is widely considered a serious threat to the traditional telecom companies. Skype was bought recently by the Internet auction firm eBay for $2.6 billion.

Direct communication is, of course, not the only way of community building. Privately owned Friendster.com Web site, started in 2002, tried to introduce young people of similar interests, or assist in finding lost friends or schoolmates. Friendster supports texts, mails, blogs, pictures, music, and videos. It serves to a large extent as a dating site, requiring its users "to be serious" and fit into certain categories. Just as Friendster had an early phase where groups of friends were drawn in, it demonstrated that after a while, there was a tendency for these groups to fade away—at least that is the experimental observation. A lot of Friendster's initiatives were taken away by MySpace.com and more recently Orkut.com. Both came later: MySpace in 2003 and Orkut in 2004. Both took a somewhat different approach, also addressing younger people as well, but allowing a lot of room for personal development.

MySpace now has fourteen million viewers a month. It is already the sixth-most visited site worldwide according to Alexa.com (after four search engines—Yahoo!, MSN, Google, and Baidu.com—and the Chinese instant messenger service qq.com).[18] This got the attention of the traditional media—as the time spent before a screen, whether a computer or television screen, seems to be limited for all users. Quite understandably, in 2005 Rupert Murdoch's News Corporation bought MySpace for $580 million. Orkut, number ten on the Internet hit list of late 2006, is closely affiliated with Google (its entrepreneurial birthplace). Both are now among the most visited Web sites, topped only by search engines. Obviously, communities are of prime interest to Internet users of any age.

Community seeking and building is not restricted to youth. Common hobbies or interests are motives for people at any age. Flickr.com and YouTube.com are two more examples of large communities spreading over the Internet. Flickr allows picture galleries together with social networking, chat, groups, and photo ratings by tags. The tagging system is a breakthrough when compared to other photo-upload sites. Flickr does not use a fixed classification system for its exhibited objects but a flexible description of tag words by the authors and users. Though these tags may not necessarily constitute the most systematic index system, they form clusters of related photos quite well.

Bottom-up tags seriously challenge traditional and fixed top-down classifications. Flickr is not only number thirty-nine on in the Alexa list of the most visited sites worldwide but is number nine in terms of the fastest-growing U.S. sites.

Video server YouTube, founded in 2005, bears the motto "broadcast yourself." It lets its users upload, view, and share video and music clips or video blogs, and is estimated to download a hundred million videos per day. Google's acquisition of YouTube for the sum of $1.65 billion in stock has increased the amount of Google-controlled video streams by a factor of ten. In November 2006, YouTube was the eighth most visited site worldwide according to Alexa. Obviously it offers some replacement to traditional television while also cross-referencing television material. Both NBC and CBS were irritated by the use of unlicensed video clips on YouTube, but settled out of court, probably because they are contemplating the use of the YouTube platform as an advertising field for television.

Both Flickr and YouTube made clever use of a community experience in combination with advertising, demonstrating that new Internet media may be introduced that overcome traditional mass media, allowing the user to feel part of an active process whenever they want to do so.

Global versus Local

The Pizza Hut experiment revealed some poor choices for global network ordering and distribution. Pizza is a local product by its very nature, despite Pizza Hut's universal recipes. The Internet as a global network had in 1994 restricted means to identify the local place of a user. Internet protocol numbers give some hint as to the sender's and receiver's physical place, but the Internet's address space is not precisely coupled to physical spaces. This changes by allowing maps and aerial surface images to be accessed by Web sites through suitable Application Programming Interfaces. Although maps and satellite images have been accessible via the Net for some time, two Google services made them available for mesh-up: Google Earth and Google Map.

A typical application like Fon.com that establishes wireless Internet access worldwide by allowing subscribers to log on to privately operated wireless local area networks, adds users by showing them a Google map of their home area and then asking them to pinpoint the exact place of their shared access point (a small box at the telephone or cable outlet). This results in an accurate and up-to-date map of global access points— much better than any other description.

The Web that started as a worldwide network, a "global digitized village" to paraphrase Marshall McLuhan, can also have value as a local neighborhood. Once again, networked communities add value to the Internet—this time as real neighbors in virtual space.

Market and Agora

Though the modern Internet may have started in 1994 with the expansion of a formerly academic network to commercial enterprises, it is still an open question whether the Net has become a *market* or shopping mall, where "an immense accumulation of commodities" and services are to be sold and bought, or if the Internet is to be modeled along the classical *agora*, a plaza, or "the street," as a common public space where politics, games, culture, banks, and markets are situated adjacent to each other. Certainly the Internet as a service must be paid for, but this payment may be secured in many ways. Selling and buying is only one aspect of such an agora, attractive enough to establish and permeate the existence of the Net. The transfer of money and goods is only one aspect of exchange between people among many others. Between the commercial use of advertising to the noncommercial uses of discussion groups and blogs, the global Net is a space for learning, play, work, entertainment, and many other not yet invented purposes. As the Internet grows steadily as marketplace, it becomes more obvious that this market depends on a stable infrastructure. But the infrastructure is much like its material counterparts of streets and places, which are not necessarily only spaces for trading goods. Its value increases with additional noncommercial community uses, much as the ambiance of a city is not defined primarily by its shopping malls. The building of communities, global and local, is arguably the most important promise of an Internet to be inhabited by citizens who may sometimes be clients and consumers, but remain at all times users of this twenty-first-century century global infrastructure.

Notes

1. Helmut F. Spinner, *Die Wissensordnung* (Opladen: Leske + budrich, 1994), 48f.; Robert K. Merton, "The Normative Structure of Science," in *The Sociology of Science* (Chicago: University of Chicago Press, 1974), 267ff.

2. Mod.ki was a moderated German newsgroup on *Künstliche Intelligenz* (artificial intelligence).

3. Michael Hauben, Ronda Hauben, and Thomas Truscott, *Netizens: On the History and Impact of Usenet and the Internet.* (Los Alamitos: Wiley–IEEE Computer Society Press, 1997).

4. This is, however, not the oldest news; a newsgroup from April 1980 can be found at ⟨http://communication.ucsd.edu/bjones/Usenet.Hist/Nethist/0061.html⟩. Older ones may also exist.

5. Laurence A. Canter and Martha S. Siegel, *How to Make a Fortune on the Information Superhighway: Everyone's Guerrilla Guide to Marketing on the Internet and Other On-line Services* (New York: Harper-Collins, 1994).

6. *Computer Underground Digest* 6, no. 83 (September 21, 1994). ⟨http://venus.soci.niu.edu/~cudigest/CUDS6/cud6.83⟩.

7. Ronald J. Vetter, Chris Spell, and Charles Ward, "Mosaic, HTML, and the World Wide Web," *IEEE Computer* 27, no. 10 (October 1994).

8. Taken from a survey of one hundred thousand servers, available at ⟨http://news.netcraft.com/archives/web_server_survey.html⟩ (accessed November 11, 2006).

9. Hotline was published in 1996, and gradually closed down after management disputes.

10. Peter Biddle, Paul England, Marcus Peinado, and Bryan Willman, "The Darknet and the Future of Content Distribution," available at ⟨http://msl1.mit.edu/ESD10/docs/darknet5.pdf⟩.

11. ⟨http://slashdot.org⟩ ("aitch tee tee pee colon slash slash slash dot dot org").

12. "Free software is a matter of liberty, not price. To understand the concept, you should think of free as in free speech, not as in free beer" (⟨http://www.gnu.org/philosophy/free-sw.html⟩ [accessed November 5, 2006]).

13. ⟨http://www.fbi.gov/majcases/fraud/internetschemes.htm⟩ (accessed November 11, 2006).

14. In Germany, the sale or even use of fascist symbols, flags, and uniforms is forbidden and punishable by prison sentences of up to three years. Similar laws hold for Austria. In these cases, free speech is not considered to be the dominant value. The denial of the gas chambers at Auschwitz is liable to prosecution in many European countries, among them Austria, Belgium, France, Germany, Luxembourg, Switzerland, and the Netherlands—but not in the United States or United Kingdom.

15. Konrad Zuse, *Der Computer—mein Lebenswerk* (Berlin: Springer, 1977).

16. Larry Sanger, former editor in chief for Nupedia and the person who baptized the new project "Wikipedia," has recently proposed a new edited encyclopedia project called "Citizendium" (initially as a fork to Wikipedia) at the Wizard of OZ conference, Berlin, September 15, 2006 (⟨http://www.wizards-of-os.org/⟩).

17. ⟨http://www.alexaradar.com/⟩.

18. Data from November 2006.

17 Conclusions

William Aspray and Paul E. Ceruzzi

When the journal *Annals of the History of Computing* was founded in the late 1970s, its editorial board established a rule that it would not publish accounts of computer history unless the events described occurred more than fifteen years before publication.[1] The contributors to this volume, many of whom have also contributed to that journal, have obviously relaxed that rule. Of all the chapters, perhaps only chapter 2 would satisfy the *Annals'* fifteen-year rule; some would not even satisfy a fifteen-day rule! When we began this project, we knew that we would be unable to get historical distance from the events of the Internet's commercialization. But we decided to proceed anyway. The topic is too important. And for the past two decades the pace of computing history has sped up so much that the *Annals'* rule now seems quaint, and it may even have lost its utility. We have taken on this study mainly because we believe that we bring to the topic a set of scholars' tools that enable us to examine more critically the events in question, even if they occurred so recently. The popular and financial press and news media, not to mention numerous online journals and blogs, are all busy reporting these topics. With our training, we hope to provide added value to these accounts, by drawing on comparisons with the past and using the levers of economic, business history, and other related disciplines. Perhaps the events of the next few years will quickly render our work obsolete, but the topics of each chapter all point to a need for further study, and we would consider this book a success if others take up these topics and elaborate, support, or refute what we have said here.

The preceding chapters cover a wide range of activities and events. Of all the themes they introduce, the one that stands above the rest is that of Internet entrepreneurs searching for a viable business model upon which to base their offerings and with which they can make money. Wolfgang Coy's chapter emphasizes the noncommercial forces that have driven the Internet, but it does no disservice to his observation to note that nearly every other chapter portrays a tortuous journey through a set of challenges and opportunities to make money, somehow. During the Internet bubble in the late 1990s, many believed that the laws of economics no longer held. This was not in fact true, as Carl Shapiro and Hal Varian show in their book, *Information Rules*.[2] David

Kirsch's chapter reminds us of the legendary stories of dot-com startups hoping to make a fortune selling or delivering products that no one wanted; but the advent of the Internet presented a host of new opportunities that remain to this day.

The Internet did change the cost to market one's products or services to a global audience. That cost has become much lower. One can spend a lot of money marketing one's Web site, but many sites have arisen and flourished through word-of-mouth (the new buzzword is "viral") marketing, simply because they offer a specialized service that many people want. In the heart of lower Manhattan one can still find specialized shops that cater to niche markets, although not as many as there were twenty years ago. Similar neighborhoods exist in London, Tokyo, and Hong Kong. Few other places in the world offer such variety. It is not that no one outside New York City or London wants what these shops sell, it is that specialized shops did not have the resources to market themselves widely and transactions were often carried out in person, creating a geographic limitation on their market. They could only survive in the densest of cities. Now they can market themselves to the world—and many do. Likewise, the cost of setting up servers to handle classified advertising, as Craigslist does, is not trivial, but it is much lower than the corresponding costs of buying classified ads in local newspapers. Blogs are a lot cheaper than mimeographing and mass-mailing pamphlets, or printing and handing out broadsides, even if a blog makes it no easier to write well. In their chapter on news media, Christine Ogan and Randall Beam show how threatening this cost structure has already become to traditional news organizations, and how a search is ongoing for a business model that will enable them to survive.

Related to this opportunity is the ability of well-designed Web sites to transform the advertising world. Advertising has always been a black art. How many people remember the slogan "Where's the beef?" but do not remember what product was being advertised? The Internet offers a way to target ads carefully to those who might respond to them, and to let advertisers know precisely whether their ad campaign is working or not. Tom Haigh shows how Google, drawing on some earlier work, refined these techniques to a high art, making it such a successful site. Google's success partly answers the question about the search for a viable business model: it is a reliance on targeted advertising revenue. We are reluctant to draw that conclusion, however, as this phenomenon has occurred so recently. Here is an instance where one needs historical perspective. James Cortada relates a similar story of how the Internet has affected what might be called the internal marketing for businesses, namely the way suppliers and customers learn of each other's needs and offerings, in the design and manufacture of a product, before it is marketed to the consumer.

It is no surprise that these opportunities were accompanied by challenges to traditional ways of doing business, as Jeffrey Yost, Nathan Ensmenger, and other authors have shown. It is not only the obvious example of newspapers that are threatened.

Threats are found wherever one finds people or professions that mediate between producer and consumer. They are also found in the internal workings of a business, as Martin Campbell-Kelly, Daniel Garcia-Swartz, and James Cortada show. Doing business anywhere requires a level of trust, and the Internet is no exception. When there is no face-to-face contact, when one places an order with a credit card and the order is processed by a server with no human input, how is trust maintained? We have seen that the giants of Internet commerce, eBay and Amazon, go to great lengths to prevent fraud and to gain the trust of their users. Theirs is an ongoing battle, never to end. As Tom Haigh shows, the prevalence of "spam," "phishing," viruses, and denial-of-service attacks are serious threats. He further reveals how these have been enabled and are hard to eliminate because of technical decisions made long ago, when the Internet and Web were being designed. Ward Hanson's chapter on the response of bricks and mortar establishments illustrates further the many challenges that retail stores had to face, trying to meet the challenge of the Web while also facing the more immediate and tangible threat of "big-box" stores like Wal-Mart entering their territory.

One of the most fascinating aspects of Internet commerce to come out of these chapters is the way users selectively adopt, or reject, the new medium. Nathan Ensmenger shows how doctors use the Web but eschew e-mail, while university professors do the opposite. At the extremes are technophiles who adopt every new thing and neo-Luddites who reject it, but most people find themselves like the professionals in Ensmenger's chapter, carefully if unconsciously weighing the advantages and threats to using new ways of communicating and doing business. Jeff Yost's analysis of the travel industry illustrates another facet of this phenomenon: how in the face of booking trips on the Web, customers continue to use travel agents for specialized services, like ecotourism, the high commissions for which offset the agents' loss of traditional bookings. These two chapters further reveal how Internet commerce, while on one hand threatening those people who made a living mediating between producers and consumers—for example, an airline and travelers—on the other hand gives rise to *new* intermediaries such as Orbitz.com, which did not exist in the pre-Web era. The companies that supply software over the Internet, described by Campbell-Kelly and Garcia-Swartz, may also be considered a variant of this phenomenon of the rise of new intermediaries.

The Internet has not overturned economic laws, but it has put an enormous strain on the traditional body of case, constitutional, and international law. Issues that had been worked out over decades, even centuries, in the United States and in world courts have suddenly been thrown into chaos. These include, at minimum, issues of copyright, trademarks, contracts, the First Amendment, and jurisdiction. William Aspray's chapter on music file sharing reveals not only how chaotic things have become, but also how quickly the Internet overthrew methods of distribution and payment that had evolved over many decades. Blaise Cronin's chapter on pornography likewise reveals the chaos of an industry that, although shadowy and secretive, had nevertheless built a stable relationship with the courts and law-enforcement agencies as to what it could and could not sell, in what neighborhoods it could open shops, and to what ages of customers it could market its products.[3] All of that is thrown into question. Add to this mix the existence of off-shore Web sites, in countries that are nearly impossible to locate on a map, where purveyors of music, pornography, gambling, and so forth are welcome. This is the reverse side of the coin of the tiny shop in Tokyo's crowded electronics district selling things that one cannot buy anywhere else—now it is an Internet supplier in a tiny, inaccessible country selling things to customers all over the world.

While preparing this volume over the summer of 2006, the editors and many of the chapter authors met for a three-day workshop in Munich to discuss the project. The issue of how much this story is an exclusively American one, and how little it includes European contributions, naturally came up. There are obvious restrictions to the scope of this study, it is only a beginning, and we did not feel able to include a European or international perspective. However, the emergence of a commercialized Internet is an international story, and we hope that future scholars will bring that to light.

In writing any history one must avoid the fallacy that because things turned out a certain way, they had to turn out that way. Shane Greenstein's chapter reveals the unique nature of the U.S. telecommunications industry, in which the voice telephone network was controlled by a regulated monopoly (AT&T), but it was not under direct government control, as was the case in many European nations. That allowed Internet Service Providers to emerge in between the rules of telecom regulation—something that could not have happened as easily elsewhere. Is this a post hoc explanation? Perhaps, but if AT&T had been either a government agency, or on the other had been a completely deregulated business, cyberspace would operate very differently today. One could also note the reverence to which Americans hold freedom of "the Press," guaranteed by the First Amendment, although in practice its application is complicated.[4] The United Kingdom, for example, has no equivalent—might this also be a factor? Some countries, such as Germany, have strict privacy laws that prevent the gathering of customer data that is so dear to Internet advertisers; other countries, such as China, have almost opposite notions of privacy—placing the needs of the government over the privacy of individuals. This is a subject for further research. Another topic we do not cover in depth is the way the Internet may be used by the poorer countries of the world. A Web site such as Amazon, whose profits come in part from affluent Americans making impulse purchases with discretionary income, will not translate well. Yet why could a site such as eBay not allow the citizens of poorer nations to obtain basic goods at cheap prices? Atsushi Akera's chapter, on the use of the Web to build communities, speaks in part to this question, but much more work needs to be done on this topic. One major factor that gives the commercial Internet its American stamp is suggested in Paul Ceruzzi's and Tom Haigh's chapters. That is the roots of the Internet in projects sponsored by the U.S. Defense Department's Advanced Research Projects Agency (ARPA). Ceruzzi describes the ascendance of the TCP/IP protocol over the internationally supported Open Systems Interconnect (OSI) protocols, in part due to ARPA influence. He begins with the myth of the Internet's invention as a way of communicating during a nuclear war. A more subtle version of that story is that the modern Internet reflects the United States' desire to maintain and monitor global information flows, as the world's remaining superpower.[5] From this perspective, it is no surprise that much of Internet management takes place in northern Virginia, near the Pentagon, or that the U.S. Commerce Department resisted efforts to place the Domain Name System and its root servers under United Nations jurisdiction.

What Next?

The authors of this study are no better than any other group in their ability, or lack thereof, to predict the future. To return to the scholars' "toolkit," we offer an ability to place the Internet's commercialization in the context of similar ways that other media evolved and found a viable business model in the early twentieth century. The history of commercial radio and television is especially relevant, as scholars have explored how those media developed a relationship with advertisers, entered into contracts with musicians, writers, artists, and actors, how they threaded a regulatory maze with local and national governments, and so on.[6] One of Marshall McLuhan's most memorable observations was how new media typically use the content of an older media in their formative years. Commercial Web pages seem to combine the features of glossy magazines, MTV, and other older media. But clearly the Web is finding its own place.

As this is being written, the Internet seems to be heading toward social networks, or to use the buzzword, "Web 2.0." Newly dominant sites include MySpace, YouTube, and Friendster. Atsushi Akera's chapter describes this phenomenon's deep roots, of which few visitors to these sites are aware. It is possible that social networks will indeed become the dominant paradigm of the Internet in the coming decade. The Internet is also seeing a new level of sophisticated games and programs that take role-playing to a deeper dimension: Second Life and Live Journal are in the news as of this writing. As before, the creators of these sites are searching for a viable business model, even if at first some of them are not interested in making money at all. For these sites, the question is not "Where's the beef?" but "Where's the money?" These sites may not last—already there are reports that teens are abandoning MySpace.[7] But Akera's chapter, and the discussion of AOL and CompuServe by Ceruzzi and Haigh, suggest that these trends are not a flash in the pan, but rather are integral to the fundamental notion of computer-mediated communication networks, which began before the Internet was invented.

One can also look to science fiction writers as well as to the more visionary computer scientists for a glimpse of the future: books by Vernor Vinge, David Gelertner, or Ray Kurzweil. The similarity of sites like Second Life to Gelertner's "Mirrorworlds," which he described in 1992, is uncanny.[8] For now, we must leave that speculation to those who do it better, with one caveat that brings us back to the theme of this book: namely, a search for a viable business model, one that exploits the advantages of networking while remaining mindful of the basic laws of economics. Some of those laws might include the basic human need for communication and community, the use of communications technologies to satisfy those needs as well as to bring buyers and sellers together, the fundamental need of a business to find a customer, the well-known process of "creative destruction" that accompanies new technologies, and the observation attributed to P. T. Barnum that "there is a sucker born every minute."

Shapiro and Varian's study of *Information Rules* reminds us of what we have already mentioned: that the laws of economics are still valid, although the coefficients of the variables that express those laws have changed. Those who lived through the dot com collapse of 2000–2001 should have learned that lesson, and the proponents of "Web 2.0" may have to learn it anew. In any event, the coming years promise to be as exciting and dynamic as those just past. The authors of this volume will consider our work a success, not if we predict the future—we cannot—but if we provide a framework for understanding how the future of Internet-based commerce will unfold.

Notes

1. Since 1992 the *IEEE Annals of the History of Computing*. In the first issue, the late Bernie Galler, editor in chief, stated, "To make sure that our material can be placed in some historical perspective, we stipulate that it be concerned with events and developments that occurred at least fifteen years before publication, which is, of course, an intentionally moving target." (Vol. 1 # 1, July 1979, p. 4).

2. Carl Shapiro and Hal Varian, *Information Rules: A Strategic Guide to the Network Economy* (Boston, Harvard Business School Press, 1998).

3. As well-known as the "Where's the beef?" phrase is Supreme Court Justice Potter Stewart's definition of pornography as "I know it when I see it."

4. The term "the Press" is a synecdoche for news gathering and dissemination, again subject to interpretation by the courts.

5. Compare the current role of the United States to the United Kingdom and undersea telegraphs, in Daniel Headrick, *The Invisible Weapon: Telecommunications and International Politics, 1951–1945* (New York, Oxford University Press, 1991).

6. For example, Susan Douglas, *Inventing American Broadcasting* (Baltimore, Johns Hopkins University Press, 1987).

7. This observation is based in part on newspaper accounts as well as a personal observation of one of the author's teenage children. Teenagers seem to feel about MySpace the way Yogi Berra felt about a restaurant in New York: "Nobody goes there any more; it got too crowded."

8. David Gelertner, *Mirror Worlds: or the Day Software Puts the Universe in a Shoebox ... How it will Happen and What it Will Mean* (New York, Oxford University Press, 1992).

Contributors

Atsushi Akera is an assistant professor in the department of science and technology studies at Rensselaer Polytechnic Institute. He conducts research on the early history of scientific and technical computing in the United States and the history of invention and innovation. He has recently published *Calculating a Natural World: Scientists, Engineers, and Computers During the Rise of U.S. Cold War Research* (MIT Press, 2006).

William Aspray is Rudy Professor of Informatics at Indiana University in Bloomington. He conducts research on the history, policy, and social study of information technology. Recent books include *Globalization and the Offshoring of Software* (ACM, 2006, coedited with Moshe Vardi and Frank Mayadas) and *Women and Information Technology: Research on the Reasons for Under-representation* (coedited with Joanne McGrath Cohoon, MIT Press, 2006), and a mass-market history of the computer, *Computer: A History of the Information Machine* (coauthored with Martin Campbell-Kelly, Westview, 2004).

Randal A. Beam is an associate professor of communication at the University of Washington in Seattle. His research focuses on media management and the sociology of news. Recent publications include *The American Journalist in the 21st Century*, a book that he coauthored with David H. Weaver, Bonnie J. Brownlee, Paul S. Voakes, and G. Cleveland Wilhoit; and "Quantitative Methods in Media Management and Economics," in the *Handbook of Media Management and Economics*.

Martin Campbell-Kelly is a professor in the Department of Computer Science at Warwick University and a consultant with LECG. He has broad interests in the history of computing, and has recently undertaken studies of the software industry and associated intellectual property issues. His publications include *From Airline Reservations to Sonic the Hedgehog: A History of the Software Industry* (MIT Press, 2003) and *Computer: A History of the Information Machine* (2nd edition, Westview, 2004, coauthored with William Aspray).

Paul E. Ceruzzi is curator of aerospace electronics and computing at the Smithsonian Institution's National Air and Space Museum in Washington, DC. Since joining the Smithsonian he has worked on a number of exhibitions relating to the history of space exploration, most recently "The Global Positioning System: A New Constellation." In addition to museum work, he has published numerous articles and papers on modern technology in popular and scholarly journals, and he is an associate editor of the journal *IEEE Annals of the History of Computing*. Among his recent books are *Smithsonian Landmarks in the History of Digital Computing* (1994, with Peggy Kidwell); *A History of Modern Computing* (MIT Press, 1998), and *Beyond the Limits: Flight Enters the Computer Age (1989)*. *Internet Alley: High Technology in Tysons Corner*, 1945–2005 (MIT Press) will appear in 2008.

James W. Cortada is a member of the IBM Institute for Business Values. He conducts research on the history and management of information technology, with particular emphasis on its effects in business and American society at large. Recent publications include *Making the Information Society* (Pearson, 2002), which discusses the role of information in American society, and a three-volume study of how computing was used in over forty industries over the past half century, *The Digital Hand* (Oxford University Press, 2004–2007).

Blaise Cronin is the Rudy Professor of Information Science at Indiana University, Bloomington, where he has been dean of the School of Library and Information Science since 1991. His principal research interests are scholarly communication, citation analysis, and scientometrics. He is editor of the *Annual Review of Information Science and Technology*. Recent books include *The Hand of Science: Academic Writing and its Rewards* (Scarecrow, 2005).

Wolfgang Coy is professor of informatics and dean of the faculty for mathematics and science at Humboldt-Universität zu Berlin. He is the German representative on the Computers and Society technical committee of the International Federation for Information Processing (IFIP). His research focuses on computers and society, computers and culture, and digital media. Recent publications include two volumes of *HyperKult* (with Christoph Tholen and Martin Warnke), and articles on the history of digital storage, the status of the information society, and intellectual property.

Nathan Ensmenger is assistant professor of history and sociology of science at the University of Pennsylvania. His research addresses the intersection of information technology, organizational theory, and the sociology of work and professions. Recent publications include "Letting the 'Computer Boys' Take Over: Technology and the Politics of Organizational Transformation," *International Review of Social History* (2003) and "The 'Question of Professionalism' in the Computing Fields," *IEEE Annals of the History of Computing* (2001). He is currently working on a book on the history of software development.

Daniel D. Garcia-Swartz is a senior managing economist in the Chicago office of LECG. He holds a joint PhD in economics and history from The University of Chicago. For many years he has been conducting research on the economics of high-technology industries—among them, electronic payment instruments, computer software, the Internet and e-commerce, and embedded computing components. His publications include "Economic Perspectives on the History of the Computer Timesharing Industry," *IEEE Annals of the History of Computing* (with Martin Campbell-Kelly, 2007), "Open Source and Proprietary Software: The Search for a Profitable Middle Ground" (with Chris Nosko and Anne Layne-Farrar, at http://www.techcentralstation.com, April 2005), and "The Failure of E-Commerce Businesses: A Surprise or Not?" *European Business Organization Law Review* (with David Evans and Bryan Martin-Keating, 2002).

Brent Goldfarb is an assistant professor of management and entrepreneurship at the University of Maryland's Robert H. Smith School of Business. Goldfarb's research focuses on how the production and exchange of technology differs from more traditional economic goods, with emphasis on the implications of the role of startups in the economy. He focuses on such questions as how do markets and employer policies affect incentives to discover new commercially valuable technologies, and when is it best to commercialize them through new technology-based firms? Why do radical technologies appear to be the domain of startups? And how big was the dot-com boom?

Shane Greenstein is the Elinor and Wendell Hobbs Professor of Management and Strategy at the Kellogg School of Management, Northwestern University. He has published numerous articles on the growth of commercial Internet access networks, the industrial economics of computing platforms, and changes in communications policy. He has been a regular columnist and essayist for *IEEE Micro* since 1995. Recent books include *Diamonds are Forever, Computers are Not* (Imperial College Press, 2004), and an edited volume *Standards and Public Policy* (with Victor Stango, Cambridge University Press, 2006).

Thomas Haigh is an assistant professor in the School of Information Studies of the University of Wisconsin-Milwaukee and a partner in the Haigh Group, providing historical interviewing and research services. His research focuses on the history of the use and management of information technology in business, on the history of software technologies and the software industry, and on knowledge practices in online communities. He edits the biographies department of *IEEE Annals of the History of Computing* and chairs the Special Interest Group on Computers, Information, and Society of the Society for the History of Technology. Haigh has published articles on the scientific office management movement of the 1910s and '20s, the early history of data processing in corporate America, the management information systems movement of the 1960s, the creation of the packaged software industry, the trade association ADAPSO and its role in the evolution of the software and services industry, the origins of the database management system, and word processing technologies of the 1970s.

Ward Hanson is a member of the Stanford Institute for Economic Policy Research, where he is a fellow and the policy forum director. He analyzes the economics and marketing of new technology. Other areas of interest include the role of competition, policy issues involving the evolving role of interactivity, and optimal product line pricing. As a pioneer in studying the commercialization and impact of the Internet, he has published research articles, a leading text, and created online courses on Internet marketing.

David Kirsch is assistant professor of management and entrepreneurship at the University of Maryland's Robert H. Smith School of Business. His research interests include industry emergence, technological choice, technological failure, and the role of entrepreneurship in the emergence of new industries. He has published on the automobile industry, including *The Electric Vehicle and the Burden of History* (Rutgers University Press, 2000) and articles in *Business History Review* and *Technology and Culture*. Kirsch is also interested in methodological problems associated with historical scholarship in the digital age. With the support of grants from the Alfred P. Sloan Foundation and the Library of Congress, he is building a digital archive of the dot-com era that will preserve at-risk, digital content about business and culture during the late 1990s.

Christine Ogan is professor of journalism and informatics at Indiana University. She has been researching and teaching about information technologies for thirty years. Currently she is teaching a course titled the Internet and the Future of the Media. Her most recent publications include "Communication Technology and Global Change," in C. Lin and D. Atkin, eds., *Communication Technology and Social Change: Theory, Effects and Applications* (Lawrence Erlbaum, 2006); "Gender Differences among Students in Computer Science and Applied Information Technology," (with Jean Robinson, Manju Ahuja, and Susan Herring) in Joanne McGrath Cohoon and William Aspray, eds., *Women and Information Technology: Research on the Reasons for Under-representation* (MIT Press, 2006); and "Confessions, Revelation and Storytelling: Patterns of Use on a Popular Turkish Web Site," (with Kursat Cagiltay) in *New Media & Society*, August, 2006.

Jeffrey R. Yost is associate director of the Charles Babbage Institute at the University of Minnesota. He conducts research on the business, social, and cultural history of computing, software, and networking. Recent publications include *The Computer Industry* (Greenwood, 2005); "Computers and the Internet: Braiding Irony, Paradox and Possibility" in C. Pursell, ed. *A Companion to American Technology* (Blackwell, 2005); and "Maximization and Marginalization: A Brief Examination of the History and Historiography of the Computer Services Industry" *Entreprises et Histoire* (November 2005).

Index

Abbate, Janet, 3, 106, 110
ABC (television network), 298
About.com, 289
Academia, 374, 541–542. *See also* Universities
Academic addresses, 33–34
Academic researchers, 107, 147
Acceptable Use Policy, 27, 28, 29, 31, 34, 38, 543
Access, Internet, 394; among U.S. households, 98n. 31; availability of, 69–70, 223; first advertised prices for commercialized, 49; locations lacking, 70; and telemedicine, 358
Account information, spammers' access to, 121
Ackman, Dan, 513, 515
Active Desktop, 141
ActiveX programs, 145
Adams, Rick, 28
Address, e-mail, 33, 116, 120. *See also* Domain name analysis
Adolescents: and Napster, 469; use of media by, 304
AdSense, Google's, 186
Adult entertainment industry, 501, 503, 504; corporatization of, 521; customization in, 505–506; Internet in, 492, 532; in mainstream media, 495, 506; major makeover of, 521–522; major players in, 522–532; media substitution in, 499–502; statistics for, 508–515; structure of, 520–522
Advanced Audio Coding (AAC), 459–460

Advanced Network and Services (ANS), 53, 65, 66
Advanced Research Projects Agency (ARPA), U.S. Dept. of Defense, 3, 9, 107, 111, 561; Information Processing Techniques Office of, 318; and Internet development, 22; as source of protocols, 31; and TCP/IP, 11–12
Advertising, 293, 558; commercial use of Internet for, 554; on community-based sites, 553; decline in magazine, 286; to finance Internet, 548; of media organizations, 296; search-related, 297
Advertising, Internet, 119, 160, 170–174, 171, 184, 295; and customer profiling, 172; and dot-com crash, 180; early mass, 543–544; Google's, 184; with instant messaging programs, 123; Internet search, 191; on navigation sites, 159; pay-per-click, 173, 186; preferential treatment in, 173; "pushed," 296; rising revenues in, 185; syndicated, 191; via spam, 121; on Web, 170–174, 186; Yahoo's first, 166
Aimster, 470
Airline reservations industry, 317, 320–321, 322–323, 400. *See also* Travel reservations industry
Airlines, direct sales model of, 319
Ajax (Asynchronous JavaScript and XML), 146, 216
Akimbo, 299
Alcatel, in early server market, 59

Alexander, Kelly, 479

Alex.com, 181, 552

Alibris, 397

Allaire, ColdFusion produced by, 134

Allstate, 375

AltaVista, 159, 167, 169, 174, 181, 185, 190

Amazon.com, 133, 191, 233, 260, 315, 397, 412, 435; and customers' buying patterns, 411; and fraud, 559; Get Big Fast approach of, 261, 263–265, 272; origins of, 343n1, 408, 410; and social dynamics, 426

American Airlines, 318, 319

American Express, 318

American Idol franchise, 279

American Lawyer Media, 429

American Library Association, 461

AmericanSingles, 436

America Online (AOL), 17, 37, 38, 67, 177; advertising policy of, 177; as bulletin board operator, 57; chat rooms of, 17, 122; community orientation of, 426; competition with Microsoft, 99n45; converted services of, 427; customer services of, 188–189; early e-mail services offered by, 113; e-mail gateways of, 114; emerging operations of, 65, 130; flat-rate pricing of, 75; full Internet access offered by, 130; instant messaging applications, 68; Internet Explorer of, 68, 140; Kayak investment of, 324; in market for home Internet users, 131; market share of, 99n43; and media organizations, 280; Netscape acquired by, 143, 157n110, 544; Nullsoft acquired by, 464; online advertising revenues of, 172; pricing scheme of of, 427; restructuring of, 88; software promotion by, 154n73; travel service of, 322; in U.S. market, 119; volunteer network of, 427

Anderson, Chris, 244

Annals of the History of Computing, 557

Anticircumvention legislation, 461

Antipornography legislation, 501

Antispam industry, 121–122, 215

Antitrust law, enforcement agencies for, 56–57

Anycasting technique, 37

AOL Europe, 468

AOL Instant Messenger (AIM), 122–123, 552

AOL Time Warner, 143, 180

Apache software, 63, 70, 131–132, 544

Apollo, United Airlines, 318, 322

Apollo Group, Univ. of Phoenix owned by, 371

Apple Computer, 332; AAC of, 435, 460; iChat software of, 123; iPods sold by, 410; iTunes of, 469; iTunes Store of, 474–476, 546; record companies signed on with, 474

Application Service Providers (ASPs), 209–213; customization service provided by, 212–213; hosting services of, 211; and Internet bubble, 212; pure-play, 210–211

Arbitron research, 287

Architext, 170

ARPANET, 4, 9, 10, 11, 13, 26, 38, 107, 110; access to, 15; compared with commercialized Internet, 13–14; design for, 107; early history of, 105, 497–498; experimental connections of, 21; newsgroups from, 542; original nodes of, 368; replaced by Internet, 111; as research network, 21; TCP/IP protocols of, 12; transition from, 23; unsolicited commercial e-mail sent on, 120

ARTISTdirect, 468

Ask Jeeves service, 170, 174

ASP Industry Consortium, 212

Assets-Based Community Development (ABCD), 442–443

Assortment, for Internet retailers, 244–245, 248, 256n36

Asynchronous JavaScript + XML (Ajax), 146, 216

AT&T, 10, 87; court-ordered breakup of, 24–25; early e-mail services of, 113; emerging operations of, 65; monopoly of, 560; national fiber network of, 54; restructuring of, 88; as retail Internet provider, 69; as tier-1 provider, 66; Worldnet service of, 65, 75

Auction-hosting site, 246. *See also* eBay

Auction houses, online, 234, 246, 248, 257n52, 549

Auction system, 173, 327–328
Audience: decline in print, 293; estimating size of, 286–287; and news creation, 300–301
Audio Home Recording Act (AHRA) (1992), 456, 457, 460, 466
Australian Record Industry Association (ARIA), 472
Authentication, problem of, 363
Automatic Data Processing (ADP), 202–203, 219
Automatic Payrolls, Inc., 203
Automobile industry, 271, 376, 395, 398

Backbone firms, 12, 66, 69, 80; as commercial resource, 28; NSF-supported, 25; overbuilding of, 71; public and private, 52; reselling capacity of, 74; run by commercial entities, 38; speeds upgraded for, 23–24
Backbone Network Service, 31
Baidu Chinese search engine, 159, 552
Ballmer, Steve, 142, 156n101
Bands, and fan sites on Internet, 479
Bandwidth, 543
Banking industry, 549; online, 400, 408, 419n14; outsourcing in, 402; telecommunication use in, 393
Bank of America, E-Loan investment of, 330
Baran, Paul, 11
Barthes, Roland, 503
Baud, use of term, 15
BBC, audience participation area of, 302
Beam-It service, 465
Beastie Boys, 458–459
Beate Uhse Ag., and pornography, 502
Bebo.com, 304
Bell, C. Gordon, 18, 23
Bell Labs, AT&T, 10, 19
Bellovin, Steve, 542
Benioff, Marc, 201, 214
Bennett, Frederick, 371
Berkeley Software Distribution (4.2BSD), 19
Berners-Lee, Tim, 61, 62, 124, 125, 126, 132, 133, 162–163, 165, 167–168, 191

Bertelsmann Music Group (BMG), 467–468, 469
Best Buy, 249, 258n64
Betamax Case, Sony, 456–457
Bezos, Jeff, 233, 260, 263, 264, 265, 432
Bhatia, Sabeer, 118
Bibliometrics, 183
Bicycle messenger services, 342
Big Blue, 317, 332
Biotech firms, B2B of, 400
BITNET, 20, 21, 25, 38, 112
BitTorrent, 452, 546
Blackberry, 85
Blacksburg Electronic Village, 424, 426
Bloch, Eric, 23
Blogs, 120, 303, 304, 498; cost structure for, 558; journalistic value of, 301
BN.com, 468
Board games, early, 550
Boczkowski, Pablo, 281, 282
Boles, David, 469
Boo.com, 261
Book dealers, secondhand, 397. See also Amazon.com
BookLine, purchased by AOL, 131
Boston, Internet diffusion in, 58
Boston Globe, 290
Boucher, Rep. Rick, 29–30, 465
Bowie, David, 458
Bowman, Shayne, 300, 303
Brand, Stewart, 14, 16, 18, 426
Brands, 330; marketing value of, 341; online retailing, 233, 243
Braun, Lloyd, 301
Brin, Sergey, 182, 185
Broadband connections, 35, 64–65, 455–456; for AOL's customers, 188; availability of, 394; and computer games, 550; home, 545; of U.S. households, 98n32
Broadband Internet access firms, 81, 88, 89
Broadband servers: emergence of, 49, 81–83
Broadcast media: business models of, 295–296; impact of Web on, 297; online, 282, 291

Broadcast news, decline of, 285

Broadvision, 211

Broadwatch, 54

Brokerage industry, 398; declining number of brokers in, 407, 420n30; effect of Internet on, 405–408; first online Internet-based services in, 407; fixed commissions in, 406; rise in number of customers for, 406–407; transaction-based offerings of, 415; work-flow redesign in, 392

Brokerage services, 423, 430–435

Brown, James, 239, 252

Brown, John Seely, 371, 374

Browsers, 136–138; Mosaic; Navigator; diffusion of, 64; early, 126, 128; improvement of, 76; line mode text-only, 153n. 54; Mozilla-based, 145; Netscape's share in market, 144, 157n114; for online retail, 212; software, advances in, 130; Unix version of, 144; wars, 138–144, 176; widespread use of, 127, 129–131. *See also* Internet Explorer

Bryan, J. Stewart III, 290

Buddy list, 99n44, 122

Bug Juice, 467

Bulletin board operators, 55–58, 58

Bulletin board system (BBS), 112, 542; and regulatory policy, 56; sharing of files on, 468

Buma/Stemra (Dutch music publisher), 471

Bundling, of media organizations, 296–297

BurnITFIRST.com, 467

Bush, Vannevar, 162

Bush (George W.) administration, 90, 142, 157n105, 503

Business directories, 173

Businesses: connection approaches for, 221; and e-commerce, 252; and "free" Internet, 547; information-dependent, 403; information retrieval in, 160; internal operations of, 401–404, 558; and Internet, 148; Internet channels added by, 395; and Internet in, 85–86; and Internet infrastructure, 131; and Internet of, 392; new online opportunity for, 409–412; online customers of, 400; proprietary networks used by, 393–394; storefronts *vs.* Internet models of, 341; time-sharing access of, 225n20; Web sites of, 163. *See also* Entrepreneurs; Industry

Business models, 292, 391, 412–414

Business Plan Archive (BPA), 247n17, 266

Business process outsourcing, 339

Business Software Alliance, 546

Business-to-business market, 299

Cable Act (1992), 460

Cable companies, 77, 81, 88, 98n32

Cable technology, 64, 221, 222, 229n73

Cable television, 284, 298, 305

Cactus Data Shield, 485n53

Call centers, 336, 338, 339

Campbell-Kelly, Martin, 417, 559

Canter, Laurence A., 120, 543

Capital markets, and commercial Internet, 269. *See also* Market, Internet

Capitol Records, early Internet use of, 458–459

CarFinance.com, 330

Cars.com, 293

Case, Steve, 17, 33, 427

Case management services, 361–362

Catalog retailers, 243, 395

CBS (television network), 298

CCITT (international federation of telecommunications carriers), 114

CDNow, 233, 253n2, 468

CD-ROMs, interactive, 459

Cell phones, 398; text messaging on, 123; video presentations created by, 300; Web-enabled, 253

Centrino program, 84

Cerf, Vint, 25, 27, 37

CERN, 124, 125, 126, 165

Chain stores, 233–234, 235, 249

Chandler, Alfred D. Jr., 398, 399, 416

Charter, Dennis, 549

Chat rooms, 17, 122, 402, 498

Chaum, David, 549

Chemical industry, B2B of, 399

Chemistry.com, 437

Chess, computer, 550

Chiariglione, Leonardo, 462

ChicagoTribune, 280

Child Online Protection Act (1998), 519

Children's Internet Protection Act, 519

China, 337–338, 341, 343, 560

Chinese instant messenger service, 552

Chomskytorrent.org, 546

Christmas season, and business, 263–264

CICS, 417

Cinemania film review archive, 136

Circuit City, 249, 258n64

Cisco (equipment supplier), 59, 86, 413

Citicorp, 401

"Citizendium," 555n16

Citizen journalists, 300, 304

Clark, Jim, 61, 354, 365

Clarke, Ian, 473

Classroom, computers in, 371, 372, 373–374, 375. *See also* Universities

Clerical positions, decline in numbers of, 403

Cleveland Free-Net, 424

Click fraud, 186–187

Clinton administration, 62, 79–80, 408–409

Club Jenna, Inc., 522, 529, 530

CNET, 176

CNN (television network), 279, 280, 285, 288, 290

Coase, Ronald, 343n3

Cohen, Bram, 545

ColdFusion, 134

Cold war, and origins of Internet, 9

College credit, online courses for, 374

College professors, Internet use of, 368–369. *See also* Professors

Comcast, AT&T cable assets acquired by, 81

Comedy Central, 298

Commerce, business-to-business or B2B, 399–400. *See also* e-commerce

Commerce, U.S. Dept. of, 13, 36–37

Commercial Internet eXchange (CIX), 29, 32, 63

Commercialization: of early Free-Nets, 425; Internet's, 4, 148, 557

Committee 802, 83–84

Common carrier law, tradition of, 57

Common Gateway Interface (CGI), 128

Communication industry, and Internet, 54, 58

Communication networks, computer-mediated, 562. *See also* Networks

Communications, 299, 416–417. *See also* Media; Telecommunications

Communications Decency Act (1996), 519

Communities, online, 120; brokerage services, 430–436; building, 554, 560; community networking services, 440–443; conversational, 438; converted services, 427–430; discursive, 547; game, 549–550; global *vs.* local, 553–554; origins of, 424–426; social networking services, 436–440; virtual sexual, 496

Community informatics, 423, 441, 443

Community networking services, 423, 440–443. *See also* Networks

Community seeking, 552

Community Technology Centers (CTC), 440–442

Compact discs (CDs), 135, 335, 454, 476

Compaq Corp., 332–333, 333

Competition, 48, 49; and FCC, 90; on global scale, 405; and pricing practices, 64; in voice telephony, 79

Competitive Local Exchange Carriers (CLECs), 77, 78, 90

Compression systems, 468. *See also* MP3

CompUSA, 333

CompuServe, 15, 27, 67, 113, 114, 427; bought by AOL, 68, 177; as bulletin board operator, 57; Internet Explorer distributed by, 140

Computer Fraud and Abuse Act, 25–26

Computer industry: browser wars in, 138–139; and demand for software, 203; disruptive technologies in, 217; and Internet music boom, 477; mainstream of, 129; Windows 95 in, 134

Computer literacy, 403

Computer manufacturing firms, Intranet used by, 401

Computer networks, 9, 10, 147. *See also* Networks

Computer reservation systems (CRSs), 317, 318

Computers, 110, 121, 208, 358. *See also* Personal computer business

Computer Sciences Corp. (CSC), 203, 209

Computer utility, 201, 204, 209–213, 220

ComScore Media Metrix, 185, 516

Concentric Systems, 211

Conferencing, Web-based, 402

Connection speed, in 1960s and 1970s, 221–222

Consignment selling, 234, 247

Consumer confidence, 16

Consumer relationship management (CRM), 315

Consumers, Internet, 250, 306, 426; customers, and augmented retailing, 253; increasing numbers of, 413; "new hybrid," 249; and online pricing, 255n33; and online retailing, 244; pornography, 496, 499, 517–518

Content providers, 413

Content Scramble System (CSS), 462

Control Video Corp., 16, 17

Converted services, 423, 427–429

Cookies, introduction of, 133

Coopersmith, Jonathan, 495, 501

Copyright infringement, 452, 457, 469, 470, 473

Copyright law, 456, 466

Copyright protection: in cyberspace, 373; and Internet, 410

Copyright violation: and home broadband access, 545–546; and music file sharing, 5

Corel, 138, 140

Corio, 210, 211, 212

Corley, Eric, 462

Corning, 86, 411

Corporate culture, 505

Corporate record indexing, market for, 194n49

Corporate search products, development of, 189

Corporate software products, 203–204

Corporations, 202, 203. *See also* Businesses

Correspondence schools, 373

Costs, sales volume as driver of, 256n46. *See also* Pricing

Counsel Connect, 428–429, 430

Counterculture, origins of cyberculture out of, 482n22

Courses, online-only, 374

Courseware, electronic, 373

Cox (cable television firm), 81

Craig's List, 293, 300, 558

Creative Commons, 551

Credit card transactions, 16, 395, 549, 559

CRM. *See* Customer relationship management

CRSs. *See* Computer reservation systems

CSC. *See* Computer Sciences Corp.

CSNET, 19–20, 22, 38

CTSS system, at MIT, 150n.12

Current TV, 301

Customer relationship management (CRM) software, 211, 214

Customers: in brokerage industry, 406–407; eBay's, 433; on Internet, 400, 415; profiling of, 172; reviews and ratings of, 322, 425–426

CyberCash, 171

Cybereducation, 370–375, 373

Cyberintermediaries, 321, 322–324, 326, 329–330

Cybersell, 120

Cyberspace, 16

"Darknet," 546

DASnet, 18

DAT. *See* Digital audio tape

Databases, 14, 169

Data network, commercial, 15

Data processing service industry, batch data processing in, 202, 203

Data rate, 221–222, 229n74

Datek, 407

Dating sites, online, 436, 437, 438

Davies, Donald, 11

DeConcini, Dennis, 457

DeCSS program, 462

"Deep Blue," 550

"Deep web," 161

Defense, U.S. Dept. of (DOD), 21, 111, 401, 420n31. *See also* Advanced Research Projects Agency

Defense Advanced Research Projects Agency (DARPA), 39n2, 62

Defense industry firms, Internet adopted by, 394, 418n6

Delivery services, effect of Web on, 342

Dell, Michael, 333, 334

Dell Computer, 316; business model of, 334; in China, 337–338; competition with Gateway of, 334; computer design of, 333, 334; direct model and logistics capabilities of, 336; direct sales of, 239; in India, 336; mass-customization strategy of, 334; Olympic product of, 333–334; Wi-Fi products of, 84

Delta Airlines, 319

Demand: and backbone capacity, 74; broadband, 83; for Internet traffic, 71, 74; and pricing practices, 74

Democracy: semiotic, 302, 452, 480, 489n91; and voluntary association, 440

Democratization: of innovation, 498; of pornography, 507–508

Department stores, 233–234. *See also* Retailers

Design: retail sales in site, 242; social mechanisms behind, 107

Desktop operating system market, 141

Deviance, sociology of, 434

Dialog search site, 161, 172

Dial-up access, 35, 49; costs of, 74–75, 221; data rate for, 222, 229n73; modems for, 229n71; origins of market for, 92; and UUCP programs, 20

Diamond Multimedia, 466

Digital Audio Music, 465

Digital audio tape (DAT) recorders, 455, 456

Digital Audio Tape Recorder Act (1991), 457

Digital camera business, 340

Digital communication technologies, 287

Digital content, emergence of, 546

Digital Equipment Corp. (DEC): ALL-IN-1 system of, 113; DECnet of, 19; PDP-10s of, 40n16; VAX computers of, 21; VT-100 of, 13; X.400 capabilities of, 114

Digital film sharing, 452

Digital Millennium Copyright Act (DMCA) (1998), 453, 460, 461–462, 463, 466, 473

Digital minidisc players, 457

Digital photo kiosks, 340, 341

Digital Playground, 521

Digital subscriber lincs (DSL), 64, 83, 456; asymmetrical protocol for, 548; costs of, 221; data rate for, 222, 229n73; maximum coverage radius for, 82, 102n72

Digital technologies: and recording industry, 453–456; and television networks, 298

Dimmick, John, 283, 284

Directory services, 159, 165–167, 173

Discursive communities, 547

Discussion groups, noncommercial uses of, 554

Disintermediation, 321, 375, 378, 498

Disk operating system (DOS), 60

Distribution: impact of Internet on, 559–560; and online retail, 244; for SaaS, 218

DLJdirect, 406

Doctors, Internet use by, 559. *See also* Physicians

"Docuverse," 162

Domain name analysis, of pornography market, 499

Domain Name System (DNS), 32–38, 61, 125, 561

Domino, web mail offered by, 119

Dorner, Steve, 117

DOS operating system, 39n10

Dot-com boom, 5, 51, 94n. 4, 159, 160, 174

Dot-com bust, 78, 80, 86, 87, 91, 94n4, 179–
181, 269, 562

Dot-com era, 6, 179, 268–269; business history
of, 261; business planning documents from,
266; firms, 269–271, 272–273; follies of, 164–
165; GBF in, 262–265; startups, 558; stock
valuations during, 186; survival of firms in,
261–262; venture creation during, 265–267

Doubleclick.com, 172, 186

Drucker, Peter F., 416

DrugEmporium.com, 375, 376

DSL. *See* Digital subscriber lines

Dulles Airport, and DNS registry, 37

DVD Forum, 462

DVDs, 462, 495

Earthlink, 65

Earth Station Five file sharing system, 473

easySABRE, 319

eBay, 191, 233, 246–248, 261, 315, 411, 430,
431, 443; business model of, 432; community
orientation of, 426, 432–433; and complementary suppliers, 248; fees set by, 246,
247 *fig.*; and fraud, 559; history of, 431;
and Meetup.com, 439; PayPal purchased by,
548; rating scale of, 435; revenues of, 431;
Skype purchased by, 552; transitioning of,
246

e-Chemicals, emergence of, 408

e-commerce, 86, 233, 256n48, 264, 405, 430,
514; business sales in, 252; determination of
success in, 263; discriminatory laws on,
101n65; disintermediation in, 375; mortgage
brokers in, 329; multichannel retailing in,
249–252; online auctions in, 246; potential
scale of, 238; small *vs.* large firms in, 247; and
social dynamics, 426; within supply chain,
238; transition to, 241; travel reservations
industry in, 316–317

Economy: e-commerce in, 397; goods and
services in, 397, 405; integrated multichannel
retailing in, 249–252; Internet in, 413; online

retail sales in, 237–239, *240*; retail sector in,
235, 236, 237; successful Internet-based
enterprises in, 412; Wal-Mart expansion in,
239–240

Ecotourism, 325, 559

Edmond, A., 516

eDonkey, 452

Education: impact of Internet on higher, 368;
and Internet, 305, 352; Internet-based, 371;
online, 372, 373. *See also* Academia;
Universities

Educational management organizations, 373

Educational reform, and World Wide Web,
370–371

eGovernment Portals, 179

Egreetings Network, 468

eHarmony, 436, 437

E-health, 352–356, 358–367

Electronic commerce. *See* e-commerce

Electronic Data Interchange (EDI), 392, 405,
414–415, 415

Electronic Data Systems (EDS), 209

Electronic digital computer, early goals for, 10.
See also Computers

Electronic Frontier Foundation, 463

Electronic information industry, personal
computer in, 14. *See also* Information

Electronic mail. *see* E-mail

Electronic publishing systems, 159, 160

Electronic university, 371

Ellis, Jim, 542

Ellison, Larry, 138, 211

Elm (public domain product), 111

E-Loan, 330, 331

e-mail, 18, 55, 109, 114; of ARPANET, 110;
bulk, 121; demand for, 105; dramatic increase
of, 116; early commercial, 113; Internet as
gateway for, 85, 111–115; limitations of, 110;
for masses, 115–116; Microsoft application
for, 60; and person-to-person communication, 110–111; physician aversion to,
362; university faculty use of, 369–370,

385n97; unstructured nature of, 363; volume of, 116; weaknesses of, 120
e-mail address formats, 115. *See also* Domain name system
Emdeon, 365, 366
EMI, online albums of, 467, 485n50
Employees: economic effects of Internet on, 404–405; and intranet use, 403, 419n19
eMusic, 466, 477
Encarta encyclopedia, 136
Encryption technologies: and average Internet users, 363; for DVDs, 462
Engelbart, Douglas, 15
Enterprise e-mail and calendaring systems, market for, 118
Enterprise Resources Planning, 189
"Enterprise search," 189
Enterprise software products industry, growth of, 210
Entertainment: online auction houses providing, 248–249; shopping as, 245
Entertainment industry, 461–463, 504. *See also* Adult entertainment industry
Entrepreneurs, 49, 260, 272, 273, 439
Entropy file sharing system, 473
E-rate program, 78
Eritmo.com, 468
E-tail sales, 405
Ethernet network, 26, 38, 223
Ethernet standard, 83
Ethical issues, in online advertising, 292
e-tickets, 320
eToys.com, 261, 263, 315
E*Trade Financial, 407, 408, 430
Eudora, 117, 119, 122, 151n32
Event ticketing, 239, *240*
Excel, 135
Excite, Inc., 119, 159, 164, 173, 178, 181, 190, 322; emergence of, 170; liquidation of, 180; public offerings of, 174
Exodus, 211
Expedia, 321–322, 323, 326

Facebook, 147, 299, 303, 304, 437–438, 438, 441
Faculty, university, 374. *See also* Professors
FairPlay antipiracy technology, 475
FanGlobe, 468
Fanning, Shawn, 433, 468, 469
Farechase, 324
FastTrack, 471, 472
Faxes, 112, 342
Federal Communications Commission (FCC), 24; and bulletin boards, 56; in enhanced service markets, 80; and exploratory investments, 89–90
Fees, original e-mail, 116. *See also* Costs; Pricing
Felten, Edward, 463, 477
Fiber-optic connections, 23, 24, 222, 229n73
FidoNet, 112, 114, 150n18
File sharing: based on FastTrack, 471; and music industry, 451–452; Napster and, 468–473; peer-to-peer, 470, 473; privatization of, 480; and record sales, 476; technology of centralized, 451. *See also* Music file sharing
File transfer protocol (FTP), 125, 168, 458, 468
Film, 480, 492
Film industry, 340–341; and circumvention, 460–461; pornography in, 499; and streaming, 460; and video file sharing, 451–452. *See also* Adult entertainment industry, business strategies of
Filo, David, 166
Financial industries, early Internet use of, 397–398
Financial press, 129
Financial Times (London), 290
First mover advantage (FMA), 262–263
First Virtual (start-up firm), 171
Fisher, William III, 282, 302, 480, 497
Flickr.com, 191, 300, 552
Floppy hard drive, 455
FlowerNet, 341

Flynt, Larry, 503–504
Folksonomy, 191, 498
Forrester Research, 320, 329, 518
ForSaleByOwner.com, 289
Foucault, Michel, 496
401 (k) accounts, 406, 407–408
Foxfire browser, 145
Fox television network, 279, 298
Frankel, Justin, 455, 464, 470
Freedom to Innovate Network, 144
Freenet file sharing system, 473
Free speech, 547, 549, 555n14
Free Speech Coalition, 509
Frequent-flier plans, 319, 322
Friedman, Thomas L., 338
Friendster.com, 304, 437, 438, 439, 552, 561
Friis, Janus, 471
FrontBridge (antispam service), 215
Front Page (Web site creation tool), 135
Fuzzball software, 41n32

Galbraith, John Kenneth, 262
Galileo, 322
Gambling online, 497
Games, computer, 16, 480, 549–550
Garageband.com, 466
Gates, Bill, 25, 135, 147, 378, 416; Internet
  Tidal Wave memo of, 166; retirement from
  Microsoft management of, 156n101
Gateway, Inc., 334–335
Gateway, use of term, 195n54
Gateway device, 18, 21, 239
General Electric, 410
General Motors, proprietary EDI network of,
  393
Genie, 67, 113, 114
Genuity, 79, 87, 88
GeoCities, 468, 486n57
Germany: Internet conference in, 5–66;
  privacy in, 560
Gerstner, Louis, 59
Get Big Fast (GBF) strategy, 261–265, 272
GetMusic.com, 467

Glaser, Rob, 459
Global access points, up-to-date map of, 553
Global community, Internet as, 549
Global Crossing, 87, 88
Global delivery systems (GDSs), CRS
  enterprises as, 324
"Global digitized village," 553
Globalization, 338, 442
Glossbrenner, Alfred, 14, 16, 18, 28, 35
Gmail, 119, 123, 146
Gnutella, 464, 470–471, 545
Go.com Web portal, 177, 315
Goodmusic, 466
Google, 159, 160, 182–183, 190, 191, 315,
  552; advertising of, 184, 186; business
  operations of, 411; corporate culture of, 185;
  dominant position of, 187; Earth satellite
  image browser, 187; financing of, 548; free
  e-mail service of, 119; information retrieval
  capabilities of, 185; Maps, 146, 187, 216; as
  Microsoft competition, 147; Office, 215;
  PageRank algorithm of, 183–184; Print book
  indexing project, 188; public offering of, 185;
  revenues of, 189; search appliance of, 184,
  189–190; services offered by, 187–188; service
  to China of, 479; Spreadsheets, 147, 216;
  Usenet absorbed by, 21; video store, 299;
  YouTube acquired by, 553
Goolsbee, Austan, 243
Gopher, 125–127, 168
Gore, Vice Pres. Al, 30, 31, 78, 301, 461
GoTo.com, 173
Government agencies, and Internet, 402, 408–
  409
Government Systems, Inc. (GSI), 36
govWorks, 179
Gracenote, 467
Grateful Dead, the, 458
Great Plains, Microsoft's acquisition of, 218
Greenberg, Eric, 259
Grokster file sharing program, 471, 472–473
Gross, Bill, 173
Grove, Andy, 416

GTE, emerging operations of, 65
Guardian Unlimited (British), 302
Guba, 299
Gurstein, Michael, 443

Hafner, Katie, 3
Handy, Charles, 416
Hard drives, personal computer, 455
Hardware, 12
Harris Interactive, 284
Hawkins, Gordon, 509, 513, 520
Hayes Communications, 222
HBO, 180
HCL Technologies, 339
Health Care Financing Administration (HCFA), 361
Health care industry, 354, 364; case management services in, 361–362; interactive communication, 352, 353; selective Internet use in, 352; technological innovation in, 356, 415; third-party benefits providers in, 361; waste in, 353, 365
Healtheon, 354, 365
Health informatics, 352, 353
Health information: portals, 365; protected, 363
Health Insurance Portability and Accountability Act (HIPAA), 363
Heiferman, Scott, 439
Hewlett-Packard, 114, 336, 466
High Performance Computing Act (1991), 30, 36
High-speed connections, 67, 82, 222–223, 229n73
HIPAA. *See* Health Insurance Portability and Accountability Act
Hirsch, Steven, 531
History, Internet, 107, 108, 109–110, 562n1
Home-based businesses, 403
Home computer revolution, 109
Home Depot, 377
Home sales, commissions in, 331, 348n83
Hosted antispam service, 215

Hosting services, 67, 211
HOSTS.TXT file, 33
HotBot search service, 170
Hotels: adult movies in, 502; and bundling of travel packages, 322; Internet access in, 223; web site development of, 395
Hotline, MP3 distribution of, 545
Hotmail, 118, 119, 187, 189, 213
Hot spot locations, 223
HotWired, 170, 171, 194n38, 426
Households, U.S., with Internet access, 47, *48*
H&R Block, 376, 411
Hummer Windblad, 469
Hun's Yellow Pages, 506
Hurricane Katrina, and Web activity, 300
Hypercard system, 162
Hypertext, 162, 190
Hypertext markup language (HTML), 124–125, 163; early support for, 61; and Web systems, 127–128
Hypertext transfer protocol (http), 124–125, 132, 148

IBM Global Network, emerging operations of, 65
IBM (International Business Machine), 53; in Chinese personal computer market, 337; civilian air traffic control system of, 430; internal applications of, 401; and Internet, 392, 413; as ISP for business, 59; Notes, 118; and NSF, 24, 53; PC, 39n10, 332–333; PROFs system of, 113; proprietary protocols of, 23; RS/6000 computer, 37; SABRE developed by, 317–318; and SAGE, 317; Systems Network Architecture of, 19; System/370 mainframes, 20; TCP/IP used by, 24; as tier-1 provider, 66; unbundling decision of, 205
ICANN. *See* Internet Corporation for Assigned Names and Numbers
iChatAV, 552
ICQ (I seek you), 68, 122–123
Illinois, Univ. of, and Mosaic, 128
Inclusion, Internet's philosophy of, 33

Income, and Internet, 305

Incumbent Local Exchange Carrier (ILEC), 77

Independent software vendors (ISVs), impact of SaaS on, 217–219

Indexes, for Web directories, 165–166, 167

India, 336, 339

Industry portals, 178–179

INFONET network, 209–210

Informatics: community, 423, 441, 443; health, 352, 353; social, 424, 429, 438–439

Information: contracted-for services, 399; cross-channel flow of, 249; health-related, 365–366; and market's balance of power, 408; offshoring, 317; for online shoppers, 241–243; Wikipedia-type, 304

Information and Communications Technologies (ICT), 423

Information ecosystems, 303

Information Infrastructure Task Force (IITF), 461

Information Processing Techniques Office (IPTO), ARPA, 10, 11

Information retrieval systems, 160–161, 162

Information science, 160

Information Segment, 413

Information services, 424; brokerage services, 430–436; community-oriented, 440–443, 444; converted services, 427–430; development of online, 161; niche markets for, 444; origins of, 424–426; social networking services, 436–440

Information society, 109

"Information superhighway," 30, 109, 544

Information technology, 52, 391; for auto industry, 393; automating functional activity in, 86; and brokerage industry, 407; employees in, 404; heavy users of, 86; incremental adoption of, 417; investment in, 85; offshoring of, 338, 339, 349n108; and private sector, 399; proprietary, 367; and social contexts, 429; transformation of, 201; VC-backed, 266–267; Wal-Mart's use of, 239

Infoseek, 172, 174, 177, 181, 190, 194n. 42

Infosys, 339

Infrastructure markets, investment in, 49

In-house computing, *vs.* time-sharing services, 208, 226n22, 226n23. *See also* Intranets

Inktomi, 170, 174, 185

Innovation: democratization of, 498; effect on market structure of, 51; and indispensable intermediaries, 379; and regulatory policy, 93; user-driven, 148; Web's reliance on, 190. *See also* Technological innovation

Inova Health Systems, 357

Instant messaging, 99n44, 122; adoption of, 124; AOL's version of, 68; as business tool, 123–124

Institute for Scientific Information, 183

Institute of Electrical and Electronics Engineers, 76, 83

Insurance, employer-provided health, 366–367

Insurance industry, 395, 401, 402

Integrated Media Systems, 459

Integrated services digital network (ISDN), 65

Intel Corp., 84, 332, 334

Intellectual property, academic *vs.* commercial understanding of, 549

Intellectual property law, and noncommercial home recording, 456

Interactive books, market for, 136

Intercarrier compensation issues, 80

Interconnection: of networks, 112; technology of, 57

Interface Message Processor, 108

Intermediaries, indispensable, 352, 375–379

International Conference on Computer Communication (Washington, D.C.), 1972, 11

International perspective, on Internet, 560

Internet, 5, 18: academic, 54, 106; beginnings of, 21; in business, 391–392, 396, 399–400; in China, 337; commercial development of, 4–5, 61, 245, 260, 379; and community, 443, 541–542; diffusion of, 239, 285; early history of, 497–498; economic effects of, 404–405; economics of providing information on, 170;

effects on public sector of, 408–409; expansion in U.S., 393; forces driving, 557; free, 547–549; governance of, 13, 36, 38, 561; incremental adoption of, 417; for internal operations, 404; mass media's use of, 299; and media businesses, 279; monitoring deployment of, 393, 418n2; and music business, 476–480; music files distributed on, 454; and niche theory, 284; number of Americans connected to, 229n69, 229n74; overseas expansion of, 32; paid, 296; and personal computer business, 332–338; and pornography, 496–497, 502, 532; precommercial, 148; privatization of, 47, 51–52, 54, 58, 63–64; realtors and mortgage brokers in, 326–331; resistance to, 351; slow business acceptance of, 414; and social change, 305; structure of, 52; and TCP/IP creation of, 107; threat of, 342; traditional applications of, 5; and travel agencies, 317–326; use of term, 18. *See also* Access; Protocols

Internet Activities Board (IAB), 62

Internet adoption, stages of, 393–399

Internet bubble, 29, 557. *See also* Dot-com bubble; Dot-com era

Internet Corporation for Assigned Names and Numbers (ICANN), 12, 36, 37, 38, 62

Internet Engineering Task Force (IETF), 62, 76

Internet Explorer, 99n45, 134, 136, 544; competition with Netscape of, 143; first release of, 138; Microsoft's support for, 141; versions of, 68, 139, 144, 146

Internet fraud, 547–548

Internet music companies, 466–467. *See also* Music industry

Internet navigation services, development of, 190. *See also* Navigation industry

Internet Network Information Center (InterNIC), 35, 36

Internet payment firms, 171

Internet protocol, voice-over, 87. *See also* Protocols, Internet

Internet-related firms, public offerings of, 165

Internet Relay Chat, 122

Internet Revolution, 129

Internet service providers (ISPs), 15, 65, 560; academic, 52–53; and CIX, 63; connectivity of, 52, 53; distribution of, 72, 73; earliest advertisements for, 54; and FCC policies, 77; first commercial, 53, 95n. 9; geographic coverage of, 69; home service first offered by, 75; increased reliability of, 80; local, 52; location of, 70; mom-and-pop, 65, 68; nonprofit regional networks replaced by, 38; origins of, 12; and peering, 31; phone numbers supported by, 65; and POP capabilities, 116; pricing policies of, 64, 76; and regulatory norms of bulletin board industry, 57; regulatory umbrella for, 80; and routine network operations, 67; smaller-scale, 38; Web sites of, 394–395

Internet Society, 12, 62

Internet Tax Freedom Act (1998), 79, 101n65

Internet Underground Music Archive, 458, 545

Interoperability, 20, 26

Intranets, 403, 404–405, 419n18

Inventory replenishment, 86

iPods, 295, 410, 435, 474, 475

IPv6, 87

IRA accounts, 407–408

ISPs. *See* Internet service providers

iTunes, 410, 475

iTunes Music Store, 434, 473–476

iXL, 260

Jabber instant messaging standard, 123

Jaffee, David, 374

James, David, 531

Jameson, Jenna, 507, 529

Java, 137–138, 140, 147

JavaScript, 137

Java System Web Server, Sun's, 544

JD Uniphase, 86

Jennings, Dennis, 22

JetBlue, 324

Jobs, Steve, 474

Johansen, Jon, 462
Johnson, David, 428–429
Johnson Commission, 520
Journalism, 291–292, 300, 301, 304, 315
Joy, Bill, 19
Jupiter Research, 263, 264

Kahn, Robert, 11, 25, 37
Kamali, Peter, 439
Katz, Jon, 281
Kawasaki, 375
Kay, Alan, 14, 163
Kayak (Internet site), 324
Kazaa, 433, 435, 471–472, 545
Kearby, Gerry, 459
Kelly, Jill, 507
Kendrick, Walter, 501, 503
Killebrew, Kenneth, 292
Kimsey, Jim, 17
Kleiner Perkin, 183
Kling, Rob, 429, 430, 498
Knight Ridder, 294
Knowledge, sociology of, 369
Kodak, and digital camera technology, 340
Krol, Ed, 126, 129, 130

Labor costs, 402, 404–405
Language, and Internet-based interaction,
  418n9. See also Communication
Larson, Christopher, 330
Lasercomb v. Reynolds, 461
Lastminute.com, 322
Law, impact of Internet on, 559
Leavitt, Mike, 373
Lebda, Doug, 330
Legacy systems, and Internet business, 415
Legal case opinions, searchable database of,
  428
Lehman, Bruce, 461
LendingTree, 330, 331
Lenk, Toby, 263
Lenova (Chinese firm), 337
Lessig, Lawrence, 106, 428, 429, 430

Letters, e-mail as replacement for, 109, 551–
  552
Level3, 87
Levi Strauss & Co., withdrawal from e-
  commerce of, 375, 376
Lexis, creation of, 161
LexisNexis, 161, 172, 427, 428–429, 494,
  495
Lexmark, 463
Libraries, 78, 127, 462, 541
Licklider, J. C. R., 10
Lifeco, 318
LimeWire, 433, 471, 545
Linux, 63, 218, 462
Liquid Audio, 459, 467
Listen.com, 467, 468
Listservs, 20, 112–113
Live Journal, 561
Local area networks (LANs), 26
"Long tail," 191, 244
LookSmart, 166–167
"Look-to-book" ratio, in airline industry, 320
Lord, Rob, 458, 464
Los Angeles Times, 302
Lotus, 113, 206, 214, 332
Lotus Notes, 119, 214, 401, 430
L-systems, and demand for software, 203
Lucent, 59, 86
Luini, Jon, 458
Lycos, 119, 159, 164, 167, 178, 181, 190; and
  dot-com crash, 180; public offerings of, 174
Lycos Europe, 468
LycoShop, 178
Lynx browser, 127

Mac computer, 552
MacCoun, Robert, 479–480
Macintosh, and Foxfire browsers, 145
Macy's, comparison matrix used by, 258n64
Madster, 470
Magazine industry, 294
Magazines, 283, 286, 293, 494, 495
Magellan directory, 166, 174

Mail-order houses, 233–234, 405, 544

Mainframes, 13, 392, 202, 208, 223

Malpractice suits, and e-mail, 362

Management consulting firms, outsourcing of, 402–403

Manufacturing, 397, 405; e-commerce in, 397; exit rates in, 271; work-flow redesign in, 392. *See also* Industry

MapQuest, 187

Market, Internet access, 62, 92, 554; commercial viability of, 54; competition in, 48; early evolution of, 47, 51–52; and enterprise computing, 86; fast growth of, 55, 70, 80; innovation in, 48, 49; new businesses in, 50; new users in, 68; retail access providers in, 51; revenues in, 93n1; transition of bulletin board operators to, 57; U.S., 64; and venture-funded entrepreneurial firms, 63–64. *See also* Access, Internet

Market Watch, 289

Mark's Bookmark, 506

Martha's Girls, 521

Massachusetts Institute of Technology (MIT), 128, 371

Mass media, 279; and asymmetrical digital subscriber line protocol, 548; Internet as extension of, 543; on Web, 304

Mass media online, 280–284

Match.com, 426, 436, 437

McClatchy (newspaper co.), 294

MCI, 24, 53; Backbone Network Service of, 31; early e-mail services of, 113; headquarters for, 37; Internet services offered by, 59; MFS absorbed by, 32; national fiber network of, 54; network connection of, 27; and NSF, 24, 53; as tier-1 provider, 66

MCI Mail, 27, 28, 114

MCI-WorldCom, 87

McKinley (start-up), 166

McLuhan, Marshall, 553, 561

McNair, Brian, 494

McPheters, Rebecca, 286

Mead Data Central, 428

Media: business models in, 561; changes in, 299–300; and cultural changes, 292; early Internet in, 25–26; international, 129; and Internet technology firms, 265; interpersonal, 304; narrowcasting of content in, 305; participatory, 301

Media businesses, 287; early Internet use of, 397–398; electronics in, 416; estimating audience size in, 286–287; and Internet, 279; and niche theory, 283; online *vs.* off-line, 284–287; organizational transformation of, 287–292

Media General, integrated operations of, 290

Media Metrix Research, 516

Media news, and Internet, 304–305

Media organizations, 280; business models for, 292–299; and cultural changes, 287–288, 291–292; and interactive capability of Internet, 281; multimedia platforms of, 302; online strategies of, 288; "platforms" of, 290; structural changes in, 287, 288

Medical claims, electronic processing of, 366

Medical practice, influence of Internet on, 354–355

Medical records, 352, 353, 362

Medicare, Medicaid, and State Childrens' Health Insurance Program Benefits Improvement Act (2001), 361

Medicine, Internet-based transformation of, 354. *See also* Health care industry; Telemedicine

Meeker, Mary, 55

Meeker, Matt, 439

Meese Commission, 520

Meetup.com, 436, 439, 440, 443

MessageLabs, 215

Metadata, 191

Metallica (heavy metal band), 478–479

Metasearch business strategy, 323

Metasearch firms, impact of, 324

Metcalfe, Bob, 83

Metcalfe's Law, 40n25

Metropolitan Access Exchange (MAE), 32

Metropolitan Fiber Systems (MFS), 32
Michigan Education Research Information
  Triad (MERIT), 24, 27, 53
Micropayment firms, 171
Microprocessors, development of, 332
Microsoft, 25, 138–139, 144, 206, 340;
  antitrust case against, 60, 144, 469; and
  antitrust law, 141–142, 156n95; brain trust
  for, 97n24; and commercialization of
  Internet, 148; competition with Netscape of,
  155n85; e-mail software of, 118; European
  Union's case against, 143–143; and formation
  of MSNBC, 282; Hotmail acquired by, 213;
  and impact of SaaS, 218; instant messaging
  networks of, 123; Internet Platform and Tools
  Division of, 139; Internet strategy of, 60–61;
  MS-DOS operating system of, 163; msn.com
  of, 176; MSN of, 159, 181; near monopoly of,
  334; online advertising revenues of, 172;
  online information and entertainment
  provided by, 135; Outlook package of, 113,
  117; in PC market, 136–137; and proprietary
  standards, 60; sued by Sun, 156n91; and
  Web, 134–136; Web site searches of, 172–
  173; Windows CE of, 85; Windows Live
  initiative of, 189; Windows 95 of, 60; X.400
  capabilities of, 114. See also Internet Explorer;
Microsoft Exchange, 113, 118, 119
Microsoft Network (MSN), 55, 67, 68, 189, 552
Microsoft Office, 134, 135, 215, 216
Microsoft Outlook, 121
Microsoft Plus! pack, 134
Microsoft's Xbox, 550
Midbar, 468
Military, intranet used by, 401
Miller, Matthew, 529, 530
Miller test, 493
MILNET, 13, 21, 35, 107
Min, Jean, 300
Mindspring, 65
Minicomputers, and time-sharing industry,
  223
"Mirror sites," 168

MITRE Corp., 317
MIT survey, of Web servers, 165
Mjuice, 467, 477
Mobile phone companies, deals with record
  industry of, 475. See also Cell phones
Modems, 75, 221, 222, 229nn 71, 73
Modularity, in auction economics, 246
Mokapetris, Paul, 33
Monster.com, 261, 293
Moore's Law, 23, 351
Morpheus file sharing system, 472
Morris, Robert T., 25, 26
Mortgage brokers, 326, 329, 330, 331
Mosaic, 61, 127, 128, 129, 131, 544
Motion Picture Association of America
  (MPAA), 460, 462, 473
Motley Fool personal finance guide, 170
Movie industry, and illegal copying, 546. See
  also Film industry
Movielink, 299
Movies: adult, 500, 506; pirated, 420n35. See
  also Film industry
Moving Picture Experts Group (MPEG-1), 455
Mozilla, 145, 544
MPAA. See Motion Picture Association of
  America
MP3 (compressed audio format), 434; creation
  of standard for, 454–455; illegal files in, 468;
  Internet music business, 464–468; and Liquid
  Audio, 459–460
MP3.com, 459, 464, 465
MP3 players, 462, 464
MSN. See Microsoft Network
MSNBC cable news channel, 136, 279, 282
MSN Group, 303, 304
MTV (television network), 298, 434, 458,
  483n23
Multiforce concept, 220
Multihoming, 66
Multiple Listing Service (MLS), 326–327, 377,
  378
Multipurpose Internet Mail Extensions
  standard, 111

Murdoch, Rupert, 249, 295, 304

Music: bands, and fan sites on Internet, 479; Internet propagation and sharing of, 451, 545; pirated, 420n35; traditional distribution system for, 452–453

Music, sharing: early legal foundations for, 456–458; early uses of Internet for, 458–460; experiments with, 460; legal swapping in, 480, 488n89; technologies for, 451

MusicCity, 472

Music file sharing, 433, 559, 476, 479, 487n76

Musicians, 458, 478. *See also* Recording artists

Music industry: anticircumvention in, 463; exposés of, 481n5; file sharing and, 451–452; and illegal copying, 546; impact of Internet on, 476–480; Internet experiments of, 467–468; iPods in, 410; and MP3 technology, 464–468; SDMI of, 462. *See also* Film industry; Record industry

Music locker services, 465, 468

Music Owners' Licensing Rights Act, 465

Music subscription services, 459

Mutual funds, professionally managed, 408

MySpace, 289, 298, 437, 441, 552, 561, 563n7; functions of, 295; growth of, 303; long-term implications of, 303–304

Nando Times, 281

Napster, 430, 433, 434, 435, 445n20, 477, 545; creation of, 468–469; and file sharing, 468–473; immediate popularity of, 469; intellectual property of, 486n59; and Internet music business, 464; and Metallica case, 478–479; music distribution system of, 545; online file sharing system of, 459; used to share music, 451; users of, 479

Napster 2, 469

NASA, and telemedicine, 357

NASDAQ exchange, 179, 265, 406, 523

National Association of Realtors (NAR), 327, 377, 378, 379

National Broadband Co., 289

National Center for Atmospheric Research, 22

National Center for Supercomputer Applications (NCSA), 61, 127, 544

National Commission on Cost of Higher Education, 371

National Music Publisher's Association, 477

National Research Council Report, 515

National Science Foundation (NSF), 13, 22–32, 52

Navigation industry, 159, 172, 190, 242

Navigator browser, 117, 123, 136, 334; first release of, 129; upgrades of, 144; version 2.0 of, 137

NBC (television network), 282, 289, 298

Nelson, Ted, 162, 163, 170, 190

Netcenter, 175

Netcom, emerging operations of, 65

"Netiquette," 120, 543, 548

NetMeeting, 135

Net neutrality, 89, 148

Netscape Communications Corp., 54, 60, 61, 167, 365, 544; AOL acquisition of, 100n46, 143, 157n110; browser business of, 137, 145; and commercialization of Internet, 148; competition with Microsoft, 155n85; Enterprise Server of, 544; and HTML support, 61; http cookies introduced by, 133; initial public offering of, 55, 130; Open Directory Project acquired by, 167; Secure Sockets Layer of, 132. *See also* Navigator browser

NetSuite, 213

Network Access Points (NAPs), 31, 32

Network Control Protocol (NCP), transition from, 21

Network operations, 10; cooperation in, 52; emergence of routine, 64–67; idea of host of, 220

Networks, 107; academic, 19–21, 541–542; for-profit private, 19; high-bandwidth home, 109; interconnectivity of, 38, 112; regional, 52; social, 551–553; television, 285–286

Network Solutions, Inc. (NSI), 36, 37

New Frontier Media, Inc., 505, 514, 522, 527–529

News, digital access to, 480. *See also* Broadcast
  media; Media
News Corp., 279, 289, 297
Newsgroup messages, 121
News organizations, cost structure for, 558
Newspapers: adult entertainment in, 494, 495;
  audience for, 294; decline of, 285, 293, 294;
  and niche theory, 283; online, 281, 282, 291;
  staff reductions of, 292–293; Web portals
  and, 178; Web sites of, 133, 295–296
NewsRecord (Greensboro, NC), 302
Newsweek, 286
New York City, Internet diffusion in, 58
New York Stock Exchange, 525
New York Times, 161, 290, 291
New York Times Co., and Web potential, 288
New York Times Electronic Media Co., 280
Nexis Information service, 428
NeXT computers, 124
NFM Inc., sales of, 528
Niche markets, 191, 558
Niche theory, 283
Nielsen/NetRatings, 303, 323, 516
Nintendo machines, 550
Noble, David, 372, 373
Nortel, 59, 86
North American Industry Classification
  (NAIC), 235, 413
Northern Light, 172
NorthWestNet, 27
Norwest, 330
Novell Netware, 59, 113, 114, 118
NSFNET (National Science Foundation
  Network), 4, 14, 15, 38, 41n33, 97n26, 107
NH, Inc., 515
Nullsoft, 464, 470
Nupedia Internet project, 551
NYSERNET, 27

Oasis, 479
Obscenity, litmus test for, 493
Obscenity and Pornography, U.S. Commission
  on, 520
OfficeMax, B2B of, 399

Offshoring, 338, 339–340, 340
OhmyNews, 300
Oh Yeon-ho, 300
Olim brothers, 233
Omidyar, Pierre, 233, 431, 432, 439
OnDemand (Siebel's CRM), 214
Online communities. *See* Communities, online
On Line Guitar Archive, 477
Online services: health-related information,
  365; home market for, 55; information, 80;
  during 1980s, 114
Online time, and contractual limits, 75
Onsale.com, 257n49
Oodle (classified advertising site), 293
Open access, wikis and, 550–551
Open Archive, 551
OpenCourseWare, 371, 551
Open Directory Project, 167, 182, 193n. 24
Openness, Internet's philosophy of, 26, 33
Open System Interconnection (OSI), 31, 114,
  151n24, 561
Open Text, 174, 189
Open TV, Spyglass acquired by, 154n. 65
Oracle, 204, 211, 217, 218–219
Orbitz.com, 321, 322, 323, 559
O'Reilly, Tim, 146
Organizational structure, and Internet, 416
Orkut.com, 437, 438, 552
Orlikowski, Wanda, 430
OS/2 Netware, 113
Outing, Steve, 288
Outlook Express, 117, 152n. 33
Outsourcing, 252, 334, 339, 402
Overture, 173, 174, 184, 185
Ovitz, Michael, 469–470

Package delivery services, 392
Packet switching, 11, 15
Page, Larry, 182, 185
PageRank algorithm, Google's, 183–184
Palm, 85
Pape, Jeff, 259
Parcel tracking, 392
Path dependency, 399

Pathfinder, 176, 180, 288, 291
Patient-physician communication, 359, 360
PatroNet, 459
Patterson, Jeff, 458
Paulson, Ken, 290
Pawlowski, Janina, 330
Payment methods, 545, 549, 559–560
PayPal, 171, 431, 432, 548
PayPal Buyer Protection Plan, 433
"Pay-per-click" model, 173
Pay2See, 171
PC Flowers, 233, 253n1
PDP-10 computer, 23
PDP-11 computer, 23, 42n51
Peeps Republic, 467
Pentium chips, 130
PeopleSoft, 211
Perdue, Lewis, 513
Performance Systems International, 28
Performing rights organizations, 452
Personal ads, 436
Personal computer business, 136, 332, 333, 334, 336, 337–338
Personal computer gaming industry, 546
Personal computers: advent of inexpensive, 14; availability of, 38; costs of, 140; demand for, 332, 334; development of processors for, 455; Internet e-mail on, 117; pricing of, 207, 226n25; and SaaS, 202; storage devices for, 455; and time-sharing industry, 206, 208–209, 223, 224
Personal computer services, 14, 15–16, 32
Personal computer software industry, 206
Personal electronics industry, and Internet music boom, 477
Petroleum companies, proprietary networks of, 394
Pets.com, 261, 315, 342
Pew Internet and American Life Project, 284, 478, 480
Pew Internet survey, 285
Pharmaceutical industry, 342, 367, 400
Philips, and introduction of DAT recorders, 457

Phishing, 26, 121, 559
Phoenix, Univ. of, 371, 374, 385n102
Photo compact discs, 340
Photography industry, film development segments of, 340. *See also* Film industry
PHP (open-source package), 134
Physicians: continuing medical education for, 354; and HIPAA requirements, 363; Internet use by, 352, 355; and reimbursement Internet-based activities, 355; use of e-mail by, 359–360, 364
Picard, Robert, 296, 299
"Pick, pack, and ship" model, 244, 250
Pittman, Bob, 458
Pizza Hut, 544, 553
PlaceWare, 218
PlanetRX.com, 342
Playboy Enterprises, Inc., 522, 524–525, 526, 527
Plugged In, 441, 443
Podcasts, 287
PointCast, 297
Point of presences, 71
Point-of-sale tracking, 86
Political blogs, 300
Politics, and origins of Internet, 12
POP (Post Office Protocol), 116
Porn actors, 505, 506–507, 529
"Porn chic," 518
Pornography, 493, 504, 559–560, 562n3; access to, 494, 519; "alternatives," 506; amateur, 508; antipornography legislation, 501; availability of, 67; and censorship laws, 494; child, 503, 519; consumers of, 517–518; contradictory attitudes to, 503; corporatization of, 504–507; democratization of, 507–508; do-it-yourself, 507; domestication of, 502–504; in history, 491; impact of Internet on, 560; increasing popularity of, 532–533; Internet, 5, 58, 497; and law enforcement, 504; mass consumer market for, 500; new forms of spectatorship in, 492; and new media, 492; nonmainstream forms of, 497; in public discourse, 518; "reality," 497;

Pornography (cont.)

representations in, 492–493; retail business, 501–502; social construction of, 494; sociotechnical analysis of, 498; traffic volume for, 516, 519; U.S. television/cable revenues in, 518, 519; video revenues in, 517, 519. *See also* Adult entertainment industry

Pornography, U.S. Dept. of Justice's Commission on, 520

Pornography industry: aggregate revenues in, 510; infrastructure providers in, 520; Internet revenues in, 511–513; media substitution in, 499–502; niche market suppliers in, 532; preference for VHS over Betamax in, 495; problems for, 498–499; revenues in, 510–513, 515; size of, 500, 501; Web sites in, 514–515

Portable People Meters, 287

Portal industry, 181

Postcasts, 295

Postel, Jonathan, 33, 36, 37

Postini, 215

Powells, 397

PowerPC chips, 130

PowerPoint, 38, 135

Preferential treatment, with Web advertising, 173

Press, the: business, 265, 266, 272; financial, 129; freedom of, 560; trade, 315; use of term, 562n4. *See also* Media

Pressplay (online music store), 469, 473

Preview Travel, 322

Priceline.com, 321, 432

Pricing, 64; of academic ISPs, 53, 95n7; for broadband, 82; for dial-up ISPs, 74–75; fees set by transaction, 246, 247; flat-rate, 51, 75; Internet's impact on retail, 243, 255n33; for local telephone calls, 56; and online retail sales, 243–244, 255n33; and peak load issues, 75; and regulatory decisions, 50; in time-sharing experience, 221

Print media, 280–282, 293, 295–296

Privacy: patient, 363; and social informatics, 438–439

Private Gold (pay-television channel), 527

Private Media Group (PMG), Inc., 500, 508, 519, 522, 523–524

Private networks, and Internet business, 415

Private sector, adoption of Internet by, 393

Private Spice (pay-television channel), 527

Privatization plan, of NSF, 52

Processing services industry, 209

Processor capability, increase in, 455

Prodigy, 17, 18, 57, 67, 113, 131, 140

"Productivity paradox," 430

Product providers. Web sites of, 394–395

Professionals, Internet use by, 559

Professors, university: Internet use by, 352, 375, 559; and online instructional technology, 370

PROFS, 401, 417

Programming, open-source, 551

Progressive Insurance, 394

Progressive Networks, 459

Proprietary surveys, 437

"Prosumer," pornographic, 507

Protocols, Internet, 105–106, 147–148, 149n2; for browsers, 129–131; design in, 19, 106; for e-mail, 111; instant messaging, 122–124; Internet e-mail software, 116–118; for packet switching, 11; publicly available, 108; spam, 120–122; for Webmail, 118–120; World Wide Web, 124–128

Providers. *See* Internet service providers

PSINet, 28, 29, 65, 87, 88

Public discourse, pornography in, 518. *See also* Communities

Public offerings, of new Internet ventures, 71

Public policies, 76–77, 78–79, 79

Public relations, to finance Internet, 548

Public sector, effects of internet on, 408–409

Public sphere, sexualization of, 496

Publishers, business model for, 296–297

Publishing: off-line, 171; on Web, 160–165, 171, 186

Purchasing, diffusion of, 239, 254n18

Putnam, Robert, 439

Qmail, 118
Qualcomm, 117
Quantum Computer Services, 17
Quicken, 331
QuickenMortgage, 330
Qwerty keyboard, 110
Qwest, 87, 88

Radio, 282, 287; commercial history of, 561;
digital access to, 480; Internet, 543; and
niche theory, 283; satellite, 286, 287; and
technological innovation, 459
Radio frequency identification (RFID) tags,
inventory-tracking, 408, 420n31
Radiology, in telemedicine, 361
Radio Webcasts, 282
Raleigh News and Observer, 281, 289
Rbuy, Inc., 328
Reader's Digest, LookSmart launched by, 166–
167
Read-Only Memory-Basic Input-Output System
code (ROM-BIOS), 332
RealAudio, 459
Real estate brokers, 377–379; businesses, 326,
329; commissions in, 331; electronic auctions
in, 327–328; Internet strategies in, 327; MLS
in, 326–327; mortgage rates in, 328, 329
Really Simple Syndication (RSS), 295
RealNetworks, 460, 462, 467
RealOne media player, 467
RealPlayer, 467
Record industry: and Apple, 474; and
circumvention, 460–461; and DAT
technology, 457–458; distribution in, 452–
453; DMCA and, 477; and Internet, 410;
and iPod use, 435; legal resources of, 462–
463; and MP3.com, 465; and MP3 players,
466; and MTV, 458; and music file sharing,
434; Napster case and, 469; as oligopoly,
453; online albums of, 467; sales in, 476;
and streaming, 460; and technological
innovation, 453–456; and Webcasting music,
282

Record Industry Association of America
(RIAA), 457, 469, 471–472, 482n15, 545
Recording artists: costs of, 452–453; early
Internet use of, 458–459; and illegal file
sharing, 478; revenues for, 452
Regulation: in adult entertainment industry,
524; of community sites, 542; in Internet
access market, 50; and investment trends,
92–93; political, 543; of pornography, 519–
520
Reid, Brian, 542
Reimbursement, in telemedicine, 360–361
Religiosity, and pornography, 494
Rental car businesses, and bundling of travel
packages, 322
Request for Comments (RFC), 19
Research and development, offshoring of, 340
Research community, and private
development, 61. *See also* Academic research;
Professors
Research in Motion, 84–85
Retailers: big-box, 234, 239; e-commerce
capabilities of large national, 252; and Inter-
net services, 69, 239; and online commerce,
234; online operations of, 250, 251, 252; and
online transition of customers, 253; pure-play,
234, 253n5; small independents, 233–234
Retailers, online: auctions used by, 248–249;
start-up costs for, 245
Retail industries: Web site development in,
395; work-flow redesign in, 392
Retailing codes, NAIC, 235
Retailing functions, 240–243, 244, 245
Retailing techniques, online, 233
Retail sales: e-commerce in, 397; online
accounts as percentage of, 234
Retail sales, online: classification of, 239, *240*;
growth of, 237–238; modular auctions, 245–
249; user reviews in, 242; value of, 397, 405
Retail sector, 236; characteristics of, 235–237;
consumption expenditures in, 235, 237,
254n8; direct sales methods in, 237; spending
in, 237

RetroRaunch, 521
Revenues, *vs.* market share, 413, 421n39
Rhapsody, 473
Rheingold, Howard, 424, 426
RIAA. *See* Record Industry Association of
America
Riffage.com, 467, 468
RightNow, 214
Rio portable player, 466
Roberts, Larry, 10, 15
Robertson, Michael, 459, 464
RocketMail, 119
Rolling Stone, 434
Rollins, Kevin, 336
ROM-BIOS. *See* Read-Only Memory-Basic
Input-Output System
Root Zone File, 38
Rose, 508
Rothenbuhler, Eric, 283
Routers, 12, 21
Roxio, 434–435, 469
RT computers, IBM, 24, 41n37
Rural access, 70–71, 100n51, 100n52
Ryobi, 375, 377

SAGE. *See* Semi-Automatic Ground
Environment
Sales, online, 237, 240, 395, 419n10. *See also*
Retail sales
Salesforce.com, 201, 213, 214, 220
Sales tax, and online retail, 243–244
Samuel, Arthur, 550
San Francisco Chronicle, 280
San Francisco Examiner, 280
Santa Cruz Operation, Inc. (SCO), 544
SAP, 204, 213, 217, 219
Sargent (software vendor), 211
Satellite connections, 221, 222, 229n73
Satyam, 339
SBC, mergers of, 90–91
Schmidt, Eric, 416
Schneier, Bruce, 549

Schools, 78, 371. *See also* Classroom;
Universities
Schrader, William L., 28
Science and Technology Act (1992), 29–30
Science Applications International
Corporation (SAIC), 36
Science fiction, 562
Scient Corp., 259–260, 261, 272, 273n1
Scientific research, sharing community in, 541
SCO Group, 544
Scour Exchange (SX), 469
SDMI. *See* Secure Digital Music Initiative
Seagate, 455
Sealey, Peter, 331
Search engines, 159, 170; after dot-com crash,
182; effectiveness of, 190; and online
advertising, 172; for online retail, 242; as
portals, 181; spamming of, 182; usage charges
for, 172. *See also* Google
Search Engine Watch, 182
Search goods, information requirements of,
242, 255n28
Sears, 17, 257n58
Seattle Community Network, 424
Second Life, 561
Secure Digital Music Initiative (SDMI), 462,
463, 484n36
Secure Sockets Layer (SSL), 132
Securities industry, online trading in, 407
Security, Internet, 16, 32, 548
Security First Network Bank, 408–410
Security technologies, circumvention of, 460–
462
Sega Enterprises v. Accolade, 461
Selling, over Web, 133. *See also* Retail sales
Semantic Web, 191
Semi-Automatic Business Research
Environment (SABRE) system, 317, 318, 319
Semi-Automatic Ground Environment (SAGE)
radar, 317
Semiotic democracy, 302, 452, 480, 489n91
Sendmail, 111, 118

Sequoia Capital, 183

Serial copy management system (SCMS), 457, 482n16

Seriff, Marc, 17

Server software, early, 131

Service bureaus, 202, 205

Service providers. *See* Internet service providers

Seventeen magazine, 286

Sex industry, 491; and changing mores, 437, 494; and Internet, 501; legal, 491; on stock market, 506. *See also* Pornography industry

Sexploitation films, 499

Sextracker, 516

Sexualization, of public sphere, 496

Sforce platform, 220

Shannon, Claude, 549

Shapiro, Carl, 413–414, 562

Sharman Networks, 471, 472

Shelf space, for Internet retailers, 244–245

Sherman, Chris, 182

Shop.org, survey of, 250

Shoppers, 240–241, 249, 257n53, 257n54. *See also* Consumers

Shoppers, online, 241–245

Shopping: electronic, 405; online, 238–239, 245, 251–252; as recreational activity, 245

Shopping assistant, in online retailing, 251

Shopping.com, 432

SHOUTcast MP3 streaming media server, 464

"Shovelware," 281

Sidgmore, John, 29

Siebel Systems, 211, 213

Siegel, Martha S., 120, 543

Silicon Graphics, 365

Silicon Valley, Internet diffusion in, 58

Simple Mail Transfer Protocol (SMTP), 106, 111, 115, 120, 124, 148

Skylink Technologies, 463

Skype, 123, 432, 471, 552

Slashdot, 547

Slate (online magazine), 136, 289

Sloan Foundation, 266

Small-to-midsize business (SMB) sector, 213

Smartmodem, 222

Smith, Cyrus R., 317

Smith, Jack, 118

Smith, R. Blair, 317

Social change, and shaping of Internet, 305–306

Social networking, 303–304, 423, 436, 437, 439, 551–553

Social networks, development of, 561. *See also* Communities

Society, place of Internet in, 28

Sociology: of deviance, 434; of knowledge, 369; symbolic interactionist tradition in, 436

Sociosexual realm, of Internet, 533

Sociotechnical analysis, 498

Software: from custom written to standardized, 210; of early Internet, 12; free, 555n12; independent of hardware, 203; Internet e-mail, 116–118; licensing of, 128; middleware server for, 134; social, 551–553; sold on Internet, 545; terminal-emulation, 129

Software, on-demand, 213, 214–216, 217

Software as Service (SaaS), 201, 213; impact on independent software vendors of, 217–219; and power/software locus, 220; and service bureaus, 203; survival of, 209–213; *vs.* computer utility idea, 201

Software industry, 204; anticipation of Internet in, 59–60; contractors, 203, emergence of, 206; Google in, 147; piracy in, 434

Software package, pre-Internet, 417

Software producers, and illegal copying, 546

Sony, 456, 457, 469, 550

SoundScan, 476

Source Telecomputing Corp., The, 15, 16

Southwest Airlines, 324

Spam, 120, 121; blocking of, 121–122; costs of, 153n47; early eradication programs for, 215; first case of, 543; prevalence of, 559; volume of, 121; vulnerability to, 115

Specialized services, attraction of, 558

Spider, in Web indexing systems, 169

Spink, J., 514, 516

Spoofing, 26

Spreadsheet programs, 215

Sprint, 27, 54, 65, 66, 87

Spyglass, 128, 154n65

St. Petersburg Times, 280

Stallman, Richard, 547

Starbucks, Wi-Fi access at, 83

StarOffice, 215

Starr Report, 502

Startup.com (documentary film), 179

Start-ups, Internet, 209, 272; investment in, 177; perceived value of, 273; retail, 234, 250; venture-backed dot-com retail entrants, 250; and venture capitalists, 180

State Farm, intranet used by, 401

State government agencies, internal intranet applications of, 402

Stevenson, Veronis Suhler, 286

Stewart, Justice Potter, 493, 562n3

Storage devices, for personal computers, 455

Streambox, 462

StreamCast, 472

Strossen, Nadine, 491

Students, and Internet, 368. *See also* Universities

Students Against Facebook News Feed, 438

Subscription business, 172, 296, 473

Subscription sites, in pornography industry, 510

Sullivan, Danny, 182

Sulzberger, Arthur, 291

Sun, 139, 156n91

Sunburst Farms, 341

Sunflower Travel Agency, transformation of, 325

Sun Microsystems, 137, 215

Supercomputers, 13, 22

Supply-side issues, 82

SURANET, 27

Symantec, 215

Synchronous Optical Network (SONET), 31–32

"Sysops," 112

System Development Corp. (SDC) and SAGE, 317

Tagged.com, 304

Tallinn, Jaan, 471

Tata Consulting, 339

Tauzin-Dingle Bill (2002), 90

Taxation, and provision of Internet access, 79

Tax savings, and online retail, 243–244

Taylor, Robert, 11

TCP/IP. *See* Transmission-Control Protocol/Internet Protocol

Teaching resources, online, 372

Technological innovation: and adult entertainment business, 501; and business practices, 415; and health care industry, 415; investment in, 126; management and, 394; and medical practice, 356; and organized workplace, 429–430; and pornography, 494–496; and recording industry, 453–456; and unemployment, 370

Technology: adoption of new, 351; and business, 416; and "creative destruction," 562; DAT recorder, 456; digital *vs.* analogue, 453, 454; interconnection, 57; Internet, 4; offshoring, 317; peer-to-peer, 545; relationship between information and communication, 492; and retail sales, 242; and sex industry, 491; social shaping of, 106, 497–499; and success of Web, 128; of web mail, 119. *See also* Information technology

Teenagers, 469, 437–438, 563n7. *See also* Adolescents

Telecom Meltdown, 87, 91

Telecommunications: in brokerage industry, 406; business use of, 393, 417; and emergence of Internet, 109; and global competition, 405; merger policy for, 79; and outsourcing, 402

Telecommunications Act (1996), 47, 50, 77, 79, 89, 90

Telecommunications industry, U.S., 560

Telecommunications policy, and fostering exploration, 90

Telehealth, 352

Telemedicine, 352, 355, 356; benefits of, 353; cost-effectiveness of, 357; and e-mail, 362; and federal programs, 361; lack of widespread access to, 367; National Library of Medicine projects in, 358; radiology in, 361; reimbursement in, 360–361; "store and forward" feature of, 357–358, 361

Telephone directories, on Web, 127

Telephone industry: and bulletin board operators, 55–58; and computer II, 56; and dial-up ISP business, 69; and DLS services, 82; and DSL services, 91; household subscriptions of, 98n32; and ISPs, 95n. 6; merger of local, 88–89; U.S., 80

Telephone networks, regulating interconnection to, 101n. 63

Telephony systems, 79, 123, 551–552

Telepsychiatry system, 356

Teletext, 281

Television: broadcast news on, 285; cable, 284, 298, 305; commercial history of, 561; culture, 302; digital access to, 480; early Web ventures of, 282; impact of Web on, 297; and niche theory, 283; reality, 303

Television industry, 271, 297–298

Tel-Save telephone firm, 178

Terra Networks, 179

Terrorist attacks, September 11, 2001, 87, 319, 326

Texas Instruments, 323, 340

Text messaging, 84–85

Text-searching companies, 173–174

The Globe (dot-com firm), 261

TheStreet.com, e-commerce index of, 265

The WELL. *See* Whole Earth 'Lectronic Link

Thinking Machines, 169

Third-party benefit providers, 361, 367

Thomas, Sean, 497

3Com, in early server market, 59

Ticketing, online sales of, 239, *240*

Tijuana bibles, 501

Time, 12, 286, 288, 291

Time-sharing, computer, 204, 221

Time-sharing industry, 202, 204–206; competition faced by, 207, 226n25; destruction of, 218; growth of, 207–208; interactivity of, 205; and other computing models, 223; resurrected, 220, 221; Tymshare in, 205; typical customer of, 225n20

Time-sharing services: instant messaging and, 122; operating systems for, 110, 150n12; *vs.* in-house computing, 208, 226n22, 226n23

Times Mirror Co., 290

Time Warner, 81, 143, 176, 180

Tivo, 451

Toffler, Alvin, 109, 507, 508

TopTenReviews.com, 516

Tower Records, 377

TownOnline, 441

Toyota, 375

TP/4 protocol, 114

Trade, and asymmetrical digital subscriber line protocol, 548

Trading, online, 407

Transaction costs: economics of, 316, 343n3, 377

Transactions, of media organizations, 296

Transit services, 66, 71

Transmission capacity, redundant investment in, 87

Transmission-Control Protocol/Internet Protocol (TCP/IP), 11, 12, 106, 148, 561; commercial sales of, 28; compatible files, 60; flexibility of, 109; NSF adoption of, 22; universal use of, 13

Transmission costs, 76

Transparency, lack of, 33

Transportation, 395, 405

Trans World Airline (TWA), 318

Travel agencies, 317, 324–325, 559; decline in numbers of, 320–321; and online travel packages, 323; reduced dependence on, 400; transformation of, 325. *See also* Travel reservations industry

Travelocity, 321, 322, 323, 325

Travel reservations industry: commissions in, 319–320; consolidation in, 318; growth in, 316–317; and Internet, 325–326, 342–343; online, 321, 324; revenues in, 318–319; specialization in, 325

Tribune Co., 289, 290, 294

Trilian program, 123

Troland, Thomas R., 294

True.com, 437

Truscott, Tom, 542

Turing, Alan M., 549

Turner, Fred, 425

Tuvalu, domain names registered in, 35

Twang This!, 467

Tymnet, 15

Tymshare, 205–206, 208

Uhse, Beate, 502, 523

Unemployment, technologically driven, 373

United Airlines, 318

United Kingdom, privacy in, 560

Universal Music Group, 465

Universities: adoption of Internet by, 394, 418n6; connected through regional networks, 27; and early Internet, 368; e-mail used in, 369–370, 385n97; virtual, 373

Unix, 19, 20, 21, 62, 111, 126, 542

Unix Netware, 113

Unix-to-Unix CoPy program (UUCP), 20, 21, 99n40

Upstartle, 216

URL (uniform resource locator), 125

U.S. News and World Report, 280, 286

Usage charges, for search engines, 172

USA Today, 280, 288, 290

Usenet, 20–21, 25, 113, 542, 543–544

Usenet newsgroups, AOL incorporation of, 131

Usenix conference, 1980, 542

Users of Internet, 9–10, 68, 284

USinternet-working (USi), 210, 211, 212

UUCPNET, 113, 114

UUNet, 21, 28–29, 29, 66

Vault Corporation v. Quaid Software, 460

VAX computer, 23

Venture capital, 30, 261, 354

Venture creation, during dot-com era, 260, 265–267

Venture Economics (database), 267, 274n21

Venture investors, decision making of, 267

Verisign, 38

Verity (Web site search pioneer), 189

Verizon, 24, 79, 90

Vermeer Technologies, 97n22, 135

Veronica, creation of, 168

VHS videocassette recorder standard, 110

Video cassette recorders (VCRs), 456

Videoconferencing, 357, 543, 552

Video equipment, for telemedicine, 356–357

Videos: adult, 505; neighborhood stores for, 341; on-demand, 543

Video telephony, 552

Videotex, 280–281, 281, 296

Viola browser, 126

Virtual community centers, 295

"Virtual reality spaces," 137

Virtual red-light zone, 503

"Virtual Sex," 521

Viruses, computer, 26, 559

Virus writers, 215

Vivid Entertainment Group, Inc., 522, 531–532

Vixie, Paul, 37

VM mainframe operating systems, 417

Vodcasts, 295

Voice traffic, *vs.* packet-switched data traffic, 25

Vongo, 299

von Hippel, Eric, 146, 498

von Meister, William, 16

Voyeur sites, Web-based, 494

Wainwright, Julie, 264

WAIS. *See* Wide Area Information Server

Walled garden approach, of AOL, 68

Wall Street Journal, 289, 290, 296

Wal-Mart, 234, 235, 239, 254n20, 559; employees of, 254n13; and local retailers, 240; and retail changes, 254n19; RFIDS used by, 420n31

Warner American Express Satellite Co., 458

Warner Brothers Records, 377

Washington Post Co., 288, 289

Washington University, WebCrawler of, 169

Waste file sharing systems, 464

Watchravesringkan, Kittichai, 436

Watermarking process, 481n8

Wayback Machine, 269

Web, 147–148, 148; dynamically generated pages on, 127; early design of, 168; firms adversely impacted by, 316; free, 296; interactive use of, 551; online newspapers on, 282; and personal computer business, 332–338; pornography sites on, 509–510, 514–515; selling over, 133; television industry's adoption of, 298

Web 2.0, 146, 561

Web browsers: home page for, 175, 195n57; and media businesses, 279–280; role of, 120. *See also* Browsers

Web-browsing, Internet adoption for, 85

Webcams, in telemedicine, 357–358

"Web crawler," 169

WebCrawler (navigation service), 174

Web directories, 165–167

Webex, 214–215, 216

"Webisodes," 291

Web mail, 118–120, 152n. 41

WebMD.com, 354, 364, 365, 366

Web-native applications, 213, 215–216, 217

Web navigation industry, 132, 185–190

Web pages, 76, 161

Web portals, 159, 174–179; and dot-com crash, 179–180; end of, 180, 190; Go.com, 177; msn.com, 176; national, 178; Netcenter,

175–176; origins of, 164; and partnership deals, 178; personalization features of, 181; popularity of, 179; services offered by, 175; specialist, 178; use of term, 174–175; Yahoo, 191

Web publishing industry, 132, 171, 176, 186

Web search, development of, 167–170

Web services, 219–220, 246

WebSideStory, 214

Web sites, 124; airline, 319, 320; business development of, 393, 394; catalog-based, 250, 251; experts for, 404; most visited, 175, 552–553; off-shore, 560; retail, 250, 251; store-based, 250, 251; sudden proliferation of, 169

Webvan, 261, 315, 342

Western Governors' Virtual University Project, 373

Western Union, 10

Westinghouse, early computer network of, 150n11

West Publishing Co., 428

Whitman, Meg, 432

Whole Earth Catalog, 425–426, 431, 433

The Whole Earth 'Lectronic Link (The WELL), 425, 426, 427, 440, 458

Wholesale industry, 392, 397, 405

Wiatt, Ted, 334–335

Wide Area Information Server (WAIS), 168–169, 193n30

Wi-Fi Alliance, 84, 85

Wi-Fi mode, 83

Wikipedia, 146, 300, 304, 551

Wikis, 120, 551

Williams, Linda, 87, 504

Willis, Chris, 300, 303

Winamp, 464

Winamp MP3, 455

Windows, Microsoft, and Foxfire browsers, 145; Hypertext in, 162; redesign of, 146; target markets for, 97n. 23

Windows 95, 60, 134

Windows 98, 141

Windows Live Desktop, 152n. 34

Windows Vista operating system, Windows
  Mail, 117
Windows XP, 145
Windows X version, of Mosaic, 127
Winplay3, 455
Wipro Technologies, 339
Wired magazine, 281, 434
Wireless access, 49, 83–85
Wireless Ethernet Compatibility Alliance
  (WECA), 84
Women, Web sites for, 498
Wood, David, 260
Word, Web pages created in, 135
WordPerfect word processor, 138
Word processing, 147, 215
Wordstar, MicroPro's, 332
Workplace, and technological innovation,
  373, 430
WorldCom, 69, 79, 88
World Information Property Organization
  (WIPO), 461
World (ISP), 38
World Summit on Information Society (2005),
  UN-sponsored, 36
World Wide Web, 12; accessing, 129–130;
  appeal of, 126; as business platform, 131–
  134; commercialization of, 163; creation of,
  106, 124; demand for, 105; dynamic page
  generation on, 133; and early browsers, 127;
  early free publicity for, 166; as electronic
  publishing system, 160–165; spread of, 129;
  standards of, 124–125
World Wide Web Consortium, 62, 76, 133
World Wide Web Virtual Library Subject
  Catalog, of CERN, 165
Worm, early appearance of, 25–26
WrestlingGear.com, 259, 260, 261, 272
Writers, Internet-related demands on, 291–292
Wulf, William, 30

Xanadu global hypertext network, 162, 170
X-rated products, media substitution for, 500–
  501

X.400 standard, 114, 115, 120
X.500 standard, 114, 115

Yahoo!, 119, 159, 166, 261, 315, 552; adult
  content sold by, 520; and dot-com crash,
  181; Farechase acquired by, 324; geographic
  spread of, 70; initial public offering of, 166;
  instant messaging of, 68, 123; online
  advertising revenues of, 172; Overture
  purchased by, 185, 186; search technology
  of, 185; in U.S. market, 119; user-generated
  content of, 301; in Web directory field, 166–
  167
Yahoo Internet Life (print magazine), 166
Yahoo Mail, 119
Yang, Jerry, 166
Yellow Pages, 173
YHOO, 179
Y2K, 403
York University, automated instruction at, 372
Yost, Jeffrey, 377, 378, 395, 558, 559
Young people, media use of, 303. *See also*
  Adolescents; Teenagers
YouTube, 146, 147, 191, 552, 553, 561

Zennstrom, Niklas, 471
Zimmerman, Phil, 549
Zimring, Franklin, 509, 513, 520
Zuse, Konrad, 549, 550